UNIVERSITY OF STRATHCLYDE

30125 00804481 9

D1759582

**Books are to be returned on or before
the last date below.**

DUE

1 3 SEP 2007

LIBREX —

ANDERSONIAN LIBRARY
★
WITHDRAWN
FROM
LIBRARY
STOCK
★
UNIVERSITY OF STRATHCLYDE

Nanobiotechnology II

Edited by
Chad A. Mirkin and
Christof M. Niemeyer

1807–2007 Knowledge for Generations

Each generation has its unique needs and aspirations. When Charles Wiley first opened his small printing shop in lower Manhattan in 1807, it was a generation of boundless potential searching for an identity. And we were there, helping to define a new American literary tradition. Over half a century later, in the midst of the Second Industrial Revolution, it was a generation focused on building the future. Once again, we were there, supplying the critical scientific, technical, and engineering knowledge that helped frame the world. Throughout the 20th Century, and into the new millennium, nations began to reach out beyond their own borders and a new international community was born. Wiley was there, expanding its operations around the world to enable a global exchange of ideas, opinions, and know-how.

For 200 years, Wiley has been an integral part of each generation's journey, enabling the flow of information and understanding necessary to meet their needs and fulfill their aspirations. Today, bold new technologies are changing the way we live and learn. Wiley will be there, providing you the must-have knowledge you need to imagine new worlds, new possibilities, and new opportunities.

Generations come and go, but you can always count on Wiley to provide you the knowledge you need, when and where you need it!

William J. Pesce
President and Chief Executive Officer

Peter Booth Wiley
Chairman of the Board

Nanobiotechnology II

More Concepts and Applications

Edited by
Chad A. Mirkin and Christof M. Niemeyer

WILEY-VCH Verlag GmbH & Co. KGaA

The Editors

Prof. Dr. Chad A. Mirkin
Department of Chemistry
International Institute for Nanotechnology
Northwestern University
2145 Sheridan Road
Evanston, IL 60201
USA

Prof. Dr. Christof M. Niemeyer
Department of Chemistry
University of Dortmund
Chair of Biological and
Chemical Microstructuring
Otto-Hahn-Str. 6
44227 Dortmund
Germany

Cover
Cover images prepared
by Matthew Banholzer
and Rafael Vega

■ All books published by Wiley-VCH are
carefully produced. Nevertheless, authors,
editors, and publisher do not warrant the
information contained in these books,
including this book, to be free of errors.
Readers are advised to keep in mind that
statements, data, illustrations, procedural
details or other items may inadvertently be
inaccurate.

Library of Congress Card No.: applied for

**Bibliographic information published by
the Deutsche Nationalbibliothek**
Die Deutsche Nationalbibliothek lists this
publication in the Deutsche National-
bibliografie; detailed bibliographic data are
available in the Internet at ⟨http://dnb.d-nb.de⟩.

© 2007 WILEY-VCH Verlag GmbH & Co.
KGaA, Weinheim

All rights reserved (including those of
translation into other languages). No part of
this book may be reproduced in any form –
by photoprinting, microfilm, or any other
means – nor transmitted or translated
into a machine language without written
permission from the publishers. Registered
names, trademarks, etc. used in this book,
even when not specifically marked as such,
are not to be considered unprotected by law.

Printed in the Federal Republic of Germany
Printed on acid-free paper

Typesetting Asco Typesetters, Hong Kong
Printing betz-druck GmbH, Darmstadt
Binding Litges & Dopf GmbH, Heppenheim
Cover Design Adam-Design, Weinheim
Wiley Bicentennial Logo Richard J. Pacifico

ISBN 978-3-527-31673-1

D
660·6
NAN

Contents

Nanobiotechnology II. Edited by Chad A. Mirkin and Christof M. Niemeyer
Copyright © 2007 WILEY-VCH Verlag GmbH & Co. KGaA, Weinheim
ISBN: 978-3-527-31673-1

Preface

The broad field of nanotechnology has undergone explosive growth and development over the past five years. In fact, no field in the history of science has experienced more interest or larger government investment. Indeed, by the end of 2006, the worldwide government and private sector investment in nanotechnology is projected to be approximately $9 billion. The enthusiasm researchers have for this field is fueled by: 1) the desire to determine the unusual chemical and physical properties of nanostructures, which are often quite different from the bulk materials from which they derive, and 2) the potential to use such properties in the development of novel and useful devices and materials that can impact and, perhaps even transform, many aspects of modern life.

The subfield known as Nanobiotechnology holds some of the greatest promise. This highly interdisciplinary field, which draws upon contributions from chemistry, physics, biology, materials science, medicine and many forms of engineering, focuses on several important areas of research. Some of these include: 1) the development of methods for building nanostructures and nanostructured materials out of biological or biologically inspired components such as oligonucleotides, proteins, viruses, and cells; these structures are intended for both biological and abiological uses, 2) the utilization of synthetic nanomaterials to regulate and monitor important biological processes, and 3) the development of synthetic and soft matter compatible surface analytical tools for building nanostructures important in both biology and medicine. Advances in this field offer novel and potentially useful approaches to building functional structures including computational tools, energy generation, conversion and storage materials, powerful optical devices, and new detection and therapeutic modalities. Indeed, advances in Nanobiotechnology have the potential to revolutionize the way the medical community approaches modern disease management.

Although the field is still embryonic, major strides have been made. Powerful new forms of signal amplification have been realized for both DNA and protein based detection systems. Indeed, the first commercial molecular diagnostic systems that rely upon nanoparticle probes are expected to be available in 2007. Biological labels based upon nanocrystals are commercially available and used routinely for research purposes in laboratories worldwide. Many new nanomaterials have boosted the efficacy and viability of several powerful pharmaceutical agents.

Nanobiotechnology II. Edited by Chad A. Mirkin and Christof M. Niemeyer
Copyright © 2007 WILEY-VCH Verlag GmbH & Co. KGaA, Weinheim
ISBN: 978-3-527-31673-1

Nanofabrication tools that allow one to routinely build oligonucleotide, protein, and other biorelevant nanostructures on surfaces with extraordinary precision have evolved to the point of commercialization and widespread use. These examples represent only a small part of this expansive field but are realized potential and serve as motivators for future developments within it.

In 2004, we edited the book *Nanobiotechnology: Concepts, Applications and Perspectives* which was intended to provide a systematic and comprehensive framework of specific research topics in Nanobiotechnology. Due to the great success of this first volume, *Nanobiotechnology II – More Concepts and Applications* now follows the notion of its precursor by combining contributions from bioorganic and bioinorganic chemistry, molecular and cell biology, materials science and bioanalytics to cover the entire scope of current and future developments in Nanobiotechnology. The collection of articles in this volume again emphasizes the high degree of interdisciplinarity necessarily implemented in the joint-venture of biotechnology and nano-sciences. During the selection of potential chapters for this volume we took into account, on the one hand, the progress by which particular areas had developed in the past three years. Because this occurred primarily in two areas, namely the development of nanoparticle science and applications as well as in the refinement of scanning probe microscopy related methods, the majority of the chapters are concerned with these issues. On the other hand, additional topics not yet covered in the first volume were identified, thus leading to contributions from the area of small molecule- and peptide-based self-assembly (chapters 1 and 2), the use of nanomaterials for medicinal applications (section 3), and the utilization of biomolecular machinery to create hybrid devices with mechanical functionalities (section 4).

The current volume is divided into four main sections. Section I (Chapters 1–6) concerns novel principles in self-assembly and nanoparticle-based systems. In Chapter 1, Mary S. Gin, Emily G. Schmidt, and Pinaki Talukdar provide an overview of artificial transmembrane channels, attainable by organic synthesis and the assembly of small molecule building blocks. This synthetic approach to ion channels, initially aimed at elucidating the minimal structural requirements for ion flow across a membrane, nowadays is focusing on the development of synthetic channels that are gated, thus providing a means to control whether the channels are open or closed. Such artificial signal transduction could lead to novel sensing and therapeutic applications. The self-assembly of small molecules also represents the underlying theme of Chapter 2, written by Maxim G. Ryadnov and Derek N. Woolfson. They summarize current efforts to build nanoscopic and mesoscopic supramolecular structures from short oligopeptides comprising the α-helical coiled-coil folding motif. Examples of nanostructures and materials made from such coiled-coil building blocks include programmable nanoscale linkers, molecular switches, and fibrous and gel-forming materials which may be useful for the production of peptide-polymer hybrids combining the advantages of both natural and synthetic polymers. These structures also could lead to the design of peptide-based switches that may render hybrid networks more controllable and increase sensitivity and responses to local environments.

The notion of the first two chapters is connected with the area of nanoparticle research in chapter 3, where Erik Dujardin and Stephen Mann illustrate how unraveling the specific interactions between bio-derived templates and inorganic materials not only yields a better understanding of natural hybrid materials but also inspires new methods for developing the potential of biological molecules, superstructures and organisms as self-assembling agents for materials fabrication. In particular, the chapter describes the use of various types of bio-related molecules, ranging from biopolymers, peptides, oligonucleotides to the complex biological architecture of proteins, viruses and even living organisms, for the synthesis and assembly of organized nanoparticle-based structures and materials. Two additional chapters deal with the conversed approach, that is, the modification of nanoparticles with biomolecules to add functionality to the inorganic components. In Chapter 4, Rochelle R. Arvizo, Mrinmoy De, and Vincent M. Rotello describe recent developments involving protein functionalized nanoparticles. These conjugates, which are produced either by covalent or non-covalent coupling strategies, hold potential for the creation of novel materials and devices in the biosensing and catalysis fields. The combination of proteins and nanoparticles also opens up novel approaches to the synthesis of nanoparticles, as summarized in Chapter 6, written by Ronan Baron, Bilha Willner, and Itamar Willner. They describe the application of biocatalysts, enzymes, such as oxidases and hydrolases, as active components for the synthesis and enlargement of nanoparticles and for biocatalytic growth of nanoparticles, mediated by specific enzyme reactions. This concept has strong implications in biosensor design, and the nanoparticle-enzyme hybrid systems also can be used as biocatalytic inks for the generation of metallic nanowires, and thus, bioelectronic devices.

The self-assembly behavior of another class of biomolecular recognition elements is described in Chapter 5. There, Thomas H. LaBean, Kurt V. Gothelf, and John H. Reif summarize the current state-of-the-art of self-assembling DNA nanostructures for patterned assembly of (macro)molecules and nanoparticles. This field of research, which was initially covered in the previous volume of Nanobiotechnology, has evolved significantly within the past three years. A large number of groups are actively conducting research on such DNA-based nanoarchitectures. Although it is not yet clear whether commercially relevant applications of DNA scaffolds will ever be realized, the basic exploration of design principles based on the predictable Watson-Crick interaction of oligonucleotides opens up long term perspectives in studying novel nanoelectronic structures, sensing mechanisms, and materials research.

The increasing importance of nanostructures in analytical applications is reflected in the seven chapters of section II (Chapters 7–13). Developments of nanoparticle-based technologies are described in the first three chapters. As shown by Joseph Wang, the large number of inorganic ions incorporated within nanoparticles can be employed for signal amplification by electrochemical means. This approach offers unique opportunities for electronic transduction of biomolecular interactions, and thus, for measuring protein and nucleic acid analytes (Chapter 7). The peculiar optical properties of semiconductor nanoparticles are also used for bioanalytical purposes. Hedi Mattoussi and colleagues describe in their contribution the latest developments in this area (Chapter 8). This quantum dot biolabeling is rapidly moving towards sophisticated applications in cell and tissue imaging as well as in FRET-based immuno-assays, thereby posing great demands on the chemical and structural integrity of the hybrid probes. A more fundamental approach combining nanoparticle technologies and spectroscopy is described in Chapter 9 by Richard P. Van Duyne and colleagues. The localized surface plasmon resonance, occurring in optically coupled nanoparticles, coupled with the size, shape, material and local dielectric environment dependence of the nanostructures, forms the basis for a novel class of biosensors.

While the above mentioned chapters have the "bottom-up" assembly of functional nanostructures in common, the following four chapters of section 2 take advantage of micro- and nanosized probe structures fabricated by conventional "top-down" methodologies. The developments of micromechanical cantilever array sensors for bioanalytical assays are described by Hans Peter Lang, Martin Hegner, and Christoph Gerber in Chapter 10. The cantilever arrays respond mechanically to changes in external parameters, like temperature or molecule adsorption, and thus, can be used to monitor binding events and chemical reactions occurring at the sensors' surfaces. In Chapter 12, James R. Heath reviews work on nanotube-based sensors, which enable the label-free detection of biomarkers for cancer and other diseases. It is pointed out here that the establishment of viable carbon nanotube or semiconductor nanowire devices for routine diagnostics will require a high-throughput fabrication method with an extraordinary high level of integration of nanoelectronics, microfluidics, chemistry, and biology.

Chapter 11, written by Harald Fuchs and colleagues, reports on uses for shear force-controlled scanning ion conductance microscopy. By using a local probe that is sensitive to ion conductance in an electrolyte solution, gentle scanning of delicate biological surfaces, allows one to obtain well resolved images of fine surface structures, such as of membrane proteins on living cells. In Chapter 13, Chad Mirkin and colleagues report on the preparation of arrays of nanoscale features of biomolecular compounds by using dip-pen nanolithography. Arrays with features on the nanometer length scale not only open up the opportunity to study many biological structures at the single particle level, but they also allow one to contemplate the creation of a combinatorial library, for instance, a complex protein array, underneath a single cell. This would open new possibilities for studying important fundamental biological processes, such as cell-surface recognition, adhesion, differentiation, growth, proliferation, and apoptosis.

Section III (Chapters 14–19) of this volume concerns the use of nanostructures in medicinal applications. The six chapters focus on three major topics: the development of nanoparticle-based drug delivery systems, the use of nanoparticles for imaging, and the design of scaffolds for tissue engineering. Chapter 14, written by Rudy Juliano, gives an introductory overview on biological barriers to nanocarrier-mediated delivery of therapeutic and imaging agents. This chapter also provides a brief assessment of the toxicology of nanomaterials, a subject which has currently initiated widespread discussion because it is anticipated that nano(bio)-

technology will be a key aspect of the future economy. With respect to the development of suitable carrier systems, Larken E. Euliss, Julie A. DuPont, and Joseph M. DeSimone in Chapter 15 summarize work on the development of biocompatible organic nanoparticles, in particular, by means of top-down fabrication techniques, so called lithographic imprinting, adapted from the electronics industry. This method enables the inexpensive fabrication of monodisperse particles of various size and shape from a large variety of matrix materials, which have great potential as functionalized carriers for applications in nanomedicine. An alternative class of particles is described in Chapter 16, where Thommey P. Thomas and colleagues report on poly(amidoamine) dendrimer-based multifunctional nanoparticles as a tumor targeting platform. The biocompatible dendrimer macromolecules act as carriers of molecules for delivery into tumor cells and can achieve increased drug effectiveness at significantly lower toxicity as compared to the free drug.

With respect to clinical imaging techniques, Young-wook Jun, Jae-Hyun Lee, and Jinwoo Cheon review current work on the development of magnetic nanoparticle-based contrast agents for molecular magnetic resonance imaging in Chapter 17. This research is aimed towards advances in cancer diagnosis because nanoparticle contrast agents promise *in vivo* diagnosis of early staged cancer with sub-millimeter dimension, and might help to unravel fundamental biological processes, such as *in vivo* pathways of cell evolution, cell differentiations, and cell-to-cell interactions. A different class of nanoparticles is described in Chapter 19 by Samuel A. Wickline and colleagues. They have developed nanoparticles comprised of perfluorocarbon materials which are biologically and metabolically inert, chemically stable, and non-toxic. These nanoparticles have been employed for molecular imaging with MRI and targeted drug delivery.

Chapter 18 of this section, written by Robert L. Langer and colleagues reviews methodologies for generating two- and three-dimensional scaffold architectures for tissue engineering. The authors analyze the use of micro- and nanoscale engineering techniques for controlling and studying cell-cell, cell-substrate and cell-soluble factor interactions as well as for fabricating organs with controlled architecture and resolution.

Section IV of this volume is devoted to one of the most innovative topics of Nanobiotechnology which concerns the fabrication of hybrid devices using organic and inorganic structures equipped with parts of Nature's biomolecular machinery. To facilitate an introduction to natural molecular nanomotors, Manfred Schliwa describes in Chapter 20 how these fascinating molecular machines are built from amino acids and how they convert chemical energy into mechanical motion. In Chapter 21, Carlo D. Montemagno and colleagues summarize current approaches to fabricate biologically inspired hybrid nanodevices. In particular, two lines of work are shown, protein-based mechanical devices and cellular power generation devices, which both have in common the theme of combining biological molecules with synthetic host structures.

Similar to the first volume, the purpose of *Nanobiotechnology II – More Concepts and Applications* is to provide both a broad survey of the field as well as instruction and inspiration to scientists at all levels from novices to those intimately engaged in this new and exciting field of research. To this end, the current state-of-the-art of the above described topics has been accumulated by international renowned experts in their fields. Each of the chapters consists of three sections, (i) an overview which gives a comprehensive but still condensed survey on the specific topic, (ii) a methods section which points the reader to the most important techniques relevant for the specific topic discussed, and (iii) an outlook discussing academic and commercial applications as well as experimental challenges to be solved.

We are most grateful to the authors for providing this collection of high quality manuscripts. We also would like to thank Dr. Sabine Sturm and the production team of Wiley-VCH for continuous and dedicated help during the production of this book.

Evanston and Dortmund, November 2006
Chad A. Mirkin
Christof M. Niemeyer

List of Contributors

Boris Anczykowski
nanoAnalytics GmbH
Heisenbergstr. 11
48149 Münster
Germany

Rochelle R. Arvizo
Department of Chemistry
University of Massachusetts
710 North Pleasant St.
Amherst, MA 01003
USA

James R. Baker, Jr.
Michigan Nanotechnology
Institute for Medicine and
Biological Sciences
University of Michigan
Rm 9220C, MSRB III
Ann Arbor, MI 48109
USA

Ronan Baron
Institute of Chemistry
The Hebrew University of
Jerusalem
Jerusalem 91904
Israel

Matthias Böcker
Center for Nanotechnology (CeNTech)
and Institute of Physics
University of Münster
Heisenbergstr. 11
48149 Münster
Germany

Shelton D. Caruthers
Department of Medicine and
Biomedical Engineering
Washington University School of
Medicine
4320 Forest Park Ave.
St. Louis, MO 63108
USA

Jinwoo Cheon
Department of Chemistry and Nano-
Medical National Core Research
Center
Yonsei University
134 Sinchon-dong
Seodaemun-gu
120-749 Seoul
South Korea

Aaron R. Clapp
Division of Optical Sciences
US Naval Research Laboratory
Washington, DC 20375-5320
USA

Nanobiotechnology II. Edited by Chad A. Mirkin and Christof M. Niemeyer
Copyright © 2007 WILEY-VCH Verlag GmbH & Co. KGaA, Weinheim
ISBN: 978-3-527-31673-1

Mrinmoy De
Department of Chemistry
University of Massachusetts
710 North Pleasant St.
Amherst, MA 01003
USA

Joseph M. DeSimone
Department of Chemistry
University of North Carolina at
Chapel Hill
Chapel Hill, NC 27599
USA

Julie A. DuPont
Department of Chemistry
University of North Carolina at
Chapel Hill
Chapel Hill, NC 27599
USA

Erik Dujardin
NanoSciences Group
CEMES, CNRS UPR 8011
BP 94347
29 rue Jeanne Marvig
31055 Toulouse Cedex 4
France

Eric Dy
Department of Bioengineering
University of California, Los
Angeles
420 Westwood Plaza
Los Angeles, CA 90095-1600
USA

Larken E. Euliss
Department of Chemistry
University of North Carolina at
Chapel Hill
Chapel Hill, NC 27599
USA

Harald Fuchs
Center for Nanotechnology (CeNTech)
and Institute of Physics
University of Münster
Heisenbergstr. 11
48149 Münster
Germany

Christoph Gerber
Institute of Physics
University of Basel
Klingelbergstrasse 82
4056 Basel
Switzerland

Mary S. Gin
Department of Chemistry
University of Illinois at Urbana-
Champaign
600 S. Mathews Ave.
Urbana, IL 61801
USA

Charles A. Goessmann
Department of Chemistry
University of Massachusetts
710 North Pleasant St.
Amherst, MA 01003
USA

Kurt V. Gothelf
Department of Chemistry
Aarhus University
Langelandsgade 140
8000 Aarhus C
Denmark

W. Paige Hall
Department of Chemistry
Northwestern University
2145 Sheridan Road
Evanston, IL 60208-3113
USA

James R. Heath
Caltech Chemistry MC 127-72
and the NanoSystems Biology
Cancer Center
1200 East California Blvd.
Pasadena, CA 91125
USA

Martin Hegner
Institute of Physics
University of Basel
Klingelbergstrasse 82
4056 Basel
Switzerland

Rudy Juliano
Department of Pharmacology
School of Medicine
University of North Carolina at
Chapel Hill
Chapel Hill, NC 27599
USA

Young-wook Jun
Department of Chemistry and
Nano-Medical National Core
Research Center
Yonsei University
134 Sinchon-dong
Seodaemun-gu
120-749 Seoul
South Korea

Joseph J. Kakkassery
Department of Chemistry
International Institute for
Nanotechnology
Northwestern University
2145 Sheridan Road
Evanston, IL 60208
USA

Jeffrey M. Karp
Massachusetts Institute of Technology
77 Massachusetts Ave.
Cambridge, MA 02139-4307
USA

Ali Khademhosseini
Massachusetts Institute of Technology
65 Landsdowne St.
Cambridge, MA 02139
USA

Thomas H. LaBean
Departments of Computer Science and
Chemistry
Duke University
Durham, NC 27708
USA

Hans Peter Lang
Institute of Physics
University of Basel
Klingelbergstrasse 82
4056 Basel
Switzerland

Robert Langer
Massachusetts Institute of
Technology
77 Massachusetts Ave.
Cambridge, MA 02139-4307
USA

Gregory M. Lanza
Department of Medicine and
Biomedical Engineering
Washington University School of
Medicine
4320 Forest Park Ave., Campus
Box 8215
St. Louis, MO 63108
USA

Jae-Hyun Lee
Department of Chemistry and
Nano-Medical National Core
Research Center
Yonsei University
134 Sinchon-dong
Seodaemun-gu
120-749 Seoul
South Korea

Yibo Ling
Massachusetts Institute of
Technology
65 Landsdowne St.
Cambridge, MA 02139
USA

Istvan J. Majoros
Michigan Nanotechnology
Institute for Medicine and
Biological Sciences
University of Michigan
109 Zina Pitcher Place
Ann Arbor, MI 48109
USA

Stephen Mann
School of Chemistry
University of Bristol
Bristol BS8 1TS
UK

Hedi Mattoussi
Division of Optical Sciences
US Naval Research Laboratory
Washington, DC 20375-5320
USA

Igor L. Medintz
Division of Optical Sciences
US Naval Research Laboratory
Washington, DC 20375-5320
USA

Chad A. Mirkin
Department of Chemistry
International Institute for
Nanotechnology
Northwestern University
2145 Sheridan Road
Evanston, IL 60208
USA

Carlo D. Montemagno
Department of Bioengineering
University of California, Los Angeles
420 Westwood Plaza
Los Angeles, CA 90095-1600
USA

Andrzej Myc
Michigan Nanotechnology Institute for
Medicine and Biological Sciences
University of Michigan
109 Zina Pitcher Place
Ann Arbor, MI 48109
USA

Christof Niemeyer
Chair of Biological and Chemical
Microstructuring
University of Dortmund
Department of Chemistry
Otto-Hahn-Str. 6
44227 Dortmund
Germany

Jordan Patti
Department of Bioengineering
University of California, Los Angeles
420 Westwood Plaza
Los Angeles, CA 90095-1600
USA

Thomas Pons
Division of Optical Sciences
US Naval Research Laboratory
Washington, DC 20375-5320
USA

John H. Reif
Department of Computer Science
Duke University
Durham, NC 27708
USA

Vincent M. Rotello
Department of Chemistry
University of Massachusetts
710 North Pleasant St.
Amherst, MA 01003
USA

Maxim G. Ryadnov
School of Chemistry
University of Bristol
Cantock's Close
Bristol BS8 1TS
UK

Khalid Salaita
Department of Chemistry
Northwestern University
2145 Sheridan Road
Evanston, IL 60208
USA

Tilman E. Schäffer
Center for Nanotechnology (CeNTech)
and Institute of Physics
University of Münster
Heisenbergstr. 11
48149 Münster
Germany

Manfred Schliwa
Institute for Cell Biology
University of Munich
Schillerstr. 42
80336 Munich
Germany

Emily G. Schmidt
Department of Chemistry
University of Illinois at Urbana-
Champaign
600 S. Mathews Ave
Urbana, IL 61801
USA

Leif J. Sherry
Department of Chemistry
Northwestern University
2145 Sheridan Road
Evanston, IL 60208-3113
USA

Rameshwer Shukla
Michigan Nanotechnology Institute for
Medicine and Biological Sciences
University of Michigan
109 Zina Pitcher Place
Ann Arbor, MI 48109
USA

Pinaki Talukdar
Department of Chemistry
University of Illinois
600 S. Mathews Ave, Box 31-5
Urbana, IL 61801
USA

Thommey P. Thomas
Michigan Nanotechnology Institute for
Medicine and Biological Sciences
University of Michigan
109 Zina Pitcher Place
Ann Arbor, MI 48109
USA

Richard P. Van Duyne
Department of Chemistry
Northwestern University
2145 Sheridan Road
Evanston, IL 60208-3113
USA

Rafael A. Vega
Department of Chemistry and
International Institute for
Nanotechnology
Northwestern University
2145 Sheridan Road
Evanston, IL 60208
USA

Joseph Wang
Biodesign Institute
Center for Bioelectronics and
Biosensors
Box 875801
Arizona State University
Tempe, AZ 85387-5801
USA

David Wendell
Department of Bioengineering
University of California, Los
Angeles
420 Westwood Plaza
Los Angeles, CA 90095-1600
USA

Samuel A. Wickline
Department of Medicine, Physics,
Biomedical Engineering and Cell
Biology & Physiology
Washington University School of
Medicine
4320 Forest Park Ave.
St. Louis, MO 63108
USA

Katherine A. Willets
Department of Chemistry
Northwestern University
2145 Sheridan Road
Evanston, IL 60208-3113
USA

Bilha Willner
Institute of Chemistry
The Hebrew University of Jerusalem
Jerusalem 91904
Israel

Itamar Willner
Institute of Chemistry
The Hebrew University of Jerusalem
Jerusalem 91904
Israel

Patrick M. Winter
Department of Medicine and
Biomedical Engineering
Washington University School of
Medicine
4320 Forest Park Ave.
St. Louis, MO 63108
USA

Derek N. Woolfson
Department of Biochemistry
University of Bristol
Cantock's Close
Bristol BS8 1TS
UK

Xiaoyu Zhang
Department of Chemistry
Northwestern University
2145 Sheridan Road
Evanston, IL 60208-3113
USA

Jing Zhao
Department of Chemistry
Northwestern University
2145 Sheridan Road
Evanston, IL 60208-3113
USA

Part I
Self-Assembly and Nanoparticles: Novel Principles

Nanobiotechnology II. Edited by Chad A. Mirkin and Christof M. Niemeyer
Copyright © 2007 WILEY-VCH Verlag GmbH & Co. KGaA, Weinheim
ISBN: 978-3-527-31673-1

1
Self-Assembled Artificial Transmembrane Ion Channels

Mary S. Gin, Emily G. Schmidt, and Pinaki Talukdar

1.1
Overview

Natural ion channels are large, complex proteins that span lipid membranes and allow ions to pass in and out of cells. These multimeric channel assemblies are capable of performing the complex tasks of opening and closing in response to specific signals (gating) and allowing only certain ions to pass through (selectivity). This controlled transport of ions is essential for the regulation of both intracellular ion concentration and the transmembrane potential. Ion channels play an important role in many biological processes, including sensory transduction, cell proliferation, and blood-pressure regulation; abnormally functioning channels have been implicated in causing a number of diseases [1]. In addition to large ion channel proteins, peptides (e.g., gramicidin A) and natural small-molecule antibiotics (e.g., amphotericin B and nystatin) form ion channels in lipid membranes.

Due to the complexity of channel proteins, numerous research groups have during recent years been striving to develop artificial analogues [2–7]. Initially, the synthetic approach to ion channels was aimed at elucidating the minimal structural requirements for ion flow across a membrane. However, more recently the focus has shifted to the development of synthetic channels that are gated, providing a means of controlling whether the channels are open or closed. Such artificial signal transduction could have broad applications to nanoscale device technology. In this chapter we will present examples of some strategies used in the development of artificial ion channels, and describe some techniques commonly used to evaluate their function.

1.1.1
Non-Gated Channels

The artificial ion channels described to date can be divided into two major classes. The first class consists of non-gated channels, which are synthetic compounds

Nanobiotechnology II. Edited by Chad A. Mirkin and Christof M. Niemeyer
Copyright © 2007 WILEY-VCH Verlag GmbH & Co. KGaA, Weinheim
ISBN: 978-3-527-31673-1

that simply form transmembrane pores. The second class comprises gated channels that incorporate a means of regulating ion flow across a membrane. A number of strategies have been used in the development of non-gated artificial channels, ranging from the assembly of monomers to form transmembrane pores to the use of single molecules capable of spanning the entire thickness of a lipid bilayer. The monomeric channels generally have more well-defined structures than those formed through aggregation.

1.1.1.1 Aggregates

For ion channels produced through the aggregation of amphiphilic molecules, monomers must first assemble in each leaflet of a lipid bilayer to form a pore with a hydrophilic interior. When aggregates in each leaflet of the bilayer align, a transmembrane channel is formed (Figure 1.1A). Examples of amphiphilic molecules that display this behavior include an oligoether-ammonium/dialkyl phosphate ion pair **1** [8] and a sterol-polyamine conjugate **2** [9] (Figure 1.1B). Artificial ion channels have also been generated through the stacking of cyclic monomers. Both cyclic β^3-peptides **3** [10] and D,L-α-peptides **4** [11] form transmembrane pores. The activities of these peptide-based aggregate ion channels are comparable to that of the natural channel-forming peptide gramicidin A.

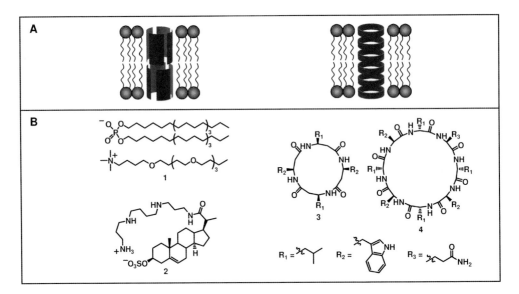

Fig. 1.1 (A) Schematic representations of artificial ion channels assembled through the aggregation of amphiphilic monomers and the stacking of cyclic peptides. (B) Compounds that aggregate to form transmembrane ion channels.

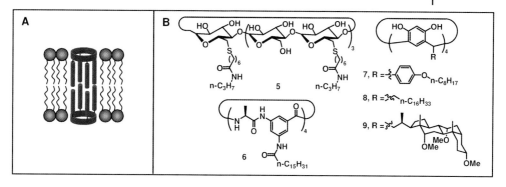

Fig. 1.2 (A) Schematic representation of a transmembrane channel formed through the dimerization of pore-forming monomers. (B) Compounds that form ion channels through dimerization.

1.1.1.2 Half-Channel Dimers

A common approach to designing synthetic ion channels has been to functionalize a pore-forming macrocycle with lipophilic groups such as alkyl chains or cholic acid. When these molecules insert into each leaflet of the bilayer and align, the macrocycles act as pores at each membrane surface, while the lipophilic groups serve as channel walls (Figure 1.2A). A variety of macrocycles have been utilized in the construction of half-channel molecules, including β-cyclodextrin **5** [12], cyclic peptides **6** [13], and resorcinarenes **7–9** [14–17] (Figure 1.2B).

1.1.1.3 Monomolecular Channels

Using a similar strategy to that described above for the assembly of half-channel dimers, a monomolecular channel **10** has been reported that comprises β-cyclodextrin with oligobutylene glycol chains attached to one face [18] (Figure 1.3A,B). In this case, the macrocycle provides a pore at the surface of the membrane, but the chains are sufficiently long so that a single molecule spans the entire thickness of the bilayer. This monomolecular channel was reported to have a Na^+ transport activity that was 36% that of gramicidin A.

Alternatively, monomolecular ion channels have been designed such that a single macrocycle resides near the center of the bilayer, while the attached lipophilic chains radiate outward toward the membrane surfaces (Figure 1.3A). Examples of molecules reported to function in this manner include a β-cyclodextrin with oligoethers attached to both the primary and secondary faces **11** [19], a calixarene-cholic acid conjugate **12** [20], as well as crown ethers functionalized with cholesterol **13** [21], bola-amphiphiles **14** [22], or oligoethers **15** [23] (Figure 1.3B). The activity of the calixarene-cholic acid conjugate **12** was found to be approximately 73% that of the channel-forming antibiotic amphotericin B.

Artificial single-molecule ion channels that incorporate multiple pore-forming crown ether macrocycles include a peptide-crown ether conjugate **16** [24] and a

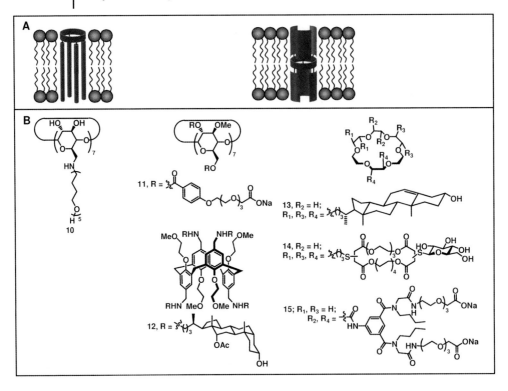

Fig. 1.3 (A) Schematic representations of monomolecular ion channels incorporating a single macrocycle. (B) Structures of monomolecular ion channels that incorporate a single macrocycle.

tris(macrocycle) hydraphile channel **17** [25] (Figure 1.4A,B). Attaching crown ethers to a helical peptide scaffold provided a channel that allowed ions to pass through a series of macrocycles as they traversed the membrane. In the hydraphile channel, distal crown ethers are thought to serve as pore openings at the membrane surfaces, while the central azacrown ether stabilizes ions as they pass across the bilayer. It was reported that the hydraphile channel was 28% as active as gramicidin D.

Although common, the incorporation of macrocycles is not a prerequisite for monomolecular channel formation. An amphiphilic molecule **18** incorporating a number of lysine and cholic acid groups as well as a *p*-phenylene diamine linker served as a monomeric transmembrane channel [26] (Figure 1.4A,B).

1.1.2
Gated Channels

While the previous examples demonstrate that artificial channels can promote transmembrane ion transport, they do not provide a means of controlling

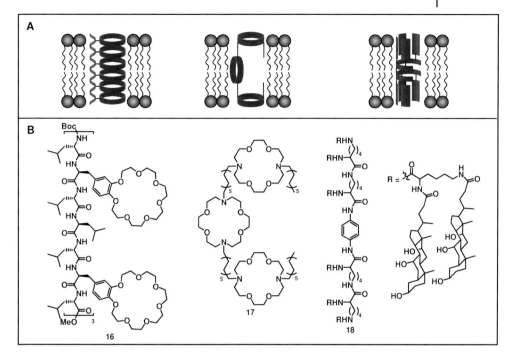

Fig. 1.4 (A) Schematic representations of monomolecular ion channels.
(B) Compounds that serve as monomolecular ion channels.

whether the channel is open or closed. Signal-activated synthetic channels bring the field one step closer to mimicking the function of natural ion channels, as active channels are only formed in the presence of a specific signal. As with natural channels, a variety of methods can be used to gate these synthetic analogues, including light, voltage, and ligand activation.

1.1.2.1 Light-Gated Channels
Although not a stimulus for natural channels, light has been used to control transmembrane ion transport through a synthetic channel. This was accomplished by incorporating an azobenzene group into an oligoether carboxylate-alkylammonium ion pair **19** [27] (Figure 1.5B). With a *trans*-azobenzene unit present, the ion pair aggregates promoted transmembrane ion transport (Figure 1.5A). However, upon isomerization to the *cis*-azobenzene, single channel currents were no longer detected, indicating channel blockage.

1.1.2.2 Voltage-Gated Channels
An early example of a voltage-gated channel relied on the use of an alkylammonium-oligoether phosphate ion pair **20** with an overall negative charge (Figure 1.6B). These ion pairs assemble into half-channels in each leaflet of a

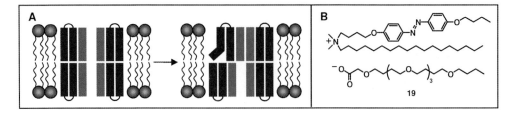

Fig. 1.5 (A) Schematic representation of a light-gated ion channel.
(B) Structure of ion pairs that assemble into a light-gated ion channel.

bilayer (Figure 1.6A). When aggregates in each leaflet align to form trans-membrane channels, there are typically unequal numbers of negatively charged monomers in each half channel, resulting in an overall molecular dipole [28]. Depending on the orientation of this molecular dipole, an applied voltage either stabilizes or destabilizes the assemblies, providing voltage-dependent ion transport.

Similar voltage-gated channels were constructed using membrane-spanning monomers with molecular dipoles (Figure 1.6A). Both, a bis-macrocycle bolaamphiphile **21** with a carboxylic acid and a succinic acid on opposite ends [29] and a bis-cholic acid compound **22** with a carboxylic acid on one end and a phosphoric acid group on the other [30], assemble into voltage-gated channels (Figure 1.6B).

The use of charged monomers is not a prerequisite for achieving voltage-gated transport through artificial channels. Using a peptide-dialkylamine conjugate **23** that dimerizes in lipid bilayers, a chloride-selective channel was developed that demonstrated voltage-dependent gating [31] (Figure 1.7A,B). A second example of a channel incorporating uncharged monomers utilizes tripeptide-functionalized *p*-octiphenyl rods with a methoxy group on one end and a methyl

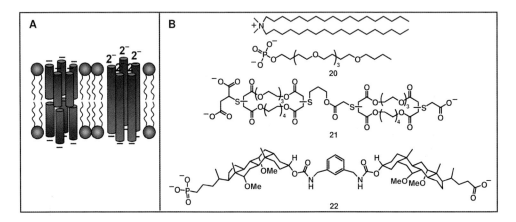

Fig. 1.6 (A) Schematic representations of voltage-gated ion channels.
(B) Compounds that assemble into voltage-gated ion channels.

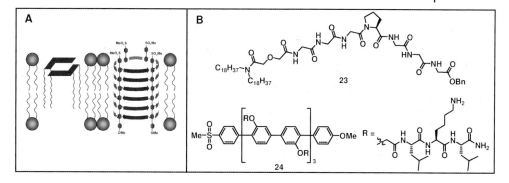

Fig. 1.7 (A) Schematic representations of voltage-gated ion channels.
(B) Uncharged compounds that assemble into voltage-gated ion
channels.

sulfone on the other **24** [32]. These *p*-octiphenyl rods with axial dipoles displayed
voltage-dependent β-barrel assembly.

1.1.2.3 Ligand-Gated Channels

A number of ligand-gated artificial ion channels have been reported. In an early
example of a ligand-activated channel, polyhistidine and copper ions were utilized
to organize transmembrane assemblies of iminodiacetate-functionalized oligo-
phenylenes **25** [32] (Figure 1.8A,B). This assembly was found to have a K$^+$ transport activity comparable to that of amphotericin B.

A second example of ligand gating relies on the formation of charge-transfer
complexes to open the channel [33]. In this case, *p*-octiphenylene rods functionalized with naphthalenediimide groups **26** initially assembled into π-helices which
act as closed ion channels (Figure 1.9A,B). Upon the addition of a dialkoxynaphthalene **27** that intercalated between the naphthalenediimide groups, charge-

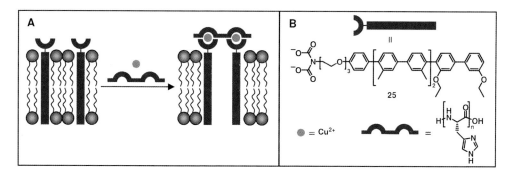

Fig. 1.8 (A) Schematic representation of a ligand-gated ion channel.
(B) Components of the ligand-gated ion channel.

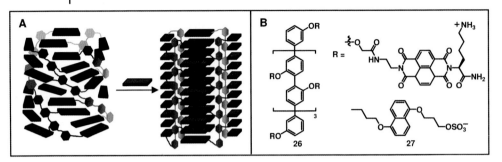

Fig. 1.9 (A) Schematic representation of a ligand-gated ion channel. The intercalation of dialkoxynaphthalene molecules (red) between naphthalenediimide groups (blue) causes the channel to open. (B) Components of the ligand-gated ion channel.

transfer complexes formed, leading to untwisting of the assemblies and an opening of the channels.

The addition of a ligand may not only lead to the formation of open channels; rather, it can also cause the blockage of artificial ion channels. This type of blockage gating has been demonstrated with a cucurbituril-based channel that is blocked by acetylcholine [34] as well as with a β-barrel pore blocked by polyglutamate [35].

1.2
Methods

As natural ion channels act in cell membranes, cell membrane mimics are used to assess the activity of artificial ion channels. Either planar lipid bilayers or spherical lipid bilayers called vesicles, or liposomes, are utilized in ion transport experiments.

1.2.1
Planar Bilayers

Planar bilayer clamp studies provide a means of establishing that a synthetic compound acts as a transmembrane ion channel [36, 37]. The set-up for these experiments involves preparing a bilayer membrane across a small hole in a hydrophobic partition between two chambers containing an electrolyte solution (Figure 1.10A). An electric potential is established across the lipid bilayer by inserting electrodes into the solution chambers; the current passing between these electrodes is then monitored using a bilayer clamp instrument. As the bilayer itself acts as a good insulator, step changes in the conductance represent ion transport through transmembrane channels.

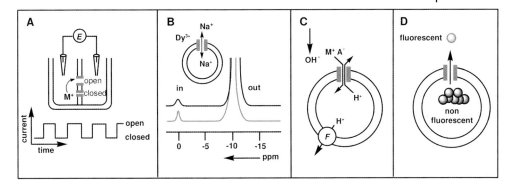

Fig. 1.10 (A) Schematic representations of the apparatus used in planar bilayer experiments and the type of data generated. (B) Depiction of the exchange of Na$^+$ ions inside vesicles and how it affects the ^{23}Na NMR spectra. (C) Representation of the flow of ions in and out of vesicles during a pH-stat experiment. (D) Depiction of an experiment using a concentration-sensitive fluorescent dye to monitor ion transport.

1.2.2
Vesicles

Vesicles, or spherical lipid bilayers enclosing an aqueous space, are also used to assess the ability of synthetic compounds to act as artificial ion channels. Vesicles can be prepared by a number of different methods, including sonication, extrusion, and detergent dialysis [38]. Dynamic light scattering and electron microscopy allow the size distribution and morphology of vesicles to be assessed. Common techniques used to monitor ion transport across vesicle bilayers include ^{23}Na NMR, pH-stat, pH- or environment-sensitive fluorescent dyes, and ion-selective electrodes.

1.2.2.1 ^{23}Na NMR
For ^{23}Na NMR experiments [39–42], large unilamellar vesicles are prepared in a NaCl solution. Addition of a dysprosium tripolyphosphate shift reagent changes the chemical shift of the Na$^+$ in the external solution [43]. In the presence of an active channel, the Na$^+$ ions inside and outside the vesicles exchange, leading to line broadening of the ^{23}Na NMR signals (Figure 1.10B). This line broadening is directly proportional to the rate of transmembrane Na$^+$ transport.

1.2.2.2 pH-Stat
For a typical pH-stat experiment [44], vesicles are prepared in a pH 6.6 buffer and the external solution is subsequently adjusted to pH 7.6. Following the addition of the proton carrier carbonyl cyanide 4-(trifluoromethoxy)phenylhydrazone (FCCP) to ensure rapid proton transport across the vesicle membranes, a metal sulfate solution is added to establish an opposing cation gradient (Figure 1.10C).

Upon addition of the channel, proton efflux occurs and a solution of base is added to maintain the pH at 7.6. The amount of base needed to maintain the pH is related to the activity of the channel.

1.2.2.3 Fluorescence

A variety of fluorescent dyes can be entrapped in vesicles to provide information regarding the activity of ion channels (Figure 1.10D). Fluorescent probes utilized include the pH-sensitive dye 8-hydroxypyrene-1,3,6-trisulfonic acid trisodium salt (HPTS) [45–47], the concentration-sensitive dye 5(6)-carboxyfluorescein (CF) [48, 49], the potential-sensitive dye safranin O [50], and the fluorophore/quencher pair 8-aminonaphthalene-1,3,6-trisulfonate (ANTS)/*p*-xylenebis(pyridinium) bromide (DPX) [51].

1.2.2.4 Ion-Selective Electrodes

Channel activity can also be assessed using an ion-selective electrode to monitor the amount of a certain ion (such as Na^+ or Cl^-) released from vesicles [52, 53].

1.3
Outlook

Significant progress has been made in the field of artificial ion channels since the first example was reported in 1982. To date, a variety of strategies have been utilized in the construction of ion channels that mimic the function of those found in nature. These artificial channels have found applications in molecular recognition [54], sensing enzymatic reactions [32], as artificial enzymes [55], and in biosensors [56]. In addition, a few synthetic channels have exhibited antibacterial activity [57, 58].

Despite this progress, certain obstacles remain in the drive to achieve truly biomimetic ion transport. One issue that must be addressed is the regulation of transmembrane ion transport through these synthetic channels. While considerable progress has been made in the development of gated channels, there is still a need for artificial channels with well-defined structures that can be opened and closed repeatedly, and in a reliable manner.

References

1 Ashcroft, F. M. (2000) *Ion Channels and Disease*. Academic Press, San Diego, CA.
2 Matile, S., Som, A., Sordé, N. (2004) Recent synthetic ion channels and pores. *Tetrahedron* **60**, 6405–6435.
3 Gokel, G. W., Mukhopadhyay, A. (2001) Synthetic models of cation-conducting channels. *Chem. Soc. Rev.* **30**, 274–286.
4 Koert, U. (2004) Synthetic ion channels. *Bioorg. Med. Chem.* **12**, 1277–1350.
5 Fyles, T. M., van Straaten-Nijenhuis, W. F. (1996) Ion channel models. In: Reinhoudt, N. D. (Ed.), *Comprehensive*

Supramolecular Chemistry. Elsevier, Oxford, Vol. 10, pp. 53–77.

6 Hector, R. S., Gin, M. S. (2005) Signal-triggered transmembrane ion transport through synthetic channels. *Supramol. Chem.* **17**, 129–134.

7 Koert, U., Al-Momani, L., Pfeifer, J. R. (2004) Synthetic ion channels. *Synthesis* 1129–1146.

8 Kobuke, Y., Morita, K. (1998) Artificial ion channels of amphiphilic ion pairs composed of oligoether-ammonium and hydrophobic carboxylate or phosphates. *Inorg. Chim. Acta* **283**, 167–174.

9 Merritt, M., Lanier, M., Deng, G., Regen, S. L. (1998) Sterol-polyamine conjugates as synthetic ionophores. *J. Am. Chem. Soc.* **120**, 8494–8501.

10 Clark, T. D., Buehler, L. K., Ghadiri, M. R. (1998) Self-assembling cyclic β^3-peptide nanotubes as artificial transmembrane ion channels. *J. Am. Chem. Soc.* **120**, 651–656.

11 Ghadiri, M. R., Granja, J. R., Buehler, L. K. (1994) Artificial transmembrane ion channels from self-assembling peptide nanotubes. *Nature* **369**, 301–304.

12 Tabushi, I., Kuroda, Y., Yokota, K. (1982) A,B,D,F-tetrasubstituted β-cyclodextrin as artificial channel compound. *Tetrahedron Lett.* **23**, 4601–4604.

13 Ishida, H., Qi, Z., Sokabe, M., Donowaki, K., Inoue, Y. (2001) Molecular design and synthesis of artificial ion channels based on cyclic peptides containing unnatural amino acids. *J. Org. Chem.* **66**, 2978–2989.

14 Wright, A. J., Matthews, S. E., Fischer, W. B., Beer, P. D. (2001) Novel resorcin[4]arenes as potassium-selective ion-channel and transporter mimics. *Chem. Eur. J.* **7**, 3474–3481.

15 Chen, W.-H., Nishikawa, M., Tan, S.-D., Yamamura, M., Satake, A., Kobuke, Y. (2004) Tetracyano-resorcin[4]arene ion channel shows pH dependent conductivity change. *Chem. Commun.* 872–873.

16 Tanaka, Y., Kobuke, Y., Sokabe, M. (1995) A non-peptidic ion channel with K^+ selectivity. *Angew. Chem. Int. Ed.* **34**, 693–694.

17 Yoshino, N., Satake, A., Kobuke, Y. (2001) An artificial ion channel formed by a macrocyclic resorcin[4]arene with amphiphilic cholic acid ether groups. *Angew. Chem. Int. Ed.* **40**, 457–459.

18 Madhavan, N., Robert, E. C., Gin, M. S. (2005) A highly active anion-selective aminocyclodextrin ion channel. *Angew. Chem. Int. Ed.* **44**, 7584–7587.

19 Pregel, M. J., Jullien, L., Canceill, J., Lacombe, L., Lehn, J.-M. (1995) Channel-type molecular structures. Part 4. Transmembrane transport of alkali-metal ions by 'bouquet' molecules. *J. Chem. Soc. Perkin Trans. 2*, 417–426.

20 Maulucci, N., De Riccardis, F., Botta, C. B., Casapullo, A., Cressina, E., Fregonese, M., Tecilla, P., Izzo, I. (2005) Calix[4]arene-cholic acid conjugates: a new class of efficient synthetic ionophores. *Chem. Commun.* 1354–1356.

21 Pechulis, A. D., Thompson, R. J., Fojtik, J. P., Schwartz, H. M., Lisek, C. A., Frye, L. L. (1997) The design, synthesis and transmembrane transport studies of a biomimetic sterol-based ion channel. *Bioorg. Med. Chem.* **5**, 1893–1901.

22 Fyles, T. M., James, T. D., Kaye, K. C. (1993) Activities and modes of action of artificial ion channel mimics. *J. Am. Chem. Soc.* **115**, 12315–12321.

23 Jullien, L., Lehn, J.-M. (1988) The "chundle" approach to molecular channels: synthesis of a macrocycle-based molecular bundle. *Tetrahedron Lett.* **29**, 3803–3806.

24 Otis, F., Voyer, N., Polidori, A., Pucci, B. (2006) End group engineering of artificial ion channels. *New J. Chem.* **30**, 185–190.

25 Murillo, O., Watanabe, S., Nakano, A., Gokel, G. W. (1995) Synthetic models for transmembrane channels: structural variations that alter cation flux. *J. Am. Chem. Soc.* **117**, 7665–7679.

26 Chen, W.-H., Shao, X.-B., Regen, S. L. (2005) Poly(choloyl)-based amphiphiles as pore-forming agents: transport-active monomers by design. *J. Am. Chem. Soc.* **127**, 12727–12735.

27 Kobuke, Y., Ohgoshi, A. (2000) Supramolecular ion channel containing *trans*-azobenzene for photocontrol of ionic fluxes. *Colloids Surf., A* **169**, 187–197.

28 Kobuke, Y., Ueda, K., Sokabe, M. (1995) Totally synthetic voltage dependent ion channel. *Chem. Lett.* 435–436.

29 Fyles, T. M., Loock, D., Zhou, X. (1998) A voltage-gated ion channel based on a bis-macrocyclic bolaamphiphile. *J. Am. Chem. Soc.* **120**, 2997–3003.

30 Goto, C., Yamamura, M., Satake, A., Kobuke, Y. (2001) Artificial ion channels showing rectified current behavior. *J. Am. Chem. Soc.* **123**, 12152–12159.

31 Gokel, G. W., Schlesinger, P. H., Djedovič, N. K., Ferdani, R., Harder, E. C., Hu, J., Leevy, W. M., Pajewska, J., Pajewski, R., Weber, M. E. (2004) Functional, synthetic organic chemical models of cellular ion channels. *Bioorg. Med. Chem.* **12**, 1291–1304.

32 Sakai, N., Mareda, J., Matile, S. (2005) Rigid-rod molecules in biomembrane models: from hydrogen-bonded chains to synthetic multifunctional pores. *Acc. Chem. Res.* **38**, 79–87.

33 Talukdar, P., Bollot, G., Mareda, J., Sakai, N., Matile, S. (2005) Ligand-gated synthetic ion channels. *Chem. Eur. J.* **11**, 6525–6532.

34 Jeon, Y. J., Kim, H., Jon, S., Selvapalam, N., Oh, D. H., Seo, I., Park, C.-S., Jung, S. R., Koh, D.-S., Kim, K. (2004) Artificial ion channel formed by cucurbit[*n*]uril derivatives with a carbonyl group fringed portal reminiscent of the selectivity filter of K$^+$ channels. *J. Am. Chem. Soc.* **126**, 15944–15945.

35 Gorteau, V., Bollot, G., Mareda, J., Pasini, D., Tran, D.-H., Lazar, A. N.,

Coleman, A. W., Sakai, N., Matile, S. (2005) Synthetic multifunctional pores that open and close in response to chemical stimulation. *Bioorg. Med. Chem.* **13**, 5171–5180.

36 Fyles, T. M., Loock, D., van Straaten-Nijenhuis, W. F., Zhou, X. (1996) Pores formed by bis-macrocyclic bola-amphiphiles in vesicle and planar bilayer membranes. *J. Org. Chem.* **61**, 8866–8874.

37 Murillo, O., Suzuki, I., Abel, E., Murray, C. L., Meadows, E. S., Jin, T., Gokel, G. W. (1997) Synthetic transmembrane channels: functional characterization using solubility calculations, transport studies, and substituent effects. *J. Am. Chem. Soc.* **119**, 5540–5549.

38 Torchilin, V. P., Weissig, V. (2003) *Liposomes.* Oxford University Press, New York, NY.

39 Riddell, F. G., Hayer, M. K. (1985) The monensin-mediated transport of sodium ions through phospholipid bilayers studied by ^{23}Na-NMR spectroscopy. *Biochim. Biophys. Acta* **817**, 313–317.

40 Espínola, C. G., Delgado, M., Martín, J. D. (2000) Synthetic flux-promoting polyether models: cation flux dependence on polyoxyethylene chain length. *Isr. J. Chem.* **40**, 279–288.

41 Murillo, O., Watanabe, S., Nakano, A., Gokel, G. W. (1995) Synthetic models for transmembrane channels: structural variations that alter cation flux. *J. Am. Chem. Soc.* **117**, 7665–7679.

42 Pregel, M. J., Jullien, L., Lehn, J.-M. (1992) Towards artificial ion channels: transport of alkali metal ions across liposomal membranes by "bouquet" molecules. *Angew. Chem. Int. Ed.* **31**, 1637–1640.

43 Gupta, R. K., Gupta, P. (1982) Direct observation of resolved resonances from intra- and extracellular sodium-23 ions in NMR studies of intact cells and tissues using dysprosium(III)tri-polyphosphate as paramagnetic shift reagent. *J. Magn. Reson.* **47**, 344–350.

44 Fyles, T. M., James, T. D., Kaye, K. C. (1993) Activities and modes of action

of artificial ion channel mimics. *J. Am. Chem. Soc.* **115**, 12315–12321.

45 Kano, K., Fendler, J. H. (1978) Pyranine as a sensitive pH probe for liposome interiors and surfaces: pH gradients across phospholipid vesicles. *Biochim. Biophys. Acta* **509**, 289–299.

46 Clement, N. R., Gould, J. M. (1981) Pyranine (8-hydroxy-1,3,6-pyrenetrisulfonate) as a probe of internal aqueous hydrogen ion concentration in phospholipid vesicles. *Biochemistry* **20**, 1534–1538.

47 Menger, F. M., Davis, D. S., Persichetti, R. A., Lee, J.-J. (1990) Synthetic flux-promoting compounds. Exceeding the ion-transporting ability of gramicidin. *J. Am. Chem. Soc.* **112**, 2451–2452.

48 Carmichael, V. E., Dutton, P. J., Fyles, T. M., James, T. D., Swan, J. A., Zojaji, M. (1989) Biomimetic ion transport: a functional model of a unimolecular ion channel. *J. Am. Chem. Soc.* **111**, 767–769.

49 Das, G., Onouchi, H., Yashima, E., Sakai, N., Matile, S. (2002) Binding of organic anions by synthetic supramolecular metallopores with internal Mg^{2+}-aspartate complexes. *ChemBioChem* **3**, 1089–1096.

50 Sakai, N., Gerard, D., Matile, S. (2001) Electrostatics of cell membrane recognition: structure and activity of neutral and cationic rigid push-pull rods in isoelectric, anionic, and polarized lipid bilayer membranes. *J. Am. Chem. Soc.* **123**, 2517–2524.

51 Das, G., Matile, S. (2002) Transmembrane pores formed by synthetic p-octiphenyl β-barrels with internal carboxylate clusters: regulation of ion transport by pH and Mg^{2+}-complexed 8-aminonaphthalene-1,3,6-trisulfonate. *Proc. Natl. Acad. Sci. USA* **99**, 5183–5188.

52 Weber, M. E., Schlesinger, P. H., Gokel, G. W. (2005) Dynamic

assessment of bilayer thickness by varying phospholipids and hydraphile synthetic channel chain lengths. *J. Am. Chem. Soc.* **127**, 636–642.

53 Pajewski, R., Ferdani, R., Pajewska, J., Djedovič, N., Schlesinger, P. H., Gokel, G. W. (2005) Evidence for dimer formation by an amphiphilic heptapeptide that mediates chloride and carboxyfluorescein release from liposomes. *Org. Biomol. Chem.* **3**, 619–625.

54 Baudry, Y., Bollot, G., Gorteau, V., Litvinchuk, S., Mareda, J., Nishihara, M., Pasini, D., Perret, F., Ronan, D., Sakai, N., Shah, M. R., Som, A., Sordé, N., Talukdar, P., Tran, D.-H., Matile, S. (2006) Molecular recognition by synthetic multifunctional pores in practice: are structural studies really helpful? *Adv. Funct. Mater.* **16**, 169–179.

55 Som, A., Matile, S. (2002) Rigid-rod β-barrel ion channels with internal "cascade blue" cofactors – catalysis of amide, carbonate, and ester hydrolysis. *Eur. J. Org. Chem.* **22**, 3874–3883.

56 Husaru, L., Schulze, R., Setiner, G., Wolff, T., Habicher, W. D., Salzer, R. (2005) Potential analytical applications of gated artificial ion channels. *Anal. Bioanal. Chem.* **382**, 1882–1888.

57 Fernandez-Lopez, S., Kim, H.-S., Choi, E. C., Delgado, M., Granja, J. R., Khasanov, A., Kraehenbuehl, K., Long, G., Weinberger, D. A., Wilcoxen, K. M., Ghadiri, M. R. (2001) Antibacterial agents based on the cyclic D,L-α-peptide architecture. *Nature* **412**, 452–455.

58 Leevy, W. M., Gokel, M. R., Hughes-Strange, G. B., Schlesinger, P. H., Gokel, G. W. (2005) Structure and medium effects on hydraphile synthetic ion channel toxicity to the bacterium *E. coli*. *New J. Chem.* **29**, 205–209.

2
Self-Assembling Nanostructures from Coiled-Coil Peptides

Maxim G. Ryadnov and Derek N. Woolfson

2.1
Background and Overview

2.1.1
Introduction: Peptides in Self-Assembly

The past decade has witnessed growing interest in employing biomolecules in the bottom-up assembly of increasingly complex nanostructures and materials [1]. In Nature, molecular self-assembly is a ubiquitous process in which various inter-molecular forces cooperate to guide the organization of biomolecules into higher levels of structural hierarchy [2]. This organization ranges from polymers that only become ordered in the presence of their target molecules or surfaces, through the spontaneous folding of linear polymer chains to precisely defined three-dimensional structures, to the association and further assembly of such folded structures into functioning assemblies and materials [1]. The majority of these processes occur at the nanoscale, employing biomolecules such as carbohy-drates, lipids, nucleic acids, and polypeptides. The resulting structures are *nano-structures* which, in some cases, assemble further to form nano-to-mesoscale com-plexes and materials. The forces involved are non-covalent interactions, including hydrogen and ionic bonding, the hydrophobic effect, and van der Waals' interac-tions. Individually, such interactions are negligible compared with covalent bonds and even with respect to ambient thermal energy. However, through the forma-tion of extended cooperative networks, non-covalent bonds can direct and cement thermodynamically stable structures. Moreover, unlike most molecules formed by covalent bonds, the assembly of such structures is readily reversed and may even be transient. As a result, assembled biological supramolecules respond to a vari-ety of external stimuli. In turn, this affords considerable control in biomolecular assembly (and disassembly) processes.

The above is the essence of natural molecular recognition and molecular self-assembly, and the nanobiotechnologist has a good deal to learn from it. Specifically – and of immediate interest to this chapter – there are now reliable rules that link the chemistry of biological molecules to their three-dimensional

Nanobiotechnology II. Edited by Chad A. Mirkin and Christof M. Niemeyer
Copyright © 2007 WILEY-VCH Verlag GmbH & Co. KGaA, Weinheim
ISBN: 978-3-527-31673-1

structures and mechanisms of assembly. Furthermore, these rules can be harnessed in the design of supramolecular complexes and materials. This so-called "bioinspiration" has largely been taken from natural nucleic acid and protein systems. Moreover, certain of these self-associating nucleic acids and proteins are built on relatively straightforward patterns of non-covalent interactions that are well understood. This facilitates using self-assembly to access otherwise synthetically inaccessible supramolecules, and in some cases leads to the creation of unprecedented supramolecular arrangements [3, 4]. For example, the chemical synthesis of protein-based suprastructures such as viruses, ribosomes or even simpler enzyme complexes is not yet achievable, whilst Nature accomplishes it seemingly effortlessly through non-covalent biomolecular recognition [2]. Nonetheless, Nature offers a clear guide on the route to take as it often constructs its rich set of complex protein architectures from simpler and often-repeated structural units. These smaller protein units are generally termed "peptides", and so-called "peptide-folding motifs" have unparalleled potential in synthetic polymer self-assembly. First, they provide explicit selectivity and specificity in supramolecular assembly, as observed in natural protein structures, which can be understood in terms of relationships or rules that link their primary chemistry to their 3D structures. Second, peptides are also accessible through synthetic chemistry [5, 6].

Like proteins, peptides are polymers of amino acids. Peptides can be synthesized as precisely defined, monodispersed polymers from a great diversity of amino-acid monomers – which can be either natural or synthetic – to present a wide variety of chemistries along the polymer chain. In some cases, the relationship between a peptide sequence (i.e., the order of amino acids along the polypeptide chain, read from the amino to carboxy terminus) and a structure is well established allowing clear, precise and elegant designs of 3D objects [7]. Finally, peptides can be prepared using developed chemical techniques.

Currently, many peptide-folding motifs are known, including those for collagen triple helices [8], zinc-finger motifs [9] and, most recently, β-amyloid fibrils [10]. Design rules are available to differing degrees for these and other motifs. Over the past two decades, however, one motif stands out for its utility in peptide-design and engineering studies – namely, the α-helical coiled coil – which is one of the most abundant and versatile protein-folding motifs in biological systems [11, 12]. The folding and assembly of coiled coils has been comprehensively studied, and clear rules are currently available for the design of coiled coil-based structures *de novo* [13]. Furthermore, because coiled coils are essentially motifs for protein–protein recognition and association, they represent an ideal building block for constructing peptide- and protein-based nanoscale objects. Herein, we review the latest achievements in the design and engineering of nanostructures and materials using the α-helical coiled coil.

2.1.2
Coiled-Coil Peptides as Building Blocks in Supramolecular Design

Coiled coils are rod-like bundles of α-helices (Figure 2.1) [11]. The unifying characteristic of all coiled-coil structures is the interdigitation of side chains from

A

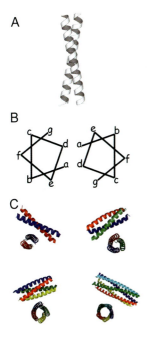

B

C

Fig. 2.1 (A) A dimeric coiled-coil superhelix. (B) These structures are characterized by heptad patterns, **abcdefg**, in their sequences, shown here configured onto helical wheels. (C) Amino acids at **a** and **d** form hydrophobic seams that cement helix–helix interactions. Different oligomerization states of coiled coils are observed in Nature (PDB entries 2ZTA, 1IJ3, 2BNI, 1T8Z).

neighboring helices in the cores of the bundles, which is referred to as "knobs-into-holes" packing [14, 15]. This packing arrangement, and the resulting association of helices, is set up by patterns of predominantly hydrophobic residues along each polypeptide chain. The most common (sometimes termed "canonical") pattern has hydrophobic residues spaced alternately three and four residues apart. The resulting seven-residue repeat is referred to as the "heptad repeat", and is usually denoted **abcdefg** with hydrophobic residues occupying positions **a** and **d**. This pattern sets up a hydrophobic seam on each α-helix. Because the average spacing of hydrophobic residues along the sequence (3.5 residues) falls short of one complete turn of an α-helix (3.6 residues), the seam has a left-handed twist with respect to the right-handed α-helix. Therefore, in order to maximize burial of the hydrophobic seams, the helices associate into bundles with left-handed helix-crossing angles. Finally, the remaining – **c**, **d**, **e**, **f**, and **g** – positions of the heptad repeat are often occupied by polar side chains, which make the individual coiled-coil helices amphipathic, and provide the resulting coiled-coil bundles with water-soluble surfaces. These basic features of coiled-coil association are illustrated in Figure 2.1.

In Nature, coiled-coil assembly is further complicated as individual bundles can have up to five helices (Figure 2.1C), the exact number being defined largely by the nature of the hydrophobic residues at the *a/d* interface [16]; higher-order arrangements are also possible [17], and coiled coils can also be either hetero- or homo-typic [18]. The saving grace, however, is that amino-acid preferences – and particularly those at the *a*, *d*, *e*, and *g* sites of the heptad repeats – determine the number and type of helices in the bundles, and some of these have been gleaned through analyses of, and experiments on, natural systems [13]. For example, a common coiled-coil motif often exploited in design and engineering studies is the *leucine zipper* (Figure 2.1A). Leucine-zipper sequences predominantly have the hydrophobic amino acid leucine in their *d* sites, which directs the assembly of parallel dimers (Figure 2.1A,B) [19, 20].

Such sequence-to-structure relationships in coiled coils provide "rules of thumb" for the prescriptive design of polypeptide chains that can be assembled from the bottom up into specific structures [13]. Furthermore, as each heptad repeat spans approximately 1 nm, the polypeptide chain can (in principle) be metered out to make struts, linkers, and fibers of precisely defined length. These features, in combination with modern methods for the synthesis of peptides, offer considerable potential and power to nanoscientists to assemble nanoscale objects. Indeed, in our view this is only rivaled by possibilities in the assembly of DNA-based objects [3, 21].

2.1.3
Coiled-Coil Design in General

The field of peptide and protein engineering and design is considerable [7]. Within this, arguably, the design of coiled coil-based motifs and structures has been the most successful over the past decade. This success can largely be attributed to: (i) the development of rules that link coiled-coil sequence and structure; and (ii) the straightforward synthesis of coiled-coil peptides. Foregoing studies of coiled-coil design include the assembly of straightforward homodimer, trimer, tetramer and pentamers; the rational design of heterodimers, trimers and tetramers; self-replicating systems, and metal- and heme-binding coiled-coil systems [7, 22]. Although not directly the subject of this chapter, this background provides a firm platform on which more ambitious coiled coil-based designs such as nanoscale linkers, peptide-based fibrous and gel materials, and components of peptide-based switches, can be built.

2.2
Methods and Examples

2.2.1
Ternary Coiled-Coil Assemblies and Nanoscale-Linker Systems

Some of the most illustrative and visually convincing examples of coiled coil-based nanostructures are the two- and three-dimensional networks of metal clus-

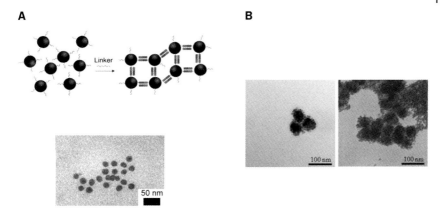

Fig. 2.2 (A) Gold nanoparticles assemblies mediated by a "belt-and-braces" nanoscale linker, with adjacent particles uniformly separated at 7 nm. (B) Satellite networks assembled from coiled coil-derivatized nanoparticles. (Adapted with permission from Refs. [23, 24]).

ters assembled with nanoscale precision from peptide-decorated gold nanoparticles (NPs) [23, 24]. To our knowledge, there are only two examples of such supramolecular arrangements based on coiled-coil linkers, though both are inspired by a larger body of work based on DNA-linker systems [25].

The first example is from our own laboratory [23]. The name of the linker – "belt-and-braces" – reflects the mode of assembly in which one peptide, "the belt", templates the co-assembly of its two half-sized "braces" (Figure 2.2A). Individually, the peptides are unstructured, but form a novel two-stranded coiled-coil assembly when mixed: the belt acts as a six-heptad template for the assembly of the two three-heptad brace peptides. The braces are directed to opposite ends of the belt using different combinations of hydrophobic, electrostatic and hydrogen-bonding interactions created through specific amino-acid placements at the *a, d, e* and *g* sites of the heptad repeats. This allows specific termini of each brace to be tagged with thiol-containing cysteine residues which, in turn, are used to bind to the cargo. The distance spanned by the assembled linker is predicted to be approximately 7 nm (6 nm for the six-heptad coiled coil plus some for the Cys-containing tags). The assembly and this spacing are confirmed by derivatization with gold NPs and the formation of mesoscopic metal networks (Figure 2.2A).

Although DNA-based linkers are better developed than those using peptides [25], peptide-based assemblies offer a number of potential advantages to this area. For instance, the peptide-based assemblies can be engineered for stability over a broader range of conditions; peptides can be functionalized readily, and they are amenable to large-scale production through the construction and expression of synthetic genes in bacterial hosts. This latter point is partly illustrated by the other report by Stevens et al. [24] where longer linkers were bacterially expressed. These studies also add another interesting example of peptide-directed

NP assemblies: namely, these authors use two sizes of NP – 8.5 and 53 nm – to produce satellite structures (Figure 2.2B). These two sets of NPs are each modified with a different complementary peptide; thus, when mixed, satellite structures resembling those organized by oligonucleotide linkers [26] are observed (Figure 2.2B).

2.2.2
Fibers Assembled Using Linear Peptides

With good reason, the field of bioinspired fibrous assemblies has been dominated over the past decade by β-amyloid and related structures. One reason for this concerted effort is the role of β-amyloid in the so-called protein-misfolding diseases, or the amyloidoses, such as Alzheimer's disease [27]. Potential applications of β-amyloid fibers in nanobiotechnology are increasingly being recognized [10]. In β-amyloid fibers, the polypeptide chains form β-strands which run perpendicular to the long axis of the fibers; these strands stack and hydrogen bond to form β-sheets that run along the long axis, and the β-sheet fibrils subsequently multimerize to form thickened fibers approximately 10 nm thick [28]. Whilst the successful rational designs of amyloid-like systems have been increasingly reported, many examples of amyloids are produced by the misfolding of natural peptides and proteins. This field has a large associated literature. Moreover, both the natural and designed variants are widely reviewed elsewhere [29]. For these reasons – and for those mentioned in the Introduction – here we focus on assemblies based on peptides designed to adopt α-helical conformations. As described in the previous section, several groups have succeeded in decorating self-assembled amyloid-like fibers and tubes based on β-structured peptides and misfolded protein with gold NPs, and some of these have been used to produce silver nanowires [30–32].

There are now several α-helix-based (and specifically coiled coil-based) designs that produce novel fibrous structures spanning the nano-to-mesoscale regimes. For example, Kojima et al. [33] describe a three-heptad coiled coil that forms helix bundles and propagates fibers 5–10 nm wide and several microns long. Though the fibers are not characterized in detail, this is the first report of such a system. The second example comes from our own group. We applied the concept of "sticky end assembly" from recombinant DNA technology to peptides for the first time to guide the longitudinal assembly of sticky-ended heterodimers into extremely long coiled coils (Figure 2.3A,B). The resulting fibers are ~50 nm thick and tens of microns long (Figure 2.3C); these structures represent the ordered co-assembly of approximately one million monomer units [34]. This approach has now been applied independently or adopted by other groups to assemble other coiled coil-based fibers [35–37], and those based on different protein-folding motifs such as the collagen triple helix [8]. This second area is gathering momentum, which is important for two reasons. First, the construction of self-assembling collagens presents a fresh set of challenges in synthesis and assembly. Second, the development of fibrous materials based on biologically relevant collagen peptides would have clear applications in tissue engineering. Unfortunately, this area is beyond the immediate scope of this chapter, and consequently the reader is referred to the recent primary literature [8, 38].

We refer to our assemblies as a self-assembling fiber (SAF) system. This comprises two complementary peptides that form a staggered heterodimer with oppositely charged "sticky ends" (Figure 2.3A). Unlike sticky-ended DNA assembly, however, it transpires that this design propagates the intended assembly not only longitudinally (Figure 2.3B) but also laterally, and this results in thickened fibers (Figure 2.3C). (Note: a dimeric coiled-coil fiber would be expected to be ~2 nm in diameter, and the fibers produced by the SAF peptides are ~50 nm thick.) As discussed below, we believe that this thickening process is inevitable; indeed, it is observed in the other designed fibrous systems.

The hallmark of the SAF system – and a key distinguishing feature from other peptide-based fibrous assemblies – is that it is a binary system. In other words, by using the aforementioned rules for coiled-coil design, two complementary (but different) peptides are encouraged to co-assemble to form the sticky-ended heterodimer. Moreover, the isolated individual peptides are unfolded, and it is only when they are mixed that the heterodimer is formed and fibriollogenesis ensues. This design produces novel features to the fibers at the nanoscale. Specifically, the fibers are polar [40], and they show considerable internal and external order on the nanoscale [39].

The binary design also allows for considerable control over the assembly process. For example, this has allowed the assembly of novel structures from the bottom up by introducing *special* peptides during the assembly of the *standard* linear SAF peptides [13, 41–43]. A later (but similar) example of coiled coil-based fiber assembly is described by Potekhin et al. as a general model for designing *n*-stranded coiled-coil ropes [35]. These authors propose that the repetition of identical heptad repeats in a peptide presents a high probability of mismatched or axially staggered α-helices during coiled-coil assembly. In turn, this should result in the formation of fibrils, which are indeed observed experimentally (Figure 2.4). This assumes the shift between adjacent helices to be equivalent to an integral number of the repeats which, in turn, gives rise to a general equation for the most favorable length of a helix to form an *n*-stranded rope: $(n \times 7 - 1)$, where n is the number of strands, 7 is the number of residues in the heptad repeat, and 1 is a residue spacer for the head-to-tail packing of the helices. Although, according to the equation the simplest design would be for two-stranded ropes using 13-residue peptides, the authors describe a longer 34-mer, which should form a five-stranded coiled coil (Figure 2.4B), arguing that this would be more stable than an assembly based on a smaller peptide. (Note that this is different from the SAF designs in which the offset between associated helices is set explicitly by rational design. Specifically, by complementary polar residues position in the helix–helix interface of the sticky-ended heterodimer [34].)

The 34-residue peptide described by Potekhin et al. forms fibrils 2.5 nm wide with the helices oriented along the rope, which is consistent with a five-stranded design without lateral association, as designed (Figure 2.4C). Curiously, however,

A

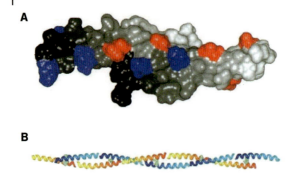

B

C

Fig. 2.3 The Self-Assembling Fiber (SAF) design. (A) A coiled-coil heterodimer designed to assemble offset and leave oppositely charged sticky ends (acidic shown in red, basic in blue). (B) Propagation of the dimer into a coiled coil protofibril. (C) Electron micrographs of matured fibers resulting from the bundling of coiled-coil protofibrils. (Adapted with permission from Refs. [34, 39]).

A

B

C   **D**

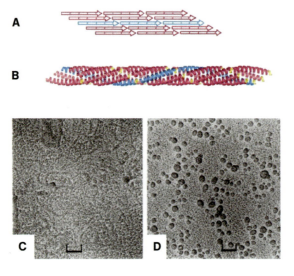

Fig. 2.4

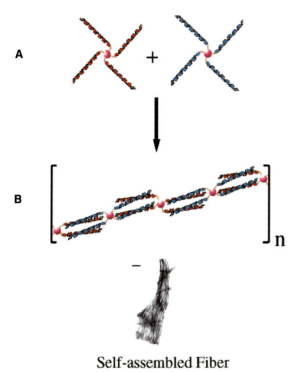

Self-assembled Fiber

Fig. 2.5 Tetramerized leucine-zipper peptides tethered to synthetic hubs (A), which co-assemble to form α-helical fibers (B). The scale bar for the electron microscopy image is 1 mμ. (Adapted with permission from Ref. [46]).

depending on pH, the peptide also associates into spherical assemblies of 10–15 nm diameter (Figure 2.4D). This design represents a very interesting model and holds potential for controlling the morphology of suprastructures, along with introducing desired functions. A more-recent fiber design by Potekhin et al. produces structures with a range of thicknesses [44]. Most recently, Zimenkov et al. presented a design for a single-peptide leucine zipper-based sequence [36] where,

Fig. 2.4 A fiber system based on a pentameric coiled-coil design. Staggered helices are depicted as arrows in (A) and modeled as ribbons in (B). One contiguous strand is shown in blue. The N-termini are shown in yellow to follow the head-to-tail alignment of the helices with each strand. Electron micrographs of fibrils (C) and spherical particles (D) for by these designed peptides. Scale bars = 100 nm. (Adapted with permission from Ref. [35]).

again, the resulting fibrillar structures were thicker (20–50 nm) than the intended two-stranded coiled-coil ropes.

2.2.3
Fibers Assembled Using Protein Fragments and Nonlinear Peptide Building Blocks

As shown recently by several groups [13, 45, 46], leucine zippers make excellent building blocks for different supramolecular nanostructures. For example, Zhou et al. described leucine-zipper peptides displayed on dendritic hubs that assemble to form fibrous structures [46]. The system comprises two complementary peptide sequences, each tetramerized as star-like dendrimers (Figure 2.5A).

The individual tetramers are α-helical to some extent on their own, but this does not lead to supramolecular assemblies. However, when mixed 1:1 the two dendrimers irreversibly form helical fibrils, which rapidly aggregate to give larger fibrous precipitates (Figure 2.5B). Intriguingly, the dendrimers sustain their pre-adopted helical structure upon aggregation, producing another and different example of α-helical fibers.

Nonlinear constructions not only provide the bulk material for supramolecular assemblies, but can also be used as supplements to program assembly. For instance, we have designed orthogonal leucine-zipper constructs that are complementary to the basic linear building blocks of the aforementioned SAF system. These introduce various morphological changes into the otherwise straight fibers [13, 41]. The characterizing feature of these constructs is that they combine multiple copies of identical or different complementary half-sized peptides into one molecule in head-to-head or tail-to-tail fashions. These *special peptides* are intended to combine with the *standard* linear SAF peptides and direct convergent and divergent modes of SAF assembly to produce nonlinear fibers (Figure 2.6). As the figure indicates, different fiber morphologies are observed depending on the precise special construct added [41]. This offers a route for controlling and programming supramolecular hierarchies on a rational basis [13].

In another topological design, Ogihara et al. use coiled coils to propagate the assembly of domain-swapped proteins [45]. Three-dimensional domain swapping in proteins is the exchange of structural subdomains between monomers, leading to oligomerization. As a result, the protein monomers become entwined in the oligomer, but most of the non-covalent interactions in the domain-swapped structure resemble those in the natively folded monomers. The starting point for Ogihara's studies is a monomeric three-helix bundle with an up-down-down topology of helices, which is derived from a previously designed coiled coil [47]. As shown in Figure 2.7A, helical regions I/II of the first monomer and region III′ of a second monomer associate to form a structural unit. This leaves regions I′/II′ and III available as "sticky ends" for further assembly into filaments. Indeed, the resulting suprastructures are characteristic of mesoscopic fibers with orientated subunits along the filament axis, and also the aforementioned lateral association of the filaments into larger fibers (Figure 2.7B).

Most recently, Lazar et al. showed that engineered helix-turn-helix fragments from the apolipoprotein A-I form extended and thickened fibers. A key difference

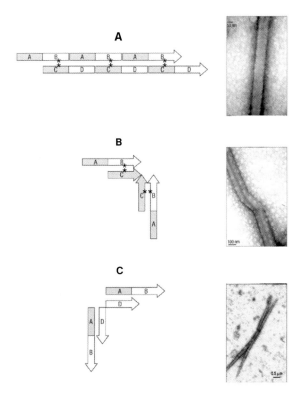

Fig. 2.6 Constructing nonlinear peptide-based fibers. (A) Standard SAFs. (B) Convergent and (C) divergent modes of assembly using special, nonlinear peptides to supplement the standard SAF peptides. (Adapted with permission from Ref. [41]).

between these fibers and those described above is that the α-helices are oriented perpendicular to the long axes of the fibers [48].

Finally in this section, Yeates and co-workers used natural protein–protein oligomerization units to build discrete nanoscale objects and fibers. A full discussion of these exciting investigations is beyond the scope of this chapter, but can be found in reports and reviews provided by Yeates's group [4].

2.2.4
Summary: Pros and Cons of Peptide-Based Assembly of Nanofibers

The preceding sections of this chapter have introduced the concepts and examples of peptide-based, self-assembling fibers. We have focused on constructions using α-helically structured peptides, although fibers have been generated by many others using β-amyloid-like structures, collagen triple helices and larger natural protein units. Thus, the field of peptide and protein-based fibrous materials is burgeoning. There is also considerable potential for such structures in

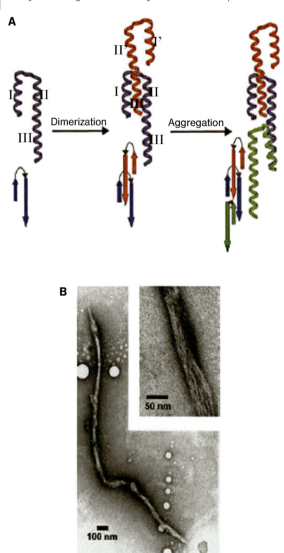

Fig. 2.7 A coiled coil-based domain-swapped system. The assembly of a monomeric three-helix bundle with up-down-down topology (A) into mesoscopic fibers (A and B). (Adapted with permission from Ref. [45]).

nanotechnology through more traditional biotechnology applications, ranging from templates for inorganic materials and nanothin wires, to biocompatible scaffolds for cell culture, including wound healing and tissue engineering. These and other applications have been reviewed recently elsewhere [5, 6].

One intriguing feature of most – if not all – of the peptide-based fibers discussed here and elsewhere is that the intended fibrils assemble further to form thickened, matured fibers. The extent of this coarsening varies from system to system; for example, with β-amyloid-like assemblies the matured fibers tend to be of the order of 10 nm thick, which implies the packing of a small number of protofibrils, whereas in our own SAF system the matured fibers can be up to 70 nm thick and involve hundreds of two-stranded coiled-coil fibrils.

Some of the questions that arise from these observations are:

1. What drives the thickening process(es)?
2. Is the thickening limited, and can it be controlled?
3. Is the thickening a failing or a drawback of the design procedures and choice of building blocks?
4. Is the converse of question 3 true – is thickening actually an advantage?

First, thickening – at least to some extent – may well be inevitable for fibers of the type we have described [34]. This is because identical chemical moieties will be presented regularly on the outer faces of the nascent fibrils through helical repetition of identical building blocks. In such arrangements, if two such moieties have some affinity for each other, however small, they will tend to zipper up via complementary patterns along the lengths of the fibrils. In this way, fibrils will be brought together to create thickened fibers via an avidity effect. With regard to limiting and controlling this process, much consideration has been made for both the amyloid and coiled-coil systems. For example, if the geometry of the fibrils being recruited to fibers is compromised at all, then coarsening will be arrested when the deformation energy outweighs the affinity. Whether thickening is a failing or an advantage of these systems is first, a matter of opinion, and second, is dependent on the particular applications in mind for the fibers. For instance, thickening may be a disadvantage when creating templates for nanothin wires, but an advantage for building a stable scaffold for tissue engineering where nanoscale precision is less important.

This subject of fiber maturation has been addressed by several groups investigating coiled coil-based fibrous systems. For example, the design of nanofilaments and nanoropes reported by Wagner et al. [37] is of a leucine-zipper peptide with a similar mode of assembly as described for the SAF system. The aim in part, however, was to address the issue of thickening by using non-covalent interactions concomitantly to favor axial and to disfavor (or at least regulate) lateral modes in the assembly of fibers. The resulting fibrous structures are indeed thinner compared to the SAF designs, but they are also much shorter. Heterogeneity in the length of nanoropes, and also some of the less-regular morphologies observed, are possibly related to poor inter-associations and stabilities of the coiled-coil protofilaments. In addition, disulfide bonds formed by oxidized cysteine residues in the sequence may contribute to the polymerization effect and partial misfolding. As noted by others, the latter can lead to the creation of less-well ordered aggregates and so-called fractal patterns of assembly, irrespective of the peptide se-quences and conformations [49–52].

Most recently, we have shown that the thickness, stability, and other properties of the SAFs can each be engineered to some extent through rational redesign [39]. Briefly, the first-generation SAFs – which comprise two complementary four-heptad leucine zippers – form fibrils that rapidly assemble to matured fibers ∼45 nm thick and tens of microns long [34]. In a second-generation SAF design, further thickening is promoted through additional favorable electrostatic interac-tions on the fibril surfaces. This results in the production of thicker (∼70 nm) fibers, and also introduces a number of interesting new features, notably that the fibers are stabilized, they are internally more ordered on the nanoscale, and they are polar [39, 40]. In a third-generation redesign, the initial step in fibrillo-genesis (i.e., formation of the heterodimer) is promoted by increasing the lengths of the interacting SAF peptides to five heptads each; otherwise, the design is as for the second-generation SAFs. Interestingly, the resulting fibers are more stable than their second-generation counterparts and maintain the nanoscale order, but they are slightly thinner at ∼60 nm on average.

A final intriguing observation from the redesigned SAFs is that the second- and third-generation fibers (but not the first-generation) show striations in negative-stain transmission electron microscopy reminiscent of patterns observed in nat-ural fibrous assemblies such as collagen and fibrin [53, 54]. For the synthetic SAF system, the distance between striations matches the expected lengths of the com-ponent peptides folded into α-helices and aligned with the long axes of the fibers (Figure 2.8). This suggests: (i) nanoscale order in the fibers above that originally designed; and (ii) that these features can be tuned by using peptides of different lengths [39].

2.2.5
Assembling More-Complex Matrices Using Peptide Assemblies as Linker Struts

Coiled coils have also been proposed as components and building blocks in the assembly of gels [55–57] and fibrillar matrices [13, 37]. With the latter in mind in particular – since coiled coils share similar dimensions and morphologies with fibrous components of the extracellular matrix (ECM) – designed α-helical fibrillar matrices represent a promising route to the engineering of synthetic ECM analogues, although as yet this suggestion needs to be supported by experi-mental investigations.

2.2.5.1 Programmed Matrices Assembled Exclusively from Coiled-Coil Building Blocks
As described above, the SAF system can be supplemented with special peptides to create more-complex suprastructures. Most recently, this has included assemblies that resemble natural fibrillar networks [13], where the fibers are themselves con-sidered as the building blocks or struts of the matrices, and the special peptides as the shapers and connectors [13, 41]. We have observed various fiber and matrix

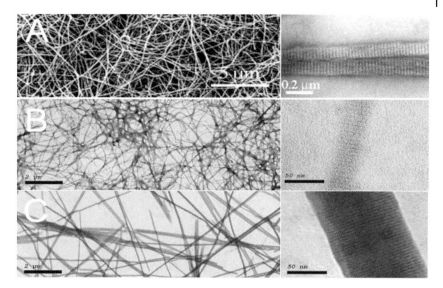

Fig. 2.8 Electron micrographs of natural (fibrin) and designed (SAF) fibers with characteristic band patterns. (A) Fibrin fibers striated with a repeat of 22.5 nm, (B) and (C) two SAF generations with striations separated at 5.3 and 4.2 nm, respectively. (Adapted with permission from Refs. [39, 53]).

morphologies depending on the special peptide added (Figure 2.9). Again, as these assemblies are peptide-based and founded on non-covalent interactions, they are biocompatible and reversible – which is attractive for creating remodelable matrices for *in-vivo* applications. However, stability remains an issue here as cell-culture media and biological fluids each present challenges to the self-assembly of peptide systems cemented by a small number of non-covalent forces [39].

There are many other examples of artificial matrices based on fully synthetic, polymer-peptide and completely peptidic systems, which are not limited by the stability problems that currently face some of the coiled coil-based systems. Thus, various designs focused on other structural motifs such as *β*-structures are at present at a more-advanced stage of development [58, 59].

2.2.5.2 Synthetic Polymer-Coiled-Coil Hybrids

Coiled coils, whether native or designed, can undergo explicit and drastic stability, conformational and topological transitions in response to external stimuli (temperature, pH, ionic strength, etc.). In combination with the aforementioned ideas for self-assembled matrices, coiled coils represent an attractive option for the development of responsive suprastructures and materials.

For example, coiled-coil domains have been utilized in the construction of stimuli-sensitive polymer hydrogels [55–57]. Petka et al. described reversible hy-

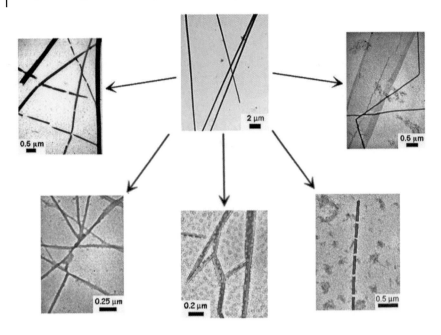

Fig. 2.9 Morphologies introduced by different nonlinear special peptides in the SAF background. (Adapted with permission from Refs. [13, 41]).

drogels incorporating self-assembling leucine-zipper sequences [55], the authors' aim being to design a polymolecular system with two key properties for gel formation. These were: (i) strong interchain interactions of polymers to form junction points in the molecular network; and (ii) solubility of the chains to allow the assembled system to swell in water, rather than to precipitate. In addition, a gel-sol transition was prescribed. The assembly that satisfied these criteria comprised leucine-zipper units to provide the interchain interactions, and polyelectrolyte domains based on peptide-PEG conjugate; the peptide component of the latter is based on alanine and glycine to provide water-solubility, but there was no fixed secondary structure that would allow the formation of a highly swollen hydrogel (Figure 2.10A). Finally, cysteine residues are included to produce disulfide cross-links and to provide additional strength to the gels, once formed.

The aforementioned reversibility of coiled-coil formation is postulated to allow control over gelation and viscoelastic properties. Indeed, by changing pH and temperature, it is possible to tune the gel-solution transition of the hydrogels (Figure 2.10A). In addition, the erosion rate of such gels (which is important for many applications) can be controlled: Shen et al. show that rapid surface dissolution is largely attributed to stronger intramolecular coiled-coil interactions during the formation of networks, and that these can be suppressed by changing the network topology to provide more-stable materials [56].

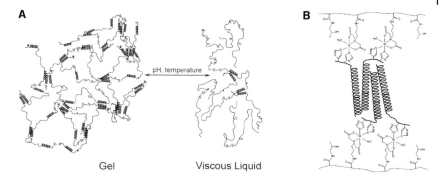

Fig. 2.10 Reversible hydrogels. (A) A co-polymer of coiled-coil domains and Ala-Gly-PEG chains self-assembles into extended networks, resulting in gelation. (B) Hydrogels formed via metal complexation between coiled-coil domains and polymer chains. (Adapted with permission from Refs. [55, 57]).

Wang et al. provide another example of environment-sensitive hydrogels based on coiled coils [57]. In this design, a tetrameric coiled coil participates in one side of a transition-metal complex, with a synthetic polymer forming the other side. The polymers are modified with iminodiacetate groups to provide partial binding of Ni^{2+}, while the coiled-coil domains bound the metal via peptidic histidine tags (Figure 2.10B). One particular proposed feature of this hydrogel system is that the use of metal complexation may open a route to monitoring hydrogels *in vivo* by using, for example, magnetic resonance imaging.

This area of biological-synthetic polymer hybrids is growing rapidly, and offers the potential for combining the functionality and synthetic accessibility of polymer chemistry with the specificity in structural organization afforded by natural biopolymers such as peptides [60, 61].

2.2.6
Key Techniques

The promise that coiled-coil motifs hold in advancing materials science and nano-biotechnology is facilitated by the synthetic accessibility of coiled-coil peptides. Such sequences are fairly straightforward to prepare, using conventional peptide synthesis techniques. The most popular protocols fall under the umbrella of the so-called Fmoc-strategy used in solid-phase peptide synthesis (SPPS) [62]. SPPS has modest requirements in terms of chemistry, skills and costs, and has already been developed as a fully automated technique that is readily available to non-specialists. Automatic synthesizers provide the necessary speed and efficiency for producing the peptide materials demanded by the rapidly evolving field of nano-biotechnology. Indeed, today a variety of peptide synthesizers is available commercially which can be matched to suit a range of budgets and requirements [62]. Nonetheless, other laboratories favor molecular-biology methods established to produce peptides by the expression of synthetic genes, usually in bacterial hosts [63].

The characterization of materials produced by these methods involves a combination of biophysical measurements, imaging, and diffraction techniques. For example, liquid chromatography, mass spectrometry, circular dichroism, analytical ultracentrifugation, electron microscopy, Fourier transform infrared and UV-visible spectroscopies have all proved to be "absolute musts" in characterizing peptide-based materials [64]. However, other techniques such as light scattering, surface plasmon resonance, solid-state NMR and atomic force microscopy are being used increasingly in the field [64].

2.3
Conclusions and Perspectives

One of the main aims of "bottom-up" assembly is to program chemical structures with the information necessary for their self-assembly to defined three-dimensional structures; these may then assemble further into functional entities, including new materials. As Nature provides the ultimate examples of bottom-up assembly, we can – by studying natural self-assembling systems – learn from, and be inspired by, the "master". Among the various biological macromolecules that self-assemble, polypeptides (peptides and proteins) arguably offer the richest collection of ordered and functional natural assemblies and materials. Whilst peptides and proteins are linear polymers of just 20 amino acids, this limited "chemical alphabet" is supplemented in Nature by so-called post-translational modifications. Moreover, unlike most other polymers, polypeptides fold up reproducibly to well-defined three-dimensional structures and assemblies. Such folding and assembly processes are governed by non-covalent interactions, which adds the elements of reversibility and control to the process. Taken together, these properties make peptides attractive building blocks for synthetic self-assembly and nanobiotechnology. It should also be noted that the use of peptides here has the added advantage that they are synthetically accessible, and that this facilitates their preparation, their functionalization via conjugation with synthetic molecules, and/or the inclusion of nonstandard amino acids.

Although in this chapter we have restricted our view only to peptides, the scope and achievements of this area are still too large to have been recounted fully. Consequently, we have focused on one peptide-folding motif – the α-helical coiled coil – which is sufficiently understood to allow both the design and characterization of self-assembled nanostructures and materials with considerable precision. Nanostructures and materials constructed thus far using coiled-coil building blocks include simple oligomerizing systems, nanoscale linkers, molecular switches, and fibrous and gel-forming materials and networks that bridge the nano-to-mesoscale regimes. Whenever possible we have referred to related areas such as studies involving alternative peptide-folding motifs and bio-inspired synthetic molecules, though these, regrettably, are described in much less depth than we would have liked.

Today, there is considerable potential for nanobiotechnological applications of objects and materials assembled using peptides, ranging from precisely defined nanometer spacers to the struts and cross-linkers for networks and hydrogels. Newly emerging and especially interesting areas in this field include the development of peptide-polymer hybrids which capture the advantages of both natural and synthetic polymers, and the design of peptide-based switches for incorporation into hybrid networks, thus providing the nanobiotechnologist with the ability to control responses to local environments.

References

1 Whitesides, G. M., Boncheva, M. (2002) Supramolecular chemistry and self-assembly special feature: beyond molecules: self-assembly of mesoscopic and macroscopic components. *Proc. Natl. Acad. Sci. USA* **99**, 4769–4774.

2 Pollard, T. D., Earnshaw, W. E. (2002) *Cell Biology.* W.B. Saunders, New York, NY.

3 Rothemund, P. W. K. (2006) Folding DNA to create nanoscale shapes and patterns. *Nature* **440**, 297–302.

4 Yeates, T. O., Padilla, J. E. (2002) Designing supramolecular protein assemblies. *Curr. Opin. Struct. Biol.* **12**, 464–470.

5 Fairman, R., Akerfeldt, K. S. (2005) Peptides as novel smart materials. *Curr. Opin. Struct. Biol.* **15**, 453–463.

6 Rajagopal, K., Schneider, J. P. (2004) Self-assembling peptides and proteins for nanotechnological applications. *Curr. Opin. Struct. Biol.* **14**, 480–486.

7 DeGrado, W. F., Summa, C. M., Pavone, V., Nastri, F., Lombardi, A. (1999) De novo design and structural characterisation of proteins and metalloproteins. *Annu. Rev. Biochem.* **68**, 779–819.

8 Kotch, F. W., Raines, R. T. (2006) Self-assembly of synthetic collagen triple helices. *Proc. Natl. Acad. Sci. USA* **103**, 3028–3033.

9 Laity, J. H., Lee, B. M., Wright, P. E. (2001) Zinc finger proteins: new insights into structural and functional diversity. *Curr. Opin. Struct. Biol.* **11**, 39–46.

10 Waterhouse, S. H., Gerrard, J. A. (2004) Amyloid fibrils in bionanotechnology. *Aust. J. Chem.* **57**, 519–523.

11 Lupas, A. N., Gruber, M. (2005) The structure of α-helical coiled coils. In: Parry, D. A. D., Squire, J. M. (Eds.), *Advances in Protein Chemistry,* Academic Press, Vol. 70, pp. 37–78.

12 Burkhard, P., Stetefeld, J., Strelkov, S. V. (2001) Coiled coils: a highly versatile protein folding motif. *Trends Cell Biol.* **11**, 82–88.

13 Ryadnov, M. G., Woolfson, D. N. (2005) MaP peptides: programming the self-assembly of peptide-based mesoscopic matrices. *J. Am. Chem. Soc.* **127**, 12407–12415.

14 Crick, F. H. C. (1953) The packing of α-helices: simple coiled-coils. *Acta Crystallogr.* **6**, 689–697.

15 Walshaw, J., Woolfson, D. N. (2001) SOCKET: a program for identifying and analysing coiled-coil motifs within protein structures. *J. Mol. Biol.* **307**, 1427–1450.

16 Harbury, P., Zhang, T., Kim, P., Alber, T. (1993) A switch between two-, three-, and four-stranded coiled coils in GCN4 leucine zipper mutants. *Science* **262**, 1401–1407.

17 Walshaw, J., Woolfson, D. N. (2003) Extended knobs-into-holes packing in classical and complex coiled-coil

assemblies. *J. Struct. Biol.* **144**, 349–361.

18 O'Shea, E. K., Lumb, K. J., Kim, P. S. (1993) Peptide velcro: design of a heterodimeric coiled coil. *Curr. Biol.* **3**, 658–667.

19 O'Shea, E. K., Rutkowski, R., Kim, P. S. (1989) Evidence that the leucine zipper is a coiled coil. *Science* **243**, 538–542.

20 Woolfson, D. N., Alber, T. (1995) Predicting oligomerization states of coiled coils. *Protein Sci.* **4**, 1596–1607.

21 Goodman, R. P., Schaap, I. A. T., Tardin, C. F., Erben, C. M., Berry, R. M., Schmidt, C. F., Turberfield, A. J. (2005) Rapid chiral assembly of rigid DNA building blocks for molecular nanofabrication. *Science* **310**, 1661–1665.

22 Woolfson, D. N. (2005) The design of coiled-coil structures and assemblies. In: Parry, D. A. D., Squire, J. M. (Eds.), *Advances in Protein Chemistry*, Academic Press, Vol. 70, pp. 79–112.

23 Ryadnov, M. G., Ceyhan, B., Niemeyer, C. M., Woolfson, D. N. (2003) "Belt and Braces": a peptide-based linker system of de novo design. *J. Am. Chem. Soc.* **125**, 9388–9394.

24 Stevens, M. M., Flynn, N. T., Wang, C., Tirrell, D. A., Langer, R. (2004) Coiled-coil peptide-based assembly of gold nanoparticles. *Adv. Mater.* **16**, 915–918.

25 Rosi, N. L., Mirkin, C. A. (2005) Nanostructures in biodiagnostics. *Chem. Rev.* **105**, 1547–1562.

26 Mucic, R. C., Storhoff, J. J., Mirkin, C. A., Letsinger, R. L. (1998) DNA-directed synthesis of binary nanoparticle network materials. *J. Am. Chem. Soc.* **120**, 12674–12675.

27 Caughey, B., Lansbury, P. T. (2003) Protofibrils, pores, fibrils, and neurodegeneration: separating the responsible protein aggregates from the innocent bystanders. *Annu. Rev. Neurosci.* **26**, 267–298.

28 Makin, O. S., Serpell, L. C. (2005) Structures for amyloid fibrils. *FEBS J.* **272**, 5950–5961.

29 Rochet, J.-C., Lansbury, P. T. (2000) Amyloid fibrillogenesis: themes and variations. *Curr. Opin. Struct. Biol.* **10**, 60–68.

30 Reches, M., Gazit, E. (2003) Casting metal nanowires within discrete self-assembled peptide nanotubes. *Science* **300**, 625–627.

31 Scheibel, T. (2005) Protein fibers as performance proteins: new technologies and applications. *Curr. Opin. Biotech.* **16**, 427–433.

32 Mihara, H., Matsumura, S., Takahashi, T. (2005) Construction and control of self-assembly of amyloid and fibrous peptides. *Bull. Chem. Soc. Jpn.* **78**, 572–590.

33 Kojima, S., Kuriki, Y., Yoshida, T., Yazaki, K., Miura, K. (1997) Fibril formation by an amphipathic alpha-helix-forming polypeptide produced by gene engineering. *Proc. Jpn. Acad. B Phys. Biol. Sci.* **73**, 7–11.

34 Pandya, M. J., Spooner, G. M., Sunde, M., Thorpe, J. R., Rodger, A., Woolfson, D. N. (2000) Sticky-end assembly of a designed peptide fiber provides insight into protein fibrillogenesis. *Biochemistry* **39**, 8728–8734.

35 Potekhin, S. A., Melnik, T. N., Popov, V., Lanina, N. F., Vazina, A. A., Rigler, P., Verdini, A. S., Corradin, G., Kajava, A. V. (2001) De novo design of fibrils made of short [alpha]-helical coiled coil peptides. *Chem. Biol.* **8**, 1025–1032.

36 Zimenkov, Y., Conticello, V. P., Guo, L., Thiyagarajan, P. (2004) Rational design of a nanoscale helical scaffold derived from self-assembly of a dimeric coiled coil motif. *Tetrahedron* **60**, 7237–7246.

37 Wagner, D. E., Phillips, C. L., Ali, W. M., Nybakken, G. E., Crawford, E. D., Schwab, A. D., Smith, W. F., Fairman, R. (2005) Toward the development of peptide nano-filaments and nanoropes as smart materials. *Proc. Natl. Acad. Sci. USA* **102**, 12656–12661.

38 Koide, T., Homma, D. L., Asada, S., Kitagawa, K. (2005) Self-complementary peptides for the formation of collagen-like triple helical supramolecules. *Bioorg. Med. Chem. Lett.* **15**, 5230–5233.

39 Smith, A. M., Banwell, E. F., Edwards, W. R., Pandya, M. J., Woolfson, D. N. (2006) Engineering increased stability into self-assembled protein fibers. *Adv. Funct. Mater.* **16**, 1022–1030.

40 Smith, A. M., Acquah, S. F. A., Bone, N., Kroto, H. W., Ryadnov, M. G., Stevens, M. S. P., Walton, D. R. M., Woolfson, D. N. (2005) Polar assembly in a designed protein fiber. *Angew. Chem. Int. Ed.* **44**, 325–328.

41 Ryadnov, M. G., Woolfson, D. N. (2003) Engineering the morphology of a self-assembling protein fibre. *Nat. Mater.* **2**, 329–332.

42 Ryadnov, M. G., Woolfson, D. N. (2003) Introducing branches into a self-assembling peptide fiber. *Angew. Chem. Int. Ed.* **42**, 3021–3023.

43 Ryadnov, M. G., Woolfson, D. N. (2004) Fiber recruiting peptides: noncovalent decoration of an engineered protein scaffold. *J. Am. Chem. Soc.* **126**, 7454–7455.

44 Melnik, T. N., Villard, V., Vasiliev, V., Corradin, G., Kajava, A. V., Potekhin, S. A. (2003) Shift of fibril-forming ability of the designed α-helical coiled-coil peptides into the physiological pH region. *Protein Eng.* **16**, 1125–1130.

45 Ogihara, N. L., Ghirlanda, G., Bryson, J. W., Gingery, M., DeGrado, W. F., Eisenberg, D. (2004) Design of three-dimensional domain-swapped dimers and fibrous oligomers. *Proc. Natl. Acad. Sci. USA* **98**, 1404–1409.

46 Zhou, M., Bentley, D., Ghosh, I. (2004) Helical supramolecules and fibers utilizing leucine zipper-displaying dendrimers. *J. Am. Chem. Soc.* **126**, 734–735.

47 O'Neil, K., DeGrado, W. F. (1990) A thermodynamic scale for the helix-forming tendencies of the commonly occurring amino acids. *Science* **250**, 646–651.

48 Lazar, K. L., Miller-Auer, H., Getz, G. S., Orgel, J. P. R. O., Meredith, S. C. (2005) Helix-turn-helix peptides

that form helical fibrils: turn sequences drive fibril structure. *Biochemistry* **44**, 12681–12689.

49 Prusiner, S. B. (1998) Prions. *Proc. Natl. Acad. Sci. USA* **95**, 13363–13383.

50 May, B. C. H., Govaerts, C., Prusiner, S. B., Cohen, F. E. (2004) Prions: so many fibers, so little infectivity. *Trends Biochem. Sci.* **29**, 162–165.

51 Sneer, R., Weygand, M. J., Kjaer, K., Tirrell, D. A., Rapaport, H. (2004) Parallel beta-sheet assemblies at interfaces. *ChemPhysChem* **5**, 747–750.

52 Lomander, A., Hwang, W., Zhang, S. (2005) Hierarchical self-assembly of a coiled-coil peptide into fractal structure. *Nano Lett.* **5**, 1255–1260.

53 Weisel, J. W. (2004) The mechanical properties of fibrin for basic scientists and clinicians. *Biophys. Chem.* **112**, 267–276.

54 Weisel, J. W. (2005) Fibrous proteins: coiled-coils, collagen and elastomers. In: Parry, D. A. D., Squire, J. M. (Eds.), *Advances in Protein Chemistry*, Academic Press, Vol. 70, pp. 247–299.

55 Petka, W. A., Harden, J. L., McGrath, K. P., Wirtz, D., Tirrell, D. A. (1998) Reversible hydrogels from self-assembling artificial proteins. *Science* **281**, 389–392.

56 Shen, W., Zhang, K., Kornfield, J. A., Tirrell, D. A. (2006) Tuning the erosion rate of artificial protein hydrogels through control of network topology. *Nat. Mater.* **5**, 153–158.

57 Wang, C., Stewart, R. J., Kopecek, J. (1999) Hybrid hydrogels assembled from synthetic polymers and coiled-coil protein domains. *Nature* **397**, 417–420.

58 Zhang, S. (2003) Fabrication of novel biomaterials through molecular self-assembly. *Nat. Biotechnol.* **21**, 1171–1178.

59 Lutolf, M. P., Hubbell, J. A. (2005) Synthetic biomaterials as instructive extracellular microenvironments for morphogenesis in tissue engineering. *Nat. Biotechnol.* **23**, 47–55.

60 Klok, H.-A. (2005) Biological-synthetic hybrid block copolymers: combining the best from two worlds. *J. Polym. Sci. Part A: Poly. Chem.* **43**, 1–17.

61 Vandermeulen, G. W. M., Kim, K. T., Wang, Z., Manners, I. (2006) Metallopolymer-peptide conjugates: synthesis and self-assembly of polyferrocenylsilane graft and block copolymers containing a sheet forming Gly-Ala-Gly-Ala tetrapeptide segment. *Biomacromolecules* **7**, 1005–1010.

62 Chan, W. C., White, P. D. (Eds.) (2000) *Fmoc Solid Phase Peptide Synthesis: A Practical Approach.* Oxford University Press, New York.

63 Maniatis, T., Fritsch, E. F., Sambrook, J. (1982) *Molecular Cloning: A Laboratory Manual.* Cold Spring Harbor Laboratory Press, Cold Spring Harbor, New York.

64 Nölting, B. (2005) *Methods in Modern Biophysics.* Springer-Verlag, New York.

3
Synthesis and Assembly of Nanoparticles and Nanostructures Using Bio-Derived Templates

Erik Dujardin and Stephen Mann

3.1
Introduction: Elegant Complexity

The emergence of complexity is a defining feature of our planet. Moreover, complexity is ubiquitous. It is observed in inanimate and animate matter, social systems and communication protocols, and in diverse environmental networks. In each case, these systems exhibit at a fundamental level organizational properties that emerge without the intervention of any distinct organizing agent. This notion provides a formidable challenge for scientists attempting to decipher the general and specific processes responsible for complexity, and is a source of deep inspiration for those who attempt to shape and organize matter through synthetic procedures [1]. Over the past few decades, it has been realized that morphological, structural and functional complexity is often a defining characteristic of biomineralization, which generally involves temporally and spatially dependent interactions between assembling organic scaffolds and inorganic minerals [2]. Fortunately, the overwhelming variety found in biomineralization (and Nature in general) does not necessarily arise from a cumulative complexity involving excessive complication and layering of increasing numbers of components and networks, but rather from an *elegant complexity* where a limited number of principles and building blocks are combined in a multitude of ways to produce adaptive and evolutive structures with precise functions. This second paradigm offers hope that complexity can also be achieved in synthetic materials if the key concepts and processes can be distilled from the appropriate biological archetypes. For complex nanostructures, a good starting point is to develop experimental approaches based on principles of biomineralization, such as templated nucleation, growth and directed self-assembly. Thus, in this chapter we illustrate how unraveling the specific interactions between bio-derived templates and inorganic materials not only yields a better understanding of natural hybrid materials but also inspires new methods for developing the potential of biological molecules, superstructures and organisms as self-assembling agents for materials fabrication.

Nanobiotechnology II. Edited by Chad A. Mirkin and Christof M. Niemeyer
Copyright © 2007 WILEY-VCH Verlag GmbH & Co. KGaA, Weinheim
ISBN: 978-3-527-31673-1

Within the sections of the chapter we describe the use of various types of bio-related molecules (biopolymers, peptides, oligonucleotides) for the synthesis and assembly of organized nanoparticle (NP)-based structures and materials (Section 3.2). This is followed by specific highlights of recent investigations demonstrating the enormous potential for using proteins (Section 3.3), viruses (Section 3.4) and living organisms (Section 3.5) in bionanomaterials fabrication. Finally, Section 3.6 presents a short outlook and perspective on the future of the research field.

3.2
Polysaccharides, Synthetic Peptides, and DNA

Biopolymers constitute the most abundant organic compounds in the biosphere, and include several families of biological macromolecules such as polysaccharides, polypeptides and oligonucleotides, which display a wide range of architectures, chemical functionalities and informational contents. Most of these readily available polymers can be advantageously used for the templated synthesis of intricate hybrid nanostructures when associated with inorganic mineral phases such as silica. For example, cellulose-based polysaccharides can adopt various long-range architectures that can be in-filled or replicated by inorganic mineralization (Figure 3.1). In particular, chains of cellulose-based molecules spontaneously order into iridescent cholesteric superstructures, with the consequence that these ordered mesostructures can be used as templates for the synthesis of cholesteric organic/inorganic hybrid materials [3]. The hybrid cellulose/silica materials exhibit optical textures under polarized light and ordered pores in electron microscopy that indicate that the complex nanometer-scale structure of the polysaccharide template is maintained during mineralization. Moreover, removal of the polysaccharide template then produces a silica replica consisting of systems of imprinted chiral pores with long-range order (Figure 3.1A). The high degree of structural imprinting probably arises from specific hydrogen-bonding interactions between the numerous hydroxyl groups of the biopolymer and siloxane precursors generated from the sol-gel reactions. Similar approaches have been developed using preformed cellulose NPs isolated from biomaterials such as cotton, algal membranes, or wood. These NPs are typically in the form of chiral, spindle-shaped crystalline rods, 5–20 and 100–500 nm in diameter and length, respectively, and can be ordered into nematic or cholesteric liquid crystals. Such colloidal suspensions have been used as templates for the synthesis of ordered mesoporous silica materials (Figure 3.1B) [4]. When the space between rods is infiltrated with silica precursors, a hybrid cellulose/silica nanocomposite is obtained that retains the co-alignment of the rod template upon silica condensation. Removal of the template by calcination completes the silicate cross-linking and results in silica replicas with parallel pores, 10–20 nm in diameter, separated by 10 nm-thick walls (black arrows in Figure 3.1B). Such pore systems are substantially larger than the voids associated with many mesoporous silicas produced by surfactant- or block copolymer-mediated synthesis.

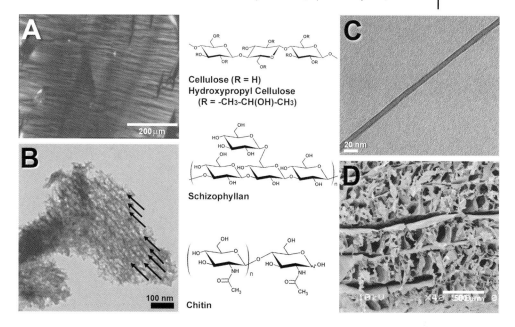

Fig. 3.1 Complex silica architectures from cellulose templates and derivatives.
(A) Porous silica from templated cholesteric solutions of hydroxypropylcellulose. Figure reproduced with permission from Ref [3].
(B) Mesoporous silica from templated cholesteric liquid crystal suspension of cellulose nanorods. The black arrows indicate 10 nm-thick parallel silica walls separating 10- to 20-nm-diameter cylindrical pores [4]; figure reproduced by permission of The Royal Society of Chemistry. (C) Silica nanofibers from replica of inner cavity of *Schizophyllan* triple helix [5]; figure reproduced by permission of The Royal Society of Chemistry. (D) Ordered chamber-like macroporous *β*-chitin-silica structure by mineralization of cuttle bone organic matrix. Figure reproduced with permission from Ref. [6]; © 2000, American Chemical Society.

Recent studies have utilized structural and chemical modifications of biological polysaccharides to prepare hybrid nanostructures. For example, *Schizophyllan*, a natural fungal poly-*β*-(1-3) glucose with one *β*-(1-6) glucose side chain every third main-chain glucose (Figure 3.1) has been used as a template for the synthesis of one-dimensional silica nanostructures [5]. When a dimethylsulfoxide solution of this polysaccharide is gradually enriched in water, *Schizophyllan* self-assembles into a triple helix with a hydrophobic inner cavity that sequesters alkylsiloxane molecules with the result that silica nanofibers (Figure 3.1C) or chains of silica NPs, depending on the ageing conditions, are produced. Other studies have used scaffolds based on *β*-chitin (Figure 3.1), which is a homopolymer of *N*-acetyl-D-glucos-2-amine (i.e., cellulose with one hydroxyl group on each monomer replaced by an acetylamino group that provides increased hydrogen bonding between adjacent polymer strands and increased material strength). In particular, the *β*-chitin matrix associated with the calcified cuttlebone of the cuttlefish (*Sepia officinalis*) was isolated by demineralization of the shell to produce an intact organic chamber-like macroporous framework that could be readily infiltrated with silica precursors to form an architecturally complex chitin/silica composite (Figure 3.1D) [6].

The above examples of using polysaccharide templates in nanostructure synthesis and fabrication highlight how a variety of complex morphologies can emerge by directed mineralization of closely related bio-derived templates. Clearly, a wealth of other biomolecule families exists that can potentially be used to generate organic/inorganic composites with well-defined structures in the nanometer to micrometer range. It should also be emphasized that silica is only one among a multitude of potential inorganic materials to be templated by biomolecules. For example, self-assembled peptide nanotubes act as direct molds for the casting of continuous metallic nanowires in their inner cavity [7], or as one-dimensional templates for the nucleation of aligned calcium phosphate crystals resulting in bone-like hybrid materials [8]. DNA is yet another type of biopolymer with a high information content that can be considered as a biomolecular template, in particular for the growth of metallic nanowires [9, 10]. This topic has been recently reviewed in detail [11, 12].

Specific chemical functionalities borne by biopolymers not only act as nucleation promoters or habit modifiers during growth but also encode information driving the self-assembly of hybrid nanostructures. Polypeptides and DNA are probably among the most promising biopolymeric templates for achieving higher-order, self-assembled architectures. In this respect, recent achievements in the development of peptide nanotubes and their use as templates are very significant for future nanobiotechnological applications [13]. Cyclic D,L-polypeptides can be triggered to self-assemble into long nanotubes with uniform diameters typically around 1 nm, with specific chemical moieties on the outer and inner surfaces of the architecture. However, extending this approach to tubes of larger diameters or non-alternating peptide sequences is limited principally by the molecular design inherent in the cyclic building blocks. In contrast, the self-organization of linear polypeptides composed of polar and non-polar blocks results in tubular structures typically with diameters of the order of 10 nm. Moreover, the morphology of these tubes can be significantly modified by introducing rigid or flexible monomers, and the desired polypeptide structure chemically functionalized by incorporating binding sites with high affinity for specific materials. For example, the use of sequences such as the histidine-rich HRE domain AHHAHHAAD, for gold [14], and HGGGHGHGGGHG or NPSSLFRYLPSD for copper [14] or silver [15, 16] deposition, respectively. Upon exposure to metal ions, and in the presence of a reducing agent, these templates are completely covered with self-assembled monodispersed metallic NPs. Significantly, a very recent report has demonstrated that highly regular, single-particle linear chains of 10 nm Au NPs could be produced by templating HRE-tagged peptide nanotubes [17].

The very rich library of peptide structures is only just now beginning to be explored with respect to the potential of designing polypeptide templates for the synthesis and self-assembly of organic/inorganic nanostructures. Moreover, this approach has the advantage of integrating morphological complexity [18, 19] with biocompatibility and bioactivity of the obtained architectures [20]. However, one shortcoming of polypeptide templates could be their relatively low level of addressability which, if not circumvented practically, would confine these materials to static rather than evolutive structures. In contrast, DNA has been studied extensively over the past decade as it provides an external stimulus via denaturation of the double strand upon heating. Indeed, DNA can act as a polyelectrolyte for the self-assembly of charged NPs by simple electrostatic interactions [21], but increasing control over the self-assembled structures has been achieved by covalently tethering DNA to NPs. The cornerstone of DNA-based NP self-assembly was established in 1996, when thiol-terminated oligonucleotides grafted onto gold NPs were shown to induce reversible aggregation, provided that two populations of gold colloids were derivatized with two different strands, mixed together, and that a free strand which was half-complementary to both of them was introduced into the mixture [22, 23]. Reversibility was achieved upon heating above the melting temperature of the three-strand double helix. The original aggregates were isotropic and compact, but several developments of this approach have led to increased complexity. Topological complexity was explored – first conceptually and lately experimentally – by N.C. Seeman and coworkers (for basic principles, see Ref. [24]). This group focused on the design and production of double-strand (ds) DNA building blocks able to pave a flat surface and even to self-assemble into precise geodesic three-dimensional nano-objects [25, 26]. By an appropriate choice of the base sequence, it was possible to attach gold NPs to a selected subset of DNA tiles, thus converting the two-dimensional ordered DNA structure into a template for the assembly of an ordered array of metallic NPs with specific symmetry [27].

Symmetry breaking in DNA-NP systems can alternatively be achieved by choosing non-spherical NPs. Oligonucleotide-functionalized gold nanorods with an aspect ratio of 5:1 showed large-scale uniaxial organization upon hybridization, and thermal reversibility of the aggregation was achieved when a direct two-strand duplex was immobilized on the nanorod surfaces [28]. Recently, we showed that these self-organized ensembles could also attain functional complexity, which is a key feature for future technologically relevant systems. For example, 100-nm mesoporous silica NPs were reversibly conjugated to 13 nm-sized gold particles via a three-strand duplexation process (Figure 3.2A) [29]. The two building blocks not only have significantly different sizes but, more importantly, have very differing chemical and physical properties. Whereas the mesoporous silica NPs can act as a reactant reservoir, adsorbate or pH adjustable surface for heterogeneous catalysis, the metallic NPs are potential catalysts or microwave-triggered localized heating elements [30]. In principle, the integration of several functionalities in DNA-linked hybrid nanostructures could lead to nanoreactors coupled to several reconfigurable shells of different metallic NPs that collectively enable a cascade of reactions to be undertaken on the molecules pre-loaded in the pore. We investigated the concept of self-assembled, reconfigurable, multilayered nanostructures by successively attaching a biotinylated derivative of the iron oxide-containing protein, ferritin, onto carbon nanotubes, followed by linking gold NPs to the

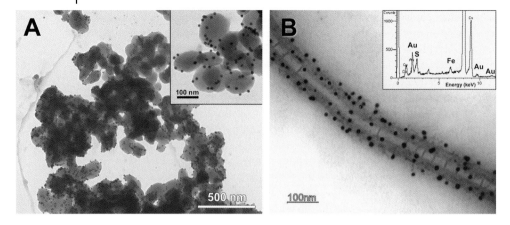

Fig. 3.2 Functional complexity in DNA-driven nanoparticle self-assembly. (A) Mesoporous silica/gold nanoparticle conjugates with satellite structure. Figure reproduced with permission from Ref. [29]. (B) Multilayer self-assembled architecture composed of carbon nanotubes, ferritin and gold nanoparticles as inorganic building blocks and double-strand DNA duplex and streptavidin–biotin as biomolecular mortars [31]; figure reproduced by permission of The Royal Society of Chemistry.

ferritin layer via a streptavidin/ds-DNA linker (Figure 3.2B) [31]. It was shown that the biomolecular bridging groups could be independently and reversibly broken to allow a sequential dismantling of the nanotube/ferritin/Au NP structure. Since the dissociation stimuli and dissociation constants between DNA duplexes and protein complexes are very different, the increased complexity in the carbon nanotube system resides not only in the types of self-assembled NPs but also in the concomitant use of different self-assembling agents. Of course, formation of isolated complex nanostructures is not the sole goal of this strategy; for example, DNA-coupling of two, or more, distinct nanoscale components can also lead to extended networks incorporating several properties, as shown recently by cross-linking of gold NPs and ferritin molecules though DNA base pairing [32].

3.3
Proteins

Proteins exhibit chemical properties similar to those of other polypeptides, but are distinguished in most cases by a precise three-dimensional (3D) folded structure and associated functional specificity. Thus, certain amino acids are exposed to the external medium and can act as binding sites for the positioning of specific ligands across a well-defined nanoscale platform. In this respect, proteins are major candidates for the templated growth of nanomaterials, and in addition can be viewed as highly evolved structuring agents for higher-order assembly of extended superstructures across a range of length scales. In some cases, protein superstructures in the form of hierarchical architectures (collagen in bone), structured gels (amelogenins in teeth), and complex macromolecular frameworks (acidic proteins in seashells) are intimately associated with the deposition of complex and functional biominerals. This close interplay between proteins and inorganic nucleation and growth involves specific interactions at the nanometer length scale, and suggests that proteins extracted from biomineralized tissues could act as templates in the controlled synthesis of NPs in the laboratory. For example, it was shown recently that a mixture of proteins extracted from the fluoroapatite shell of the brachiopod *Lingula anatina* resulted in promotion of the *in vitro* crystallization of this mineral when added to buffered calcium phosphate/fluoride metastable solutions. Regularly shaped nanocrystals were obtained because the proteins promoted the dissolution of amorphous calcium phosphate precursor NPs and appeared to induce the local ordering of fluoroapatite crystals [33].

Although generally considered within the context of promoting inorganic NP nucleation and growth or assembly, recent studies have exploited the favorable and specific interactions between the exposed amino acids of proteins and mineral surfaces to achieve the converse process in which individual protein molecules are stabilized by entrapment within ultrathin coatings of certain inorganic-based materials. This approach was demonstrated by wrapping myoglobin (Mb), hemoglobin (Hb) or glucose oxidase (GOx) with an ultrathin shell of aminopropyl-functionalized magnesium (organo)phyllosilicate (Figure 3.3A,B) [34]. The organoclay was exfoliated in water, and the disintegration yielded cationic oligomers that could be purified into stable sols. Simple electrostatic interactions between anionic peripheral moieties of the proteins and organoclays oligomers produced spheroidal core-shell hybrid NPs with diameters around 4, 8 and 6.5 nm for Mb, Hb and GOx, respectively. Significantly, the proteins and enzymes were shown to remain intact both structurally and functionally after entrapment. Similarly, a more generic protocol was developed by isolating individual proteins (Mb, Hb) within the water droplets of a microemulsion prepared in the presence of an oil phase that contained silica precursors such as tetramethoxysilane that readily hydrolyze when they come in contact with the nanoscale water droplets [35]. Thus, mineralization occurred at the water/oil interface and produced a 2- to 3 nm-thick amorphous silica shell that preserved the protein structural integrity. At low protein loadings, individual 9-nm core-shell NPs were obtained, whereas colloidal aggregation was observed at high protein loading, resulting in 20-nm raspberry-like clusters of coated proteins.

The protein/mineral core-shell nanostructure of the above examples can be inverted by using a hollow protein cage such as the iron storage protein, ferritin, as nanoreactors for producing a mineral core. Ferritin is composed of 24 polypeptide subunits forming a hollow spherical cage that is perforated by two types of ion channel. In living organisms, this structure is optimized to store and release iron. Ferritin was the first protein to be used as a biomimetic template for the synthesis of monodispersed hybrid inorganic NPs [36, 37]. The general route consists of isolating the empty protein shell (apoferritin), re-loading it with chosen

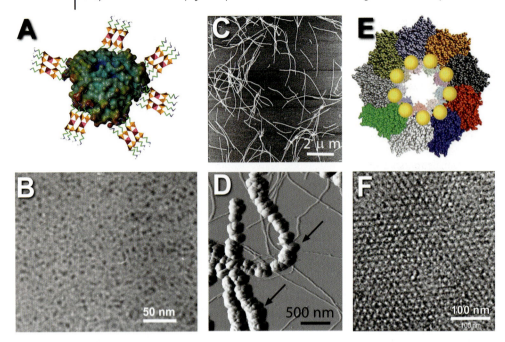

Fig. 3.3 Organizational complexity in protein-templated bioinorganic nanostructures. (A) Schematic and (B) transmission electron micrograph (TEM) of individual organoclay-wrapped myoglobin. Figures reproduced with permission from Ref. [34]. (C) Transmission electron micrograph of 10-nm-wide metal-coated amyloid fibers obtained by self-assembly of a prion determinant from *Saccharomyces cerevisiae*, the N-terminal and middle region (NM) of Sup35p. (D) Atomic force micrograph of gold-toned NM fibers. Growth of 1-D gold nanowires by successive silver and gold electrodeless plating occurs only where the protein fiber template has initially been labeled by binding 1.4-nm gold colloid (arrows). Figures reproduced with permission from Ref. [45]; © 2003, National Academy of Sciences. (E,F) Chaperonin/gold nanoparticles conjugates. (E) Schematic of a 17-nm chaperonin protein decorated with nine 1.4-nm gold nanoparticles tethered to the protein subunits. (F) TEM image of a 2-D crystal of chaperonin/gold colloid self-assembled nanostructures. Figures reproduced with permission from Ref. [51]; © 2002, Macmillan Publishing Ltd.

precursors, and subsequently inducing inorganic precipitation specifically within the 8 nm-diameter cavity. This approach produced a range of artificial ferritins with metallic, semiconductive and magnetic core structures [38]. Careful control of the mineralization steps can even lead to a fine-tuning of the filling fraction of the ferritin cavity. Indeed, a thin and continuous coating of the inner wall or complete filling of the ferritin cavity was successively demonstrated recently for cobalt oxide [39]. A similar approach has been applied to other proteins with spherical cage morphology such as lumazine synthase [40], the ferritin-like protein from *Listeria innocua* [41], or the heat shock protein from *Methanococcus jannaschii* [42, 43]. The native protein shells are usually suitable for the growth of only a limited number of materials, so that chemical modification of the peptide sub-units is required to expand the potential of this approach to a wider range of inorganic materials. Consequently, site-directed mutagenesis (see Section 3.4) is now being more readily applied to modify the peptide sequence in order to favor the intake of different inorganic precursors [43], or to promote the interaction of the mineralized protein with a given substrate while keeping the cage structure intact [44].

Significantly, protein-templated nanostructures are not necessarily restricted to spherical (isotropic) morphologies. In particular, proteins that form highly anisotropic superstructures such as microtubules or amyloid-type fibers have been identified as potential templates for the preparation of one-dimensional inorganic-organic hybrid nanostructures. For example, the N-terminal and middle region (NM) of a yeast (*Saccharomyces cerevisiae*) protein precisely forms β-sheet-rich amyloid fibers with a diameter of 9–11 nm (Figure 3.3C) [45]. These fibers are chemically stable, do not aggregate as readily as other amyloid fibers, and can be chemically modified by site-directed mutagenesis, as shown by studies in which genetically engineered NM domains containing a cysteine residue were successfully self-assembled into fibers and used as a template for the production of metallic Ag/Au nanowires with diameters of about 100 nm (Figure 3.3D).

In addition to one-dimensional superstructures, certain proteins spontaneously assemble into 2-D planar arrays that can be used for replication and patterning of inorganic NPs. Bacterial surface proteins, for example, are known to form stable, ordered arrays called S-layers that exist in a range of 2-D symmetries with lattice constants between 3 and 30 nm and are of 5–15 nm in thickness. The proteins self-organize into lattices with periodic pores with diameters between 2 and 6 nm [46]. When anchored on a substrate and exposed to ionic precursors (e.g., Cd(II)), the S-layer films selectively bind ions at specific locations of opposite charge, and nucleation is triggered at these sites when the layers are incubated with a second reactant (e.g., H_2S) [47]. This results in the direct formation of ordered 2-D arrays of NPs, such as CdS, with the interparticle distance determined by the underlying S-layer symmetry. Similarly, electron beam-induced reduction of metal precursors chemisorbed onto modified S-layers has been used to produce ordered NP arrays [48]. In principle, an extension of these approaches to a wide variety of materials should be relatively straightforward, as chemical modification of the S-layer proteins is well known. Alternatively, the charged surface of the crystalline S-layer superstructure can be used specifically to bind pre-formed inorganic NPs of opposite charge to produce extended arrays of periodically arranged NPs [49]. Perfect matching persists only as long as the NP diameter is smaller than the S-layer film lattice parameter. A more specific recognition process between protein arrays and pre-formed NPs can be obtained based on the strong binding affinity between streptavidin and biotin by derivatizing the S-layer proteins with streptavidin and inorganic NPs with biotin [50].

A similar approach has been described involving chaperonins, which comprise hollow double-ring structures composed of several heat shock proteins (HSP) in the form of 17 nm-sized aggregates that spontaneously self-assemble into 2-D crystals in a similar fashion to S-layers [51]. Chaperonins were genetically engineered to promote the self-assembly of metallic or semi-conductor NPs in two distinct ways. When a cysteine residue was placed at various solvent-exposed HSP sites, the corresponding chaperonins possess a ring of binding thiol sites with high affinity for gold or zinc. Size-selective binding of pre-formed 5 ± 3 nm-sized NPs was observed, with chaperonins grafted with a 3-nm ring of thiols Similarly, chaperonins with a 9-nm binding ring specifically ordered 10 ± 2 nm Au NPs, but not those with diameters of 5 nm or 15 nm. Similar templated arrays were obtained with core-shell CdSe-ZnS quantum dots, thus confirming the binding of the zinc-rich inorganic surface to exposed chaperonin thiols. Alternatively, very small gold NPs (1.4 nm) were covalently tethered to mutated HSP isolated subunits and self-assembly of the NP-loaded HSP then induced by the addition of ATP/Mg^{2+} to produce 2-D crystals of chaperonins in which each organized ring consisted of up to nine Au NPs around the pore perimeter (Figure 3.3E,F).

Finally, besides acting as a patterned host for NP deposition within self-assembled 2-D or 3-D architectures, proteins can play a direct structuring role in biomolecule-NP conjugates. This was demonstrated initially for metallic NPs coated with complementary antigen and antibodies [52], as well as streptavidin and biotin [53]. Proteins have also been used simultaneously as both the building unit and structuring agent; notably, the use of streptavidin to control the aggregation of biotinylated ferritin molecules and their associated superparamagnetic iron oxide NPs [31, 54].

Considering the ease with which proteins can be tailored and modified to optimize specific interactions with given materials, and the robustness of their folding and self-assembling properties, one can reasonably anticipate that proteins will provide an expanding toolbox for both directing the spontaneous organization of NPs into larger ordered structures, and conferring hybrid conjugates with multiple functionalities such as biocompatibility, chemical specificity, and responsiveness.

3.4
Viruses

Viruses occupy the frontier between inanimate and living matter. These complex adaptive structures represent the natural extension and integration of protein and biopolymer (RNA/DNA) self-assembly. Approximately 1000 viruses were identified between 1885 – when the rabies virus was first identified by L. Pasteur – and 1986. Since then, the number of types documented has greatly increased to more than 6700 in 2006, though this probably accounts for less than 1% of existing viruses. This vast diversity strongly suggests that the development of a few basic principles and methods for the use of viruses in nanomaterials science will open up countless opportunities for generating increasing complexity in synthetic architectures. Indeed, from the materials chemist's point of view, viruses can be considered as protein superstructures that offer ready-to-use chemical platforms with precisely defined topology, symmetry, and morphology. For example, spherical viruses such as chlorotic cowpea mottle virus (CCMV), bovine papillomavirus or herpes simplex virus (HSV) protect their genetic material by enclosing the nucleic acid strand within protein cages with diameters of 26, 65, or 125 nm, respectively. Filament or rod-shaped viruses either produce a long, thin protein enclosure to contain the plasmid DNA (e.g., bacteriophage M13), or coil a RNA strand within a helicoidal protein nanotubule, as exemplified by tobacco mosaic virus (TMV).

Following a protocol reminiscent of the mineralization of apoferritin, one can remove the RNA strand from the cavity of CCMV and purify the empty icosahedral virion. The resulting highly cationic inner surface consists of 1080 arginine and 540 lysine groups that facilitate the intake of anionic precursors such as aqueous tungstate ions which, on acidification, aggregate into polyoxotungstate NPs entrapped within the CCMV cavity [55, 56]. However, the mineralization of CCMV using cationic precursors is not readily achieved due to unfavorable electrostatic interactions, and other viruses may offer better flexibility. For example, TMV is constructed from 2130 identical coat proteins aligned along a right-handed RNA helix to produce a 300 nm-long cylinder with an 18-nm outer diameter and a 4-nm inner cavity. This complex object is nontoxic to humans and is readily produced in identical copies by infecting tobacco plants. A closer analysis of the spatial distribution of amino acids within the coat protein structure reveals that TMV can act as a flexible template for inorganic mineralization and NP assembly due to the presence of surfaces with opposite charge distributions. Glutamic and aspartic acids (pK_a values 1–3 and 6–8, respectively) predominantly line the inner surface, whilst lysine and arginine (pK_a values 11 and 12.5, respectively) are mostly found on the exposed outer surface and in the RNA groove, respectively [57]. When incubated at pH < 3, the cavity can be considered as neutral, whereas the outer surface is highly positively charged. Thus, the selective binding of anionic precursors such as $AuCl_4^-$ or $PtCl_4^-$ to the outer surface, followed by reaction with a reducing agent (hydrazine), resulted in the formation of virioids decorated specifically with dense arrays of metal NPs (Figure 3.4A,B) [58]. Similarly, uniform coatings of iron oxides, cadmium or lead sulfide or silica were produced on the external surface of TMV particles [57]. Conversely, at neutral pH, the outer amino acid moieties are positively charged, whereas those in the inner cavity are highly anionic. Under these conditions, the binding of Ag(I) ions along the internal channel is enhanced such that photochemical reduction resulted in the formation of linear arrays of metallic 4 nm-sized silver NPs within the TMV cavity [58]. The above principles were applied to produce metallic nanowires with diameters defined by the cavity size [59, 60]. First, TMV-attached palladium NPs were produced from $[PdCl_4]^-$ ions adsorbed on the outer surface at pH 3. Subsequently, increasing the pH to 7 resulted in infiltration of the inner cavity with nickel or cobalt ions, which were reduced in situ by the redox reaction between Pd(0) and Ni(II) or Co(II). Although the mechanistic details of this second step are unclear, it seems that dissolution of the Pd catalyst from the outer surface was coupled to the constrained growth of continuous metallic nanowires inside

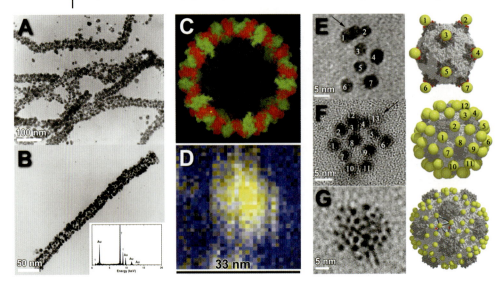

Fig. 3.4 Virus/nanoparticle architectures with morphological and functional complexity. Low- (A) and high- (B) magnification TEM micrographs showing wild-type TMVs with dense external coating of gold nanoparticles. Inset: corresponding energy-dispersive X-ray (EDX) spectrum showing Au (and Cu from supporting grid) signals. Figure reproduced with permission from Ref. [58]; © American Chemical Society. (C) Cut-away view of a ribbon diagram of the cowpea chlorotic mottle virus showing the central cavity of the protein cage. (D) Superimposed iron (yellow) and nitrogen (blue) electron energy-loss spectrum images of iron oxide-filled CCMV (common scale bar at bottom is 33 nm). The compositional maps clearly show that the mineral core is contained within the protein shell. Figure reproduced with permission from Ref. [65]. (E,F,G) Unstained TEM images of gold nanoparticles bound to isolated mutant CPMV virus (left) and corresponding models (right). (E) The BC mutant can accommodate one 5-nm gold nanoparticle per fivefold axis. (F) The EF mutant offers five accessible binding sites for each of the twelve fivefold axes. (G) Forty-two of the 120 possible sites of the DM mutant are observed to be occupied by 2-nm nanoparticles. Figure reproduced with permission from Ref. [74]; © American Chemical Society.

the cavity, with diameters precisely matching the available 4-nm space of the inner channel.

As viral-directed mineralization is strongly influenced by the nature of the amino acid side groups located at specific symmetry-related sites in the assembled viral capsid superstructure, chemical derivatization, directed mutagenesis or evolutionary pressure at these sites can result in reproducible modifications designed for non-biological purposes. For example, CCMV consists of 180 identical protein subunits, each of which contributes 11 carboxylate and seven amino moieties to the outer surface of the virus. Fluorophores or oligopeptides have been coupled by direct esterification to the exposed glutamate and aspartate residues, or to lysine residues after activation by a succinimidyl ester intermediate [61]. Similarly, thiol-selective chemical agents (e.g., ethyl mercury phosphate or

5-maleimidofluorescein) were used to derivatize cysteine sites in wild-type cowpea mosaic virus (CPMV) [62]. This technique was extended to the covalent coupling of entire proteins to cysteine or lysine residues of the CPMV interior by derivatizing them with the appropriate ditopic crosslinker [63]. A single tyrosine site per protein subunit of the icosahedral virus, bacteriophage MS2, was also targeted to develop a new chemical protocol for the covalent modification of the inner surface of the viral cage [64].

Site-directed mutagenesis is a powerful tool that has been extensively developed for biomolecular research, and which is now being increasingly used in biotemplated materials chemistry. By using this technique it is possible to modify the amino acid sequence – and hence the chemical properties of the protein – while preserving the tertiary structure of the polypeptide subunits and higher-order self-assembly of the virus superstructure. For example, site-directed replacement of glutamate 95 and aspartate 109 residues of the TMV subunit with uncharged polar amide side significantly reduces the anionic charge associated with the 4 nm-wide inner channel of the virus particles. As the reduction by two negative charges is repeated for each of the 2130 polypeptide subunits present in the assembled virus, metal-ion binding within the inner channel is curtailed. As a consequence, the specificity for Ag NP nucleation within the TMV channel as described above for the native virus is significantly diminished [58]. Other studies have used extensive site-directed modifications to produce mutants with charge-reversed viral surfaces. For instance, the interior surface of wild-type CCMV is lined with 1620 protonated amino groups (six Arg and three Lys for each of the 180 coat proteins) to package and condense the anionic RNA viral genome, and replacement of the nine basic residues with one glutamic acid produced a mutant with 180 anionic sites on its cavity wall without affecting self-assembly of the virus architecture (Figure 3.4C). In contrast to wild-type CCMV, this mutant sequestered cationic precursors such as Fe^{II} ions, which were successfully transformed by air oxidation into 25 nm-sized iron oxide NPs within the virus cage (Figure 3.4D) [65]. Genetic engineering has also been used to insert extra peptide sequences in targeted proteins. For example, peptide sequences known to favor the mineralization of ZnS, such as Cys-Asn-Asn-Pro-Met-His-Gln-Asn-Cys or Val-Ile-Ser-Asn-His-Ala-Glu-Ser-Ser-Arg-Arg-Leu, or CdS (Ser-Leu-Thr-Pro-Leu-Thr-Thr-Ser-His-Leu-Arg-Ser), were inserted in the coat protein pVIII of M13 bacteriophage using the relevant gene in the phage genome [66]. The incubation of mutated M13 viruses with zinc or cadmium chloride, followed by exposure to aqueous Na_2S, resulted in the selective formation of densely packed crystalline ZnS or CdS NPs on the elongated viral capsid that, in some cases, could be fused into nanowires. Semiconducting hybrid ZnS–CdS or magnetic CoPt NPs, as well as FePt 1-D NP assemblies that could be transformed into freestanding crystalline nanowires upon thermal annealing, were also prepared using this approach [67].

Clearly, the control of genetic engineering techniques provides a powerful tool to modify or expand the peptide sequences of viral coat proteins in order to promote interactions with target materials. This approach not only facilitates the nucleation of selected materials at specific locations on the virus surface, but can also be used to transform viruses into encoded templates for the controlled self-assembly of pre-formed NPs. Indeed, the phage display technique can be used to identify peptide sequences that promote interactions with specific crystallographic faces of chosen metals, semiconductors, oxides, etc. [68–70]. This process is an accelerated Darwinian evolution of the initial phage population under the constraint of adhesion to the selected material. Briefly, this involved generating M13 bacteriophage libraries that contained random sequences of the gene encoding for protein III (pIII), which are located at one end of this filamentous virus. The phage was then exposed to a target material, after which the non-interacting individuals were removed by washing and the interacting ones were collected by elution under specific unbinding conditions (e.g., controlled pH). The latter were then multiplied by bacterial infection and re-exposed to the target material. This selection/elution/amplification cycle was repeated a number of times until DNA sequencing showed a limited number of pIII gene sequences that were highly specific for the inorganic material (e.g., ZnS) under consideration [66, 71]. Re-suspension of the selected evolved phage, for example in $Zn(II)/HS^-$ solutions, promoted the deposition of monodispersed ZnS nanocrystals specifically at the pIII end of the phage particles. Moreover, owing to the large aspect ratio of phage M13, concentrated solutions of the phage pack into a smectic C liquid crystalline phase that was still observed when NPs were attached to the ends of the virus, even in M13 dried films [72]. This property was exploited to form highly ordered 3-D structures of hybrid phage-ZnS components that were cast as a film which exhibited a high degree of ordering, with layers of M13 intercalated with planes of ZnS NPs. The above-described approach was significantly generalized by selecting a phage library against streptavidin (STV) in order to identify a peptide sequence able to selectively bind to any STV-conjugated NP [73]. Furthermore, by combining templated growth on engineered pVIII capsid proteins and binding NPs with designed pIII end sequences, it was possible to use M13 bacteriophage as a dual platform for the integration of two different nanomaterials while preserving the ability of the virus to self-assemble.

Increasing the regioselectivity of viral-templated nucleation and assembly of inorganic NPs necessitates that the surface distribution of certain key amino acids is highly precise. Recently, the accurate positioning of NPs on designed sites was demonstrated for the icosahedral CPMV virus [74, 75]. Wild-type CPMV was genetically engineered to present cysteine residues at selected positions on the outer surface. A first mutant (BC) had a single cysteine added to each protein subunit at a location very close to the fivefold symmetry axis, while another mutant (EF) had a single cysteine inserted per subunit as a GGCGG loop, such that the overall intercysteine distances were in the range of 7–8 nm. A third mutant (DM) had two cysteines replacing two exposed alanine and glutamic acid residues. Simple geometrical considerations indicated that, whereas only one or two 5 nm-sized gold NPs can bind per axis in the BC mutant (Figure 3.4E), the NPs can readily access all 60 binding sites in the EF mutant (Figure 3.4F). The DM mutants were specifically decorated using 2 nm-sized gold NPs that self-positioned at potentially all the 120 cysteine sites (Figure 3.4G). For the EF and DM viruses, the

NP-positioning accuracy could be probed by bridging them with rigid dithiolated conjugated molecules and subsequently observing the hybrid nanostructures by scanning tunneling microscopy [75]. In the absence of bridging molecules, or when the molecular length was smaller than the inter-particle distance, individual NPs could be distinguished on the surface of CPMV due to the increased conductance at the metallic NP sites. When adequate bridging molecules were added, the high-conductance patterns extended over areas much larger than isolated NPs, indicating that the conjugated molecules and NPs had created a conductive network that behaved as large high-conductance domains. More recently, five CPMV mutants were produced with six histidine residues at different surface-exposed locations [76]. This resulted in a pH-dependent charged environment at specific sites on the virus outer surface that could bind metal ions such as Ni(II). Using linkers terminated by an amine triacetate nickel complex (Ni-NTA), these histidine sites could be decorated with gold NPs [76], as well as luminescent CdSe-ZnS quantum dots [77].

Virus particles have also been used to direct the self-assembly of nanostructures by acting as a platform comprising site-specific ligands for antibody-induced binding of capped NPs. For example, HSV and adenovirus (ADV) induced the extended assembly of dextran-coated magnetic particles functionalized with anti-HSV or anti-ADV antibodies [78]. T2 nuclear magnetic resonance relaxation time measurements on the assemblies proved to be a rapid and very sensitive detection method for as few as five viruses per 10 μL aliquots, which is a major improvement compared to the relatively slow polymerase chain reaction-based method for viral detection.

Higher-order levels of complexity can be achieved using virus-based self-assembly of NPs, provided that the superstructures of the viral particles can be readily attained. Viruses with more regular shapes – such as TMV or CPMV – can even order into porous macroscopic crystals. Concentrated TMV dispersions in the form of nematic liquid crystals have been exploited to spatially pattern the deposition of meso-structured silica [79]. The concentrated solutions consisted of closely packed hexagonally ordered bundles of co-aligned TMV particles that were subsequently infiltrated with silica precursors such as reaction mixtures of 90% tetraethoxysilane (TEOS) and 10% aminopropyltriethoxysilane (APTES) at slightly alkaline pH. Using gentle, close-to-neutral conditions allowed the hydrophilic reaction mixture to infill with high fidelity the voids between adjacent TMV particles, with the consequence that extensive silica condensation occurred throughout the interstitial spaces of the virus liquid crystal. Removal of the virus particles by thermal degradation produced a unique form of mesoporous silica with hexagonally ordered straight cylindrical pores with diameters of around 11 nm and a wall thickness of 10 nm. In the case of CPMV, a body-centered cubic crystal is obtained with large cavities and channels between capsids that amount to 50% of the total volume. By analogy to earlier studies on inverse opal or replication of sea urchin skeletons, these crystals can be infiltrated with mineral precursors in order to fill the voids. This concept has been successfully demonstrated for metallic palladium and platinum [80]. More complex assemblies can be constructed by combining the DNA-driven self-assembly of nano-objects described in Section 3.2 with oligonucleotide-functionalized viruses. Following this concept, CPMV was derivatized with both a fluorescent tag and DNA single strand, and aggregation was obtained by adding a free complementary DNA strand to a mixture of two virus populations bearing different dyes [81]. Electron microscopy, fluorescence resonance energy transfer (FRET) measurements and DNA melting experiments proved that the two types of virus alternated within the extended aggregates, which could also be reversibly disassembled. Finally, chemical modifications of virus coat proteins have been exploited to create CPMV self-assembled nano-patterns on solid surfaces by Dip-Pen nanolithography [82, 83].

The above examples clearly indicate that virus-based approaches in nanobio-technology constitute a very versatile and yet highly specific toolkit through the application of relatively simple concepts, using just a few available templates [84]. They also specifically illustrate how biological technologies and principles can help to develop an innovative approach to bio-derived nanomaterials chemistry. The full potential of this approach necessitates the use of a wider range of viral templates and the extension of this strategy for the preparation of functional virus-NP conjugates that exhibit evolutive, adaptive and self-replicating proper-ties. Increased complexity and functionality can also be attained in bioinorganic nanosystems by replacing the viral templates with living cells and cellular super-structures. This rapidly emerging field is the focus of the following section.

3.5
Microorganisms

Biomineralization by living organisms is an essential part of the chemistry of life, and is a major source of inspiration in biomimetic materials chemistry [2]. In addition, the synthetic mineralization of cellular superstructures and living/dead organisms (*mineralized biology*) is making a significant impact in advancing the synthesis of functional materials and composites in the laboratory [85]. Consider-ing typical cell sizes, living species are well suited for structuring inorganic nano-scale building blocks across the micron and sub-micron length scales. As silica condensation in water represents a mild but versatile process, it is often used as a test system to develop the mineralization of new biotemplated synthesis. For ex-ample, threads of co-aligned multicellular bacterial filaments of *Bacillus subtilis* were infiltrated with either amorphous silica NPs or with pre-hydrolyzed siloxane reaction solutions directly after pulling from a web culture [86]. The minerali-zation of the inter-filament spaces produce hybrid threads consisting of close-packed cells embedded in a continuous silica matrix. Moreover, removal of the bacterial template by heating produced mesoporous silica fibers in the form of precise replicas of the bacterial thread superstructure.

Numerous recent reports have exploited biological processes associated with the deposition of inorganic NPs as environmentally benign routes to materials synthesis. For example, the ability of many microorganisms to induce the formation of metallic or metal oxide NPs by redox processes involving aqueous ions and complexes has attracted significant interest [87]. Indeed, reduction of the metal ions is sometimes coupled to respiration through the associated oxidation of organic molecules, thereby replacing aerobic metabolism based on dioxygen [88]. In some cases, metallic NPs are deposited within the microbial cell as a response to cytotoxic levels of metal ions [89], whereas in other systems the deposition of mineral NPs has a precise functional use. For example, magnetic iron oxide or sulfide NPs are produced in magnetotactic bacteria that navigate in the geomagnetic field towards areas of optimum oxygen partial pressure [90]. Recently, some bacteria have been shown to extrude polysaccharide fibrils that template the growth of several micron-long pseudo-crystals of akaganeite (β-FeOOH) in close vicinity to the bacterial cell wall, and which are then discarded (Figure 3.5A) [91]. The initial filaments consisted of a 3 nm-sized akaganeite 1-D crystalline core sheathed by 2 nm-diameter ferrihydrite NPs (Figure 3.5B). Interestingly, it was postulated that the oxidation of soluble Fe(II) into insoluble iron (III) oxyhydroxides produces two protons per metallic ion, and that the local increase in

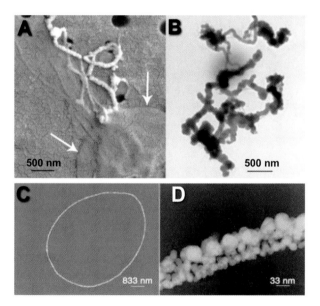

Fig. 3.5 Microbial production of inorganic nanocrystals. (A) SEM image of a cell (edge marked with white arrows) associated with iron oxide-mineralized filaments and non-mineralized fibrils. (B) TEM image of amorphous FeOOH-mineralized filaments filtered from the biofilm supernatant. Figures reproduced with permission from Ref. [91]; © 2004, American Association for the Advancement of Science. (C) TEM image of a thin cross-section of an individual *Aspergillus niger* hypha, loaded with 13-nm-diameter Au particles and then assembled with 30-nm-diameter Au particles through DNA hybridization. (D) Higher-magnification TEM image of the double-layered nanoparticle ring structure. Figure reproduced with permission from Ref. [94].

proton concentration induces a proton-motive force across the membrane wall, which increases the energy-generating potential of the cell.

Semiconductor NPs, which have been demonstrated as bio-compatible fluorescent markers in living cells [92], have recently been produced in bacterial cells [93]. For example, when incubated with cadmium chloride and sodium sulfide, *Escherichia coli* bacteria secreted spherical CdS NPs with a wurtzite crystal structure and size distribution between 2 and 5 nm. The bio-production of quantum dots depended heavily on the bacterial growth phase, with a 20-fold increase in NP formation observed in stationary-phase cells compared with cultures in the late logarithmic phase. Interestingly, no crystals were detected in mid-logarithmic-phase cells. One possibility is that high cellular contents of the thiol-containing peptide, glutathione, were responsible for the differences in NP synthesis observed at different stages of cell growth.

Although several metal/metal oxide-producing bacteria have been reported, the output in terms of a general protocol for the templated growth of monodispersed NPs or nanostructures remains to be further developed. In this regard, a recent report has described the use of the hyphae of a living fungus as a dynamic template for the self-assembly of pre-formed gold NPs [94]. For this, oligonucleotide-functionalized Au NPs were dispersed in a culture solution and spores of *Aspergillus niger* filamentous fungus added. After germination, the hyphae grew and became branched, and the Au NPs were progressively adhered to the hyphae surface by electrostatic and chemical interactions until all the NPs were assembled into mycelia pellets (Figure 3.5C). The adsorbed NPs with attached oligonucleotides could themselves serve as templates for the assembly of a second layer of different NP by standard DNA duplex formation (Figure 3.5D). Since the fungus is alive, it was also possible to grow the hyphae sequentially in a series of culture media, each containing different types of NPs to produce layers of NPs organized along the fungus tubules. By extending this approach to other fungi, it was possible to vary the template morphology to include tube-shaped hyphae with diameters ranging between 800 nm and 12 mµ. These studies highlight the possibility that many other types of living microorganisms might be recruited for the fabrication of nanoscale materials using bionanotechnological processes.

3.6
Outlook

Over the past decade, a variety of studies of biomineralization have significantly increased our understanding of the interactions between bio-organic templates and growing inorganic minerals, and how these are exploited in the synthesis of complex bioinorganic architectures often based on the integration and organization of nanoscale building blocks. In so doing, these studies have identified and promoted a number of methods that have been used to produce new organic-inorganic structures with well-defined nanoscale features. In this regard, the research field has partially moved in perspective from bio-inspired crystal engineering to bio-inspired nanoscience. Indeed, in recent years, an impressive range of bio-derived templates (biomolecules, viruses, living organisms, etc.) have been studied that constitute a growing toolkit to shape and assemble many types of organic and inorganic materials with nanometer-scale precision. Whilst further insight into the underlying principles of biomineralization is still needed, another major challenge will be to proceed beyond the direct mimicking of biological principles by developing novel templates and synthetic/assembly methods that foster new bio-derived functions such as self-replication, adaptation and evolution in nanomaterials systems.

Several potential developments indicate that exciting new science will continue to emerge in this field:

- Although bio-derived macromolecules such as polysaccharides, proteins and nucleic acids continue to offer new opportunities, new types of synthetic derivatives will significantly extend the capability to design templates that combine functionalities and properties across the range of biomolecule families. For example, a synthetic analogue of DNA such as PNA – in which a polypeptide backbone replaces the (deoxy)ribose sugar backbone – shows the same recognition and duplication properties, but is essentially uncharged and therefore exhibits a higher binding energy than DNA, as well as increased stability in low ionic strength solutions and improved ability to transfer across cell membranes. PNA is also more resistant to enzyme degradation and pH variations, while the peptide backbone might be used for interaction and recognition with other peptides [95].

- The use of dynamic and adaptive polymers as templates for NP/nanostructure synthesis and assembly could offer innovative properties for a wide range of bioinorganic hybrid materials. Indeed, "dynamers" can be built from monomeric components linked through reversible connections, making them genuinely evolutionary as they are able to modify their constitution by exchange and reshuffling of their components [96]. For example, modular reversible polymers are obtained by DNA pairing/denaturing at the ends of two autocomplementary oligonucleotides linked by different spacers [97]. The thermodynamics and kinetics of such polymers are determined by the two sequences, making complex and dynamic materials available from simple building blocks. The integration of these soft materials for nanomaterials fabrication will be a key challenge.

- The versatility and rational design of the self-assembly of isolated proteins, viruses or microorganisms will be significantly enhanced by recent developments in combinatorial chemistry, and through direct screening for specific binding abilities of molecular libraries exposed to targeted substrates [98]. For instance, the efficient binding of mannose moieties to the protein, concanavalin A, has been achieved using combinatorial libraries of dimeric carbohydrate compounds [99].

- Bio-derived routes to nanomaterials fabrication are sure to benefit from the increased availability of methods currently dedicated to biological or biochemical fields such as genotyping, proteotyping, and advanced genetic engineering.

The above comments represent just a few possible outcomes and extensions of currently developed methods for the synthesis and self-assembly of NPs and nanostructures using bio-derived templates. Recent progress, and the burgeoning of creativity reported in the literature, are strong indicators that a convergence between chemical and biological methodologies within a nanoscience context will rapidly lead to a wealth of new concepts, materials, and technologies with wide-ranging applications.

Acknowledgments

The authors thank Drs. S. A. Davis, S. R. Hall, M. Li, A. M. Patil, and W. Shenton at the University of Bristol for their key contributions to experimental studies using bio-derived templates in nanoscience. The authors acknowledge the Egide-British Council joint program "Alliance" for partial funding (Contract 07712PA).

References

1 Lehn, J. M. (2002) Toward complex matter: Supramolecular chemistry and self-organization. *Proc. Natl. Acad. Sci. USA* **99**, 4763–4768.

2 Mann, S. (2001) *Biomineralization. Principles and Concepts in Bioinorganic Materials Chemistry.* Oxford University Press, Oxford, UK.

3 Thomas, A., Antonietti, M. (2003) Silica nanocasting of simple cellulose derivatives: Towards chiral pore systems with long-range order and chiral optical coatings. *Adv. Function. Mater.* **13**, 763–766.

4 Dujardin, E., Blaseby, M., Mann, S. (2003) Synthesis of mesoporous silica by sol-gel mineralisation of cellulose nanorod nematic suspensions. *J. Mater. Chem.* **13**, 696–699.

5 Numata, M., Li, C., Bae, A. H., Kaneko, K., Kazuo, S. C., Shinkai, S. (2005) beta-1,3-Glucan polysaccharide can act as a one-dimensional host to create novel silica nanofiber structures. *Chem. Commun.* 4655–4657.

6 Ogasawara, W., Shenton, W., Davis, S. A., Mann, S. (2000) Template mineralization of ordered macroporous chitin-silica composites using a cuttlebone-derived organic matrix. *Chem. Mater.* **12**, 2835–2837.

7 Reches, M., Gazit, E. (2003) Casting metal nanowires within discrete self-assembled peptide nanotubes. *Science* **300**, 625–627.

8 Hartgerink, J. D., Beniash, E., Stupp, S. I. (2001) Self-assembly and

mineralization of peptide-amphiphile nanofibers. *Science* **294**, 1684–1688.

9 Keren, K., Krueger, M., Gilad, R., Ben-Yoseph, G., Sivan, U., Braun, E. (2002) Sequence-specific molecular lithography on single DNA molecules. *Science* **297**, 72–75.

10 Mertig, M., Ciacchi, L. C., Seidel, R., Pompe, W., De Vita, A. (2002) DNA as a selective metallization template. *Nano Lett.* **2**, 841–844.

11 Braun, E., Sivan, U. (2004) DNA-templated electronics. In: Niemeyer, C. M., Mirkin, C. A. (Eds.), *Nano-biotechnology. Concept, Applications and Perspectives*. Wiley-VCH, Weinheim, pp. 244–255.

12 Mertig, M., Pompe, W. (2004) Biomimetic fabrication of DNA-based metallic nanowires and networks. In: Niemeyer, C. M., Mirkin, C. A. (Eds.), *Nano-biotechnology. Concept, Applications and Perspectives*. Wiley-VCH, Weinheim, pp. 256–277.

13 Gao, X. Y., Matsui, H. (2005) Peptide-based nanotubes and their applications in bionanotechnology. *Adv. Mater.* **17**, 2037–2050.

14 Yu, L. T., Banerjee, I. A., Matsui, H. (2003) Direct growth of shape-controlled nanocrystals on nanotubes via biological recognition. *J. Am. Chem. Soc.* **125**, 14837–14840.

15 Banerjee, I. A., Yu, L. T., Matsui, H. (2003) Cu nanocrystal growth on peptide nanotubes by biominerali-zation: Size control of Cu nanocrystals by tuning peptide conformation. *Proc. Natl. Acad. Sci. USA* **100**, 14678–14682.

16 Naik, R. R., Stringer, S. J., Agarwal, G., Jones, S. E., Stone, M. O. (2002) Biomimetic synthesis and patterning of silver nanoparticles. *Nat. Mater.* **1**, 169–172.

17 Gao, X. Y., Djalali, R., Haboosheh, A., Samson, J., Nuraje, N., Matsui, H. (2005) Peptide nanotubes: Simple separation using size-exclusion columns and use as templates for fabricating one-dimensional single chains of an nanoparticles. *Adv. Mater.* **17**, 1753–1757.

18 Zhao, X. J., Zhang, S. G. (2004) Fabrication of molecular materials using peptide construction motifs. *Trends Biotechnol.* **22**, 470–476.

19 Lomander, A., Hwang, W. M., Zhang, S. G. (2005) Hierarchical self-assembly of a coiled-coil peptide into fractal structure. *Nano Lett.* **5**, 1255–1260.

20 Ellis-Behnke, R. G., Liang, Y. X., You, S. W., Tay, D. K. C., Zhang, S. G., So, K. F., Schneider, G. E. (2006) Nano neuro knitting: Peptide nanofiber scaffold for brain repair and axon regeneration with functional return of vision. *Proc. Natl. Acad. Sci. USA* **103**, 5054–5059.

21 Warner, M. G., Hutchison, J. E. (2003) Linear assemblies of nano-particles electrostatically organized on DNA scaffolds. *Nat. Mater.* **2**, 272–277.

22 Mirkin, C. A., Letsinger, R. L., Mucic, R. C., Storhoff, J. J. (1996) A DNA-based method for rationally assembling nanoparticles into macroscopic materials. *Nature* **382**, 607–609.

23 Alivisatos, A. P., Johnsson, K. P., Peng, X. G., Wilson, T. E., Loweth, C. J., Bruchez, M. P., Schultz, P. G. (1996) Organization of 'nanocrystal molecules' using DNA. *Nature* **382**, 609–611.

24 Seeman, N. C. (2004) Biomimetic fabrication of DNA-based metallic nanowires and networks. In: Niemeyer, C. M., Mirkin, C. A. (Eds.), *Nanobiotechnology. Concept, Applications and Perspectives*. Wiley-VCH, Weinheim, pp. 308–318.

25 Chen, J. H., Seeman, N. C. (1991) Synthesis from DNA of a molecule with the connectivity of a cube. *Nature* **350**, 631–633.

26 Zhang, Y. W., Seeman, N. C. (1994) Construction of a DNA-truncated octahedron. *J. Am. Chem. Soc.* **116**, 1661–1669.

27 Le, J. D., Pinto, Y., Seeman, N. C., Musier-Forsyth, K., Taton, T. A., Kiehl, R. A. (2004) DNA-templated self-assembly of metallic nano-

component arrays on a surface. *Nano Lett.* **4**, 2343–2347.

28 Dujardin, E., Hsin, L. B., Wang, C. R. C., Mann, S. (2001) DNA-driven self-assembly of gold nanorods. *Chem. Commun.* 1264–1265.

29 Sadasivan, S., Dujardin, E., Li, M., Johnson, C. J., Mann, S. (2005) DNA-driven assembly of mesoporous silica/gold satellite nanostructures. *Small* **1**, 103–106.

30 Hamad-Schifferli, K., Schwartz, J. J., Santos, A. T., Zhang, S. G., Jacobson, J. M. (2002) Remote electronic control of DNA hybridization through inductive coupling to an attached metal nanocrystal antenna. *Nature* **415**, 152–155.

31 Li, M., Dujardin, E., Mann, S. (2005) Programmed assembly of multi-layered protein/nanoparticle-carbon nanotube conjugates. *Chem. Commun.* 4952–4954.

32 Li, M., Mann, S. (2004) DNA-directed assembly of multifunctional nanoparticle networks using metallic and bioinorganic building blocks. *J. Mater. Chem.* **14**, 2260–2263.

33 Leveque, I., Cusack, M., Davis, S. A., Mann, S. (2004) Promotion of fluorapatite crystallization by soluble-matrix proteins from Lingula anatina shells. *Angew. Chem. Int. Ed.* **43**, 885–888.

34 Patil, A. J., Muthusamy, E., Mann, S. (2004) Synthesis and self-assembly of organoclay-wrapped biomolecules. *Angew. Chem. Int. Ed.* **43**, 4928–4933.

35 Ma, D., Li, M., Patil, A. J., Mann, S. (2004) Fabrication of protein/silica core-shell nanoparticles by microemulsion-based molecular wrapping. *Adv. Mater.* **16**, 1838–1841.

36 Meldrum, F. C., Wade, V. J., Nimmo, D. L., Heywood, B. R., Mann, S. (1991) Synthesis of inorganic nano-phase materials in supramolecular protein cages. *Nature* **349**, 684–687.

37 Meldrum, F. C., Heywood, B. R., Mann, S. (1992) Magnetoferritin – In-vitro synthesis of a novel magnetic protein. *Science* **257**, 522–523.

38 Mayes, E. L., Mann, S. (2004) Mineralization in nanostructured biocompartments: biomimetic ferritins for high-density data storage. In: Niemeyer, C. M., Mirkin, C. A. (Eds.), *Nanobiotechnology. Concept, Applications and Perspectives.* Wiley-VCH, Weinheim, pp. 278–287.

39 Kim, J. W., Choi, S. H., Lillehei, P. T., Chu, S. H., King, G. C., Watt, G. D. (2005) Cobalt oxide hollow nano-particles derived by bio-templating. *Chem. Commun.* 4101–4103.

40 Shenton, W., Mann, S., Colfen, H., Bacher, A., Fischer, M. (2001) Synthesis of nanophase iron oxide in lumazine synthase capsids. *Angew. Chem. Int. Ed.* **40**, 442–445.

41 Allen, M., Willits, D., Young, M., Douglas, T. (2003) Constrained synthesis of cobalt oxide nano-materials in the 12-subunit protein cage from *Listeria* innocua. *Inorg. Chem.* **42**, 6300–6305.

42 Varpness, Z., Peters, J. W., Young, M., Douglas, T. (2005) Biomimetic synthesis of a H-2 catalyst using a protein cage architecture. *Nano Lett.* **5**, 2306–2309.

43 Flenniken, M. L., Willits, D. A., Brumfield, S., Young, M. J., Douglas, T. (2003) The small heat shock protein cage from *Methanococcus jannaschii* is a versatile nanoscale platform for genetic and chemical modification. *Nano Lett.* **3**, 1573–1576.

44 Sano, K., Ajima, K., Iwahori, K., Yudasaka, M., Iijima, S., Yamashita, I., Shiba, K. (2005) Endowing a ferritin-like cage protein with high affinity and selectivity for certain inorganic materials. *Small* **1**, 826–832.

45 Scheibel, T., Parthasarathy, R., Sawicki, G., Lin, X. M., Jaeger, H., Lindquist, S. L. (2003) Conducting nanowires built by controlled self-assembly of amyloid fibers and selective metal deposition. *Proc. Natl. Acad. Sci. USA* **100**, 4527–4532.

46 Sleytr, U. B., Egelseer, E. M., Pum, D., Schuster, B. (2004) S-layers. In: Niemeyer, C. M., Mirkin, C. A. (Eds.), *Nanobiotechnology. Concept,*

Applications and Perspectives. Wiley-VCH, Weinheim, pp. 77–92.

47 Shenton, W., Pum, D., Sleytr, U. B., Mann, S. (1997) Synthesis of cadmium sulphide superlattices using self-assembled bacterial S-layers. *Nature* **389**, 585–587.

48 Wahl, R., Mertig, M., Raff, J., Selenska-Pobell, S., Pompe, W. (2001) Electron-beam induced formation of highly ordered palladium and platinum nanoparticle arrays on the S layer of *Bacillus sphaericus* NCTC 9602. *Adv. Mater.* **13**, 736–740.

49 Hall, S. R., Shenton, W., Engelhardt, H., Mann, S. (2001) Site-specific organization of gold nanoparticles by biomolecular templating. *Chemphyschem* **2**, 184–186.

50 Moll, D., Huber, C., Schlegel, B., Pum, D., Sleytr, U. B., Sara, M. (2002) S-layer-streptavidin fusion proteins as template for nanopatterned molecular arrays. *Proc. Natl. Acad. Sci. USA* **99**, 14646–14651.

51 McMillan, R. A., Paavola, C. D., Howard, J., Chan, S. L., Zaluzec, N. J., Trent, J. D. (2002) Ordered nanoparticle arrays formed on engineered chaperonin protein templates. *Nat. Mater.* **1**, 247–252.

52 Shenton, W., Davis, S. A., Mann, S. (1999) Directed self-assembly of nanoparticles into macroscopic materials using antibody-antigen recognition. *Adv. Mater.* **11**, 449–452.

53 Connolly, S., Fitzmaurice, D. (1999) Programmed assembly of gold nanocrystals in aqueous solution. *Adv. Mater.* **11**, 1202–1205.

54 Li, M., Wong, K. K. W., Mann, S. (1999) Organization of inorganic nanoparticles using biotin-streptavidin connectors. *Chem. Mater.* **11**, 23–26.

55 Douglas, T., Young, M. (1998) Host-guest encapsulation of materials by assembled virus protein cages. *Nature* **393**, 152–155.

56 Douglas, T., Young, M. (1999) Virus particles as templates for materials synthesis. *Adv. Mater.* **11**, 679–681.

57 Shenton, W., Douglas, T., Young, M., Stubbs, G., Mann, S. (1999) Inorganic-organic nanotube composites from template mineralization of tobacco mosaic virus. *Adv. Mater.* **11**, 253–256.

58 Dujardin, E., Peet, C., Stubbs, G., Culver, J. N., Mann, S. (2003) Organization of metallic nanoparticles using tobacco mosaic virus templates. *Nano Lett.* **3**, 413–417.

59 Knez, M., Bittner, A. M., Boes, F., Wege, C., Jeske, H., Maiss, E., Kern, K. (2003) Biotemplate synthesis of 3-nm nickel and cobalt nanowires. *Nano Lett.* **3**, 1079–1082.

60 Knez, M., Sumser, M., Bittner, A. M., Wege, C., Jeske, H., Martin, T. P., Kern, K. (2004) Spatially selective nucleation of metal clusters on the tobacco mosaic virus. *Adv. Function. Mater.* **14**, 116–124.

61 Gillitzer, E., Willits, D., Young, M., Douglas, T. (2002) Chemical modification of a viral cage for multivalent presentation. *Chem. Commun.* 2390–2391.

62 Wang, Q., Lin, T. W., Tang, L., Johnson, J. E., Finn, M. G. (2002) Icosahedral virus particles as addressable nanoscale building blocks. *Angew. Chem. Int. Ed.* **41**, 459–462.

63 Chatterji, A., Ochoa, W., Shamieh, L., Salakian, S. P., Wong, S. M., Clinton, G., Ghosh, P., Lin, T. W., Johnson, J. E. (2004) Chemical conjugation of heterologous proteins on the surface of cowpea mosaic virus. *Bioconj. Chem.* **15**, 807–813.

64 Hooker, J. M., Kovacs, E. W., Francis, M. B. (2004) Interior surface modification of bacteriophage MS2. *J. Am. Chem. Soc.* **126**, 3718–3719.

65 Douglas, T., Strable, E., Willits, D., Aitouchen, A., Libera, M., Young, M. (2002) Protein engineering of a viral cage for constrained nanomaterials synthesis. *Adv. Mater.* **14**, 415–418.

66 Mao, C. B., Flynn, C. E., Hayhurst, A., Sweeney, R., Qi, J. F., Georgiou, G., Iverson, B., Belcher, A. M. (2003) Viral assembly of oriented quantum

dot nanowires. *Proc. Natl. Acad. Sci. USA* **100**, 6946–6951.

67 Mao, C. B., Solis, D. J., Reiss, B. D., Kottmann, S. T., Sweeney, R. Y., Hayhurst, A., Georgiou, G., Iverson, B., Belcher, A. M. (2004) Virus-based toolkit for the directed synthesis of magnetic and semi-conducting nanowires. *Science* **303**, 213–217.

68 Brown, S. (1992) Engineered iron oxide-adhesion mutants of the *Escherichia coli* phage-lambda receptor. *Proc. Natl. Acad. Sci. USA* **89**, 8651–8655.

69 Brown, S. (2004) Genetic approaches to programmed assembly. In: Niemeyer, C. M., Mirkin, C. A. (Eds.), *Nanobiotechnology. Concept, Applications and Perspectives.* Wiley-VCH, Weinheim, pp. 113–125.

70 Whaley, S. R., English, D. S., Hu, E. L., Barbara, P. F., Belcher, A. M. (2000) Selection of peptides with semiconductor binding specificity for directed nanocrystal assembly. *Nature* **405**, 665–668.

71 Lee, S. W., Mao, C. B., Flynn, C. E., Belcher, A. M. (2002) Ordering of quantum dots using genetically engineered viruses. *Science* **296**, 892–895.

72 Lee, S. W., Wood, B. M., Belcher, A. M. (2003) Chiral smectic C structures of virus-based films. *Langmuir* **19**, 1592–1598.

73 Lee, S. W., Lee, S. K., Belcher, A. M. (2003) Virus-based alignment of inorganic, organic, and biological nanosized materials. *Adv. Mater.* **15**, 689–692.

74 Blum, A. S., Soto, C. M., Wilson, C. D., Cole, J. D., Kim, M., Gnade, B., Chatterji, A., Ochoa, W. F., Lin, T. W., Johnson, J. E., et al. (2004) Cowpea mosaic virus as a scaffold for 3-D patterning of gold nanoparticles. *Nano Lett.* **4**, 867–870.

75 Blum, A. S., Soto, C. M., Wilson, C. D., Brower, T. L., Pollack, S. K., Schull, T. L., Chatterji, A., Lin, T. W., Johnson, J. E., Amsinck, C., et al. (2005) An engineered virus as a scaffold for three-dimensional self-assembly on the nanoscale. *Small* **1**, 702–706.

76 Chatterji, A., Ochoa, W. F., Ueno, T., Lin, T. W., Johnson, J. E. (2005) A virus-based nanoblock with tunable electrostatic properties. *Nano Lett.* **5**, 597–602.

77 Medintz, I. L., Sapsford, K. E., Konnert, J. H., Chatterji, A., Lin, T. W., Johnson, J. E., Mattoussi, H. (2005) Decoration of discretely immobilized cowpea mosaic virus with luminescent quantum dots. *Langmuir* **21**, 5501–5510.

78 Perez, J. M., Simeone, F. J., Saeki, Y., Josephson, L., Weissleder, R. (2003) Viral-induced self-assembly of magnetic nanoparticles allows the detection of viral particles in biological media. *J. Am. Chem. Soc.* **125**, 10192–10193.

79 Fowler, C. E., Shenton, W., Stubbs, G., Mann, S. (2001) Tobacco mosaic virus liquid crystals as templates for the interior design of silica mesophases and nanoparticles. *Adv. Mater.* **13**, 1266–1269.

80 Falkner, J. C., Turner, M. E., Bosworth, J. K., Trentler, T. J., Johnson, J. E., Lin, T. W., Colvin, V. L. (2005) Virus crystals as nanocomposite scaffolds. *J. Am. Chem. Soc.* **127**, 5274–5275.

81 Strable, E., Johnson, J. E., Finn, M. G. (2004) Natural nanochemical building blocks: Icosahedral virus particles organized by attached oligo-nucleotides. *Nano Lett.* **4**, 1385–1389.

82 Cheung, C. L., Camarero, J. A., Woods, B. W., Lin, T. W., Johnson, J. E., De Yoreo, J. J. (2003) Fabrication of assembled virus nanostructures on templates of chemoselective linkers formed by scanning probe nanolithography. *J. Am. Chem. Soc.* **125**, 6848–6849.

83 Smith, J. C., Lee, K. B., Wang, Q., Finn, M. G., Johnson, J. E., Mrksich, M., Mirkin, C. A. (2003) Nano-patterning the chemospecific immobilization of cowpea mosaic virus capsid. *Nano Lett.* **3**, 883–886.

84 Flynn, C. E., Lee, S. W., Peelle, B. R., Belcher, A. M. (2003) Viruses as

vehicles for growth, organization and assembly of materials. *Acta Materialia* **51**, 5867–5880.

85 Dujardin, E., Mann, S. (2002) Bio-inspired materials chemistry. *Adv. Mater.* **14**, 775–788.

86 Davis, S. A., Burkett, S. L., Mendelson, N. H., Mann, S. (1997) Bacterial templating of ordered macrostructures in silica and silica-surfactant mesophases. *Nature* **385**, 420–423.

87 Sastry, M., Ahmad, A., Khan, A. M., Kumar, R. (2004) Microbial nano-particle production. In: Niemeyer, C. M. and Mirkin, C. A. (Eds.), *Nanobiotechnology. Concept, Applications and Perspectives.* Wiley-VCH, Weinheim, pp. 126–135.

88 Lower, S. K., Hochella, M. F., Beveridge, T. J. (2001) Bacterial recognition of mineral surfaces: Nanoscale interactions between *Shewanella* and alpha-FeOOH. *Science* **292**, 1360–1363.

89 Beveridge, T. J. and Murray, R. G. E. (1980) Sites of metal deposition in the cell wall of *Bacillus subtilis. J. Bacteriol.* **141**, 876–887.

90 Frankel, R. B., Bazylinski, D. A. (2004) Magnetosomes: Nanoscale magnetic iron minerals in bacteria. In: Niemeyer, C. M., Mirkin, C. A. (Eds.), *Nanobiotechnology. Concept, Applications and Perspectives.* Wiley-VCH, Weinheim, pp. 136–145.

91 Chan, C. S., De Stasio, G., Welch, S. A., Girasole, M., Frazer, B. H., Nesterova, M. V., Fakra, S., Banfield, J. F. (2004) Microbial polysaccharides template assembly of nanocrystal fibers. *Science* **303**, 1656–1658.

92 Dubertret, B., Skourides, P., Norris, D. J., Noireaux, V., Brivanlou, A. H., Libchaber, A. (2002) In vivo imaging of quantum dots encapsulated in phospholipid micelles. *Science* **298**, 1759–1762.

93 Sweeney, R. Y., Mao, C. B., Gao, X. X., Burt, J. L., Belcher, A. M., Georgiou, G., Iverson, B. L. (2004) Bacterial biosynthesis of cadmium sulfide nanocrystals. *Chem. Biol.* **11**, 1553–1559.

94 Li, Z., Chung, S. W., Nam, J. M., Ginger, D. S., Mirkin, C. A. (2003) Living templates for the merarchical assembly of gold nanoparticles. *Angew. Chem. Int. Ed.* **42**, 2306–2309.

95 Hyrup, B., Nielsen, P. E. (1996) Peptide nucleic acids (PNA): Synthesis, properties and potential applications. *Bioorg. Medicinal Chem.* **4**, 5–23.

96 Lehn, J. M. (2005) Dynamers: dynamic molecular and supra-molecular polymers. *Progress Polymer Sci.* **30**, 814–831.

97 Fogleman, E. A., Yount, W. C., Xu, J., Craig, S. L. (2002) Modular, well-behaved reversible polymers from DNA-based monomers. *Angew. Chem. Int. Ed.* **41**, 4026–4028.

98 Ramstrom, O., Bunyapaiboonsri, T., Lohmann, S., Lehn, J. M. (2002) Chemical biology of dynamic combinatorial libraries. *Biochim. Biophys. Acta – General Subjects* **1572**, 178–186.

99 Ramstrom, O., Lehn, J. M. (2000) In situ generation and screening of a dynamic combinatorial carbohydrate library against concanavalin A. *Chembiochem* **1**, 41–48.

4
Proteins and Nanoparticles: Covalent and Noncovalent Conjugates

Rochelle R. Arvizo, Mrinmoy De, and Vincent M. Rotello

4.1
Overview

Nanoparticles (NPs) provide a versatile tool for the integration of biological and materials systems. These organic–inorganic hybrid materials feature an inorganic core surrounded by a monolayer conjugated with organic and/or biomolecular ligands. Materials used for the cores include metals (e.g., Au, Pt, Ag, Co, FePt), semiconductors (e.g., CdSe, CdS, ZnSe, InP, PbSe), and core/shell hybrids (e.g., CdSe/ZnS, FePt/Fe$_2$O$_3$) [1]. With these systems, the composition of the core material dictates the primary physical and chemical properties of the NP, providing unique and useful intrinsic properties. Gold NPs (AuNPs), for example, are optically dense and have useful electronic and plasmonic properties, while CdSe NPs (quantum dots, QDs), are highly stable and efficient fluorophores for use as optical probes. The magnetic characteristics of iron oxide NPs make them functional probes for techniques such as magnetic resonance imaging (MRI). Taken together, the diversity of available core materials and properties make NPs pragmatic tools for numerous applications [1].

The organic monolayer of the NP is likewise important, providing the interface between the core and the surrounding environment. At the simplest level, the monolayer acts as a barrier between the NP core and the environment, effectively protecting and stabilizing the core. On a more functional level, the reactivity, solubility and interfacial interactions of NPs are dictated by the chemical nature of the monolayer periphery.

For the biological applications featured in this volume, solubility in aqueous environments can be provided by charged or polar groups at the monolayer periphery. The four general monolayer designs used to achieve water solubility are: amphiphilic ligands [2–6], silica shells [7], lipids [8], and polymers [9]. Building upon the water-soluble scaffold, the monolayer can be further tailored with more complex headgroups to modulate intermolecular interactions of the particle. These headgroups can range from simple relatively nonspecific ligands to biologically active components including peptides, proteins, and DNA.

Nanobiotechnology II. Edited by Chad A. Mirkin and Christof M. Niemeyer
Copyright © 2007 WILEY-VCH Verlag GmbH & Co. KGaA, Weinheim
ISBN: 978-3-527-31673-1

Two fundamentally different approaches can be used to conjugate proteins to NPs. The first approach uses noncovalent interactions between the particle and protein, while the second method uses direct covalent linkage of the protein to the particle surface. Both approaches have their strengths and limitations, and hence their place in the bionanotechnology "tool kit".

4.1.1
Covalent Protein-Nanoparticle Conjugates

The unique optical [10, 11], photophysical [12] and electronic [13] properties of metal and semiconductor NPs makes them ideal for biorecognition and biosensing processes. Conjugating proteins covalently to the surface of NPs allows for greater control of protein reactivity and aggregation of biomolecule-functionalized NPs. Covalent attachment is generally achieved by either coupling with active amino acid residues (e.g., $-NH_2$, $-SH$ and $-COOH$) on the surface of a protein, or by solid-phase synthesis of peptides, terminated with functional residue that can be further conjugated onto the NP surface.

Optical biosensing is one of the technologies enabled by covalent particle–protein conjugation. Two ways of transducing binding events using NP-biomolecule conjugates have been recently reported. The first method is simply to use metallic NPs as local quencher of a fluorophore, allowing the interaction between the NP and the fluorescent protein or fluorescent-tagged protein to be determined quantitatively [14]. The second method is through spectroscopic shifts generated by aggregated or conjugated metal NPs through plasmonic coupling [15, 16].

Similar to DNA systems, the specificity of enzyme–substrate interactions can be exploited for the creation of optical biosensors. Recently, Simonian et al. [17] reported a system for the detection of paraoxon, an organophosphate neurotoxin. In their study, these authors functionalized Au NPs with the enzyme organophosphate hydrolase (OPH), and the conjugates were then incubated with a fluorescent enzyme inhibitor. The fluorescence intensity of the inhibitor is sensitive to the proximity of the Au NP (Figure 4.1); when paraoxon was introduced to the OPH–NP–inhibitor conjugate mixtures, an increase in fluorescence was observed via displacement of the inhibitor by paraoxon.

Semiconductor NPs (i.e., QDs, in particular CdSe) are very useful as fluorescence labels due to their favorable intrinsic properties, including high fluorescence quantum yield, photostability, and size-dependent tunable fluorescence bands [7, 18, 19]. In a model system, effective fluorescent resonance energy transfer (FRET) was observed by the specific interaction of biotin-labeled CdSe NPs with Texas red-labeled streptavidin. The extent of the energy transfer was in proportion to the concentration of the dye-labeled protein, and provided proof of concept for an effective biosensor.

Controlled aggregation of NPs causes a shift in optical absorption in the surface plasmon resonance peak and broadening of the absorption spectrum of the NPs, reflecting the extent of aggregation. Otsuka et al. reported a lactose-

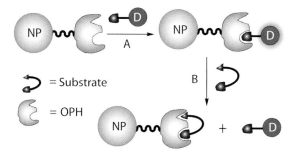

Fig. 4.1 (A) Schematic representation of Decoy **D**-Enzyme interaction for enhancement of fluorescence due to the proximity of nanogold in the absence of substrate. (B) Substrate displacement of decoy from the OPH–gold complex, leading to a decrease in the fluorescence signal from the decoy.

conjugated gold NP to target agglutinin, a bivalent lectin with D-galactose specificity [20]. As shown in Figure 4.2, the lactose on gold NPs promotes lectin-induced aggregation, leading to distinct changes in the absorption spectrum. Also, the change of color from red to purple was observed due to the aggregation. The aggregation is reversible in nature, which can be released by addition of excess galactose. Significantly, since the degree of aggregation is proportional to the

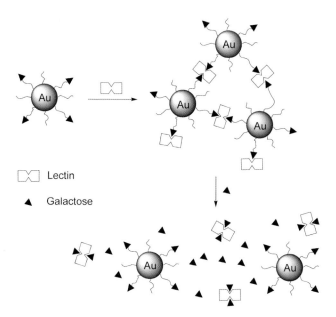

Fig. 4.2 Schematic representation of reversible lectin-induced association of gold nanoparticles modified with lactose.

lectin concentration, the target molecule could be detected quantitatively at high sensitivity. By taking advantage of the specific biotin–streptavidin interaction, Perez-Luna et al. fabricated biotinylated gold NPs and investigated their aggregation by optical properties in the presence of streptavidin [21].

The intrinsic catalytic and photoelectrochemical properties of NPs can be utilized to fabricate electronic biosensors [22, 23]. Redox enzyme functionalized NPs have been used extensively for these bioelectroanalytical systems [24]. In this system, enzyme functionalized NPs are linked through electrodes, and the concentration of substrate or specific proteins is measured by the development of electrolytic current [13, 25, 26]. The catalytic deposition of metals on biomolecule–NP hybrid labels has also been used to generate conductive domains and surfaces. This unique property of these systems allows them to be used as an electronic biosensor, as the conductivity can be measured quantitatively [27].

Electronic biosensors and biofuel cells [28, 29] can be driven by the transfer of electrons generated via biological reactions to the surface of the NP. This transfer changes the surface plasmon resonance spectra of the NP, yielding a bioelectronic/biosensing system. Based on this concept, Willner et al. reported several systems using a NP–enzyme hybrid as a sensor. In one representative example [25], the redox enzyme glucose oxidase (GOx) was connected through a single Au NP by reconstitution of the apo-flavoenzyme, apo-glucose oxidase (apo-GOx), on a 1.4 nm Au NP that was functionalized with N^6-(2-aminoethyl) flavin adenine (FAD). This enzyme–NP hybrid system was connected to the electrode by a suitable dithiol ligand. Alternatively, the FAD-functionalized NP can be assembled on the electrode first, after which the apo-GOx can be introduced (Figure 4.3A). The rate of electron transfer from the enzyme depends on dithiol linkers. Using a dithiol linker (Figure 4.3C), this system exhibits a highly efficient electrical communication with the electron-transfer turnover rate at $\sim$50 000 s^{-1}. As

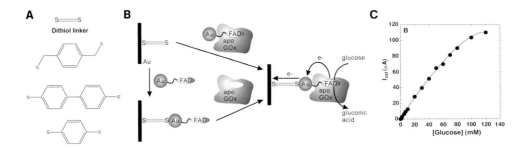

Fig. 4.3 (A) Structure of different dithiol linkers. (B) Assembly of glucose oxidase (GOx) electrode by the reconstitution of apo-enzyme on a FAD-functionalized gold nanoparticle. (C) Calibration plot of the electrocatalytic current developed by the reconstituted GOx electrode in the presence of different concentrations of glucose.

shown in Figure 4.3B, the amount of glucose present is directly proportional to the amount current produced, providing a highly efficient glucose biosensor.

This idea of "charging" NPs can be used for other catalytic application, such as the recently reported hydrogen evolution demonstrated by Yeni Astuti et al. [30]. In these studies, zinc-substituted cytochrome c (ZnCyt-c) was immobilized on metal oxide NPs. The efficient electron injection from the triplet state of ZnCyt-c to the TiO_2 NP electrode then served to generate H_2. Another example recently reported by Ipe and Niemeyer uses electrostatically attached cytochrome $450_{BS\beta}$ to cadmium sulfite (CdS) QDs via a histidine linker [31]. In the presence of hydrogen peroxide, the enzyme is able to catalyze the hydroxylation of myristic acid to hydroxymyristic acid. Since superoxide (O_2^-) and OH is generated when CdS absorbs light, this allows the enzyme to activate its catalytic machinery through scavenging the free radicals. By combining the unique properties of CdS QDs and cytochrome $450_{BS\beta}$, the authors were able to create a nanohybrid that acts as a light-switchable photocatalyst.

4.1.2
Noncovalent Protein–NP Conjugation

The simplest way to produce noncovalent NP–protein conjugates is through complementary electrostatic interactions between the NP and the protein. This can be done either on the surface of a "naked" NP (e.g., citrate-stabilized gold) or by the use of a functionalized monolayer. The use of functionalized monolayers allows the facile generation of either positively or negatively charged NPs that can bind oppositely charged proteins. Perhaps the most significant useful attribute of noncovalent protein–particle conjugation is the reversibility of appropriately chosen systems, facilitating applications in sensing and delivery. For example, Ag_2S NPs conjugated with bovine serum albumin (BSA) can be assembled and disassembled, with the change of pH causing association and dissociation of the protein [32]. Noncovalent interactions with peptides are also useful for templation of peptide assemblies to yield *de-novo* proteins [33–35]. The reversible inhibition and activation of enzymatic activity provides a further use of noncovalent electrostatic conjugation [36]. An important issue for all of these applications is the retention of native protein structure in the particle–protein conjugate. Recent studies have demonstrated that oligo (ethylene glycol) monolayer-protected NPs minimize nonspecific binding of proteins (Figure 4.5A), making these systems quite useful [36, 37].

The controlled interactions of NPs with proteins is a potential tool for both fundamental and applied biomedical investigations. In a fundamental study by Fischer et al., the use of α-chymotrypsin (ChT) with NP **1** (Figure 4.4A) revealed effective inhibition of the enzyme via a two-step process: fast reversible initial inhibition followed by a kinetically irreversible conformational change in secondary structure as determined by CD and fluorescence spectroscopy [5]. The initial binding in this system is electrostatic; the anionic monolayer of the NP is complementary to the cationic side chains surrounding the active site of ChT. Upon

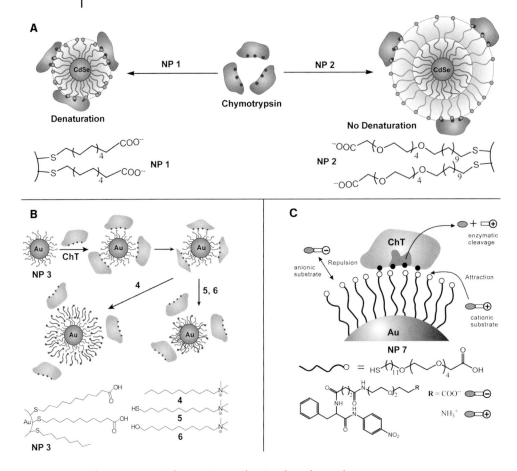

Fig. 4.4 Nanoparticle–protein complexation through complementary electrostatic interactions. (A) Structure and interaction of nanoparticles (NP) **1** and **2** with ChT. (B) ChT inhibition by NP **3** and its release mechanism from the surface of NP by addition of different surfactant (**4–6**). (C) Enzyme selectivity induced through substrate–monolayer interaction. Cationic substrates are attracted and hydrolyzed by the enzyme, but anionic substrates are repulsed by the monolayer.

binding with the NP, the active site of ChT is sterically hindered, resulting in inhibition of the protein. The second irreversible step was proposed to be the result of ChT burying its hydrophobic residues into the surface of the monolayer; this hypothesis is supported by the fact that binding without denaturation was observed for NP **2** [6].

As the nature of binding of the above NP–protein system is electrostatic in nature, surfactants can be used to restore enzyme activity. Subsequent studies conducted by Fischer et al. demonstrated that the addition of positively charged surfactants attenuated the monolayer charge (Figure 4.4B), releasing and reactivating the protein [36]. Dynamic light scattering (DLS) measurements demonstrated that NP **3** had the same hydrodynamic radius before and after surfactant addition of thiol **5** and alcohol **6**. Upon incubation with alkyl surfactant **4**, however, a bilayer is formed. Taken together, these results indicate that NP–protein interactions can be tailored to facilitate applications such as protein refolding and intracellular protein release.

Substrate-selective behavior of the NP–protein complex was observed in further investigations with NP **7** and ChT. To elucidate the role of substrate charge on the selectivity, three SPNA derivatives, each with different charges, were synthesized and analyzed [38]. Enhanced chemoselectivity of ChT activity was observed when bound to the surface of NP **7** (Figure 4.4C). The NP–ChT complex showed very low activity towards negatively charged SPNA substrate, but ∼50% and almost 100% relative activities of bound ChT to free ChT were observed towards the neutral SPNA substrate and the positively charged SPNA substrate, respectively. Considering the substrate charge together with the anionic nature of the NP monolayer, this chemoselectivity can be explained by a combination of steric hindrance and electrostatics.

The introduction of functionality onto the surface of NPs can enhance the affinity and specificity of this type of binding [39]. A report on diverse L-amino acid-terminated NPs demonstrates varying affinity towards ChT, supporting this hypothesis [39]. In addition to modulating the binding affinity to ChT, it was found that the hydrophilic side chains destabilize the structure of ChT through either competitive hydrogen-bonding or breakage of salt bridges, with denaturation much slower with hydrophobic amino acid side chains. Significantly, correlation between the hydrophobicity index of amino acid side chains and the binding affinity and denaturation rates was observed once again.

The above examples demonstrate effective binding with little or no specificity outside of electrostatic complementarity. Specificity of binding can be imparted via conjugation of the particle with biomolecular ligands. Zheng and Huang fabricated a biotin group or glutathione capped onto the surface of gold NPs protected by tri(ethylene glycol) thiols [40]. These authors were able to show specific binding of either streptavidin or glutathione-S-transferase to their respective capped NPs. In another example, Lin et al. fabricated a series of carbohydrate-capped gold NPs to explore their interaction with concanavalin A (Con A) [41]. It was shown that gold NPs functionalized with a mannose linker had a high affinity towards Con A, although the size of the NP and the mannose linker affected the interaction. By extending their investigation using mannose-conjugated gold NPs, the same authors were able to demonstrate mannose-specific adhesion FimH of type 1 pili in *Escherichia coli* [42].

Gold NPs have also been successfully tagged with covalent DNA–streptavidin conjugates. Building upon their earlier findings [27], Niemeyer et al. functionalized citrate-stabilized gold NPs with two different thiolated oligomer strands forming a difunctional DNA–gold NP [43] scheme allowing for the detection of protein antigens with enhanced sensitivity. One of the sequences on the NP is used to immobilize antibodies, while the other is used for signal amplification by means of DNA-directed assembly of multiple layers of NPs. It is suggested that this system will allow for the rapid detection of proteins, at a low cost.

4.2
Methods

The coupling and functionalization of NPs with proteins is carried out using a variety of methods, including electrostatics, ligand recognition, metal-mediated complexation, chemisorption, and covalent binding through bifunctional linkers [44–46]. These methods are summarized in the following sections.

4.2.1
General Methods for Noncovalent Protein–NP Conjugation

The simplest way to form protein–NP conjugation is through electrostatic interactions. In this case, the proteins are electrostatically attracted to oppositely charged NPs, allowing absorption. The affinity and stability of these interactions

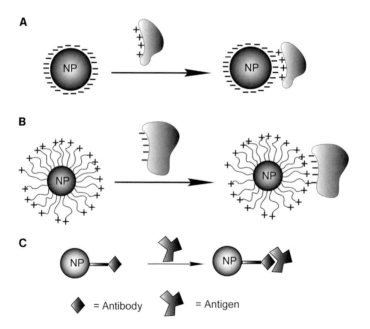

Fig. 4.5 Formation of the noncovalent biomolecule–nanoparticle (NP) conjugates. (A) Electrostatic interactions via direct absorption of the protein onto the NPs surface. (B) NP–protein conjugates formed by absorption of the protein onto the NPs monolayer. (C) Biomolecule–nanoparticle conjugates can also be assembled by antibody–antigen associations.

can be measured using various techniques such as isothermal titration calorimetry (ITC) and enzymatic activity assays. For examples of this interaction, see Section 4.1.2.

Nanoparticle–protein conjugates can also be formed through specific protein–ligand interactions. This recognition is achieved by functionalizing NPs with groups that provide specific affinity to certain proteins or oligonucleotides. For example, streptavidin-functionalized gold NPs have been used for the binding with biotinylated proteins [7], while NPs functionalized with antibodies have been used for affinity binding with their respective antigen [47]. It has been observed that the antigen–antibody binding constant on NPs is higher than in free systems [48].

Metal-mediated complexation provides a versatile method for the creation of noncovalent protein–NP conjugates. Nickel and cobalt nitrilotriacetic acid (NTA) complexes have a high affinity to histidine-tagged proteins, without nonspecific binding via metal chelation [49]. In one study, Xu et al. affixed a nickel–NTA complex onto the iron oxide shell of magnetic NPs with a dopamine anchor, and further used them to target the histidine-tagged proteins [49]. The protein–NP complex can be separated magnetically and then released by the addition of EDTA (Figure 4.6).

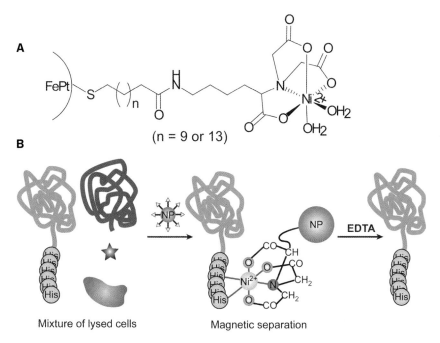

Fig. 4.6 (A) Structure of nanoparticles targeting histidine-tagged proteins. (B) Selective binding to histidine-tagged proteins and purification by magnetic separation.

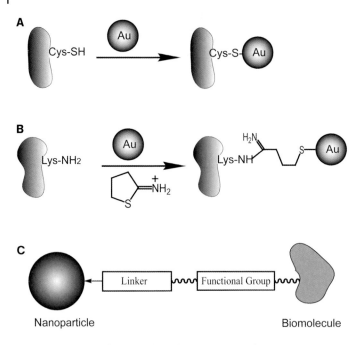

Fig. 4.7 Formation of covalent biomolecule–nanoparticle (NP) conjugates. (A) NP–protein conjugates formed by adsorption of NPs onto native thiol groups of the protein. (B) Conjugation using proteins chemically functionalized with an external thiol residue. (C) Schematic representation of covalent binding of protein through a bifunctional linker.

4.2.2
General Methods for Covalent Protein–NP Conjugation

The problems of instability and inactivation of proteins can be overcome when they are covalently conjugated to the surface of metal NPs [50]. This has been achieved in two ways: (i) chemisorption via thiol derivatives; and (ii) through the use of bifunctional linkers. The chemisorption of proteins onto the surface of NPs (which usually contain a core of Au, ZnS, CdS and CdSe/ZnS) can proceed through cysteine residues that are present in the protein (e.g., oligopeptide, serum albumin) [51], or chemically in the presence of 2-iminothiolane (Traut's reagent) [51, 52].

Bifunctional linkers are highly diverse and offer versatile covalent conjugation of proteins onto NPs. Thiols, disulfides or phosphine ligands are often used as anchor groups to bind Au, Ag, CdS, and CdSe NPs for these bifunctional linkers. Weakly absorbed molecules are displaced by the above anchor groups to further stabilize the NPs, or they are added during NP synthesis. Biological compounds are often covalently linked to NPs using bifunctional linkers via carbodiimide-mediated amidation and esterfication coupling with thiol groups [24].

4.3
Outlook

During the past few years, the development and application of protein–NP conjugates has burgeoned, mainly because the useful characteristics of NPs – such as ease of functionalization, tunable core size, and variety of core materials available – make these systems excellent scaffolds for the conjugation of proteins. There are numerous opportunities for expanding the utility of these systems. For example, in the area of noncovalent protein–particle interactions there are many unanswered questions regarding the nature of the interface, and a better understanding of this issue will provide access to a range of new diagnostics and therapeutics.

In another potential direction, proteins and NPs can self-assemble into highly organized superstructures, offering opportunities for the creation of novel materials and devices [53, 54]. Thus, an understanding of protein–NP conjugates is significant both for fundamental research and for biotechnological and nanotechnological approaches. In turn such knowledge should lead to the creation of advanced materials that can be applied to sensing, catalysis, signal transduction, transport, and other applications associated with the biomedical and/or biomaterial sciences.

References

1 Burda, C., Chen, X., Narayanan, R., El-Sayed, M. A. (2005) Chemistry and properties of nanocrystals of different shapes. *Chem. Rev.* **105**, 1025–1102.

2 Chan, W. C. W., Nie, S. (1998) Quantum dot bioconjugates for ultrasensitive nonisotopic detection. *Science* **281**, 2016–2018.

3 Uyeda, H. T., Medintz, I. L., Jaiswal, J. K., Simon, S. M., Mattoussi, H. (2005) Synthesis of compact multidentate ligands to prepare stable hydrophilic quantum dot fluorophores. *J. Am. Chem. Soc.* **127**, 3870–3878.

4 Mattoussi, H., Mauro, J. M., Goldman, E. R., Anderson, G. P., Sundar, V. C., Mikulec, F. V., Bawendi, M. G. (2000) Self-assembly of CdSe-ZnS quantum dot bioconjugates using an engineered recombinant protein. *J. Am. Chem. Soc.* **122**, 12142–12150.

5 Fischer, N. O., McIntosh, C. M., Simard, J. M., Rotello, V. M. (2002) Supramolecular chemistry and self-assembly special feature: Inhibition of chymotrypsin through surface binding using nanoparticle-based receptors. *Proc. Natl. Acad. Sci. USA* **99**, 5018–5023.

6 Hong, R., Fischer, N. O., Verma, A., Goodman, C. M., Emrick, T., Rotello, V. M. (2004) Control of protein structure and function through surface recognition by tailored nanoparticle scaffolds. *J. Am. Chem. Soc.* **126**, 739–743.

7 Bruchez, M., Jr., Moronne, M., Gin, P., Weiss, S., Alivisatos, A. P. (1998) Semiconductor nanocrystals as

fluorescent biological labels. *Science* **281**, 2013–2016.

8 Dubertret, B., Skourides, P., Norris, D. J., Noireaux, V., Brivanlou, A. H., Libchaber, A. (2002) In vivo imaging of quantum dots encapsulated in phospholipid micelles. *Science* **298**, 1759–1762.

9 Gao, X., Cui, Y., Levenson, R. M., Chung, L. W. K., and Nie, S. (2004) In vivo cancer targeting and imaging with semiconductor quantum dots. *Nat. Biotechnol.* **22**, 969–976.

10 Alvarez, M. M., Khoury, J. T., Schaaff, T. G., Shafigullin, M. N., Vezmar, I., Whetten, R. L. (1997) Optical absorption spectra of nanocrystal gold molecules. *J. Phys. Chem. B* **101**, 3706–3712.

11 Alivisatos, A. P. (1996) Perspectives on the physical chemistry of semi-conductor nanocrystals. *J. Phys. Chem.* **100**, 13226–13239.

12 Daniel, M.-C., Astruc, D. (2004) Gold nanoparticles: assembly, supra-molecular chemistry, quantum-size-related properties, and applications toward biology, catalysis, and nano-technology. *Chem. Rev.* **104**, 293–346.

13 Hicks, J. F., Miles, D. T., Murray, R. W. (2002) Quantized double-layer charging of highly monodisperse metal nanoparticles. *J. Am. Chem. Soc.* **124**, 13322–13328.

14 Dubertret, B., Calame, M., Libchaber, A. J. (2001) Single-mismatch detection using gold-quenched fluorescent oligonucleotides. *Nat. Biotechnol.* **19**, 365–370.

15 Zhang, X., Young, M. A., Lyandres, O., VanDuyne, R. P. (2005) Rapid detection of an anthrax biomarker by surface-enhanced Raman spectroscopy. *J. Am. Chem. Soc.* **127**, 4484–4489.

16 Haes, A. J., Zou, S., Schatz, G. C., VanDuyne, R. P. (2004) A nanoscale optical biosensor: the long range distance dependence of the localized surface plasmon resonance of noble metal nanoparticles. *J. Phys. Chem. B* **108**, 109–116.

17 Simonian, A. L., Good, T. A., Wang, S.-S., Wild, J. R. (2005) Nanoparticle-based optical biosensors for the direct detection of organophosphate chemical warfare agents and pesticides. *Anal. Chim. Acta* **534**, 69–77.

18 El-Sayed, M. A. (2004) Small is different: shape-, size-, and composition-dependent properties of some colloidal semiconductor nanocrystals. *Acc. Chem. Res.* **37**, 326–333.

19 Grieve, K., Mulvaney, P., Grieser, F. (2000) Synthesis and electronic properties of semiconductor nanoparticles/quantum dots. *Curr. Opinion Colloid Interface Sci.* **5**, 168–172.

20 Otsuka, H., Akiyama, Y., Nagasaki, Y., Kataoka, K. (2001) Quantitative and reversible lectin-induced association of gold nanoparticles modified with α-lactosyl-ω-mercapto-poly(ethylene glycol). *J. Am. Chem. Soc.* **123**, 8226–8230.

21 Aslan, K., Luhrs, C. C., Perez-Luna, V. H. (2004) Controlled and reversible aggregation of biotinylated gold nanoparticles with streptavidin. *J. Phys. Chem. B* **108**, 15631–15639.

22 Willard, D. M. (2003) Nanoparticles in bioanalytics. *Anal. Bioanal. Chem.* **376**, 284–286.

23 Wang, J. (2003) Nanoparticle-based electrochemical DNA detection. *Anal. Chim. Acta* **500**, 247–257.

24 Katz, E., Willner, I. (2004) Integrated nanoparticle-biomolecule hybrid systems: synthesis, properties, and applications. *Angew. Chem. Int. Ed.* **43**, 6042–6108.

25 Xiao, Y., Patolsky, F., Katz, E., Hainfeld, J. F., Willner, I. (2003) 'Plugging into enzymes': Nanowiring of redox enzymes by a gold nano-particle. *Science* **299**, 1877–1881.

26 Hernandez-Santos, D., Gonzalez-Garcia, M. B., Garcia, A. C. (2002) Metal-nanoparticles based electro-analysis. *Electroanalysis* **14**, 1225–1235.

27 Niemeyer, C. M., Ceyhan, B. (2001) DNA-directed functionalization of colloidal gold with proteins. *Angew. Chem. Int. Ed.* **40**, 3685–3688.

28 Barton, S. C., Kim, H.-H., Binyamin, G., Zhang, Y., Heller, A. (2001) The 'wired' Laccase cathode: high current density electroreduction of O_2 to water at $+0.7$ V (NHE) at pH 5. *J. Am. Chem. Soc.* **123**, 5802–5803.

29 Chen, T., Barton, S. C., Binyamin, G., Gao, Z., Zhang, Y., Kim, H.-H., Heller, A. (2001) A miniature biofuel cell. *J. Am. Chem. Soc.* **123**, 8630–8631.

30 Astuti, Y., Palomares, E., Haque, S. A., Durrant, J. R. (2005) Triplet state photosensitization of nanocrystalline metal oxide electrodes by zinc-substituted cytochrome *c*: Application to hydrogen evolution. *J. Am. Chem. Soc.* **127**, 15120–15126.

31 Ipe, B. I., Niemeyer, C. M. (2006) Nanohybrids composed of quantum dots and cytochrome P450 as photocatalysts. *Angew. Chem. Int. Ed.* **45**, 504–507.

32 Meziani, M. J., Sun, Y.-P. (2003) Protein-conjugated nanoparticles from rapid expansion of supercritical fluid solution into aqueous solution. *J. Am. Chem. Soc.* **125**, 8015–8018.

33 Pengo, P., Broxterman, Q. B., Kaptein, B., Pasquato, L., Scrimin, P. (2003). Synthesis of a stable helical peptide and grafting on gold nanoparticles. *Langmuir* **19**, 2521–2524.

34 Verma, A., Nakade, H., Simard, J. M., Rotello, V. M. (2004) Recognition and stabilization of peptide α-helices using templatable nanoparticle receptors. *J. Am. Chem. Soc.* **126**, 10806–10807.

35 Boal, A. K., Rotello, V. M. (2000) Fabrication and self-optimization of multivalent receptors on nanoparticle scaffolds. *J. Am. Chem. Soc.* **122**, 734–735.

36 Fischer, N. O., Verma, A., Goodman, C. M., Simard, J. M., Rotello, V. M. (2003) Reversible 'irreversible' inhibition of chymotrypsin using nanoparticle receptors. *J. Am. Chem. Soc.* **125**, 13387–13391.

37 Zheng, M., Davidson, F., Huang, X. (2003) Ethylene glycol monolayer protected nanoparticles for eliminating nonspecific binding with biological molecules. *J. Am. Chem. Soc.* **125**, 7790–7791.

38 Hong, R., Emrick, T., Rotello, V. M. (2004) Monolayer-controlled substrate selectivity using noncovalent enzyme-nanoparticle conjugates. *J. Am. Chem. Soc.* **126**, 13572–13573.

39 You, C.-C., De, M., Han, G., Rotello, V. M. (2005) Tunable inhibition and denaturation of α-chymotrypsin with amino acid-functionalized gold nanoparticles. *J. Am. Chem. Soc.* **127**, 12873–12881.

40 Zheng, M., Huang, X. (2004) Nanoparticles comprising a mixed monolayer for specific bindings with biomolecules. *J. Am. Chem. Soc.* **126**, 12047–12054.

41 Lin, C. C., Yeh, Y. C., Yang, C. Y., Chen, G. F., Chen, Y. C., Wu, Y. C., Chen, C. C. (2003) Quantitative analysis of multivalent interactions of carbohydrate-encapsulated gold nanoparticles with concanavalin A. *Chem. Commun.* **23**, 2920–2921.

42 Lin, C.-C., Yeh, Y.-C., Yang, C.-Y., Chen, C.-L., Chen, G.-F., Chen, C.-C., Wu, Y.-C. (2002) Selective binding of mannose-encapsulated gold nanoparticles to type 1 pili in *Escherichia coli. J. Am. Chem. Soc.* **124**, 3508–3509.

43 Hazarika, P., Ceyhan, B., Niemeyer, C. M. Sensitive detection of proteins using difunctional DNA-gold nanoparticles. *Small* **1**, 844–848.

44 Collings, A. F., Caruso, F. (1997) Biosensors: recent advances. *Reports Progress Physics* **60**, 1397–1445.

45 Rao, S. V., Anderson, K. W., Bachas, L. G. (1998) Oriented immobilization of proteins. *Microchim. Acta* **128**, 127–143.

46 Scouten, W. H., Luong, J. H. T., Stephen Brown, R. (1995) Enzyme or protein immobilization techniques for applications in biosensor design. *Trends Biotechnol.* **13**, 178–185.

47 Sondi, I., Siiman, O., Koester, S., Matijevic, E. (2000) Preparation of aminodextran-CdS nanoparticle complexes and biologically active antibody-aminodextran-CdS

nanoparticle conjugates. *Langmuir* **16**, 3107–3118.

48 Soukka, T., Harma, H., Paukkunen, J., Lovgren, T. (2001) Utilization of kinetically enhanced monovalent binding affinity by immunoassays based on multivalent nanoparticle-antibody bioconjugates. *Anal. Chem.* **73**, 2254–2260.

49 Xu, C., Xu, K., Gu, H., Zhong, X., Guo, Z., Zheng, R., Zhang, X., Xu, B. (2004) Nitrilotriacetic acid-modified magnetic nanoparticles as a general agent to bind histidine-tagged proteins. *J. Am. Chem. Soc.* **126**, 3392–3393.

50 Caruso, F. (2001) Nanoengineering of particle surfaces. *Adv. Mater.* **13**, 11–22.

51 Droz, E., Taborelli, M., Descouts, P., Wells, T. N. C., Werlen, R. C. (1996) Covalent immobilization of immuno-globulins G and *Fab'* fragments on

gold substrates for scanning force microscopy imaging in liquids. *J. Vacuum Sci. Technol. B* **14**, 1422–1426.

52 Ghosh, S. S., Kao, P. M., McCue, A. W., Chappelle, H. L. (1990) Use of maleimide-thiol coupling chemistry for efficient syntheses of oligo-nucleotide-enzyme conjugate hybridization probes. *Bioconj. Chem.* **1**, 71–76.

53 Srivastava, S., Verma, A., Frankamp, B. L., Rotello, V. M. (2005) Controlled assembly of protein-nanoparticle composites through protein surface recognition. *Adv. Mater.* **17**, 617–621.

54 Verma, A., Srivastava, S., Rotello, V. M. (2005) Modulation of the interparticle spacing and optical behavior of nanoparticle ensembles using a single protein spacer. *Chem. Mater.* **17**, 6317–6322.

5

Self-Assembling DNA Nanostructures for Patterned Molecular Assembly

Thomas H. LaBean, Kurt V. Gothelf, and John H. Reif

Abstract

This chapter describes the use of DNA for molecular-scale self-assembly. DNA nanostructures provide a versatile toolbox with which to organize nanoscale materials. The chapter commences with a discussion of DNA nanostructures, the self-assembly of various building-blocks known as DNA tiles, and how these can be made to self-assemble into two and three-dimensional lattices. Methods are then discussed for the programmed assembly of patterned and/or shaped two- and three-dimensional DNA-nanostructures, including their use to produce beautiful algorithmic assemblies displaying fractal design patterns. The resulting large DNA nanostructures provide multiple attachment sites within and between tiles for complex programmed structures, which leads to diverse possibilities for scaffolding useful constructs and templating interesting chemistries. Methods are then described for the assembly of various biomolecules and metallic nanoparticles (NPs) onto DNA nanostructures, and also of various materials using DNA nanostructures. The chapter is concluded with a discussion of the various challenges faced by DNA nanostructure self-assembly.

5.1
Introduction

Self-assembly is one of the key approaches that might enable future methods to be developed for building nanostructures and nanodevices [1]. Our current ability to form nanostructures by self-assembly is, however, quite limited compared to the power of lithographic techniques used to form solid structures in bulk material and, in particular, of electronic circuits at the nanoscale. Lithography is used to form the most complex human-constructed objects, namely microprocessors, which have been manufactured with upwards of billions of precisely patterned elements. It should be noted however, that top-down techniques such as lithography are limited in scale, whereas bottom-up methods of self-assembly are not.

Nanobiotechnology II. Edited by Chad A. Mirkin and Christof M. Niemeyer
Copyright © 2007 WILEY-VCH Verlag GmbH & Co. KGaA, Weinheim
ISBN: 978-3-527-31673-1

The motivation and inspiration to explore self-assembly can literally be found within ourselves since, as living organisms, we are the most advanced nanosystems known, far exceeding the complexity of any human-engineered nanostructure. The fantastically complex machinery of living organisms – which is composed primarily of organic molecules and polymers – is formed by, and operates by, self-assembly. Indeed, the precision and efficiency of the self-assembly process in cells derives from specific molecular interactions between proteins, DNA and RNA in particular, as well as other compounds such as lipids, carbohydrates and small molecules. The question to be considered here is whether we can use techniques inspired from the cell's self-assembly machinery to assemble artificial nanostructures? Likewise, which materials should we use as building blocks in order to direct the self-assembly process at the molecular scale?

Proteins, antibodies, peptides and their small-molecule affinity substrates are efficient and highly specific in self-assembly processes, and if only few types of specific interactions were to be required then these might be the best choice. The importance of their role in molecular biology, medicine and nanoscience cannot be underestimated, but when it comes to the individual encoding of multiple building blocks for assembly into complex nanostructures, their use may be limited due to their diverse structures and differences in the nature of their self-interactions or size. These materials have their own extensive technical literature, but this will not be discussed at this point.

In contrast, DNA nanostructures have an easily predictable secondary structure due to the well-understood properties of single-stranded DNA base-pairing and the double-helix structure of double-stranded (ds) DNA as functions of their environment (temperature and buffer solution) [2–4], and this allows software to be developed for computer-aided design of the sequences that comprise DNA nanostructures [5, 6]. There is also a well-established biotechnology for constructing DNA sequences and for executing operations on DNA sequences. Likewise, there are known techniques for attaching other molecules to specific locations on DNA sequences. Hence, DNA appears to be an ideal material for achieving complex self-assembly, and consequently this will form the basis of the present chapter.

5.2
Overview of DNA Nanostructures

The past few years have witnessed a huge number of advances in our ability to construct complex nanostructures from nucleic acid building materials [7–9]. The study of artificial DNA structures for applications in nanotechnology began during the early 1980s when Seeman sought to design and construct periodic matter and discrete objects assembled from synthetic DNA oligonucleotides [10]. He noted that simple double-helical DNA could only be used for the construction of linear assemblies, and that more complex building blocks would be required for two- and three-dimensional constructs. Seeman also noted that biological systems make use of branched base-pairing complexes such as forks (three-arm junctions) found in replicating DNA and Holliday intermediates (four-arm junctions) found in homologous recombination complexes. These natural branch junction motifs exposed a potential path toward multi-valent structural units. A *Holliday junction* is formed by four strands of DNA (two identical pairs of complementary strands) where double-helical domains meet at a branch point and exchange base-pairing partner strands. The branch junctions in recombination complexes are free to diffuse up and down the paired homologous dsDNA domains as the partners share sequence identity all along their lengths. Seeman subsequently showed that, by specifically designing sequences which were able to exchange strands at a single specified point and by breaking the sequence symmetry which allowed the branch junction to migrate [11], immobile junctions could be constructed and used in the formation of stable and rigid DNA building blocks. These building blocks (known as *DNA tiles*) – especially double-crossover (DX) complexes [12] – became the initial building blocks for the construction of periodic assemblies and the formation of the first uniform two-dimensional crystals of DNA, known as *DNA lattices* [13]. They also can be used to form long tubes [14].

A large number of distinct DNA tile types have now been designed and prototyped; some examples are shown in Figure 5.1. The high thermal stability (T_m up to at least 70 °C) of some DNA tiles, the ability to program tile-to-tile association rules via single-strand (ss) DNA sticky-ends, and the wide range of available attachment chemistries make these structures extremely useful as molecular-scale building blocks for diverse nanofabrication tasks. The process of DNA nanostructure and sequence design was laid out very effectively in a recent review article [15]. To date, all DNA tiles produced have contained double-helical DNA domains as structural members and branch junction crossovers as connectors. The use of paired crossovers greatly increases the stiffness of the tiles over that of linear dsDNA. Following the success of the DX lattices, triple-crossover (TX) tiles and their two-dimensional (2-D) uniform lattices and tubes were demonstrated [16–20]. Recently, the existence of DDX tiles containing four double helices and four crossover points has also been demonstrated [21].

Since DX, DDX and TX tiles are designed with their helices parallel and coplanar, their lattices tend to grow very well in the dimension parallel with the helix axes, but fairly poorly in the dimension perpendicular to the axis. Elimination of this problem and growth of lattices with a square aspect ratio was the primary motivation behind the design of the cross-tile [22], which allows for a uniform growth of 2-D lattices in both of these directions. Long DNA nanotubes (up to 15 μm in length) and large 2-D lattices (extending over many μm^2) have been assembled from cross-tiles (Figure 5.2). The design of the 2-D cross-tile lattices made use of an interesting corrugation method [22] for providing symmetry in both the horizontal and vertical directions, by the use of rotations and flips of neighboring cross-tiles, so as to cancel deformations that otherwise would limit lattice growth. Recently, a variant of the cross-tile [23] was developed that provides

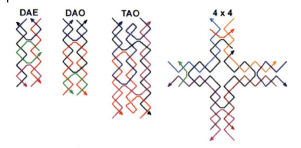

Fig. 5.1 Schematic drawings of four DNA tiles. The colored lines represent different oligonucleotide strands with arrowheads marking the 3′ ends. DAE and DAO are double crossover complexes (also known as DX), TAO is an example of a triple crossover (or TX) tile, and the 4 × 4 cross-tile is composed of four arms each of which contains a four-arm junction.

an improved symmetry in both the horizontal and vertical directions, resulting in even larger 2-D lattices (extending to millimeters).

A variety of other tile shapes have been prototyped beyond the rectangular and square tiles illustrated here. Lattices with rhombus-like units have been constructed in which the helix crossing angles are closer to the relaxed ~60° angles observed in biological Holliday junctions [24]. At least three different versions of triangular DNA tiles have been prototyped (Figure 5.3): one type which tiles the plane with triangles [25]; and two types which form hexagonal patterns [26, 27]. Such triangular lattices have not been shown to grow as large as those from rectangular and square tiles, but they may be useful for assembly applications where

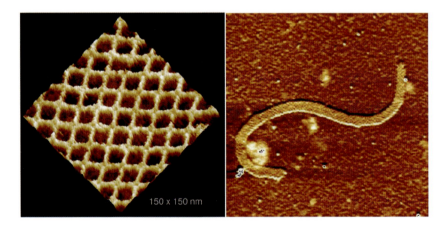

Fig. 5.2 Atomic force microscopy images of corrugated (planar) and uncorrugated (tube) versions of cross-tile lattices. The right panel is a 1 μm × 1 μm scan. (Adapted with permission from Ref. [22].)

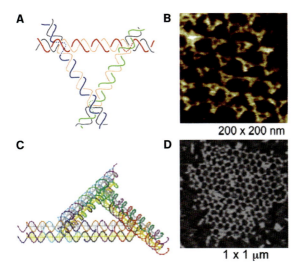

Fig. 5.3 Schematic drawings of two different triangular tiles (A, C) and AFM images of resulting 2-D lattices assembled from triangle tiles (B, D). (A,B adapted with permission from Ref. [25]; C,D adapted with permission from Ref. [26].)

slightly more structural flexibility is desired. Some interesting multilayer structures with symmetrical stacking interactions have also been demonstrated [27].

DNA tiles have also been assembled using paranemic interactions between pairs of parallel helices [28].

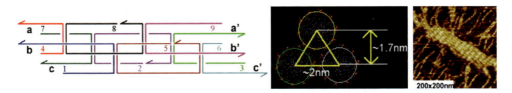

Fig. 5.4 The three-helix bundle (3HB) DNA tile shown as a schematic trace of the strands through the tile. The different colored lines represent different oligonucleotides, and arrowheads mark the 3′ ends. Six crossover points (paired vertical lines) connect the three double helices (paired horizontal lines), with two crossovers connecting each of the possible pairs of helices. The center panel is an end view of the 3HB tile to show the stacking of the helical domains. The right panel is an AFM image of a 2-D lattice formed from properly programmed 3HB tiles. (Adapted with permission from Ref. [29].)

5.3
Three-Dimensional (3-D) DNA Nanostructures

Three-dimensional building blocks and periodic matter constructed of DNA have been long-term goals of this field. In addition to forming 2-D lattices and tubes, DNA tiles can be used to form 3-D lattices. If guest molecules can be incorporated into 3-D DNA lattices, typical applications may include:

- the assembly of 3-D molecular electronics circuits and memory; and
- the molecular scaffold for structure determination of guest molecules via X-ray crystallography studies.

Non-planar tiles represent one strategy for expanding the tiling into the third dimension. DNA tiles which hold their helical domains in non-planar arrangements have been designed, and include for example a three-helix bundle (Figure 5.4) [29] and a six-helix bundle [30], although initial attempts at 3-D DNA nanostructures using these particular tiles have not yet proved successful. Recently, Seeman and colleagues [31] have demonstrated a 3-D DNA hexagonal lattice formed from a single 13-base DNA sequence that formed stacked layers of parallel helices with base-pairing between adjacent layers. However, these 3-D DNA lattices are not yet sufficiently regular to allow for their application to structure determination of guest molecules via X-ray crystallography.

Another approach is to form 3-D polyhedral DNA nanostructures which can then assembly to regular lattices. Early attempts to build a cube [32] and a truncated octahedron [33] with dsDNA edges and branch junction vertices met with some success, but the final constructs were produced in very low yields. More recently, a tetrahedral unit with short double helical edges was constructed in much higher yield [34]. Perhaps the most impressive experimental success to date in DNA-based 3-D nanostructures produced an octahedron with DX-like edges [35]. This study was noteworthy in that the 1.7 kilobase DNA strand (which folded with the help of five short oligonucleotides) was produced as a single piece by PCR-based assembly, and the octahedron was formed in sufficient yield to permit structural characterization by cryoelectron microscopy.

5.4
Programmed Patterning of DNA Nanostructures

In addition to 2-D and 3-D periodic lattices, another long-term goal of DNA self-assembly studies has been the generation of complex, non-periodic patterns on lattices. There are at least three techniques available to perform this:

1. The use of *algorithmic self-assembly*, whereby patterns are formed using a small tile set, the sticky-ends of which represent tile association rules that promote lattice formation according to the specific rules of the encoded algorithm. The first demonstrations of algorithmic self-assembly used DNA tiles to demonstrate the execution of various Boolean and arithmetic operations at the molecular scale; the computa-

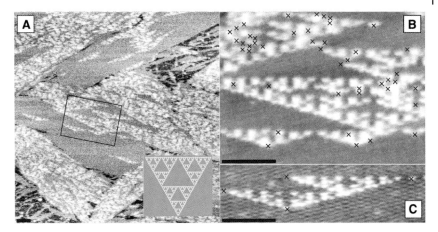

Fig. 5.5 AFM images of DX tile lattice algorithmically assembled to form fractal Sierpinski triangle patterns. (A) Bright tiles carry an additional stem-loop of DNA projected out of the tile plane which appears taller to the AFM and therefore acts as a topographic marker. The inset to panel (A) shows a schematic of the target pattern. (B) An expanded view of the boxed region in panel (A). (C) An expanded view from another section of lattice. The 'Xs' mark tiles which appear to be assembled incorrectly, based on visual inspection of the preceding (input) tiles. Scale bars = 100 nm. (Adapted with permission from Ref. [37].)

tions occurred during the assembly of linear sequences of DNA tiles that preferentially bound to each other according to computational rules [17, 36]. Subsequently, 2-D demonstrations of algorithmic self-assembly methods using DNA tiles have provided some of the most complex patterns yet demonstrated via molecular self-assembly; an example is seen in the Sierpinski triangle pattern shown in Figure 5.5 [37]. In principle, any arbitrary structure which can be specified by a set of encoded association rules can be expected to form via algorithmic self-assembly [38], albeit at some yield < 100% and with some error rate > 0%. Various schemes have been developed to reduce errors in algorithmic self-assembly, but with the side effect of increasing the size of the lattice as well as the number of tile types [39–41]. Compact error-resilient designs were later developed which provide a reduction of assembly errors without any increase in the size of the computational lattice [42, 43].

2. Another method for programmed patterning of DNA nanostructures is to use a *stepwise* or *hierarchal self-assembly*, whereby patterns are formed in multiple stages, incorporating as subcomponents patterned nanostructures formed in prior stages. A recent demonstration of this technique provided the molecular-scale self-assembly of fixed-size DNA

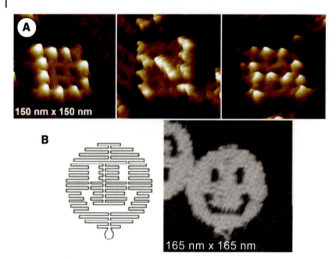

Fig. 5.6 (A) AFM images of fixed-size DNA tile lattices displaying bound streptavidin molecules (white dots) in arbitrary target patterns. (Adapted with permission from Ref. [44].) (B) Schematic drawing and AFM image of an arbitrary shape formed by folding the single-strand DNA genome of M13 bacteriophage using staple strands (short synthetic oligonucleotides). (Adapted with permission from Ref. [46].)

lattices patterned in any arbitrary way (see Figure 5.6) [44]. This study showed that, by minimizing the number of sequential steps in the assembly process, the overall yield of target structure would be maximized.

3. A final and very promising method is the use of *directed self-assembly*, whereby patterns are formed via the use of molecules that control in some way the self-assembly process so as to form the intended pattern. The directed self-assembly of one-dimensional patterned DNA lattices was first demonstrated by using scaffold strands to provide the specification of inputs to the computational assemblies mentioned above [36, 38]. The directed self-assembly of 2-D patterned DNA lattices was then demonstrated by the use of scaffold strands that were incorporated into each of the rows of a 2-D DNA lattice, allowing for the display of binary sequences as a 2-D barcode pattern that can be viewed by atomic force microscopy (AFM) imaging [45]. To date, the most impressive use of directed self-assembly has been made by Rothemund, using a 7-kb scaffold strand which folded into arbitrary 2-D shapes and patterns with the help of multiple short oligonucleotides that specify the shape and patterning (see Figure 5.6) [46].

Finite-sized arrays were also reported in another DNA system and in RNA [47, 48]. Previous DNA tiling systems all resulted in an unbounded growth of lattice, and consequently polydisperse products following annealing. These demonstrations of finite-sized arrays represent another step towards the increased control of self-assembled molecular systems.

The 2-D lattices and 3-D structures assembled from DNA and described in this and the prior sections represent interesting objects in their own right, but their real usefulness will come from their application as scaffolds and templates upon which chemistry is performed, or with which heteromaterials are organized into functioning nanodevices. Here, we will use the term *DNA-programmed patterned assembly* to denote this use of DNA lattices to organize heteromaterials. We will return to some of these applications later in the chapter.

5.5
DNA-Programmed Assembly of Biomolecules

The assembly of other biomolecules on DNA templates and arrays may prove useful for the fabrication of biomimetics and other devices, with applications such as biochips, immunoassays, biosensors, and a variety of nanopatterned materials. The logical end to the shrinking of microarrays is the self-assembled DNA nanoarray with a library of ligands distributed at addressable locations to bring analyte detection down to the single molecule level. We will return to complex DNA tiling structures momentarily, but first we examine simpler dsDNA systems.

The conjugation of DNA and streptavidin via a covalent linker was reported by Niemeyer et al. in 1994, and these conjugates were applied for DNA-programmed assembly on a macroscopic DNA array on a surface and in a nanoscale array made by aligning DNA-tagged proteins to specific positions along an oligonucleotide template [49–51]. The covalent attachment of an oligonucleotide to streptavidin provides a specific recognition domain for a complementary nucleic acid sequence. In addition, the binding capacity for four biotin molecules is utilized as biomolecular adapters for positioning biotinylated components along a nucleic acid backbone.

Besides duplex DNA structures, more complex self-assembling DNA tiling structures have been used to organize biomolecules into specific spatial patterns. DNA nanostructures covalently labeled with ligands have been shown to bind protein molecules in programmed patterns; for example, making use of the popular biotin/avidin pair, arrays of evenly spaced streptavidin molecules were assembled on a DNA tile lattice [22]. On cross-tile lattices, individual streptavidin molecules are visible as separate peaks in the AFM image (Figure 5.7). Hence, single molecule detection could be achieved on DNA nanoarrays displaying a variety of protein binding ligands.

Further design evolution of the cross-tile system to a two-tile type (A and B) tile set allowed for somewhat more complex structures and patterns [52]. In

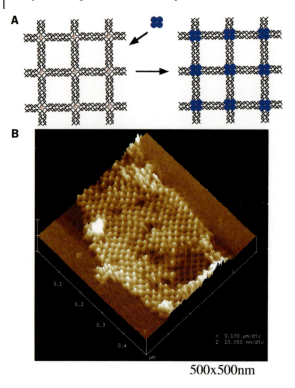

Fig. 5.7 (A) Schematic drawing of cross-tile lattice carrying a biotinylated central strand and streptavidin molecules (blue) binding to the functionalized sites. (B) AFM image showing individual streptavidin proteins at the vertices of the cross tile array. (Adapted with permission from Ref. [22].)

this study, some size control of lattice and partial addressability were demonstrated, but the display patterns were still periodic and symmetric (Figure 5.8). In ongoing experiments, finite-sized objects with independent addressing have been used to assemble a range of specifically patterned protein arrays in high yield [44].

Another exciting future use for biomolecules specifically patterned on self-assembled DNA nanostructures is the specific deposition of inorganic materials via crystal nucleation. Natural peptides and proteins have been implicated in the growth of nanopatterned silica by living organisms [53]. Peptides and RNA sequences have been artificially evolved by *in-vitro* selection to specifically bind and precipitate or crystallize various semi-conductors and metals [54, 55]. Patterning these species on 3-D DNA lattices could provide a method for bottom-up assembly and controlled deposition resulting in a wide variety of complex inorganic structures for use in nanoelectronics, photonics, and other fields.

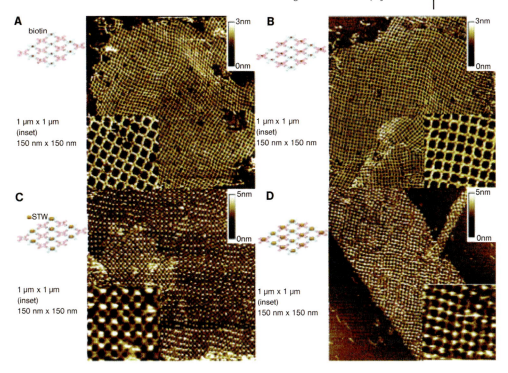

Fig. 5.8 AFM images of the programmed self-assembly of streptavidin on one-dimensional DNA nanotracks. (a,b) AFM images of bare A*B and A*B* nanotracks before streptavidin binding, respectively. Tiles marked with '*' indicate the presence of biotinylated strands. (c,d) AFM images of A*B and A*B* nanotracks after binding of streptavidin. All AFM images are 500 × 500 nm. (Adapted with permission from Ref. [52].)

5.6
DNA-Programmed Assembly of Materials

In analogy to the immobilization of DNA on a variety of solid surfaces, the conjugation of DNA and analogues with metal NPs, semiconductor NPs and polymer particles is becoming increasingly important. In particular, the DNA-directed assembly of gold NPs has been studied in much detail, primarily due to the stability and ease of preparation of gold–NP conjugates, and due to the interparticle distance-dependent plasmon resonance absorption. In 1996, both Mirkin et al. [56] and Alivisatos et al. [57] reported pioneering studies on the assembly of gold NPs by the hybridization of DNA–NP conjugates. Since then, such DNA-conjugated materials have found several applications, for example in biosensors (for a detailed overview, the reader is referred to recent excellent reviews, both in

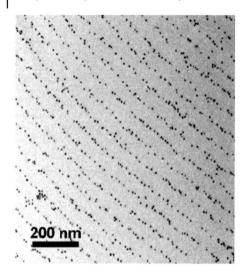

200 nm

Fig. 5.9 TEM image of gold nanoparticles organized on a self-assembled DX tile lattice using complementary base-pairing interactions. (Adapted with permission from Ref. [62].)

this book and elsewhere [58–61]). Here, the main focus will be on the assembly of materials in DNA-lattices on surfaces. Combining the ability of DNA to organize nanomaterials with the diverse and programmable structures available from self-assembling DNA tile lattice strategies has resulted in several initial steps towards the bottom-up assembly of nanomaterials that may prove valuable as electronic components. For example, a TX tile assembly was used to align a modest number of 5-nm gold particles in single and double layer queues [17]. This construction used tiles containing integral biotin-labeled DNA strands and streptavidin-bound gold particles. In another study, DX tile arrays were used to pattern 6-nm gold particles into precisely spaced rows covering micrometer scale areas (see Figure 5.9) [62]. This study featured gold NPs labeled with T_{15} oligonucleotides which base-paired with assembled DX lattice displaying single-strand A_{15} sequences hanging from certain tiles. Two recent reports by the group of Yan have demonstrated useful improvements of this method [63, 64]. The ability to organize electrically active species such as gold by using DNA points the way toward the templating of complex devices and circuits for applications in nanoelectronics.

The DNA-programmed assembly of materials other than gold NPs has also been reported, including semi-conductor NPs [65, 66], nanorods [67], mesoscale particles [68, 69], and dendrimers [70, 71]. Most of these examples are based on the linear assembly of two complementary DNA strands, leading to dimers or aggregates. Many examples of nanowires templated on DNA molecules by a variety of electrodeless deposition protocols (including the fabrication of a field effect transistor [72]) have also been reported, but these are beyond the scope of this chapter.

Carbon nanotubes represent one of the most promising materials for nanoscience due to their unique structure and mechanical and electronic properties [73], and in recent years the chemical conjugation of organic and bioorganic compounds with carbon nanotubes has been a rapidly developing field [74]. The ability to control the exact positioning of multiple carbon nanotubes by means of DNA-programmed assembly would be a major achievement in nanoscience, and indeed the conjugation of carbon nanotubes with DNA [75, 76] and with polypeptide nucleic acid (PNA) has been described [77]. In these reports the carbon nanotubes were shortened into fragments by oxidation, which resulted in carbon nanotubes fragments with carboxyl groups in the terminal positions and, to some extent, in their side walls. The covalent coupling of 5′-amino DNA-sequences or PNA to carboxyl groups at the nanotubes led to the formation of carbon nanotubes coupled with DNA or PNA sequences. Carbon nanotubes containing 12-mer PNA-sequences were annealed with dsDNA sequences containing 12-mer sticky ends and imaged by AFM [77]. In a report by Dai and colleagues [76], multi-wall carbon nanotubes (MWNTs) and single-wall carbon nanotubes (SWNTs) functionalized with 20-mer DNA sequences were annealed with complementary sequences attached to gold NPs. The resulting aggregates were deposited on mica and imaged by AFM, whereupon the images revealed the occasional interconnection of individual MWNTs by a gold NP. In a recent report by Chen and co-workers [78], the self-assembly of DNA functionalized single-walled carbon nanotubes was studied, with AFM images showing that highly branched structures were formed for the DNA-functionalized SWNTs.

In another study, SWNTs were assembled between pre-positioned metal electrodes via complementary DNA base-pairing by ssDNA on the gold electrodes (thiol-labeled oligos) and the oxidized SWNTs (3′-amino-labeled oligos) (see Figure 5.10) [79]. The electrical conductivity between the electrode pairs was shown to be highly dependent on the presence of complementary DNA on the electrodes and nanotubes. These initial investigations into carbon nanotube–DNA conjugates hold great promise for future developments in the assembly of nanotube structures with useful electronic and mechanical properties.

5.7
Laboratory Methods

The common methods utilized in DNA-based molecular assembly have been described in detail in the reports (and their supporting documents) cited throughout this chapter. An overview of the most central methods – molecular assembly by thermal annealing and the examination of 2-D nanostructures by AFM imaging – are summarized briefly in the following sections.

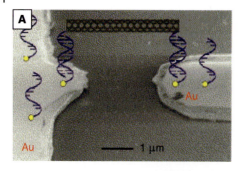

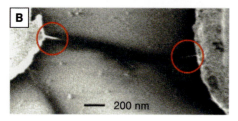

Fig. 5.10 DNA-mediated deposition of single-wall carbon nanotubes (SWNTs) between two gold electrodes. (A) Schematic representation of hybridization between complementary strands resulting in bridging the two electrodes by the SWNTs. (B) SEM image of two electrode pairs connected by SWNTs.

5.7.1
Annealing for DNA Assembly

The formation of specific hybridization complexes from designed DNA strands takes place by slow annealing. First, custom oligonucleotides are obtained from a commercial vendor and purified by polyacrylamide gel electrophoresis under denaturing conditions [16]. Complexes are formed by mixing stoichiometric quantities of the required strands (as estimated by their absorbance at 260 nm) in TAE/Mg buffer (20 mM Tris-acetate, pH 7.6, 2 mM EDTA, 12.5 mM MgCl$_2$). The concentration of DNA is typically between 1.0 and 0.05 µM (per strand) in a volume of 10–100 µL. Mixed oligonucleotide solutions are then cooled slowly from 90 °C to 20 °C to facilitate hybridization. This annealing step can be achieved by sealing the DNA solution in a microfuge tube and floating the tube in 1–2 L of water at 95 °C in a covered styrofoam box. Alternatively, a heating block or thermal cycler can be programmed to step down from 95 °C to room temperature over the course of the annealing step. Styrofoam box annealing chambers can take up to 48 h to cool down, whereas a heating block, when turned off and left on the bench, can cool in under 1 h. The duration of the annealing step depends somewhat on the complexity of the nanostructure to be formed, the number of component strands required, and the nature of the assembly process (single-step, hierarchical, etc.), with less time required for simpler structures.

5.7.2
AFM Imaging

The majority of nanostructures produced so far by DNA assembly have been 2-D in nature, and have been examined by AFM, using either wet or dry samples. For imaging under buffer, a 3–5 μL DNA sample is spotted on freshly cleaved mica and left to adsorb onto the surface for around 3 min; TAE/Mg buffer (30 μL) is then placed onto the mica. Another 30 μL of buffer is pipetted between the AFM tip and the tip-holder, so that when the tip and sample come together no air will be trapped above the tip. Imaging under buffer requires a fluid cell for the AFM. For imaging under air, a 3–5 μL DNA sample is spotted on freshly cleaved mica and left to adsorb onto the surface for around 3 min. The sample is then wicked off with filter paper, and 30 μL of distilled water is used to rinse the mica, which is then dried under a gentle stream of nitrogen. It has been found that imaging under buffer provides a higher-resolution analysis, and that "tapping mode" imaging is typically preferred over "contact mode".

5.8
Conclusions

Currently, we are learning the basics of how to form DNA nanostructures with moderately high complexity. But consider for a moment just a 2-D nanostructure patterned at the molecular scale by a binary array of pixels. Let us define the *complexity* of such a patterned nanostructure to be the total number of pixels on the array that can be set on or off, via programmed assembly. Prior review of patterned self-assembly results indicates that the experimentally achieved results for 2-D patterning using DNA self-assembly has rapidly increased in complexity over the past few years (and this rate of increase seems to be considerably higher than that experienced by lithographically based patterning over the same period). Thus, we are clearly optimistic about future improvements in the complexity of patterning using DNA self-assembly.

In the ultimate development of bottom-up nanofabrication strategies it may be possible to assemble large numbers of easily available building blocks using patterned DNA nanostructures to position a wide variety of materials. Depending on the nature and programming of the building blocks, these will self-assemble into complex nanostructures with the required properties. Such properties might include: an enzyme-like nature, electronic circuits with efficient contacts to electrodes at larger length scales, memory-storage devices, drug-delivery robots, multifunctional diagnostic devices for *in-vivo* application, or even systems capable of self-replication.

Acknowledgments

These studies were supported by NSF grants EIA-0218376, CCR-0326157, CCF-0523555, and CCF-0432038.

References

1 Whitesides, G. M., Grzybowski, B. (2002) Self-assembly at all scales. *Science* **295**, 2418–2421.

2 SantaLucia, J., Hicks, D. (2004) The thermodynamics of DNA structural motifs. *Annu. Rev. Biophys. Biomed. Struct.* **33**, 415–440.

3 Anselmi, C., DeSantis, P., Scipioni, A. (2005) Nanoscale mechanical and dynamical properties of DNA single molecules. *Biophys. Chem.* **113**, 209–221.

4 Dirks, R. M., Pierce, N. A. (2004) An algorithm for computing nucleic acid base-pairing probabilities including pseudoknots. *J. Comput. Chem.* **25**, 1295–1304.

5 Yin, P., Guo, B., Belmore, C., Palmeri, W., Winfree, E., LaBean, T. H., Reif, J. H. (2004) In: Ferretti, C., Mauri, G., Zandron, C. (Eds.), *Tenth International Meeting on DNA Based Computers (DNA10)*. Springer-Verlag, Milan, Italy, Vol. 3384.

6 Zhang, M., Sabharwal, C., Tao, W. M., Tarn, T. J., Xi, N., Li, G. Y. (2004) Interactive DNA sequence and structure design for DNA nano-applications. *IEEE Transact. NanoBiosci.* **3**, 286–292.

7 Seeman, N. C. (2003) DNA in a material world. *Nature* **421**, 427–431.

8 Seeman, N. C. (2005) Structural DNA nanotechnology: an overview. *Methods Mol. Biol.* **303**, 143–166.

9 Lee, S. H., Mao, C. (2004) DNA nanotechnology. *Biotechniques* **37**, 517–519.

10 Seeman, N. C. (1982) Nucleic-acid junctions and lattices. *J. Theoret. Biol.* **99**, 237–247.

11 Seeman, N. C. (1995) Molecular craftwork with DNA. *Chemical Intelligencer* **1**, 38–47.

12 Li, X. J., Yang, X. P., Qi, J., Seeman, N. C. (1996) Antiparallel DNA double crossover molecules as components for nanoconstruction. *J. Am. Chem. Soc.* **118**, 6131–6140.

13 Winfree, E., Liu, F. R., Wenzler, L. A., Seeman, N. C. (1998) Design and self-assembly of two-dimensional DNA crystals. *Nature* **394**, 539–544.

14 Rothemund, P. W. K., Ekani-Nkodo, A., Papadakis, N., Kumar, A., Fygenson, D. K., Winfree, E. (2004) Design and characterization of programmable DNA nanotubes. *J. Am. Chem. Soc.* **126**, 16344–16352.

15 Feldkamp, U., Niemeyer, C. M. (2006) Rational design of DNA nanoarchitectures. *Angew. Chem. Int. Ed. Engl.* **45**, 1856–1876.

16 LaBean, T. H., Yan, H., Kopatsch, J., Liu, F. R., Winfree, E., Reif, J. H., Seeman, N. C. (2000) Construction, analysis, ligation, and self-assembly of DNA triple crossover complexes. *J. Am. Chem. Soc.* **122**, 1848–1860.

17 Mao, C. D., LaBean, T. H., Reif, J. H., Seeman, N. C. (2000) Logical computation using algorithmic self-assembly of DNA triple-crossover molecules. *Nature* **407**, 493–496.

18 Li, H. Y., Park, S. H., Reif, J. H., LaBean, T. H., Yan, H. (2004) DNA-templated self-assembly of protein and nanoparticle linear arrays. *J. Am. Chem. Soc.* **126**, 418–419.

19 Liu, D., Park, S. H., Reif, J. H., LaBean, T. H. (2004) DNA nanotubes self-assembled from triple-crossover tiles as templates for conductive nanowires. *Proc. Natl. Acad. Sci. USA* **101**, 717–722.

20 Wei, B., Mi, Y. L. (2005) A new triple crossover triangle (TXT) motif for

DNA self-assembly. *Biomacromolecules* **6**, 2528–2532.

21 Reishus, D., Shaw, B., Brun, Y., Chelyapov, N., Adleman, L. (2005) Self-assembly of DNA double-double crossover complexes into high-density, doubly connected, planar structures. *J. Am. Chem. Soc.* **127**, 17590–17591.

22 Yan, H., Park, S. H., Finkelstein, G., Reif, J. H., LaBean, T. H. (2003) DNA-templated self-assembly of protein arrays and highly conductive nanowires. *Science* **301**, 1882–1884.

23 He, Y., Tian, Y., Chen, Y., Deng, Z., Ribbe, A. E., Mao, C. (2005) Sequence symmetry as a tool for designing DNA nanostructures. *Angew. Chem. Int. Ed. Engl.* **44**, 6694–6696.

24 Mao, C. D., Sun, W. Q., Seeman, N. C. (1999) Designed two-dimensional DNA Holliday junction arrays visualized by atomic force microscopy. *J. Am. Chem. Soc.* **121**, 5437–5443.

25 Liu, D., Wang, M., Deng, Z., Walulu, R., Mao, C. (2004) Tensegrity: construction of rigid DNA triangles with flexible four-arm DNA junctions. *J. Am. Chem. Soc.* **126**, 2324–2325.

26 Ding, B., Sha, R., Seeman, N. C. (2004) Pseudohexagonal 2D DNA crystals from double crossover cohesion. *J. Am. Chem. Soc.* **126**, 10230–10231.

27 Chelyapov, N., Brun, Y., Gopalkrishnan, M., Reishus, D., Shaw, B., Adleman, L. (2004) DNA triangles and self-assembled hexagonal tilings. *J. Am. Chem. Soc.* **126**, 13924–13925.

28 Shen, Z. Y., Yan, H., Wang, T., Seeman, N. C. (2004) Paranemic crossover DNA: A generalized Holliday structure with applications in nanotechnology. *J. Am. Chem. Soc.* **126**, 1666–1674.

29 Park, S. H., Barish, R., Li, H. Y., Reif, J. H., Finkelstein, G., Yan, H., LaBean, T. H. (2005) Three-helix bundle DNA tiles self-assemble into 2D lattice or 1D templates for silver nanowires. *Nano Lett.* **5**, 693–696.

30 Mathieu, F., Liao, S. P., Kopatscht, J., Wang, T., Mao, C. D., Seeman, N. C.
(2005) Six-helix bundles designed from DNA. *Nano Lett.* **5**, 661–665.

31 Paukstelis, P. J., Nowakowski, J., Birktoft, J. J., Seeman, N. C. (2004) Crystal structure of a continuous three-dimensional DNA lattice. *Chem. Biol.* **11**, 1119–1126.

32 Chen, J. H., Seeman, N. C. (1991) Synthesis from DNA of a molecule with the connectivity of a cube. *Nature* **350**, 631–633.

33 Zhang, Y. W., Seeman, N. C. (1994) Construction of a DNA-truncated octahedron. *J. Am. Chem. Soc.* **116**, 1661–1669.

34 Goodman, R. P., Berry, R. M., Turberfield, A. J. (2004) The single-step synthesis of a DNA tetrahedron. *Chem. Commun. (Camb.)* 1372–1373.

35 Shih, W. M., Quispe, J. D., Joyce, G. F. (2004) A 1.7-kilobase single-stranded DNA that folds into a nanoscale octahedron. *Nature* **427**, 618–621.

36 Yan, H., Feng, L. P., LaBean, T. H., Reif, J. H. (2003) Parallel molecular computations of pairwise exclusive or (XOR) using DNA 'String tile' self-assembly. *J. Am. Chem. Soc.* **125**, 14246–14247.

37 Rothemund, P. W. K., Papadakis, N., Winfree, E. (2004) Algorithmic self-assembly of DNA Sierpinski triangles. *PLoS Biol.* **2**, 2041–2053.

38 Winfree, E., Yang, X., Seeman, N. C. (1998) In: Landweber, L., Baum, E. (Eds.), *Second International Meeting on DNA Based Computers (DNA2)*. DIMACS, Princeton, NJ, Vol. 44, pp. 191–213.

39 Winfree, E., Bekbolatov, R. (2004) In: Chen, J., Reif, J. (Eds.), *Ninth International Meeting on DNA Based Computers (DNA9)*. Springer-Verlag, Madison, Wisconsin, Vol. 2943, pp. 126–144.

40 Chen, H. L., Goel, A. (2004) In: Ferretti, C., Mauri, G., Zandron, C. (Eds.), *Tenth International Meeting on DNA Based Computers (DNA10)*. Springer-Verlag, Milan, Italy, Vol. 3384, pp. 274–283.

41 Chen, H. L., Cheng, Q., Goel, A., Huang, M. D., de Espanes, P. M.

(2004), *Proceedings, 15th Annual ACM-SIAM Symposium on Discrete Algorithms (SODA)*, Association for Computing Machinery (ACM), New York, NY, pp. 890–899.

42 Reif, J. H., Sahu, S., Yin, P. (2006) Compact error-resilient computational DNA tiling assemblies. In: Jonoska, N. (Ed.), *Nanotechnology: Science and Computation*. Springer-Verlag, New York, pp. 79–104.

43 Sahu, S., Reif, J. H. (2006) Capabilities and limits of compact error resilience methods for algorithmic self-assembly in two and three dimensions. In: Mao, C., Yokomori, T. (Eds.), *Twelfth International Meeting on DNA Computing (DNA 12)*. Springer-Verlag, New York, pp. 1–12.

44 Park, S. H., Pistol, C., Ahn, S. J., Reif, J. H., Lebeck, A. R., Dwyer, C., LaBean, T. H. (2006) Finite-size, fully addressable DNA tile lattices formed by hierarchical assembly procedures. *Angew. Chem. Int. Ed. Engl.* **45**, 735–739.

45 Yan, H., LaBean, T. H., Feng, L. P., Reif, J. H. (2003) Directed nucleation assembly of DNA tile complexes for barcode-patterned lattices. *Proc. Natl. Acad. Sci. USA* **100**, 8103–8108.

46 Rothemund, P. W. (2006) Folding DNA to create nanoscale shapes and patterns. *Nature* **440**, 297–302.

47 Liu, Y., Ke, Y., Yan, H. (2005) Self-assembly of symmetric finite-size DNA nanoarrays. *J. Am. Chem. Soc.* **127**, 17140–17141.

48 Chworos, A., Severcan, I., Koyfman, A. Y., Weinkam, P., Oroudjev, E., Hansma, H. G., Jaeger, L. (2004) Building programmable jigsaw puzzles with RNA. *Science* **306**, 2068–2072.

49 Niemeyer, C. M. (2002) The developments of semisynthetic DNA-protein conjugates. *Trends Biotechnol.* **20**, 395–401.

50 Niemeyer, C. M., Sano, T., Smith, C. L., Cantor, C. R. (1994) Oligonucleotide-directed self-assembly of proteins: semisynthetic DNA–streptavidin hybrid molecules as connectors for the generation of macroscopic arrays and the construction of supramolecular bioconjugates. *Nucleic Acids Res.* **22**, 5530–5539.

51 Niemeyer, C. M., Burger, W., Peplies, J. (1998) Covalent DNA–streptavidin conjugates as building blocks for novel biometallic nanostructures. *Angew. Chem. Int. Ed.* **37**, 2265–2268.

52 Park, S. H., Yin, P., Liu, Y., Reif, J. H., LaBean, T. H., Yan, H. (2005) Programmable DNA self-assemblies for nanoscale organization of ligands and proteins. *Nano Lett.* **5**, 729–733.

53 Cha, J. N., Shimizu, K., Zhou, Y., Christiansen, S. C., Chmelka, B. F., Stucky, G. D., Morse, D. E. (1999) Silicatein filaments and subunits from a marine sponge direct the polymerization of silica and silicones in vitro. *Proc. Natl. Acad. Sci. USA* **96**, 361–365.

54 Whaley, S. R., English, D. S., Hu, E. L., Barbara, P. F., Belcher, A. M. (2000) Selection of peptides with semiconductor binding specificity for directed nanocrystal assembly. *Nature* **405**, 665–668.

55 Gugliotti, L. A., Feldheim, D. L., Eaton, B. E. (2004) RNA-mediated metal-metal bond formation in the synthesis of hexagonal palladium nanoparticles. *Science* **304**, 850–852.

56 Mirkin, C. A., Letsinger, R. L., Mucic, R. C., Storhoff, J. J. (1996) A DNA-based method for rationally assembling nanoparticles into macroscopic materials. *Nature* **382**, 607–609.

57 Alivisatos, A. P., Johnsson, K. P., Peng, X., Wilson, T. E., Loweth, C. J., Bruchez, M. P., Jr., Schultz, P. G. (1996) Organization of 'nanocrystal molecules' using DNA. *Nature* **382**, 609–611.

58 Rosi, N. L., Mirkin, C. A. (2005) Nanostructures in biodiagnostics. *Chem. Rev.* **105**, 1547–1562.

59 Niemeyer, C. M. (2001) Nanoparticles, proteins, and nucleic acids: biotechnology meets materials science. *Angew. Chem. Int. Ed. Engl.* **40**, 4128–4158.

60 Katz, E., Willner, I. (2004) Integrated nanoparticle-biomolecule hybrid systems: synthesis, properties, and applications. *Angew. Chem. Int. Ed. Engl.* **43**, 6042–6108.

61 Seeman, N. C. (1998) Nucleic acid nanostructures and topology. *Angew. Chem. Int. Ed. Engl.* **37**, 3220–3238.

62 Le, J. D., Pinto, Y., Seeman, N. C., Musier-Forsyth, K., Taton, T. A., Kiehl, R. A. (2004) DNA-templated self-assembly of metallic nanocomponent arrays on a surface. *Nano Lett.* **4**, 2343–2347.

63 Sharma, J., Chhabra, R., Liu, Y., Ke, Y., Yan, H. (2006) DNA-templated self-assembly of two-dimensional and periodical gold nanoparticle arrays. *Angew. Chem. Int. Ed. Engl.* **45**, 730–735.

64 Zhang, J., Liu, Y., Ke, Y., Yan, H. (2006) Periodic square-like gold nanoparticle arrays templated by self-assembled 2D DNA Nanogrids on a surface. *Nano Lett.* **6**, 248–251.

65 Mitchell, G. P., Mirkin, C. A., Letsinger, R. L. (1999) Programmed assembly of DNA functionalized quantum dots. *J. Am. Chem. Soc.* **121**, 8122–8123.

66 Parak, W. J., Gerion, D., Zanchet, D., Woerz, A. S., Pellegrino, T., Micheel, C., Williams, S. C., Seitz, M., Bruehl, R. E., Bryant, Z., et al. (2002) Conjugation of DNA to silanized colloidal semiconductor nanocrystalline quantum dots. *Chem. Mater.* **14**, 2113–2119.

67 Dujardin, E., Hsin, L. B., Wang, C. R. C., Mann, S. (2001) DNA-driven self-assembly of gold nanorods. *Chem. Commun.* 1264–1265.

68 Soto, C. M., Srinivasan, A., Ratna, B. R. (2002) Controlled assembly of mesoscale structures using DNA as molecular bridges. *J. Am. Chem. Soc.* **124**, 8508–8509.

69 Milam, V. T., Hiddessen, A. L., Crocker, J. C., Graves, D. J., Hammer, D. A. (2003) DNA-driven assembly of biodisperse, micron-sized colloids. *Langmuir* **19**, 10317–10323.

70 Choi, Y. S., Mecke, A., Orr, B. G., Holl, M. M. B., Baker, J. R. (2004) DNA-directed synthesis of generation 7 and 5 PAMAM dendrimer nanoclusters. *Nano Lett.* **4**, 391–397.

71 DeMattei, C. R., Huang, B. H., Tomalia, D. A. (2004) Designed dendrimer syntheses by self-assembly of single-site, ssDNA functionalized dendrons. *Nano Lett.* **4**, 771–777.

72 Keren, K., Krueger, M., Gilad, R., Ben-Yoseph, G., Sivan, U., Braun, E. (2002) Sequence-specific molecular lithography on single DNA molecules. *Science* **297**, 72–75.

73 Dai, H. J. (2002) Carbon nanotubes: Synthesis, integration, and properties. *Acc. Chem. Res.* **35**, 1035–1044.

74 Banerjee, S., Hemraj-Benny, T., Wong, S. S. (2005) Covalent surface chemistry of single-walled carbon nanotubes. *Adv. Mater.* **17**, 17–29.

75 Dwyer, C., Guthold, M., Falvo, M., Washburn, S., Superfine, R., Erie, D. (2002) DNA-functionalized single-walled carbon nanotubes. *Nanotechnology* **13**, 601–604.

76 Li, S. N., He, P. G., Dong, J. H., Guo, Z. X., Dai, L. M. (2005) DNA-directed self-assembling of carbon nanotubes. *J. Am. Chem. Soc.* **127**, 14–15.

77 Williams, K. A., Veenhuizen, P. T. M., de la Torre, B. G., Eritja, R., Dekker, C. (2002) Nanotechnology – Carbon nanotubes with DNA recognition. *Nature* **420**, 761–761.

78 Lu, Y. H., Yang, X. Y., Ma, Y. F., Du, F., Liu, Z. F., Chen, Y. S. (2006) Self-assembled branched nanostructures of single-walled carbon nanotubes with DNA as linkers. *Chem. Phys. Lett.* **419**, 390–393.

79 Hazani, M., Hennrich, F., Kappes, M., Naaman, R., Peled, D., Sidorov, V., Shvarts, D. (2004) DNA-mediated self-assembly of carbon nanotube-based electronic devices. *Chem. Phys. Lett.* **391**, 389–392.

6
Biocatalytic Growth of Nanoparticles for Sensors and Circuitry

Ronan Baron, Bilha Willner, and Itamar Willner

6.1
Overview

Biomolecules conjugated to metal nanoparticles (NPs) attract substantial recent interest in the rapidly developing area of nanobiotechnology [1]. The similar dimensions of metal NPs and biomolecules such as enzymes, antigens–antibodies or nucleic acids suggest that the integration of biomolecules with metal particles could yield new hybrid systems that combine the unique optical or electronic properties of metallic NPs with the selective and catalytic functions of biomolecules. Specifically, the integration of enzymes with metallic NPs could lead to new NP–enzyme biocatalytic conjugates where the metallic NPs control, or direct, the enzyme functions. Alternatively, the enzymes activities may regulate the optical or electronic features of the NPs, thus enabling the development of new bioelectronic or sensor systems. Biomolecule–metal NP hybrids are used as tags for optical sensing, and as catalytic labels for the amplified electrical, electrochemical or microgravimetric analysis of biorecognition events [2]. For example, the interparticle plasmon absorbance of gold (Au) aggregates generated in the presence of nucleic acid-functionalized Au NPs was employed to analyze DNA and single base mismatches [3]. The surface-enhanced resonance Raman scattering (SERRS) spectra of molecules associated with Ag NPs were used to amplify nucleic acid hybridization [4], and the intramolecular fluorophore-Au NP quenching within a beacon DNA structure, as well as the fluorescence reactivation upon the opening of the beacon were used to analyze DNA [5]. Also, Au NPs conjugated to biomolecules acted as labels for the optical detection of antigen–antibody [6] or DNA [7] recognition complexes on surfaces. The catalytic properties of metallic particles, that induce the deposition of different metals on the NPs by chemical means, were used for the amplification of biorecognition processes. The catalytic enlargement of Au NP labels associated with biorecognition complexes was employed to generate bridging conductive paths across electrodes, thus enabling the detection of antibody–antigen complexes [8] or DNA [9] hybridization by conductivity measurements. The deposition of metals on core metal NPs acting as

Nanobiotechnology II. Edited by Chad A. Mirkin and Christof M. Niemeyer
Copyright © 2007 WILEY-VCH Verlag GmbH & Co. KGaA, Weinheim
ISBN: 978-3-527-31673-1

labels was used to amplify biorecognition events by voltammetric stripping off of the deposited metal [10]. The catalytic enlargement of metallic NPs bound to biological complexes was also used to amplify the recognition events by following the "weight" of the linked NPs, thus permitting the sensitive microgravimetric detection of DNA using a quartz-crystal-microbalance [11]. Apart from bioanalytical applications, biological structures can be used as templates for the synthesis of well-defined complex nano-objects through the direct reduction of metallic salts to metal clusters [12], or the growth of preassembled NPs [13].

During the past few years, Sastry and colleagues [14], followed by others [15], have developed the concept of using biological entities as "reaction containers" for the synthesis of NPs. For example, the exposure of the fungus, *Verticillium* sp. to aqueous $AuCl_4^-$ resulted in the reduction of the salt to gold NPs with a diameter of ca. 20 nm formed on the cell surface, as well as intracellularly [14a]. Similarly, the fungi *Fusarium oxysporum* and *Verticillium* sp. generated, in the presence of ferric and ferrous salts, magnetite NPs [14b]. It was also shown that silver single crystals with defined shapes, such as triangles or hexagons, were generated in the *Pseudomonas stutzeri* AG259 bacterium in the presence of $AgNO_3$ [15a]. Cell extracts from the lemongrass plant also yield, in the presence of $AuCl_4^-$, single-crystalline gold nano-triangles and nano-prisms [14c]. The mechanisms of the formation of the NPs in these biological media are, however, unknown. Presumably, different biosynthetic products such as α-hydroxy carboxylic acids or reduced cofactors play an important role in the reduction of the respective salts to the NPs. The biocatalytic deposition of metallic seeds on specific cellular domains, such as surfactant polysaccharides, may then enhance the shape-controlled growth of the NPs.

In this chapter we address a new emerging area in nanobiotechnology that deals with the use of specific biomolecules and enzymes for the synthesis and growth of NPs. Special emphasis is given to the possible applications of biocatalytic-NPs conjugates for the development of sensors and the design of metallic nanopatterns.

6.1.1
Enzyme-Stimulated Synthesis of Metal Nanoparticles

The H_2O_2-mediated enlargement of Au NPs in the presence of $AuCl_4^-$ leads to a strong increase in optical absorbance [16]. Typical high-resolution transmission electron microscopy (HRTEM) images of Au NPs seeds before and after enlargement by H_2O_2 are presented in Figure 6.1.

The oxidation of glucose by the enzyme glucose oxidase (GOx) generates H_2O_2, and the latter product allows the reduction of $AuCl_4^-$ at catalytic Au NP seeds. This observation led to the development of an optical detection path for glucose oxidase activity, and for the sensing of glucose by the growth of the NPs [16]. The enlargement of the Au NP seeds was examined in solution and on surfaces. The citrate-capped Au NPs were assembled on a triethylaminopropyl siloxane (TEAS) film that was assembled on glass plates (Figure 6.2A). The GOx-

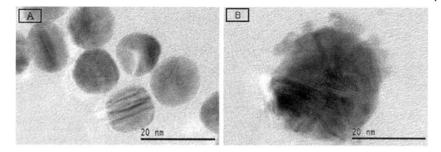

Fig. 6.1 HRTEM images of: (A) the Au NP seeds; (B) after 5 min of reaction, with $HAuCl_4$ (2×10^{-4} M), CTAC (2×10^{-3} M), and H_2O_2 (2.4×10^{-4} M). (Reprinted from Ref. [16], with permission; © American Chemical Society, 2005.)

biocatalyzed oxidation of glucose leads to the formation of H_2O_2 that acts as a reducing agent for the catalytic deposition of Au on the Au NPs associated with the glass support. Enlargement of the particles was then followed spectroscopically (Figure 6.2B). A calibration curve for the glucose in the range corresponding to 10^{-6} M to 10^{-4} M (Figure 6.2C) was extracted. It should be noted that GOx revealed a good stability in the growth solution used [2×10^{-4} M $HAuCl_4$, cetyltrimethylammonium chloride (CTAC), 2×10^{-3} M in 0.01 M phosphate buffer]. As different oxidases generate H_2O_2 upon the O_2-driven biocatalyzed oxidation of the corresponding substrates, the method can be generalized to sense a large number of substrates. Furthermore, the enzyme-stimulated catalytic deposition of metals on metal NP seeds may be extended to other biocatalysts that yield products capable of reducing metal salts. For instance, the alkaline phosphatase hydrolyzes p-aminophenol phosphate (**1**) to yield p-aminophenol (**2**), and the latter product reduces Ag^+ on Au NP seeds to form core-shell metallic structures [17].

$$H_2N\!-\!\!\bigcirc\!\!-\!OPO_3^{2-} \qquad H_2N\!-\!\!\bigcirc\!\!-\!OH$$

$$\text{(1)} \qquad\qquad\qquad \text{(2)}$$

Different bioactive o-hydroquinone derivatives such as the neurotransmitters dopamine (**3**), L-DOPA (**4**), adrenaline (**5**), or noradrenaline (**6**), were found to act as effective reducing agents of $AuCl_4^-$ to Au NPs [18]. Figure 6.3A depicts the absorbance features of the Au NPs generated by different concentrations of dopamine. As the concentration of dopamine increased, the plasmon absorbance of the Au NPs was intensified. Noteworthy is, however, the fact that although the Au NPs were enlarged as the concentration of dopamine increased, the maximum absorbance of the plasmon-band is blue-shifted, in contrast to the expectation that enlarged particles should reveal a red-shifted plasmon absorbance. The TEM images recorded upon the dopamine-induced growth of the Au NPs led

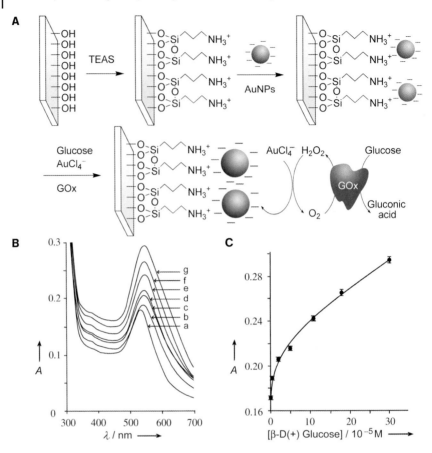

Fig. 6.2 (A) The biocatalytic enlargement of Au NPs associated with a glass support in the presence of glucose and glucose oxidase (GOx). (B) Absorbance spectra of the Au NP-functionalized glass supports upon reaction with 2×10^{-4} M HAuCl$_4$, 47 µg mL^{-1} GOx in 0.01 M phosphate buffer that includes CTAC (2×10^{-3} M), and different concentrations of β-D(+) glucose: (a) 0 M; (b) 5×10^{-6} M; (c) 2×10^{-5} M; (d) 5×10^{-5} M; (e) 1.1×10^{-4} M; (f) 1.8×10^{-4} M; (g) 3.0×10^{-4} M. The reaction time interval was 10 min at 30 °C. (C) Calibration curve corresponding to the absorbance of the enlarged Au NPs formed in the presence of glucose and measured at $\lambda = 542$ nm. (Figures 6.2B and 6.2C are adapted from Ref. [16], with permission; © American Chemical Society, 2005.)

to an understanding of the spectral absorbance features of the NPs. The increase of the concentration of dopamine, indeed, enhanced the growth of the Au NPs. However, the enlargement process generated on the Au NPs surface small Au clusters (1–2 nm) that were detached to the solution, and these acted as additional seeds for enlargement. As a result, the absorbance spectra corresponded to the enlarged particles, and to numerous small clusters/NPs that led to the blue shift of the plasmon absorbance. The spectral features enabled the extraction of a calibration curve corresponding to the optical detection of dopamine. Similar

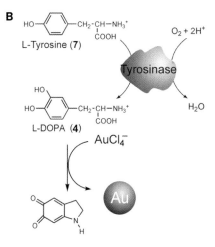

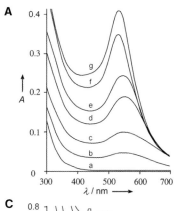

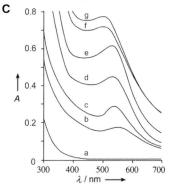

Fig. 6.3 (A) Absorbance spectra of the Au NPs formed in the presence of different concentrations of dopamine: (a) 0 M; (b) 2.5×10^{-6} M; (c) 5×10^{-6} M; (d) 8×10^{-6} M; (e) 1×10^{-5} M; (f) 1.5×10^{-5} M; (g) 2×10^{-5} M. All systems include HAuCl$_4$ (2×10^{-4} M) and CTAC (2×10^{-3} M) in 0.01 M phosphate buffer. Spectra were recorded after a fixed time-interval of 2 min. (B) Assay of tyrosinase activity through the biocatalyzed oxidation of tyrosine and the L-DOPA-mediated formation of Au NPs. (C)

Absorbance spectra of Au NPs formed by variable concentrations of tyrosinase: (a) 0 U mL^{-1}; (b) 10 U mL^{-1}; (c) 20 U mL^{-1}; (d) 30 U mL^{-1}; (e) 35 U mL^{-1}; (f) 40 U mL^{-1}; (g) 60 U mL^{-1}. All systems include tyrosine (2×10^{-4} M), HAuCl$_4$ (2×10^{-4} M), CTAC (2×10^{-3} M) in 0.01 M phosphate buffer. Spectra were recorded after a fixed time-interval of 10 min. (Adapted from Ref. [18], with permission; © American Chemical Society, 2005).

results were observed for the other neurotransmitters [18], and appropriate calibration curves for the optical sensing of the different neurotransmitters were extracted. The growth of the particle not only follows the concentration of the substrates that correspond to the respective enzymes, but may also be used to assay enzyme activities. For example, the enzyme tyrosinase is specifically expressed in melanocytes and melanoma cells, and is viewed as a specific marker for these cells [19], which makes its detection particularly interesting. Tyrosinase hydroxylates tyrosine (**7**) to L-DOPA (**4**), using O$_2$ as the oxygen source. The biocatalytically generated L-DOPA acts, however, as the reducing agent for the generation of Au NPs (Figure 6.3B). The generated Au NPs act then as optical labels for the activity of the enzyme. This is depicted in Figure 6.3C which shows the increase in the plasmon absorbance of the Au NPs at time intervals of reacting tyrosinase with tyrosine, in the presence of O_2. The tyrosinase-mediated formation of L-DOPA generates the reducing agent that yields and enlarges the Au NPs.

Enzyme inhibition was also probed using the catalytic growth of metallic NPs by enzymes. This was demonstrated for the inhibition of acetylcholine esterase (AChE), an enzyme that catalyzes the hydrolysis of acetylcholine. In fact, the development of rapid and sensitive detection protocols for AChE activity and its inhibition has important homeland security implications. The inhibition of AChE by nerve gases leads to perturbation of the nerve conduction system and to the rapid paralysis of vital functions of living systems [20]. Thus, analytical procedures for following AChE activity have significant possible applications for the detection of chemical warfare. The biocatalyzed hydrolysis of acetylthiocholine (**8**) to thiocholine (**9**), and the subsequent catalytic reduction of $AuCl_4^-$ on Au NP seeds were used to follow the AChE activity [21] (Figure 6.4A, B). The addition of paraoxon (**10**), a well-established AChE irreversible inhibitor that mimics the functions of organophosphate nerve gases, resulted in the inhibition of the Au NPs growth [21]. Figure 6.4C illustrates the inhibited growth of the Au NPs upon the increase of the concentration of the inhibitor. According to the inhibition mechanism, paraoxon (CX) leads to the phosphorylation of the active site of the biocatalyst, with concomitant release of the leaving group X (X = *p*-nitrophenol) [Eq. (1)]. The overall rate constant of the inhibition of AChE is given by Eq. (2), where [E] is the concentration of the non-inhibited enzyme, $[E_o]$ is the initial concentration of AChE, and [I] is the concentration of the inhibitor:

$$E + C\text{-}X \underset{k_{-1}}{\overset{k_1}{\rightleftharpoons}} [EC\text{-}X] \xrightarrow{k_2} EC + X \qquad \qquad (1)$$

$$\ln[E] = \ln[E_o] - k_i[I]t \qquad \qquad (2)$$

Figure 6.4C depicts the time-dependent absorbance changes of the Au NPs in the presence of variable concentrations of the inhibitor, according to Eq. (2). The value $k_i = 3.3 \times 10^5$ M min^{-1} was derived from the respective plots, consistent with the previously reported value of 2.9×10^5 M min^{-1} [22].

Redox enzymes usually lack direct electron transfer with their macroscopic environment [23]. Redox-active cofactors or small molecular oxidizing or reducing agents (e.g., O_2 or H_2S, respectively) activate the electron-transfer proteins towards the respective biotransformations. Molecular electron-transfer mediators are often used to activate the biocatalytic functions of redox proteins by a diffusional mechanism [23]. Transition metal complexes are often employed as electron-transfer mediators that make the electrical communication between the redox enzymes and their environment (e.g., electrodes), and activate the biocata-

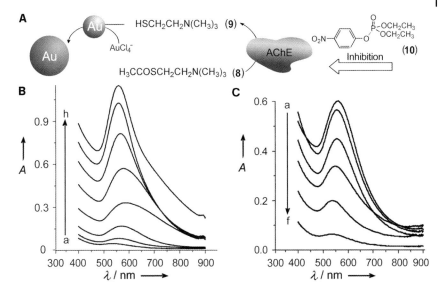

Fig. 6.4 (A) Acetylcholine esterase (AChE)-mediated growth of Au NPs, and its inhibition by a nerve-gas analogues. (B) Absorbance spectra of the Au NPs formed in the presence of AChE (0.13 U mL^{-1}), HAuCl$_4$ (1.1 × 10^{-3} M), Au NP seeds (3.6 × 10^{-8} M), and variable concentrations of acetylthiocholine (**7**): (a) 0 M; (b) 2.4 × 10^{-5} M; (c) 4.8 × 10^{-5} M; (d) 9.5 × 10^{-5} M; (e) 1.4 × 10^{-4} M; (f) 1.9 × 10^{-4} M; (g) 2.4 × 10^{-4} M; (h) 3.8 × 10^{-4} M. Spectra were recorded after 5 min of particle growth. (C) Absorbance spectra of the Au NPs formed in the presence of AChE and acetylthiocholine, in the presence of different concentrations of the inhibitor paraoxon: (a) 0 M; (b) 1.0 × 10^{-8} M; (c) 5.0 × 10^{-8} M; (d) 1.0 × 10^{-7} M; (e) 2.0 × 10^{-7} M; (f) 4 × 10^{-7} M. All systems include AChE (0.13 U mL^{-1}), HAuCl$_4$ (1.1 × 10^{-3} M), Au NP seeds (3.6 × 10^{-8} M), and acetylthiocholine (1.4 × 10^{-3} M). In all experiments AChE was incubated with the respective concentration of the inhibitor for a time-interval of 20 min, and the absorbance spectra of the enlarged Au NPs was recorded after 5 min of biocatalytic growth. (Figure 6.4B and 6.4C are adapted from Ref. [21], with permission; American Chemical Society, 2005).

lytic functions of the enzymes. As the reduction of the metal salts and the generation (or enlargement) of metal NPs is a redox process, the possibility exists of facilitating the synthesis of the NPs by electron-relay-mediated biocatalysis. This has been demonstrated with the application of Os(II) bispyridine-4-picolinic acid, [Os(bpy)$_2$PyCO$_2$H]$^{2+}$ (**11**) as an electron-transfer mediator for the activation of the biocatalyzed oxidation of glucose by GOx, and the concomitant reduction of AuCl$_4^-$ and the enlargement of Au NPs seeds [24]. The enlargement of the Au NP seeds proceeds effectively, and the resulting dimensions and absorbance properties were controlled by the concentration of glucose. The absorbance features for different concentrations of glucose are depicted in Figure 6.5.

The detailed mechanism for the enlargement of the Au NPs in the system was elucidated. During the first step, the [Os(bpy)$_2$PyCO$_2$H]$^{2+}$ reduces the Au(III)

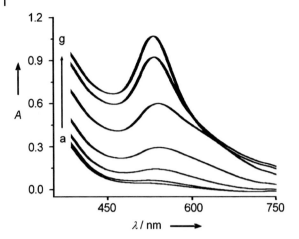

Fig. 6.5 Absorbance spectra of the Au NPs formed in the system consisting of 4.7 U mL^{-1} GOx, 1 × 10^{-4} M [Os(bpy)$_2$PyCO$_2$H]$^{2+}$, 1.5 × 10^{-3} M HAuCl$_4$, 7.5 × 10^{-3} M citrate, 2.5 × 10^{-8} M Au NP seeds, in the presence of different concentrations of glucose: (a) 0 M; (b) 7.4 × 10^{-5} M; (c) 3.7 × 10^{-4} M; (d) 7.4 × 10^{-4} M; (e) 1.5 × 10^{-3} M; (f) 3.0 × 10^{-3} M; (g) 4.4 × 10^{-3} M. Spectra recorded under air after a time interval of 10 min. (Adapted from Ref. [24], with permission; © Wiley-VCH, 2005).

salt to Au(I) [Eq. (3)], and the latter species is reduced by citrate to Au^0 that is deposited on the gold seeds, acting as catalyst [Eq. (4)]. A TEM analysis of the samples confirmed the growth mechanism. Au NP seeds of 2–3 nm were enlarged to NPs with diameters in the region of 10–30 nm, by the system that included glucose (5.9 × 10^{-3} M) with a growth time interval of 10 min.

$$2[Os(II)(bpy)_2PyCO_2H]^{2+} + AuCl_4^- \longrightarrow 2[Os(III)(bpy)_2PyCO_2H]^{3+}$$

$$+ Au(I) + 4Cl^- \tag{3}$$

$$\underset{\underset{CO_2H}{\overset{OH}{|}}}{HO_2CCH_2\text{-}C\text{-}CH_2CO_2H} + 2Au(I) \xrightarrow{Au} HO_2CCH_2\overset{O}{\overset{\|}{C}}CH_2CO_2H$$

$$+ 2Au^0 + 2H^+ + CO_2 \tag{4}$$

A similar mediated redox mechanism was used to follow the inhibition of AChE through blocking the enlargement of the Au NPs [24]. The [Os(bpy)$_2$PyCO$_2$H]$^{2+}$/citrate-mediated growth of the Au NPs was employed to follow the activity of AChE by coupling the synthesis of the particles to the AChE/choline oxidase (ChOx) enzyme cascade (Figure 6.6). The AChE-stimulated hydrolysis of acetylcholine yields choline, and the [Os(bpy)$_2$PyCO$_2$H]$^{3+}$ (**12**) -mediated oxidation of choline yields betaine and [Os(bpy)$_2$PyCO$_2$H]$^{2+}$. Then, in the presence of citrate, the resulting [Os(bpy)$_2$PyCO$_2$H]$^{2+}$ reduces AuCl$_4^-$ to Au^0 by the two-step coupled reactions, while regenerating [Os(bpy)$_2$PyCO$_2$H]$^{3+}$.

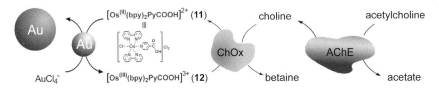

Fig. 6.6 The $[Os(bpy)_2PyCO_2H]^{2+}$-mediated growth of the Au NPs in the presence of the AChE/ChOx/acetylcholine/Au NP seeds system.

6.1.2
Enzyme-Stimulated Synthesis of Cupric Ferrocyanide Nanoparticles

The biocatalytically induced growth of cupric ferrocyanide $(Cu_2Fe(CN)_6)$ NPs was used to develop electrochemical and optical biosensing systems of glucose [25] and analogues of organophosphorus nerve agents [26]. The GOx-mediated oxidation of glucose by O_2 yields H_2O_2 and gluconic acid, whereafter the generated H_2O_2 was used to reduce ferricyanide $(Fe(CN)_6^{3-})$ to ferrocyanide $(Fe(CN)_6^{4-})$; then, in the presence of Cu^{2+} the latter complex yielded the cubic cupric ferrocyanide complex, $Cu_{3/2}Fe(CN)_6$ (Figure 6.7A). The resulting cupric ferrocyanide was analyzed by spectrophotometric or electrochemical means. The electrochemical analysis of the complex was performed by dipping a carbon paste electrode (CPE) into the reaction medium, and the electrochemical analysis of the Fe(II) content in the resulting complex structure, by the application of square-wave voltammetry. Figures 6.7B and 6.7C show, respectively, the current responses of the adsorbed cupric ferrocyanide NPs generated by different concentrations of glucose, and the corresponding calibration curve. A related procedure was employed to analyze organophosphorus pesticides and nerve gas analogues. Organophosphorus hydrolase (OPH) hydrolyzes phosphoric acid esters. Accordingly, OPH was used to hydrolyze *p*-nitrophenolphosphate. The resulting *p*-nitrophenol reduced ferricyanide to ferrocyanide to generate, in the presence of Cu^{2+} ions, the cupric ferrocyanide complex that was analyzed electrochemically.

6.1.3
Cofactor-Induced Synthesis of Metallic NPs

Many redox enzymes are coupled by the diffusional nicotinamide adenine dinucleotide (phosphate)/1,4-dihydronicotinamide adenine dinucleotide (phosphate) $(NAD(P)^+/NAD(P)H)$ cofactor system [27]. The NADH or NADPH reduced cofactors were found to enlarge Au NP seeds by the reduction of $AuCl_4^-$, thus providing a useful tool to follow many biocatalytic processes. The reduction of the Au(III) salt takes place in two steps. During the first step, the Au(III) ions are rapidly reduced to the Au(I) species [Eq. (5)]. The latter product is then reduced by NADH, in the presence of the Au NP seeds acting as catalyst, to the metal, a process that leads to the enlarged NPs [Eq. (6)] [28].

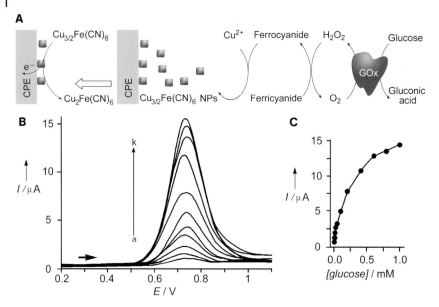

Fig. 6.7 (A) The biocatalytic formation of cupric ferrocyanide NPs in the presence of glucose, GOx, ferrocyanide and copper ions, and the electrochemical analysis of glucose by subsequent oxidation of the Fe(II) in the cupric ferrocyanide NPs at the CPE surface. (B) Square-wave voltammograms corresponding to the oxidation of enzymatically generated cupric ferrocyanide for different concentrations of glucose: (a) 0 μM; (b) 6 μM; (c) 13 μM; (d) 25 μM; (e) 50 μM; (f) 100 μM; (g) 200 μM; (h) 400 μM; (i) 600 μM; (j) 800 μM; (k) 1000 μM. (C) Corresponding calibration curve. (Figures 6.7B and 6.7C are adapted from Ref. [25], with permission; © Wiley-VCH, 2006).

$$\text{AuCl}_4^- + \text{NADH} \longrightarrow \text{Au(I)} + 4\text{Cl}^- + \text{NAD}^+ + \text{H}^+ \qquad (5)$$

$$2\text{Au(I)} + \text{NADH} \xrightarrow{\text{Au}} 2\text{Au}^0 + \text{NAD}^+ + \text{H}^+ \qquad (6)$$

The NADH-stimulated enlargement of Au NP seeds is shown graphically and microscopically in Figure 6.8. As the concentration of NADH is elevated, the absorbance of the enlarged Au NPs is intensified, indicating the formation of larger particles (Figure 6.8A). The scanning electron microscopy (SEM) images of the particles confirmed the growth of the NPs on the glass support (Figure 6.8B). The particles generated by NADH (2.7×10^{-4} M; 5.4×10^{-4} M; and 1.36×10^{-3} M) are shown in images (I), (II) and (III), respectively, that reveal the generation of Au NPs with an average diameter of 13 ± 1 nm, 40 ± 8 nm, and 20 ± 5 nm, respectively. The NPs generated by the high concentration of NADH (III) reveals a two-dimensional array of enlarged particles that touch each other, consistent with the spectral features of the surface.

The NADH-stimulated enlargement of Au NPs was used so as to amplify the thrombin-aptamer biorecognition event [29]. Au NPs, functionalized with

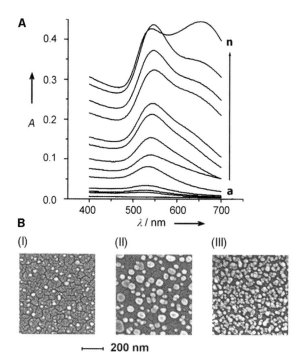

Fig. 6.8 (A) Absorbance changes of the Au NP-aminopropyl siloxane-functionalized glass slides upon enlargement with different concentrations of NADH: (a) 0 M; (b) 14×10^{-5} M; (c) 27×10^{-5} M; (d) 34×10^{-5} M; (e) 41×10^{-5} M; (f) 44×10^{-5} M; (g) 48×10^{-5} M; (h) 51×10^{-5} M; (i) 54×10^{-5} M; (j) 58×10^{-5} M; (k) 61×10^{-5} M; (l) 65×10^{-5} M; (m) 68×10^{-5} M; (n) 1.36×10^{-3} M. All systems included HAuCl$_4$ (1.8×10^{-4} M) and CTAB (7.4×10^{-2} M). (B) SEM images of NADH-enlarged Au particles generated on a Au NP-functionalized glass support using HAuCl$_4$ (1.8×10^{-4} M), CTAB (7.4×10^{-2} M) and NADH: (I) 2.7×10^{-4} M; (II) 5.4×10^{-4} M; (III) 1.36×10^{-3} M. (Adapted from Ref. [28], with permission; © Wiley-VCH, 2004).

the thrombin aptamer, were used as labels to detect thrombin on glass supports (Figure 6.9A). A glass surface was modified with the thrombin aptamer (**13**), that associates the thrombin analyte. The binding of the thrombin aptamer-functionalized Au NPs to the second binding region of thrombin, followed by NADH/AuCl$_4^-$ enlargement of the Au NPs, allowed amplification of the primary association of thrombin to the sensing surface and the spectroscopic analysis of the resulting enlarged Au NPs. Figure 6.9B shows the absorbance spectra of the enlarged particles upon analyzing different concentrations of thrombin. This method enabled the detection of thrombin with a sensitivity limit of 2 nM.

The regeneration of NADH by enzymes can be used to analyze either the activity of the enzyme, or its substrate. This was demonstrated with the analysis of lactate using lactate dehydrogenase (Figure 6.10A) [28]. Citrate-stabilized Au NP seeds were immobilized on an aminopropyl siloxane film associated with a glass

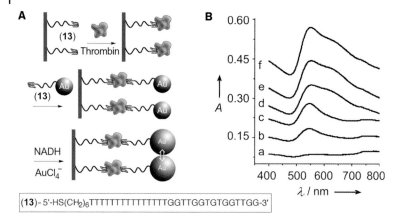

(13)= 5'-HS(CH₂)₆TTTTTTTTTTTTTTTGGTTGGTGTGGTTGG-3'

Fig. 6.9 (A) Amplified detection of thrombin on surfaces by the catalytic enlargement of thrombin aptamer-functionalized Au NPs. (B) Spectral changes of the (**13**)-functionalized glass supports upon interaction with thrombin: (a) 0 nM; (b) 2 nM; (c) 5 nM; (d) 19 nM; (e) 94 nM; (f) 167 nM. (Adapted from Ref. [29], with permission; © American Chemical Society, 2004).

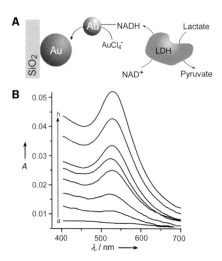

Fig. 6.10 (A) The biocatalytic enlargement of Au NPs by the NAD^+/LDH/lactate system. (B) Spectral changes of the Au NP-functionalized glass supports upon interaction with the growth solution consisting of $HAuCl_4$ (1.8×10^{-4} M), CTAB (7.4×10^{-2} M) and lactate/LDH-generated NADH formed within 30 min in the presence of variable concentrations of lactate: (a) 0 M; (b) 2.9×10^{-3} M; (c) 3.6×10^{-3} M; (d) 5.1×10^{-3} M; (e) 5.8×10^{-3} M; (f) 6.6×10^{-3} M; (g) 7.3×10^{-3} M; (h) 9.8×10^{-3} M. (Figure 6.10B is adapted from Ref. [28], with permission; © Wiley-VCH, 2004).

support, and the functionalized surface was treated with different concentrations of lactate in the presence of $AuCl_4^-$. The absorbance spectra of the enlarged NPs in the presence of different concentrations of lactate are shown in Figure 6.10B. The plasmon absorbance of the Au NPs increased as the concentration of lactate increased, and therefore the concentration of NADH, consistent with the NADH-mediated growth of the NPs. The sensitivity limit for analyzing lactate by this method was found to be 3×10^{-3} M [28].

The reduced cofactor NADH was also used to reduce Cu^{2+} ions to Cu^0 metal that was deposited on Au NPs [30]. The bioelectrocatalytic regeneration of NADH by alcohol dehydrogenase (AlcDH) and the subsequent NADH-mediated deposition of the Cu^0 metal on the Au NPs were employed for the electrochemical analysis of the enzyme substrate, ethanol (Figure 6.11A). Conductive indium-tin oxide (ITO) electrodes were modified with an aminopropyl siloxane layer on which citrate-stabilized Au NPs were immobilized. The enlargement of the Au NP seeds with copper was followed spectrophotometrically at 540 nm. Moreover, stripping voltammetry provided an electrochemical readout for the content of Cu^0 deposited by the biocatalytic process in the presence of different concentrations of ethanol. The chronocoulometric transients observed upon the stripping off of the Cu^0 generated by different concentrations of ethanol are depicted in Figure 6.11B, and the respective calibration curve is shown Figure 6.11C.

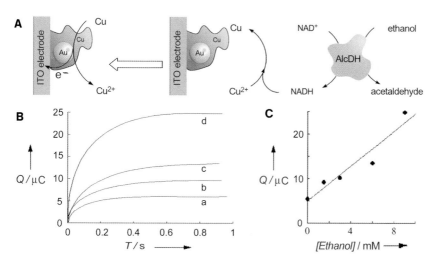

Fig. 6.11 (A) Biocatalytic enlargement of Au NP seeds associated with an electrode with Cu shells using the NAD$^+$/AlcDH/ethanol system. (B) Chronocoulometric transients corresponding to the stripping off of the Cu deposited on the electrode by different NADH concentrations generated by AlcDH in the presence of NAD$^+$ (34 mM) and different concentrations of ethanol: (a) 0 mM; (b) 1.5 mM; (c) 6 mM; (d) 9 mM. The reactions were performed in an aqueous solution for 2 h (stripping of the metal was accomplished by a potential step from -0.2 V to 0.1 V). (C) Corresponding calibration curve. (Figures 6.11B and 6.11C are adapted from Ref. [30], with permission; © Wiley-VCH, 2005.)

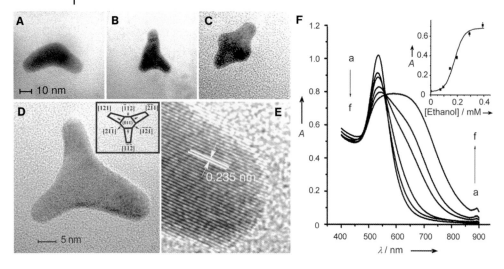

Fig. 6.12 (A–C) Typical TEM images of dipod-, tripod-, and tetrapod-shaped Au NPs formed in the presence of NADH (4.0×10^{-6} M), respectively. (D) HRTEM image of a representative tripod-shaped Au NP formed in the presence of NADH (4.0×10^{-6} M). (E) HRTEM analysis of the tripod Au NP. The inset in part (D) shows the crystal planes and the tripod growth directions extracted from the HRTEM analysis. (F) Absorbance spectra corresponding to the Au NPs formed by the biocatalytic generation of NADH in the presence of AlcDH and variable concentrations of ethanol: (a) 0 M; (b) 7.3×10^{-5} M; (c) 9.4×10^{-5} M; (d) 1.5×10^{-4} M; (e) 2.0×10^{-4} M; (f) 2.9×10^{-4} M. Inset: calibration curve corresponding to the absorbance of the shaped Au NPs at $\lambda = 680$ nm formed in the presence of different concentrations of ethanol. (Adapted from Ref. [31], with permission; © American Chemical Society, 2005.)

An interesting advance in the NADH-mediated synthesis of Au NPs was the discovery that, under special conditions, shaped NPs consisting of dipods, tripods and tetrapods can be formed [31]. It was shown that in a basic aqueous solution (pH = 11), which included ascorbate and NADH as reducing agents and cetyltrimethylammonium bromide (CTAB) as surfactant, the rapid formation of shaped Au NPs was observed. The Au NP shapes consisted of dipods (12%), tripods (45%), tetrapods (13%) and spherical particles (30%). The shaped particles were made of "arms" approximately 20–25 nm long and 2–5 mm wide (Figure 6.12). The development of the shaped particles was controlled by the concentration of NADH. At low NADH concentrations, "embryonic-type" shapes were observed, whereas at high cofactor concentrations well-shaped particles were detected (Figure 6.12A,B,C). The mechanism that leads to the formation of shaped Au NPs was discussed; in the first step, conducted at pH 11, ascorbate reduced $AuCl_4^-$ to Au NP seeds. In the second step, these particles acted as catalysts for the rapid NADH-mediated reduction of $AuCl_4^-$ to Au^0 on the seeds. HRTEM allowed the detailed analysis of the growth directions of the shaped NPs; the HRTEM images of tripod particles are shown in Figures 6.12D and 6.12E. The lattice planes exhibit an inter-planar distance of 0.235 nm that corresponds to the {111} type planes of crystalline gold. The pods, separated by 120°, revealed a crystallite orientation of [011] with growth directions of the pods of type ⟨211⟩; namely, the pods extended to the directions [1$\bar{1}\bar{2}$], [$\bar{2}\bar{1}$1], and [121] (Figure 6.12D, inset). It was suggested that the cofactor preferentially binds certain crystalline planes that dictate the growth directions of the NPs. The shaped Au NPs revealed a red-shifted plasmon absorbance band at $\lambda = 680$ nm, consistent with a longitudinal plasmon exciton in the "rod-like" structures [32]. The blue color of the shaped Au NPs differs from the red color characteristic of spherical Au NPs.

The blue color of the shaped Au NPs allowed the development of new biosensor systems for NAD^+-dependent biocatalytic transformations. This was demonstrated by using the shaped Au NPs as color labels for the analysis of ethanol in the presence of AlcDH [31]. The biocatalyzed oxidation of ethanol by AlcDH yields NADH, and the biocatalytically generated NADH acted as the active reducing agent for the formation of the blue shaped Au NPs. Figure 6.12F depicts the absorbance spectra, resulting from the gradual development of the shaped Au NPs formed in the presence of different concentrations of ethanol. The derived calibration curve corresponding to the optical analysis of ethanol by the shaped Au NPs is also shown (Figure 6.12F, inset).

6.1.4
Enzyme–Metal NP Hybrid Systems as "Inks" for the Synthesis of Metallic Nanowires

The use of biomaterials as templates for the synthesis of defined metallic nano-objects continues to attract substantial research effort. These activities represent an important direction in the developing concept of the "bottom-up" synthesis of nanostructures. For example, the association of metallic Au NPs to telomers followed by catalytic enlargement of the NPs was used to generate continuous metal wires on nucleic acid templates [33]. Similarly, the catalytic deposition of Au on Ag clusters linked to double-stranded DNA was employed to synthesize metallic nanocircuits [34]. The enlargement of Au NPs associated with actin filaments was also reported to yield micrometer-long metal nanowires that revealed ATP-driven motility and nano-transporting features, in conjunction with myosin-modified interfaces [35].

The biocatalytic growth of metal NPs may also be employed to construct nanoscale metallic circuitry. This was demonstrated by the application of metallic NP-functionalized enzymes as "biocatalytic inks" for the generation of nanowires [17]. Dip-pen nanolithography (DPN) was used as the nanopatterning tool for application of the "biocatalytic ink" on surfaces. Realizing that GOx stimulates the enlargement of Au NPs by the biocatalytic generation of H_2O_2 that reduces $AuCl_4^-$ in the presence of Au NPs as catalytic sites [16], the generation of Au NP/GOx conjugate could act as a biocatalytic ink for the synthesis of nanowires. Accordingly, the flavoenzyme GOx was functionalized with mono-N-hydroxysuccinimide-functionalized 1.4 nm Au NP to yield the AuNP/GOx conjugate with an average loading of 12 Au NPs per enzyme molecule (Figure 6.13).

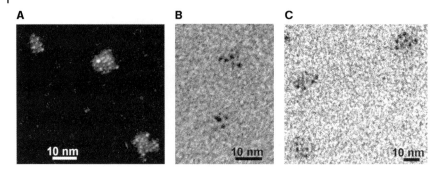

Fig. 6.13 Characterization of flavoenzymes modified with Au NPs.
(A) Scanning transmission electron microscopy (STEM) image of
AuNP-modified GOx. (B) TEM image of Au NP-modified GalOx.
(C) TEM image of Au NP-modified AlkPh. (Adapted from Ref. [17],
with permission; © Wiley-VCH, 2006.)

The resulting Au NP/GOx conjugate was then used as an "ink" for DPN patterning of "biocatalytic lines" on a Si support. In the presence of glucose, GOx
generated H_2O_2 that reduced $AuCl_4^-$ at the surface of the Au NPs (Figure
6.14A). The catalytic growth of the particles yielded micrometer-long Au-metallic
wires exhibiting heights and widths in the region of 150 nm to 250 nm, depending on the time interval used for the biocatalytic "development" of the enlarged
particles (Figure 6.14B). Another flavoenzymes, galactose oxidase, was similarly
used as a biocatalyst for the growth of Au NPs in the presence of Au NP/galactose
as "biocatalytic ink", and wires were generated in the presence of galactose (Figure 6.14C). The metallic nanowires generated by the enlargement of the Au NPs
yielded continuous metallic wires that exhibited electrical conductivity. The generation of metallic nanowires is not limited to flavoenzymes that yield H_2O_2 as reducing agent. In fact, any biocatalyst that generates a product exhibiting a reducing power to transform the metal ions to metals at a catalytic metal NP may be
used to synthesize the wires. This was demonstrated with the synthesis of Ag
nanowires using Au NP-functionalized alkaline phosphatase as a "biocatalytic
ink" (Figure 6.14D) [17]. The mechanism for the synthesis of the Ag nanowires
involves the alkaline phosphatase-mediated hydrolysis of *p*-aminophenol phosphate (**1**) to *p*-aminophenol (**2**), that reduces $AgNO_3$ to Ag on the Au NP seeds.
Continuous silver nanowires exhibiting a height and width of 20–25 nm on the
protein template were generated by this method. The procedure allowed the consecutive growth of different metallic nanowires with predesigned patterns generated by the DPN method (Figure 6.14E). First, Au NP-functionalized GOx was
used to grow the Au nanowire by the primary patterning of the Au NP/GOx conjugate, using DPN and subsequent development of the Au nanowire by glucose.
The Au nanowire was then passivated with mercaptoundecanoic acid, and the Au
NP-modified alkaline phosphatase (AlkPh/Au NPs) allowed orthogonal growth of
the silver nanowire.

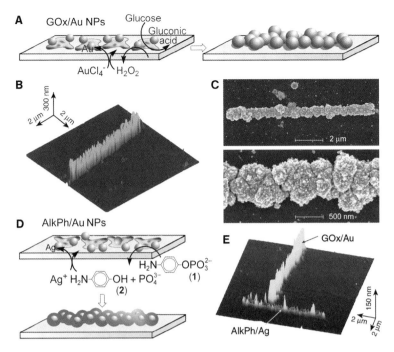

Fig. 6.14 (A) The generation of a Au nanowire by the biocatalytic enlargement of a Au NP-functionalized GOx line deposited on a silicon support by dip-pen nanolithography (DPN). (B) AFM image of the Au nanowire generated by the Au NP-functionalized GOx as "biocatalytic ink". (C) SEM images of the pattern generated by the Au NP-functionalized GalOx as "biocatalytic ink". (D) Biocatalytic enlargement of Au NP-modified AlkPh with silver, and the fabrica-tion of a silver line on a support using DPN and the "biocatalytic ink". (E) AFM image of Au and Ag nanowires generated by deposition and enlargement of the Au NP/GOx template, followed by the passivation of the Au nanowire with mercaptoundecanoic acid, and subsequent deposition of the Au NP/AlkPh template and its enlargement to the Ag nanowire. (Figures 6.14B, 6.14C and 6.14E are adapted from Ref. [17], with permission; © Wiley-VCH, 2006).

6.2
Methods

The biocatalytic growth of NPs for sensor and nanocircuitry applications involves fundamental structural characterization of the particles by different spectroscopy and microscopy methods, and the macroscopic examination of the systems for bioanalytical applications.

6.2.1
Physical Tools to Characterize the Growth of Nanoparticles and Nanowires

Absorption spectra provide a useful tool to follow the mechanism and rate of the biocatalytic growth of NPs. The plasmon absorbance of Au NPs or Ag NPs provides an effective probe to monitor the biocatalytic enlargement reaction. The red-shift in the plasmon absorbance of Au NPs indicates an increase in the dimensions of the NPs, whereas a blue-shift indicates the formation of smaller NPs or metallic nanoclusters.

TEM is used to image the morphologies of metallic NPs in solution by drying a drop of the solution on an appropriate grid. SEM is used to monitor the metallic nanoclusters on conductive supports. Atomic force microscopy (AFM) is used to image the nanostructures on surfaces.

6.2.2
General Procedure for Monitoring the Biocatalytic Enlargement of Metal NPs in Solutions

The respective enzyme is introduced into an aqueous buffer solution that includes the respective metallic salt and catalytic NP seeds (usually Au NPs). Introduction of the substrate corresponding to the enzyme initiates the biocatalyzed transformation that yields the reducing agent which activates the reduction of the metal salt on the metallic seeds, resulting in their enlargement. The extent of NP enlargement is controlled by the content of enzyme, concentration of substrate, and the time interval used for the biocatalytic enlargement process. The growth of particles is monitored spectroscopically and characterized by different microscopy methods. Changes in the absorbance features of the NPs as a function of the concentrations of the different substrates enable the extraction of appropriate calibration curves.

As a typical example, we present the conditions for growing Au NPs by glucose oxidase and glucose [16]. The growth solutions consisted of 2×10^{-4} M HAuCl$_4$ in 0.01 M phosphate buffer (pH 7.2), 2×10^{-3} M cetyltrimethylamonium chloride (CTAC) and either different concentrations of H$_2$O$_2$ or β-D(+) glucose with 47 μg mL^{-1} GOx. For the catalytic growth of the Au-NPs in solution, 3×10^{-10} M Au-NP seeds (12 ± 1 nm) were added to the growth solution. Au-NPs stabilized with citrate were prepared according to published methods [36]. The experiments were performed at ambient temperature (22 ± 2 °C), unless otherwise stated.

6.2.3
Modification of Surfaces with Metal NPs and their Biocatalytic Growth for Sensing

The surfaces (usually glass plates) are modified with an aminosiloxane thin film followed by the assembly of citrate-stabilized Au NPs on the film [37]. The surface-confined NPs act as seeds for biocatalytic enlargement. The surfaces are then interacted with a buffer solution that includes the metal salt, the respective enzyme, and the appropriate substrate. The biocatalytically generated product acts as a reducing agent that reduces the metal salts and stimulates NP growth. The absorbance features of the surfaces at different substrate concentrations enable quantitative monitoring of the substrate concentration.

As a typical example, the growth of Au NPs on a glass surface by the GOx system is described [16]. Glass slides were functionalized with 3-aminopropyltriethoxysilane with the citrate-stabilized Au-NPs (12 ± 1 nm). The gold NPs-modified glass slide was then soaked in the above-described growth solution. The absorbance features of the resulting modified slides were recorded in water.

6.2.4
Modification of Enzymes with NPs and their Use as Biocatalytic Templates for Metallic Nanocircuitry

Au NPs (1.4 nm) functionalized with a single N-hydroxysuccinimide functionality were covalently linked to glucose oxidase, galactose oxidase, or alkaline phosphatase. The resulting Au NP-functionalized enzymes are used as "biocatalytic inks" that generate templates for the synthesis of metallic nanowires. The Au-functionalized enzymes are deposited on Si surfaces using the DPN method, and the respective metallic nanowires are generated by biocatalytic enlargement of the Au NP in the presence of the respective enzyme substrate [17].

As a typical example, the growth of Ag nanowires using Au NP-functionalized alkaline phosphatase is described. Alkaline phosphatase was modified with N-hydroxysuccinimide-functionalized Au NPs (1.4 nm). The single N-hydroxy-succinimide functionality linked to the NPs ensures specific linkage of the NPs to a single enzyme unit, without crosslinking aggregation of several enzyme units. The average loading was found to be 10 NPs per enzyme molecule. The biocatalytic activity of the enzyme–NP hybrid was assessed photometrically, and the activity was found not to be altered by modification with the NPs. Biocatalytic enlargement of these particles in solution was also checked. The Au NP-modified alkaline phosphatase was then used as "biocatalytic ink" using DPN. The enzyme is deposited on a SiO_2/Si support, and the resulting surface is treated with p-nitrophenylphosphate/$AgNO_3$ as the developing solution. The enzyme-generated hydroquinone-type derivative acts as a reducing agent for the deposition of silver on the Au NPs associated with the enzyme. The height and width are dependent on the time of enlargement and the concentration of glucose, and can be modified by adjusting these parameters.

6.3
Outlook

The application of biocatalysts as active components for the synthesis of NPs and for the enlargement of metallic NPs is an emerging field in nanobiotechnology, and one which has important new implications in biosensors design. The different examples discussed in this chapter show that numerous types of enzymes such as oxidases, hydroxylases, hydrolytic proteins or $NAD(P)^+$-dependent enzymes may be employed as biocatalysts for the synthesis of metal NPs, and for the development of optical/electrochemical sensors for the respective substrates. The possibilities of immobilizing these biocatalytic systems on solid supports was illustrated, and different immobilized analytical assays of practical utility that are based on these concepts may be envisaged. The demonstration that metal NP–enzyme hybrid systems may act as biocatalytic templates (or "biocatalytic inks") for the generation of enlarged metal NPs or metallic nanowires appears to be particularly promising. In addition to the significance of this relatively simple method to fabricate predesigned nanocircuitry of controlled metal compositions, the metal NP–enzyme conjugates may have important implications in new biosensor configurations. The deposition of the metal NP–enzyme conjugates in between electrodes, and the substrate-induced enlargement of particles, may lead to biosensor devices where the conductivity between the electrodes gap provides a quantitative measure of the substrate concentration. Furthermore, the metal NP–enzyme conjugates may act as biocatalytic labels for amplifying a variety of biorecognition events such as antigen–antibody or nucleic acid–DNA interactions, and enabling the optical, electrochemical or microgravimetric readout of recognition processes [38].

The enzymatic growth of NPs may be extended to the general concept of a biocatalyzed synthesis of various materials such as semiconductor NPs (e.g., CdS, PbS) or magnetic NPs. Once such biocatalytically induced transformations have been achieved, the development of new optical [39] or photoelectrochemical [40] biosensor systems can be accomplished. Likewise, the templated growth of semiconductor nanowires, together with metallic contacting wires, is anticipated to yield new strategies for the assembly of nanodevices, such as transistors.

References

1 (a) Katz, E., Willner, I., Wang, J. (2004) Electroanalytical and bioelectroanalytical systems based on metal and semiconductor nanoparticles. *Electroanalysis* **16**, 19–44; (b) Niemeyer, C. M. (2001) Nanoparticles, proteins, and nucleic acids: biotechnology meets materials science. *Angew. Chem. Int. Ed.* **40**, 4128–4158.

2 (a) Katz, E., Willner, I. (2004) Integrated nanoparticle-biomolecule hybrid systems: synthesis, properties, and applications. *Angew. Chem. Int. Ed.* **43**, 6042–6108; (b) Medintz, I. L., Uyeda, H. T., Goldman, E. R., Mattoussi, H. (2005) Quantum dot bioconjugates for imaging, labelling and sensing. *Nat. Mater.* **4**, 435–446.

3 (a) Mirkin, C. A., Letsinger, R. L., Mucic, R. C., Storhoff, J. J. (1996) Biological synthesis of triangular gold nanoprisms. *Nature* **382**, 607–609; (b) Demers, L. M., Mirkin, C. A., Mucic, R. C., Reynolds, R. A., III, Letsinger, R. L., Elghanian, R., Viswanadham, G. (2000) A fluorescence-based method for determining the surface coverage and hybridization efficiency of thiol-capped oligonucleotides bound to gold thin films and nanoparticles. *Anal. Chem.* **72**, 5535–5541.

4 Cao, Y. W. C., Jin, R., Mirkin, C. A. (2002) Nanoparticles with Raman spectroscopic fingerprints for DNA and RNA detection. *Science* **297**, 1536–1540.

5 Dubertret, B., Calame, M., Libchaber, A. J. (2001) Single-mismatch detection using gold-quenched fluorescent oligonucleotides. *Nature Biotechnol.* **19**, 365–370.

6 Lyon, L. A., Musick, M. D., Natan, M. J. (1998) Colloidal Au-enhanced surface plasmon resonance immuno-sensing. *Anal. Chem.* **70**, 5177–5183.

7 He, L., Musick, M. D., Nicewarner, S. R., Sallinas, F. G., Benkovic, S. J., Natan, M. J., Keating, C. D. (2000) Colloidal Au-enhanced surface plasmon resonance for ultrasensitive detection of DNA hybridization. *J. Am. Chem. Soc.* **122**, 9071–9077.

8 Velev, O. D., Kaler, E. W. (1999) In situ assembly of colloidal particles into miniaturized biosensors. *Langmuir* **15**, 3693–3698.

9 (a) Park, S.-J., Taton, T. A., Mirkin, C. A. (2002) Array-based electrical detection of DNA with nanoparticle probes. *Science* **295**, 1503–1506; (b) Urban, M., Möller, R., Fritzsche, W. (2003) A paralleled readout system for an electrical DNA-hybridization assay based on a microstructured electrode array. *Rev. Sci. Instrum.* **74**, 1077–1081.

10 (a) Wang, J., Xu, D., Kawde, A.-N., Polsky, R. (2001) Metal nanoparticle-based electrochemical stripping potentiometric detection of DNA hybridization. *Anal. Chem.* **73**, 5576–5581; (b) Authier, L., Grossiord, C., Brossier, P., Limoges, B. (2001) Gold nanoparticle-based quantitative electrochemical detection of amplified human cytomegalovirus DNA using disposable microband electrodes. *Anal. Chem.* **73**, 4450–4456.

11 Willner, I., Palolsky, F., Weizmann, Y., Willner, B. (2002) Amplified detection of single-base mismatches in DNA using microgravimetric quartz-crystal-microbalance transduction. *Talanta* **56**, 847–856.

12 Knez, M., Sumser, M., Bittner, A. M., Wege, C., Geske, H., Martin, T. P., Kern, K. (2004) Spatially selective nucleation of metal clusters on the tobacco mosaic virus. *Adv. Funct. Mater.* **14**, 116–124.

13 Li, Z., Chung, S.-W., Nam, J. M., Ginger, D. S., Mirkin, C. A. (2003) Living templates for the hierarchical assembly of gold nanoparticles. *Angew. Chem. Int. Ed.* **7**, 2306–2309.

14 (a) Priyabrata, P., Ahmad, A., Mandal, D., Senapati, S., Sainkar, S. R., Khan, M. I., Ramani, R., Parischa, R., Ajayakumar, P. V., Alam, M., Sastry, M., Kumar, R. (2001) Bioreduction of AuCl₄⁻ ions by the fungus, *Verticillium* sp. and surface trapping of the gold nanoparticles formed. *Angew. Chem. Int. Ed.* **40**, 3585–3588; (b) Bharde, A., Rautaray, D., Bansal, V., Ahmad, A., Sarkar, I., Yusuf, S. M., Sanyal, M., Sastry, M. (2006) Extra-cellular biosynthesis of magnetite using fungi. *Small* **2**, 135–141; (c) Shankar, S. S., Rai, A., Ankamwar, B., Singh, A., Ahmad, A., Sastry, M. (2004) Biological synthesis of trian-gular gold nanoprisms. *Nature Mater.* **3**, 482–488; (d) Sanyal, A., Rautaray, D., Bansal, V., Ahmad, A., Sastry, M. (2005) Heavy-metal remediation by a fungus as a means of production of lead and cadmium carbonate crystals. *Langmuir* **21**, 7220–7224.

15 (a) Klaus, T., Joerger, R., Olsson, E., Granqvist, C.-G. (1999) Silver-based crystalline nanoparticles, microbially fabricated. *Proc. Natl. Acad. Sci. USA* **96**, 13611–13614; (b) Gardea-Torresdey, J. L., Parsons, J. G., Gomez, E., Peralta-Videa, J., Troiani, H. E., Santiago, P., Yacaman, M. J. (2002) Formation and growth of Au nanoparticles inside live alfalfa plants. *Nano. Lett.* **2**, 397–401; (c) Reiss, B. D., Mao, C., Solis, D. J., Ryan, K. S., Thomson, T., Belcher, A. M. (2004) Biological routes to metal alloy ferromagnetic nano-structures. *Nano. Lett.* **4**, 1127–1132.

16 Zayats, M., Baron, R., Popov, I., Willner, I. (2005) Biocatalytic growth of Au nanoparticles: from mechanis-tic aspects to biosensors design. *Nano Lett.* **5**, 21–25.

17 Basnar, B., Weizmann, Y., Cheglakov, Z., Willner, I. (2006) Synthesis of nanowires using dip-pen nanolitho-

graphy and biocatalytic inks. *Adv. Mater.* **18**, 713–718.

18 Baron, R., Zayats, M., Willner, I. (2005) Dopamine-, L-DOPA-, adrenaline-, and noradrenaline-induced growth of Au nanoparticles: assays for the detection of neuro-transmitters and of tyrosinase activity. *Anal. Chem.* **77**, 1566–1571.

19 Angeletti, C., Khomitch, V., Halaban, R., Rimm, D. L. (2004) Novel tyramide-based tyrosinase assay for the detection of melanoma cells in cytological preparations. *Diagn. Cytopathol.* **31**, 33–37.

20 (a) Pauluhn, J., Machemer, L., Kimmerle, G. (1987) Effects of inhaled cholinesterase inhibitors on bronchial tonus and on plasma and erythrocyte acetylcholine esterase activity in rats. *Toxicology* **46**, 177–190; (b) Clement, J. P. (1983) Efficacy of mono- and bispyridinium oximes versus soman, sarin and tabun poisoning in mice. *Fundam. Appl. Toxicol.* **3**, 533–535.

21 Pavlov, V., Xiao, Y., Willner, I. (2005) Inhibition of the acetycholine esterase-stimulated growth of Au nanoparticles: nanotechnology-based sensing of nerve gases. *Nano Lett.* **5**, 649–653.

22 Villatte, F., Marcel, V., Estrada-Mondaca, S., Fournier, D. (1998) Engineering sensitive acetylcholin-esterase for detection of organophos-phate and carbamate insecticides. *Biosens. Bioelec.* **13**, 157–164.

23 (a) Willner, I., Katz, E. (2000) Integration of layered redox proteins and conductive supports for bioelectronic applications. *Angew. Chem. Int. Ed.* **39**, 1180–1218; (b) Heller, A. (1990) Electrical wiring of redox enzymes. *Acc. Chem. Res.* **23**, 128–134; (c) Habermüller, L., Mosbach, M., Schuhmann, W. (2000) Electron-transfer mechanisms in amperometric biosensors. *Fresenius J. Anal. Chem.* **366**, 560–568; (d) Heller, A. (1992) Electrical connection of enzyme redox centers to electrodes. *J. Phys. Chem.* **96**, 3579–3587; (e) Gregg, B. A., Heller, A. (1991) Redox

polymer films containing enzymes. 2. Glucose oxidase containing enzyme electrodes. *J. Phys. Chem.* **95**, 5976–5980.

24 Xiao, Y., Pavlov, V., Shlyahovsky, B., Willner, I. (2005) An OsII-Bisbipyridine-4-Picolinic acid complex mediates the biocatalytic growth of Au nanoparticles: optical detection of glucose and acetylcholine esterase inhibition. *Chem. Eur. J.* **11**, 2698–2704.

25 Wang, J., Sánchez Arribas, A. (2006) Biocatalytically induced formation of cupric ferrocyanide nanoparticles and their application for electrochemical and optical biosensing of glucose. *Small* **2**, 129–134.

26 Sánchez Arribas, A., Vázquez, T., Wang, J., Mulchandani, A., Chen, W. (2005) Electrochemical and optical bioassays of nerve agents based on the organophosphorus-hydrolase mediated growth of cupric ferro-cyanide nanoparticles. *Electrochem. Commun.* **7**, 1371–1374.

27 Katakis, I., Dominguez, E. (1997) Catalytic electrooxidation of NADH for dehydrogenase amperometric biosensors. *Mikrochim. Acta* **126**, 11–32.

28 Xiao, Y., Pavlov, V., Levine, S., Niazov, T., Markovich, G., Willner, I. (2004) Catalytic growth of Au nanoparticles by NAD(P)H cofactors: optical sensors for NAD(P)$^+$-dependent biocatalyzed transformations. *Angew. Chem. Int. Ed.* **43**, 4519–4522.

29 Pavlov, V., Xiao, Y., Shlyahovsky, B., Willner, I. (2004) Aptamer-functionalized Au nanoparticles for the amplified optical detection of thrombin. *J. Am. Chem. Soc.* **126**, 11768–11769.

30 Shlyahovsky, B., Katz, E., Xiao, Y., Pavlov, V., Willner, I. (2005) Optical and electrochemical detection of NADH and of NAD$^+$-dependent biocatalyzed processes by the catalytic deposition of copper on gold nanoparticles. *Small* **1**, 213–216.

31 Xiao, Y., Shlyahovsky, B., Popov, I., Pavlov, V., Willner, I. (2005) Shape and color of Au nanoparticles follow

biocatalytic processes. *Langmuir* **21**, 5659–5662.

32 Murphy, C. J., Jana, N. R. (2002) Controlling the aspect ratio of inorganic nanorods and nanowires. *Adv. Mater.* **14**, 80–82.

33 Weizmann, Y., Patolsky, F., Popov, I., Willner, I. (2004) Telomerase-generated templates for the growing of metal nanowires. *Nano Lett.* **4**, 787–792.

34 Braun, E., Eichen, Y., Sivan, U., Ben-Yoseph, G. (1998) DNA-templated assembly and electrode attachment of a conducting silver wire. *Nature* **391**, 775–778.

35 Palolsky, F., Weizmann, Y., Willner, I. (2004) Actin-based metallic nanowires as bio-nanotransporters. *Nature Mater.* **3**, 692–695.

36 Grabar, K. C., Freeman, R. G., Hommer, M. B., Natan, M. J. (1995) Preparation and characterization of Au colloid monolayers. *Anal. Chem.* **67**, 735–743.

37 Doron, A., Katz, E., Willner, I. (1995) Organization of Au colloids as monolayer films onto ITO glass surfaces: application of the metal colloid films as base interfaces to construct redox-active monolayers. *Langmuir* **11**, 1313–1317.

38 Möller, R., Powell, R. D., Hainfeld, J. F., Fritzsche, W. (2005) Enzymatic control of metal deposition as key step for a low-background electrical detection for DNA chips. *Nano Lett.* **5**, 1475–1482.

39 Patolsky, F., Gill, R., Weizmann, Y., Mokari, J., Banin, U., Willner, I. (2003) Lighting-up the dynamics of telomerization and DNA replication by CdSe-ZnS quantum dots. *J. Am. Chem. Soc.* **125**, 13918–13919.

40 Gill, R., Patolsky, F., Katz, E., Willner, I. (2005) Electrochemical control of the photocurrent direction in intercalated DNA/CdS nanoparticle systems. *Angew. Chem. Int. Ed.* **44**, 4554–4557.

Part II
Nanostructures for Analytics

Nanobiotechnology II. Edited by Chad A. Mirkin and Christof M. Niemeyer
Copyright © 2007 WILEY-VCH Verlag GmbH & Co. KGaA, Weinheim
ISBN: 978-3-527-31673-1

7
Nanoparticles for Electrochemical Bioassays

Joseph Wang

7.1
Overview

7.1.1
Particle-Based Bioassays

The emergence of nanotechnology is opening new horizons for the application of nanoparticles (NPs) in analytical chemistry [1, 2]. In particular, NPs (such as colloidal gold and semiconductor quantum-dot NPs) are of considerable interest owing to their unique physical and chemical properties. Such properties offer excellent prospects for a wide range of biosensing applications [3, 4].

NPs hold immense promise as versatile labels for biological assays [5]. For example, metallic and inorganic NPs coupled to biomolecules such as antigens or nucleic acids have been extensively employed as active units in different biosensing systems [3, 4]. The enormous signal enhancement associated with the use of NP amplifying labels and with the formation of NP–biomolecule assemblies provides the basis for ultrasensitive electrochemical biodetection with PCR-like sensitivity [5]. Nucleic acid and antibody-functionalized NPs have been widely used as labels for the amplified optical [6] and microgravimetric [7] transduction of biorecognition events.

This chapter describes recent advances based on the use of bioconjugated NPs for the electronic transduction of biomolecular recognition events. Particular attention will be given to new signal amplification and coding strategies based on metal and semiconductor NP quantitation tags for electrochemical bioaffinity assays of DNA and proteins.

7.1.2
Electrochemical Bioaffinity Assays

Electrochemical devices have received considerable recent attention in the development of bioaffinity sensors [8–10]. Such affinity electrochemical biosensors and

Nanobiotechnology II. Edited by Chad A. Mirkin and Christof M. Niemeyer
Copyright © 2007 WILEY-VCH Verlag GmbH & Co. KGaA, Weinheim
ISBN: 978-3-527-31673-1

bioassays exploit the selective binding of certain biomolecules (e.g., antibodies, oligonucleotides) toward specific target species for triggering useful electrical signals. Electrochemical devices offer elegant routes for interfacing – at the molecular level – the biorecognition binding event and the signal-transduction element, and are uniquely qualified for meeting the size, cost, low-volume, and power requirements of decentralized biodiagnostics [10, 11]. The electrochemical transduction of binding events is commonly detected in connection with the use of enzyme- or redox labels or from other binding-induced changes in electrical parameters (e.g., conductivity or capacitance). The use of NP labels is relatively new in electrochemical detection, and offers unique opportunities for the electronic transduction of biomolecular interactions, for the biosensing of proteins and nucleic acids, for the design of novel bioelectronic devices, and for biodiagnostics in general.

7.1.3
NP-Based Electrochemical Bioaffinity Assays

Inspired by the novel use of NPs in optical bioassays [4, 12], recent studies have focused on developing analogous particle-based electrochemical routes for the detection of DNA and proteins. The majority of these studies has focused on metallic NPs and inorganic (quantum dot) nanocrystals. Additional studies have included nanowires, polymeric carrier (amplification) spheres, or magnetic (separation) beads. These NP materials offer elegant ways for interfacing biomolecular recognition events with electrochemical signal transduction, for dramatically amplifying the resulting electrical response, and for designing novel coding strategies. Most of these schemes have commonly relied on a highly sensitive electrochemical stripping measurement of the metal tag. The remarkable sensitivity of such stripping measurements is attributed to the "built-in" preconcentration step, during which the target metals are electrodeposited onto the working electrode [13]. The detection limits are thus lowered by three orders of magnitude, compared to pulse-voltammetric techniques used previously for monitoring biomolecular interactions. Such ultrasensitive electrochemical detection of metal tracers has been accomplished in connection with a variety of new and novel DNA- or protein-linked particle nanostructure networks. The successful realization of these NP-based signal amplification strategies requires proper attention to nonspecific adsorption issues.

7.1.3.1 Gold and Silver Metal Tags for Electrochemical Detection of DNA and Proteins

Several groups have developed powerful NP-based electrochemical bioaffinity assays based on capturing gold [14–16] or silver [17] NPs to the bound target, followed by acid dissolution and anodic-stripping electrochemical measurement of the solubilized metal tracer. These protocols facilitated the detection of DNA and proteins down to the picomolar and subnanomolar levels. For example, Dequaire et al. [14] demonstrated an electrochemical metalloimmunoassay based on strip-

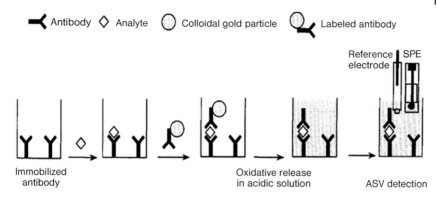

Fig. 7.1 Sandwich electrochemical immunoassay based on the use of gold nanoparticle tags and electrochemical stripping detection of the dissolved tag. ASV = anodic stripping voltammetry; SPE = screen-printed electrode. (From Ref. [14].)

ping voltammetric detection of a colloidal gold label (Figure 7.1). The same group demonstrated the applicability of the gold NP-based stripping procedure for the hybridization detection of the 406-base human cytomegalovirus DNA sequence [16]. Further sensitivity enhancement can be achieved by catalytic enlargement of the gold tag in connection with the NP-promoted precipitation of gold [15] or silver [18–20]. Combining such enlargement of the metal-particle tracers, with the effective "built-in" amplification of electrochemical stripping analysis paved the way to subpicomolar detection limits [18]. Silver enhancement relies on the chemical reduction of silver ions by hydroquinone to silver metal on the surface of the gold NPs. The silver reduction time must be controlled as a trade-off between greater enhancement and the contribution of a nonspecific background. A significant reduction of the silver staining background signals was obtained by using an indium-tin oxide (ITO) electrode possessing low silver-enhancing properties, or by modifying the gold transducer with a polyelectrolyte multilayer [19]. A simplified gold NP-based protocol was reported [21] which relied on pulse-voltammetric monitoring of the gold-oxide wave at ~1.20 V at a disposable pencil graphite electrode. A detection limit of 0.78 fmol was reported for PCR amplicons bound to the pencil electrode in connection with hybridization to oligonucleotide–NP conjugates.

The protocols described above have been based on the use of one NP reporter per one binding event. It is possible to further enhance the sensitivity by capturing multiple NPs per binding event. For example, the author's group has demonstrated an electrochemical triple-amplification hybridization assay, combining the carrier-sphere amplifying units (loaded with numerous gold NP tags) with the "built-in" preconcentration of the electrochemical stripping detection and catalytic enlargement of the multiple gold-particle tags [22]. The gold-tagged spheres were prepared by binding biotinylated metal NPs to streptavidin-coated polysty-

A Colloid Gold NP

B Enlarged Au-NP

C Multiple Au-NP on PS Carrier

D Enlarged multiple Au-NP on PS Carrier

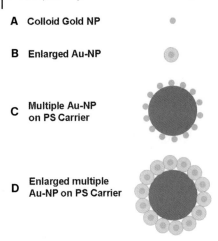

Fig. 7.2 (A–D, from top) Different generations of amplification platforms for bioelectronic detection based on gold nanoparticle (NP) tracers. (A) A single NP tag; (B) catalytic enlargement of the NP tag; (C) polymer (polystyrene, PS) carrier bead loaded with numerous gold NP tags; (D) catalytic enlargement of multiple tags on the carrier bead.

rene spheres. Scanning electron microscopy (SEM) imaging indicated a coverage of around 80 gold particles per polystyrene bead. Such a multiple-amplification route offered a dramatic enhancement of the sensitivity, and this enlargement of numerous gold NP tags (on a supporting sphere carrier) represents the fourth generation of amplification, starting with the early use of (first-generation) single gold NP tags [14–16]. These four generations of gold tags, reflecting the gradual increase in the amount of capture gold (per binding event), are displayed in Figure 7.2.

It is also possible to use gold NPs as carriers of redox markers for amplified biodetection [23]. Gold NPs covered with 6-ferrocenylhexaenthiol were used for this purpose in connection with a sandwich DNA hybridization assay. Due to the elasticity of the DNA strands, the ferrocene–Au-NP conjugates were positioned in closed proximity to the underlying electrode to allow a simple electron-transfer reaction. A detection limit of 10 amol was observed, along with linearity up to 150 nM. Applicability to PCR products related to the hepatitis B virus was reported.

It is possible also to detect DNA hybridization based on preparing the metal marker along the DNA backbone (instead of capturing it at the end of the duplex) [24]. Such a protocol relies on the DNA template-induced generation of conducting nanowires as a mode of capturing the metal tag. The use of DNA as a metallization template has evoked substantial research activity directed towards the generation of conductive nanowires and the construction of functional circuits [25], and this approach was applied to grow silver clusters on DNA templates. However, the DNA-templated growth of metal wires has not yet been exploited for detecting DNA hybridization. The new protocol (Figure 7.3) consists of the

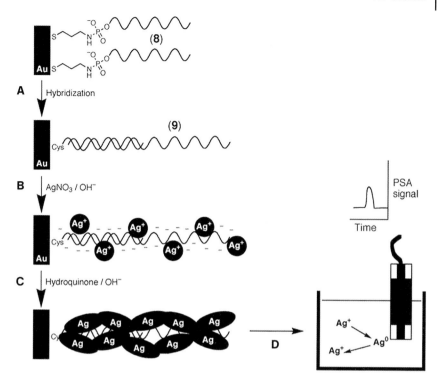

Fig. 7.3 Schematic representation of the protocol used for electrical detection of silver clusters produced along the DNA backbone. From top: (a) Formation of a self-assembled cystamine monolayer; immobilization of single-stranded (ss) DNA "probe" through the 5′-phosphate groups of ssDNA by the formation of phosphoramidate bond with the amino groups of the electrode surface; hybridization of the complementary target; (b) "loading" of the silver ion onto the DNA; (c) hydroquinone-catalyzed reduction of silver ions to form silver aggregates on the DNA backbone; (d) dissolution of the silver aggregates in an acid solution and transfer to detection cell, along with potentiometric stripping analysis (PSA) detection. (From Ref. [24].)

vectorial electrostatic "collection" of silver ions along the captured DNA target, followed by hydroquinone-induced reductive formation of silver aggregates along the DNA skeleton, along with dissolution and stripping detection of the nanoscale silver cluster.

7.1.3.2 NP-Induced Conductivity Detection

The formation of conductive domains as a result of biomolecular interactions provides an alternative and attractive route for electrochemical transduction of biorecognition events. NP-induced changes in the conductivity across a microelectrode gap were exploited by Mirkin's group for highly sensitive and selective detection of DNA hybridization [26]. Such a protocol involved capture of the NP-tagged DNA targets by probes immobilized in the gap between the two closely spaced

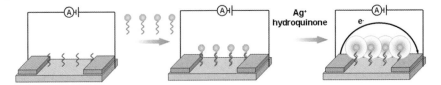

Fig. 7.4 Conductivity detection of nanoparticle-based microelectrodes arrays. Left to right: The capture of the nanoparticle-tagged DNA targets by probes confined to the gap, and subsequent silver enlargement, electrically short-circuit the gap and lead to a measurable conductivity signal. (Based on Ref. [26].)

microelectrodes, and a subsequent silver precipitation (Figure 7.4). This resulted in a conductive metal layer bridging the gap, and led to a measurable conductivity signal. Such hybridization-induced conductivity signals, associated with resistance changes across the electrode gap, offer high sensitivity down to the 0.5 picomolar level. Control of the salt concentration allowed an excellent mismatch discrimination (with a factor of 100 000:1), without thermal stringency. Similarly, Velev and Kaler [27] exploited the catalytic features of NPs for analogous conductivity immunoassays of proteins in connection with antibody-functionalized latex spheres placed between two closely spaced electrodes. A sandwich immunoassay thus led to the binding of a secondary gold-labeled antibody, followed by catalytic deposition of a silver layer "bridging" the two electrodes. The method enables the analysis of human immunoglobulin G (IgG), with a detection limit of approximately 2×10^{-13} M.

7.1.3.3 Inorganic Nanocrystal Tags: Towards Electrical Coding
Owing to their unique (tunable-electronic) properties, semiconductor (quantum dot) nanocrystals have generated considerable interest for optical DNA detection [28]. Recent activity has demonstrated the utility of such inorganic crystals for enhanced electrical detection of proteins and DNA [29–31].

Previously, the author's group reported on the detection of DNA hybridization in connection with cadmium sulfide (CdS) NP tags and electrochemical stripping measurements of the cadmium [29]. A NP-promoted cadmium precipitation was used to enlarge the NP tag and amplify the stripping DNA hybridization signal. In addition to measurements of the dissolved cadmium ion, solid-state measurements were demonstrated following a "magnetic" collection of the magnetic-bead/DNA-hybrid/CdS-tracer assembly onto a screen-printed electrode transducer. Such a protocol couples the amplification features of NP/polynucleotide assemblies and highly sensitive potentiometric stripping detection of cadmium, with an effective magnetic isolation of the duplex. The low detection limit (100 fmol) was coupled to good precision (RSD = 6%). A substantially enhanced signal was obtained by encapsulating multiple CdS NPs into the host bead, or by loading onto carbon-nanotube carriers [30].

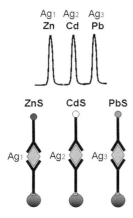

Fig. 7.5 Use of different quantum dot tracers for electrical detection of multiple protein targets. Lower diagram: Sandwich immunoassay leading to capture of the nanocrystal-linked secondary antibodies. Upper diagram: Stripping voltammogram for a solution containing dissolved ZnS, CdS and PbS nanoparticle tracers, corresponding to the three protein targets (Ag_1–Ag_2). (Based on Ref. [32].)

One interesting aspect of these inorganic nanocrystals is the possibility of combining different semiconductor tags, linked to different antibodies or DNA probes, for the simultaneous analysis of different protein or DNA targets [31, 32]. Such nanocrystals offer an electrodiverse population of electrical tags as needed for multiplexed bioanalysis. Four encoding NPs (CdS, ZnS, CuS and PbS) were thus used to differentiate the signals of four proteins or DNA targets in connection with a sandwich immunoassay and hybridization assay, respectively, along with stripping voltammetry of the corresponding metals. Each recognition event yielded a distinct voltammetric peak, the size and position of which reflected the level and identity, respectively, of the corresponding antigen or DNA target (Figure 7.5). Conducting massively parallel assays (in microwells of microtiter plates or by using multi-channel microchips, with each microwell or channel carrying out multiple measurements) could thus lead to a high-throughput analysis of proteins or nucleic acids.

Recent efforts in our laboratory have aimed at developing large particle-based libraries for electrical coding, based on the judicious design of encoded "identification" spheres [33] or striped metal microrods [34]. By incorporating different predetermined levels (or lengths) of multiple metal markers, such beads or rods can lead to a large number of recognizable voltammetric signatures, and hence to a reliable identification of a large number of biomolecules. For example, multimetal cylindrical particles can be prepared by the template-directed electrochemical synthesis, by plating indium, zinc, bismuth, and copper onto a porous membrane template. Capping the rod with a gold end facilitates its functionalization with a thiolated oligonucleotide probe. Each nanowire thus yields a characteristic multi-peak voltammogram, the peak potentials and current intensities of which reflect the identity of the corresponding DNA target. In this way, thousands of usable codes could be generated in connection with five to six different potentials and four to five different current intensities. In addition to powerful bioassays, such "identification beads" hold great promise for the identification of counterfeit products and related authenticity testing. The template-directed electrochemical route can also be used for preparing micrometer-long metal microrod tags for ultransensitive detection [34]. The linear relationship between the charge passed during the preparation and the size of the resulting microrod allows tailoring of the sensitivity of the electrical DNA assay. For example, microrods prepared by plating of indium into the pores of a host membrane offered a greatly lower detection limit (250 zmol) compared to common bioassays based on spherical NP labels. Indium offers a very attractive electrochemical stripping behavior, and is not normally present in biological samples or reagents. Solid-state chronopotentiometric measurements of the indium microrod have been realized through a "magnetic" collection of the DNA-linked particle assembly onto a screen-printed electrode transducer.

The coding and amplification features of inorganic nanocrystals have been shown to be extremely useful for monitoring aptamer–protein interactions. Very recently, the author's group described a highly sensitive and selective simultaneous bioelectronic displacement assay of several proteins in connection with a self-assembled monolayer of several thiolated aptamers conjugated to proteins carrying different inorganic nanocrystals [35]. Electrochemical stripping detection of the non-displaced nanocrystal tracers resulted in a remarkably low (attomole) detection limit, that is significantly lower than those of common aptamer biosensors (Figure 7.6). Unlike the two-step sandwich assays used in early quantum dot-based electronic hybridization or immunoassays [31, 32], this aptamer biosensor protocol relies on a single-step displacement protocol. Aptamers offer great

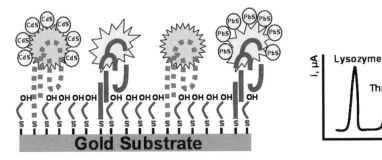

Fig. 7.6 Aptamer/quantum dot (QD)-based dual-analyte biosensor, involving displacement of the tagged proteins by the target analytes. The protocol involves the co-immobilization of several thiolated aptamers, along with binding of the corresponding QD-tagged proteins on a gold surface, addition of the protein sample, and monitoring the displacement through electrochemical detection of the remaining nanocrystals. The voltammogram on the right shows the Cd and Pb peaks corresponding to the two proteins. (Based on Ref. [35].)

promise for sensitive displacement assays as the tagged protein has a significantly lower affinity to the aptamer compared to the unmodified analyte.

Metal NPs have also been shown useful for electrical coding of single nucleotide polymorphisms (SNP) [36]. Such a procedure relies on the hybridization of monobase-modified gold NPs with the mismatched bases. The binding event leads to changes in the gold oxide peak, and offers great promise for coding all mutational changes. Analogous SNP coding protocols based on different inorganic nanocrystals are currently being examined in our laboratory [37]. This protocol involves the addition of ZnS, CdS, PbS and CuS crystals linked to adenosine, cytidine, guanosine and thymidine mononucleotides, respectively. Each mutation captures (via base pairing) different nanocrystal-mononucleotide conjugates, to yield a distinct electronic fingerprint.

7.1.3.4 Use of Magnetic Beads

Several of the above-described protocols [15, 18, 38] have combined the inherent signal amplification of electrochemical stripping analysis with an effective discrimination against non-target biomolecules. In addition to efficient isolation of the duplex, magnetic spheres can open the door for elegant approaches to trigger and control the electrochemical detection of DNA and proteins [39, 40].

For example, an attractive magnetic triggering of the electrical DNA detection has been realized through a "magnetic" collection of the magnetic-bead/DNA-hybrid/metal-tracer assembly onto a screen-printed electrode transducer that allowed direct electrical contact of the silver precipitate [39]. Such a bioassay involved the hybridization of a target oligonucleotide to probe-coated magnetic beads, followed by binding of the streptavidin-coated gold NPs to the captured target, catalytic silver precipitation on the gold-particle tags, a magnetic "collection" of the DNA-linked particle assembly, and solid-state stripping detection (Figure 7.7). The magnetic "collection" route greatly simplifies the electrical detection of metal tracers as it eliminates the acid dissolution step.

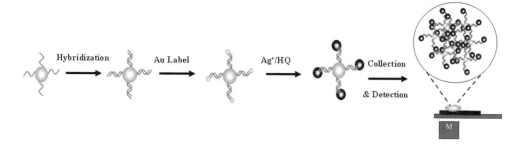

Fig. 7.7 Schematic of the magnetically induced solid-state electrochemical detection of DNA hybridization. The assay involves introduction of the probe-coated magnetic beads, the hybridization event (with the biotinylated target), capture of the streptavidin-gold particles, catalytic silver deposition on the gold nanoparticle tags, and positioning of an external magnet (M) under the electrode to attract the particle-DNA assembly and effect solid-state chrono-potentiometric detection. (From Ref. [39].)

Magnetic beads have also been used for triggering the electron-transfer reactions of DNA [40]. Changing the position of the magnet (below thick film electrodes) was thus used for switching "on/off" the DNA oxidation (through attraction and removal of DNA functionalized-magnetic spheres). The process was reversed and repeated upon switching the position of the magnet, with and without oxidation signals in the presence and absence of the magnetic field, respectively. Such magnetic triggering of the DNA oxidation holds great promise for DNA arrays (based on closely spaced electrodes and guanine-free, inosine-substituted probes). Willner and coworkers described an amplified detection of viral DNA and of single-base mismatches using oligonucleotide-functionalized magnetic beads and an electrochemiluminescence (ECL) detection [41]. The magnetic attraction of the labeled magnetic particles and their rotation on the electrode surface was used to amplify the ECL signal. The ability of external magnetic fields to control other bioelectrochemical processes, such as biocatalytic transformations of redox enzymes was also documented by Willner's group [42].

It is possible also to use magnetic spheres as tags for DNA hybridization detection in connection to stripping voltammetric measurements of their iron content [43]. A related protocol, developed in the same study, involved probes labeled with gold-coated iron core-shell NPs. In both cases, the captured iron-containing NPs were dissolved following the hybridization, and the released iron was quantified by adsorptive stripping voltammetry in the presence of the 1-nitroso-2-naphthol ligand and a bromate catalyst. Core-shell copper-gold NP tags were also shown useful by coupling the favorable electrochemical behavior of the copper core with the attractive surface modification properties of the gold shell [44].

7.1.3.5 Ultrasensitive Particle-Based Assays Based on Multiple Amplification Schemes

Several amplification processes, such as catalytic enlargement of the metal tag and its electrolytic preconcentration onto the electrode surface, have been discussed previously. These protocols have been based on the use of one reporter per one binding event, but it is possible to further enhance the sensitivity by employing multiple tracers per binding event. This can be accomplished in connection with polymeric microbeads carrying multiple redox tracers externally (on their surface) or internally (via encapsulation; Figure 7.8). Coupled with additional amplification units and processes, such bead-based multi-amplification protocols meet the high sensitivity demands of electrochemical affinity biosensors.

As indicated earlier, gold NPs can be used as carriers of multiple redox (ferrocene) markers for amplified biodetection down to the 10 amol level [23]. Similarly, a dramatic sensitivity enhancement can be achieved in sandwich bioassays involving the capture of polymeric spheres loaded with numerous gold NP tags [22], in manner described earlier in Section 7.1.3.1.

The internal encapsulation of electroactive tracers within polymeric carrier beads offers an attractive alternative to their external loading. For example, a re-

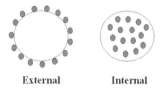

External   Internal

Fig. 7.8 Polymeric "carrier" bead amplifying units based on loading numerous redox tags either externally (on their outer surface) or internally (via encapsulation).

markably sensitive electrochemical detection of DNA hybridization was reported based on polystyrene beads impregnated with a ferrocene marker [45]. The resulting "electroactive beads" are capable of carrying a huge number of the ferrocene-carboxaldehyde tag molecules, and hence offer a remarkable amplification of single hybridization events. This allowed the chronopotentiometric detection of target DNA down to the 5.1×10^{-21} mol level ($\sim$31 000 molecules) in connection with 20 min hybridization and the "release" of the marker in an organic medium. This dramatic signal amplification advantage is combined with an excellent discrimination against a large excess (10^7-fold) of non-target nucleic acids. The DNA-linked particle assembly that resulted from the hybridization event is shown in Figure 7.9; this image indicates that the 10-μm electroactive beads are cross-linked to the smaller ($\sim$0.8 μm) magnetic spheres through the DNA hybrid. Current efforts are aimed at encapsulating different ferrocene markers within the polystyrene host beads in connection with multi-target DNA detection. Other marker encapsulation routes hold great promise for electrochemical bioassays. Particularly attractive are the recently developed nanoencapsulated microcrystalline particles, prepared using a layer-by-layer (LBL) assembly technique, that offer large marker/biomolecule ratios and superamplified bioassays [46, 47].

Electroactive (FCA) bead

Magnetic bead

50 μm

Fig. 7.9 The use of "electroactive" particles for ultrasensitive DNA detection, based on polystyrene beads impregnated with a redox marker. Scanning electron micrograph of the resulting DNA-linked particle assembly. FCA = ferrocenecarboxaldehyde. (From Ref. [45].)

7.2
Methods

The majority of the above-described studies have focused on electrochemical stripping transduction of the dissolved NPs, although some reports demonstrated the solid-state detection of the captured particles (without their dissolution). Most of these schemes have commonly relied on a highly sensitive electrochemical stripping transduction of the biorecognition event. Stripping analysis is a two-step technique. The first (deposition) step involves the electrolytic deposition of a small portion of the metal ions in solution into the working electrode to preconcentrate the metals. This is followed by the stripping (measurement) step, which involves dissolution (stripping) of the deposit. As the metals are preconcentrated into the electrode by a factor of approximately 1000, the detection limits are lowered by three orders of magnitude compared to solution-phase voltammetric measurements. Hence, four to six metals can be measured simultaneously in various matrices at concentration levels down to 10^{-10} M, utilizing relatively inexpensive instrumentation. Less sensitive – but faster – pulse-voltammetric techniques (e.g., square-wave voltammetry or differential pulse voltammetry) can also be used for monitoring the dissolved NPs or nanocrystals.

Control of the surface chemistry and coverage is essential for assuring high reactivity, orientation/accessibility, and stability of the surface-bound probe, as well as for avoiding nonspecific binding/adsorption events. Surface-blocking steps should thus be employed to avoid the amplification of background signals (associated with nonspecific adsorption of the NP amplifiers). In a typical bioassay the DNA probe, antibody or aptamer is immobilized onto the surface of the working electrode, on the walls of polymeric microwells, or onto functionalized magnetic beads. This can accomplished in connection with streptavidin-coated magnetic beads [15], via adsorption onto the walls of polystyrene microwells [14, 16], or through the use of a highly dense mixed monolayer of alkanethiols on the gold surface [35]. The latter procedure involves a hydrophilic 6-mercapto-1-hexanol component that provides the blocking action essential for minimizing nonspecific adsorption effects. Electropolymerization represents another attractive means of localizing the probes on small electrode surfaces, as is desirable for the fabrication of high-density arrays.

In the case of functionalized magnetic spheres, the binding event is followed by a magnetic separation, and transfer of the sandwich probe/target-tagged-probe system to an electrochemical cell where the NPs are detected voltammetrically. Such magnetic isolation of the duplex offers the effective removal of unwanted (non-hybridized) constituents. A similar "two-surface" (recognition and transduction) approach can be accomplished by using gold electrodes (coated with a dense monolayer of the probe) along with carbon detection electrodes.

Proper attention should be given also to conjugating the NP tags. The preparation of DNA-functionalized gold NPs or CdS nanocrystals often involves the use of thiolate-terminated oligonucleotides. Antibodies can be conjugated to NP tracers through common bifunctional linkers (e.g., carbodiimide), coupled to terminal groups on the functionalized nanocrystal.

7.3
Outlook

Nanotechnology offers unique opportunities for designing powerful electrochemical bioassays and biosensors, and adds a new dimension to such assays and devices. The ability to tailor the composition, size and shape of nanomaterials is expected to lead to entirely new types of electrochemical sensors. The above-described studies demonstrate the broad potential of bioconjugated NPs for the electronic transduction of biomolecular recognition events. Recently, the use of NPs in electrochemical bioassays has expanded dramatically, and will surely continue in this vein. Indeed, metal and semiconductor NPs, one-dimensional nanotubes and nanowires have rapidly become attractive labels for bioaffinity assays, offering unique signal amplification and multiplexing capabilities. The coupling of different NP-based amplification platforms and amplification processes have dramatically enhanced the intensity of the analytical signal and have led to the development of ultrasensitive bioassays for proteins and nucleic acids. The enormous amplification afforded by such NP-based schemes opens up the possibility of detecting disease markers, infectious agents or biothreat agents that cannot be measured by conventional methods. These highly sensitive biodetection schemes might provide an early detection of diseases, or perhaps a warning of a terrorist attack, using ultrasensitive bioelectroanalytical protocols unachievable with standard electrochemical methods.

One critical requirement for the successful realization of ultrasensitive NP-based bioelectronic assays is the ability to minimize the nonspecific binding of coexisting biomolecules. Thus, proper attention should be given to the surface chemistry and to the washing step as desired for minimizing nonspecific adsorption of the NP amplifiers. The combination of high sensitivity, specificity, and multi-target detection capabilities permits NP-based electronic bioaffinity assays to rival the most advanced optical protocols. Consequently, such nanomaterials-based bioassays and devices are expected to have a major impact upon clinical diagnostics, environmental monitoring, security surveillance, or for ensuring food safety. The application of such systems to the monitoring of other biomolecular interactions is also expected in the near future. In this regard, preliminary data from nanocrystal-based monitoring of lectin-glycan interactions have shown much encouragement [48]. A recently reported much-improved detection limit for ion-selective electrodes [49] offers great promise for the use of these potentiometric devices in the ultrasensitive monitoring of biomolecular interactions in relation to NP amplifiers (e.g., silver or lead sulfide). Future innovative research is expected to lead to advanced particle-based electrochemical biodetection strategies which, when coupled with other major technological advances, will result in effective, easy-to-use, hand-held portable devices and chip-based array formats for protein and DNA diagnostics.

Acknowledgments

The author gratefully acknowledges the financial support provided by the National Science Foundation (Grant No. CHE 0506529) and NIH (R01A 1056047-01 and R01 EP 0002189).

References

1 Niemeyer, C. M. (2001) Nanoparticles, proteins, and nucleic acids: biotechnology meets materials science. *Angew. Chem. Int. Ed.* **40**, 4128.

2 Alivisatos, P. (2004) The use of nanocrystals in biological detection. *Nat. Biotechnol.* **22**, 47.

3 Katz, E., Willner, I. (2004) Integrated nanoparticle-biomolecule hybrid systems: Synthesis, properties, and applications. *Angew. Chem. Int. Ed.* **43**, 6042.

4 Rosi, N. L., Mirkin, C. A. (2005) Nanostructures in biodiagnostics. *Chem. Rev.* **105**, 1547.

5 Wang, J. (2005) Nanomaterial-based amplified transduction of biomolecular interactions. *Small* **1**, 1036.

6 Storhoff, J. J., Elghanian, R., Mucic, R. C., Mirkin, C. A., Letsinger, R. L. (1998) One-pot colorimetric differentiation of polynucleotides with single base imperfections using gold nanoparticle probes. *J. Am. Chem. Soc.* **120**, 1959.

7 Willner, I., Patolsky, F., Weizmann, Y., Willner, B. (2002) Amplified detection of single-base mismatches in DNA using micro gravimetric quartz-crystal-microbalance transduction. *Talanta* **56**, 847.

8 Mikkelsen, S. R. (1996) Electrochemical biosensors for DNA sequence detection. *Electroanalysis* **8**, 15.

9 Skladal, P. (1997) Advances in electrochemical immunosensors. *Electroanalysis* **9**, 737.

10 Palecek, E., Fojta, M. (2001) Detecting DNA hybridization and damage. *Anal. Chem.* **73**, 75A.

11 Wang, J. (2002) Electrochemical nucleic acid biosensors. *Anal. Chim. Acta* **469**, 63.

12 Taton, T. A., Mirkin, C. A., Letsinger, R. L. (2000) Scanometric DNA array detection with nanoparticle probes. *Science* **289**, 1757.

13 Wang, J. (1985) *Stripping Analysis.* VCH, New York.

14 Dequaire, M., Degrand, C., Limoges, B. (2000) An electrochemical metalloimmunoassay based on a colloidal gold label. *Anal. Chem.* **72**, 5521.

15 Wang, J., Xu, D., Kawde, A., Polsky, R. (2001) Metal nanoparticle-based electrochemical stripping potentiometric detection of DNA hybridization. *Anal. Chem.* **73**, 5576.

16 Authier, L., Grossiord, C., Berssier, P., Limoges, B. (2001) Gold nanoparticle-based quantitative electrochemical detection of amplified human cytomegalovirus DNA using disposable microband electrodes. *Anal. Chem.* **73**, 4450.

17 Cai, H., Xu, Y., Zhu, N., He, P., Fang, Y. (2002) An electrochemical DNA hybridization detection assay based on a silver nanoparticle label. *Analyst* **127**, 803.

18 Wang, J., Polsky, R., Danke, X. (2001) Silver-enhanced colloidal gold electrical detection of DNA hybridization. *Langmuir* **17**, 5739.

19 Lee, T. M. H., Li, L. L., Hsing, I. M. (2003) Enhanced electrochemical detection of DNA hybridization based on electrode-surface modification. *Langmuir* **19**, 4338.

20 Cai, H., Wang, Y., He, P., Fang, Y. (2002) Electrochemical detection of DNA hybridization based on silver-enhanced gold nanoparticle label. *Anal. Chim. Acta* **469**, 165.

21 Ozsoz, M., Erdem, A., Kerman, K., Okzan, D., Tugrul, B., Topcuoglo, N., Ekren, H., Taylan, M. (2003) Electrochemical genosensor based on colloidal gold nanoparticles for the detection of factor V eiden mutation using disposable pencil graphite electrodes. *Anal. Chem.* **75**, 2181.

22 Kawde, A., Wang, J. (2004) Amplified electrical transduction of DNA hybridization based on polymeric beads loaded with multiple gold nano-particles tags. *Electroanalysis* **16**, 101.

23 Wang, J., Li, J., Baca, A. J., Hu, J., Zhou, F., Yan, W., Pang, D. W. (2003) Amplified voltammetric detection of DNA hybridization via oxidation of ferrocene caps on gold nanoparticle/streptavidin conjugates. *Anal. Chem.* **75**, 3941.

24 Wang, J., Rincon, O., Polsky, R., Dominguez, E. (2003) Electrochemical detection of DNA hybridization based on DNA-templated assembly of silver Cluster. *Electrochem. Commun.* **5**, 83.

25 Braun, E., Eichen, Y., Sivan, U., Ben-Yoseph, G. (1998) DNA-templated assembly and electrode attachment of a conducting silver wire. *Nature* **391**, 777.

26 Park, S., Taton, T. A., Mirkin, C. A. (2002) Array-based electrical detection of DNA with nanoparticle probes. *Science* **295**, 1503.

27 Velev, O. D., Kaler, E. W. (1999) In situ assembly of colloidal particles into miniaturized biosensors. *Langmuir* **15**, 3693.

28 Han, M., Gao, X., Su, J., Nie, S. (2001) Quantum-dot-tagged micro-beads for multiplexed optical coding of biomolecules. *Nat. Biotechnol.* **19**, 631.

29 Wang, J., Liu, G., Polsky, R. (2002) Electrochemical stripping detection of DNA hybridization based on CdS nanoparticle tags. *Electrochem. Commun.* **4**, 819.

30 Wang, J., Liu, G., Jan, R., Zhu, Q. (2003) Electrochemical detection of DNA hybridization based on carbon-nanotubes loaded with CdS tags. *Electrochem. Commun.* **5**, 1000.

31 Wang, J., Liu, G., Merkoçi, A. (2003) Electrochemical coding technology for simultaneous detection of multiple DNA targets. *J. Am. Chem. Soc.* **125**, 3214.

32 Liu, G., Wang, J., Kim, J., Jan, M., Collins, G. (2004) Electrochemical coding for multiplexed immuno-assays of proteins. *Anal. Chem.* **76**, 7126.

33 Wang, J., Liu, G., Rivas, G. (2003) Encoded beads for electrochemical identification. *Anal. Chem.* **75**, 4667.

34 Wang, J., Liu, G., Zhou, J. (2003) Indium microrod tags for electrical detection of DNA hybridization. *Anal. Chem.* **75**, 6218.

35 Hansen, J., Wang, J., Kawde, A., Xiang, Y., Gothelf, K. V., Collins, G. (2006) Quantum-dot/aptamer-based ultrasensitive multi-analyte electro-chemical biosensor. *J. Am. Chem. Soc.* **128**, 2228.

36 Kerman, K., Saito, M., Morita, Y., Takamura, Y., Ozsoz, M., Tamiya, E. (2004) Electrochemical coding of single-nucleotide polymorphisms by monobase-modified gold nanoparti-cles. *Anal. Chem.* **76**, 1877.

37 Wang, J., Lee, T., Liu, G. (2005) Nanocrystal-based bioelectronics coding of SNP. *J. Am. Chem. Soc.* **127**, 38.

38 Palecek, E., Fojta, M., Jelen, F. (2002) New approaches in the development of DNA sensors: hybridization and electrochemical detection of DNA and RNA at two different surfaces. *Bioelectrochemistry* **56**, 85.

39 Wang, J., Xu, D., Polsky, R. (2002) Magnetically-induced solid-state electrochemical detection of DNA hybridization. *J. Am. Chem. Soc.* **124**, 4208.

40 Wang, J., Kawde, A. (2002) Magnetic-field stimulated DNA oxidation. *Electrochem. Commun.* **4**, 349.

41 Patolsky, F., Weizmann, Y., Katz, E., Willner, I. (2003) Magnetically amplified DNA assays (MADA): Sensing of viral DNA and single-base mismatches by using nucleic acid modified magnetic particles. *Angew. Chem. Int. Ed.* **42**, 2372.

42 Hirsch, R., Katz, E., Willner, I. (2000) Magneto-switchable bioelectrocatalysis. *J. Am. Chem. Soc.* **122**, 12053.

43 Wang, J., Liu, G. D., Merkoci, A. (2003) Particle-based detection of DNA hybridization using electrochemical stripping measurements of an iron tracer. *Anal. Chim. Acta* **482**, 149.

44 Cai, H., Zhu, N., Ping, Y., He, P., Fang, Y. (2003) Cu:Au alloy nanoparticle as oligonucleotides labels for electrochemical stripping detection of DNA hybridization. *Biosensors Bioelectronics* **18**, 1311.

45 Wang, J., Polsky, P., Merkoci, A., Turner, K. (2003) Electroactive beads for ultrasensitive DNA detection. *Langmuir* **19**, 989.

46 Trau, D., Yang, W., Seydack, M., Carusu, F., Renneberg, R. (2002) Nanoencapsulated microcrystalline particles for superamplified biochemical assays. *Anal. Chem.* **74**, 5480.

47 Mak, W. C., Cheung, K. Y., Trau, D., Warsinke, A., Scheller, F., Renneberg, R. (2005) Electrochemical bioassay utilizing encapsulated electrochemical active microcrystal biolabels. *Anal. Chem.* **77**, 2835.

48 Dai, Z., Kawde, A., Xiang, Y., La Belle, J., Gerlach, J., Bhavanandan, V. P., Joshi, J., Wang, J. (2006) Nanoparticle-based bioelectronic sensing of glycan-lectin interactions, *J. Am. Chem. Soc.* **128**, 100018.

49 Bakker, E., Buhlmann, P., Pretsch, E. (1999) Polymer membrane ion-selective electrodes – what are the limits? *Electroanalysis* **11**, 915.

8
Luminescent Semiconductor Quantum Dots in Biology

Thomas Pons, Aaron R. Clapp, Igor L. Medintz, and Hedi Mattoussi

8.1
Overview

During the past two decades there has been a steady and growing interest in semiconductor nanocrystals (both colloidal and self-assembled) [1, 2]. Such interest, which initially was motivated by a pure academic desire to better understand the unique physical properties of these materials, has more recently been supplemented by an explosion in technological developments in this area [1, 3–6]. Potential applications in devices such as absorption filters [7, 8], light-emitting diodes [9–12], photovoltaic cells [13, 14] and, more recently, in bio-inspired applications, have spearheaded the drive to expand our understanding of their properties and to design chemical methods for preparing new and better materials.

Biological tagging using conventional fluorophores and fluorescent proteins to develop immunoassays, disease diagnosis, and cell and tissue imaging is a ubiquitous approach in biotechnology [15–18]. However, virtually all available organic dyes and fluorescent proteins have inherent limitations, such as narrow excitation spectral windows, broad red-tailing photoluminescence (PL) spectra, and low resistance to photodegradation [5, 6]. Semiconductor nanocrystals (e.g., CdSe-ZnS core-shell quantum dots; QDs) offer several unique properties in comparison, and promise significant advantages in certain bioanalytical and imaging applications [5, 6, 19]. Colloidal QDs such as those made of ZnSe, CdS, CdSe, CdTe, and PbSe emit light over a wide range of wavelengths in the visible and near infrared (IR) [1, 5, 6, 20–25]. In addition, their broad absorption envelope allows simultaneous excitation of several different colors of QDs with a single wavelength, making them naturally suitable for multiplexing applications.

QDs conjugated with biomolecular receptors (including proteins, peptides, DNA) can be used in applications such as detection of soluble substances, imaging, and diagnostics applications. Thus, a successful integration of QDs into any biotechnological application will necessitate a thorough understanding of these hybrid systems. The use of QDs in biology has particularly expanded during the

Nanobiotechnology II. Edited by Chad A. Mirkin and Christof M. Niemeyer
Copyright © 2007 WILEY-VCH Verlag GmbH & Co. KGaA, Weinheim
ISBN: 978-3-527-31673-1

past five years, with the number of biologically driven reports increasing from a few in 2000 to a few hundreds in 2005. This clearly reflects the promise of such technology and its ability to potentially gain new understanding of a variety of outstanding issues and problems.

In this chapter, we review the progress made in this particular field and provide a short history outlining what we believe are the most important bio-related developments. We also provide a critical assessment of the progress made and the problems encountered. In addition, the currently available techniques to prepare biocompatible QDs will be discussed, along with advantages and limitations associated with those techniques. Finally, an outlook is offered on where the field should be heading in the near future, and the potential hurdles which might need to be overcome.

8.1.1
QD Bioconjugates in Cell and Tissue Imaging

The use of luminescent QDs in cellular and *in-vivo* imaging has, in particular, attracted a tremendous interest [5, 6]. This is driven by the potential benefits of properties such as strong resistance to photo- and chemical degradation, multicolor imaging capacity, and high one- and two-photon excitation cross-sections for use in probing cellular processes, protein tracking and deep tissue imaging with reduced auto-fluorescence. The high photobleaching threshold of these materials also allows sample imaging to be collected over long periods of time. In order to harness some of the potential benefits offered by QDs, however, they must first be delivered inside the cell cytoplasm.

Several methods have been used to deliver QDs across the membrane of live cells and into the intracellular medium, with each presenting its advantages and drawbacks. *Endocytosis* of QDs allows parallel labeling of large cell populations using a "natural" and relatively benign process. While QDs capped with carboxy-terminated ligands have been shown to be nonspecifically endocytosed [26], conjugation with cell-penetrating peptides [27, 28] or molecules targeting cell surface receptors (e.g., folic acid [29], epidermal growth factor; EGF [30], transferrin [31]) were found to greatly improve the delivery efficiency, with reduced incubation times required. As QD-bioconjugates delivered by this mechanism largely remain sequestered within endosomes [28], this prohibits their subsequent targeting to subcellular compartments, or their use for sensing cytoplasmic processes. There is a slight benefit to this feature, however, as this isolates the nanoparticles (NPs) from the intracellular machinery and may potentially benefit long-term live cell imaging, by slowing down the possible cytotoxic activity of QDs. Electroporation and co-incubation with transfection reagents such as cationic lipids have also been shown to provide an effective tool for labeling cell cultures, though the reagents are often delivered as aggregates [32, 33]. This has not prevented studies from focusing on cell viability/toxicity [33], but more sophisticated investigations requiring well-dispersed single QDs are not yet possible using this approach. A recent report indicates that individual QDs can be delivered into the cytoplasm via osmotic lysis of pinocytic vesicles, and used to follow the movement of single kinesin motors [34]. It is not clear, however, whether this technique could be used to deliver large numbers of QDs to obtain stronger fluorescence signals. Thus far, microinjection is the only reported technique which has allowed a homogeneous cytoplasmic delivery of NPs and their conjugates [32, 35]. QDs conjugated with specific peptides (namely the family of cell-penetrating peptides derived from the HIV TAT protein motif) could be directed to specific subcellular compartments, such as the nucleus and mitochondria [32]. However, this technique is tedious and requires cells to be injected one at a time. A general conclusion of these studies is that the nature of the QD coating plays a crucial role in their intracellular stability to prevent aggregate formation [5, 6].

When successfully conjugated to surface receptors, QDs are more readily available for the effective labeling of cells. In particular, such conjugates have brought new insights into the motion of individual receptors in live cells [30, 36]. The high signal and photostability of single QDs allow the tracking of individual proteins for up to 20 minutes, compared to a few seconds for organic dyes [36]. In addition, their relatively small size does not hinder their movement in confined spaces, compared to larger metallic or latex particles, where displacement can be substantially reduced. For example, QDs were used to monitor the diffusion of glycine receptors at the surface of neuronal cells, and revealed different diffusion kinetics depending on the receptor localization with respect to the synapse [36]. In another study, QDs conjugated to EGF revealed a previously unreported retrograde transport of a receptor to the cell body [30] (Figure 8.1). By using a combination of EGF-QDs and visible fluorescent protein-tagged receptors, the authors investigated the relationship between the receptor cellular fate and its dimerization and interactions with other receptors. Another important difference between QDs and organic dyes for single molecule imaging lies in the blinking phenomenon, whereby the QD photoemission constantly switches between "bright" and "dark" states. Whilst this phenomenon provides a means of distinguishing single QDs from aggregates, it also represents a limitation, as QDs become spontaneously and momentarily unavailable for imaging. For example, one group utilized this blinking property to unequivocally identify single QD-conjugates (in comparison to small aggregates) and follow their migration with time [36].

The results of several studies have suggested that QDs might also be valuable in-vivo fluorescent labels. QDs were microinjected into a single Xenopus embryonic cell and used to follow its early stage division and development for several days. In particular, the QDs were shown to be confined to the progeny of this cell, and their presence did not cause any deleterious effects on cell growth and division [37]. In another study, five tumor cell populations labeled with different QD colors, using cationic encapsulation, were injected into the tail vein of mice and tracked (by multi-photon fluorescence microscopy) as they were extravasated into the lung tissues [33]. Again, no difference was observed between labeled and unlabeled cells. As the emission of QDs can be tuned via composition and size, nanocrystals that emit in the relative tissue-transparent region [near-infrared (NIR) region; ∼800–1100 nm) offer increased imaging depth and background

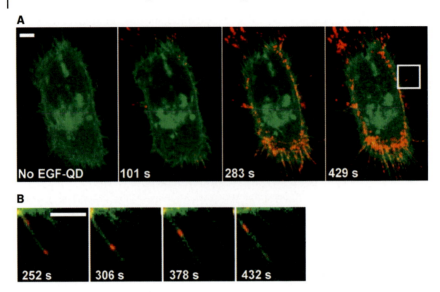

Fig. 8.1 Retrograde transport of EGF-QDs (red) on filipodia. (A) A431 cell expressing receptor erbB3-mCitrine (green); maximum intensity projection of four 0.5 μm confocal sections as a function of time. (B) Magnified image of filipodium indicated in the last panel of (A), showing uniform migration of the EGF-QDs toward the cell body with a velocity of ~10 nm s^{-1}. All scale bars = 5 μm. (Reproduced from Ref. [30].)

reduction. Indeed, the injection of NIR QDs permitted real-time mapping of sentinel lymph nodes in animal cancer surgery (Figure 8.2). Moreover, the study benefited from the prolonged imaging capacity offered by QDs, and from their relatively small hydrodynamic sizes which were optimal for lymph node retention [38]. Ultimately, this demonstration might potentially result in significant improvements in cancer surgery techniques.

Despite the remarkable progress made in developing the effective use of QDs to image cells and cellular processes, there remain several issues to be addressed, understood, and eventually solved. One major issue is the unavailability of a simple and general method that allows effective translocation of QDs across the cell membrane. Fixed cells are often easier to label than live cells because their membrane can be permeabilized without compromising the cellular architecture. This allows the specific labeling of a wide range of targets, by functionalizing the QDs with antibodies targeting cell-surface receptors, cytoskeleton components, or nuclear antigens [39], or with DNA for *in-situ* hybridization [40, 41]. The same level of success has not been achieved with live cells.

The second issue is related to toxicity, which has somewhat limited the enthusiasm for QD use despite the large potential, and this problem must be solved if

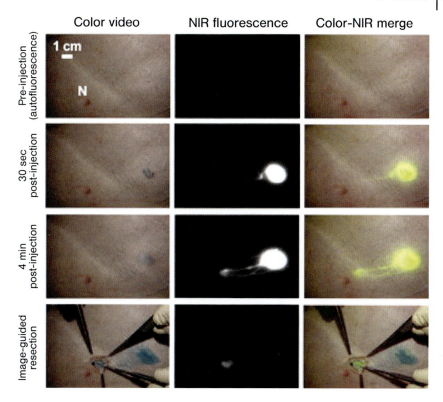

Fig. 8.2 Near-infrared (NIR) QD sentinel lymph node mapping in a pig. Images of the surgical field in a pig injected intradermally with 400 pmol of NIR QDs in the right groin. Four time points are shown from top to bottom: before injection (autofluorescence); at 30 s after injection; at 4 min after injection; and during image-guided resection.

Images are shown for each time point by color video (left), NIR fluorescence (middle), and color-NIR merge (right). To create the merged image, the NIR fluorescence image was pseudocolored lime green and superimposed on the color video image. (Reproduced from Ref. [38].)

QDs are to find wider use among *in-vivo* studies. This issue is based on concern that most commonly used QD cores are composed of metals that exhibit toxic effects (including Cd, Se, Hg, and Te ions). Leakage of such ions into the surrounding tissue and culture is thought to interfere with tissue and cell health and functions. However, these effects can be reduced by overcoating the core with other less-toxic metals and by adding a protective hydrophilic coating (e.g., a polymer shell). The development of high-quality QDs based on less-toxic materials is therefore highly desirable.

Another unresolved issue remains the clearance of the QDs from the body if used in live animals. The processing of injected QDs by organisms remains poorly understood; preliminary studies have shown that QDs may be cleared relatively quickly from the blood circulation, but they potentially accumulate in a variety of organs (e.g., liver, bone marrow, spleen), depending on their size and coating [42, 43].

8.1.2
Quantum Dots in Immuno- and FRET-Based Assays

Quantum dots also bring significant advantages to immunoassays, where their spectral characteristics allow the simultaneous detection of several targets in multiplexed sensing schemes. The use of QD–antibody conjugates has been demonstrated in direct and sandwich fluoroimmunoassays [44], and a simultaneous detection of four different toxins in single wells of a microtiter plate was realized with four different colored QDs [45]. Mixed toxin samples were first applied to capture antibody-coated wells, followed by exposure to specific antibodies coupled to different colored QDs. The resulting PL spectrum was then analyzed by linear spectral deconvolution to identify the toxin sample composition.

Fluorescence resonance energy transfer (FRET) between QD donors and organic dye-labeled biomolecules is particularly interesting as it allows modulation of the QD PL signal as a function of the target concentration, thus creating "active" fluorescence probes. Non-radiative energy transfer between colloidal QDs is well-documented in close-packed films [46, 47]. Subsequent investigations have since shown QDs to be excellent donors for various bio-inspired FRET-based studies. In particular, their broad absorption spectra in the blue/UV region allows excitation far from the acceptor absorption spectrum, and this reduces its direct excitation contribution to the PL signal. At the same time, their narrow PL spectra provide tunable spectral overlap and simplify data analysis. Furthermore, a QD can act as both energy donor and nano-scaffold for conjugating several acceptors, providing increased FRET efficiencies [48].

Different sensing schemes have been recently developed that utilize QDs as FRET donors. The most straightforward approach consists of labeling the target with a dye acceptor. Because the FRET process is on the nanometer scale, it only occurs when the target binds to the QD–receptor conjugate. For example, QDs conjugated to oligonucleotides have been used to monitor DNA replication in solution using FRET for signal transduction [49]. CdSe-ZnS QDs stabilized with mercaptopropionic acid were attached to a thiol-terminated DNA primer and incubated with the complementary sequence to induce hybridization. The addition of polymerase mixed with a complementary DNA sequence labeled with Texas Red allowed real-time monitoring of the dynamics of DNA replication through changes in the QD and dye emission signals. Replication of the DNA brings the dye progressively closer to the QD, resulting in improved FRET efficiency between the QD and proximal Texas Red acceptors. The obtained data indicate that the replication process is effectively complete after about 1 hour.

An alternative, recently developed sensing scheme employed FRET to carry out solution-phase monitoring of proteolytic enzyme activity. QDs were self-assembled with dye-labeled modular peptides engineered to include a polyhistidine tract for conjugation to CdSe-ZnS QDs capped with dihydrolipoic acid (DHLA), a protease-specific cleavage sequence, and a terminal cysteine group for dye labeling (Figure 8.3) [50]. In the absence of any proteases, the conjugation of several copies of dye-labeled peptides to the QDs resulted in a pronounced rate of FRET, which manifested as a loss of QD emission. Added protease was seen to cleave the peptide, causing the dye to diffuse away from the QD, thereby altering the rate of energy transfer and restoring the QD signal. Proteolytic activity was thus monitored for a variety of proteases, yielding quantitative kinetic parameters and mechanisms of enzymatic inhibition. These conjugates were also employed to detect the inhibition of protease activity in the presence of active inhibitors [50]. Proteases targeted in this demonstration included thrombin, collagenase, chymotrypsin and caspase-1, all with clinical or analytical use.

Another strategy utilizes FRET to detect competitive binding between a target and a dye-labeled analogue. In one demonstration, a bioconjugate specific for the disaccharide maltose was developed using DHLA-capped QDs self-assembled with maltose-binding protein (MBP) tagged with a terminal polyhistidine tract (MBP-His) [51]. An analogue sugar (β-cyclodextrin) was chemically attached to an organic quenching dye, which was then pre-bound to the MBP binding pocket prior to assembly of the bioconjugate. The close proximity of the dye to the QD donor created a favorable condition for efficient energy transfer, and this resulted in effective quenching of the QD PL. The addition of maltose (MBP substrate) to the solution competitively displaced β-cyclodextrin away from the MBP sugar-binding site and a substantial (concentration-dependent) recovery of the QD emission was measured. The response of the nanosensing assembly allowed direct measurement of the MBP-maltose dissociation constant, K_D, and showed that MBP effectively retains its native function when attached to the nanocrystals. Building on these results, a FRET-based sensing assembly targeting the explosive trinitrotoluene (TNT) in solution was also developed using a similar rationale [52]. Efficient rates of FRET combined with specificity for TNT was achieved using a single-chain antibody fragment (much smaller in size than full antibodies) that preferentially binds TNT (TNB2-45). As with the maltose sensor, an analogue substrate (TNB) covalently labeled with a quencher dye (BHQ10) was pre-bound to the antibody fragment prior to the assay. Due to favorable spectral overlap and small separation distance between the donor and acceptor, efficient energy transfer between the QD and proximal dye was measured. As TNT was added to the solution, it competed for binding to the antibody fragment and displaced the TNB-BHQ10 analogue; this altered the FRET interactions and translated to QD fluorescence recovery with increasing TNT concentration. Discrimination of the sensor was verified by testing the assembly against other analogues, and data showed that specificity for TNT was retained [52].

The utilization of DNA molecular beacons coupled to QDs has also been demonstrated [53, 54]. In the absence of a target, the DNA molecular beacons form a hairpin structure, bringing the dye-labeled end in close proximity to the QD and resulting in efficient QD quenching due to FRET interactions. In the presence of the target, the DNA strand undergoes a conformational change (to an open struc-

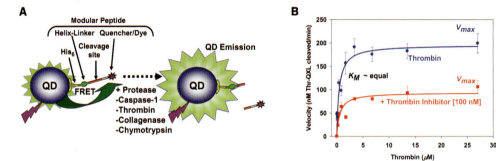

Fig. 8.3 (A) Schematic diagram of the self-assembled QD-peptide nanosensors; for clarity, only one peptide is shown. Dye-labeled modular peptides containing appropriate cleavage sequences are self-assembled onto the QD. FRET from the QD to the proximal acceptors quenches the QD PL. Specific protease cleaves the peptide and alters FRET signature. (B) Results of assaying a constant amount of QD-thrombin-quencher peptide substrate versus increasing thrombin concentration in the absence (blue) and presence (red) of 100 nM thrombin inhibitor. (Adapted from Ref. [50].)

ture) that increases the QD-dye acceptor separation distance, restoring the QD PL signal. For example, a QD-aptamer biosensor based on this concept has been demonstrated in the detection of thrombin, an important blood-clotting protein [54]. In these studies, the FRET efficiency measured for the QD-to-one-acceptor complex was smaller than those measured with organic dye donors, due to the large QD size, and this was compensated for by conjugating several molecular beacons per QD.

Finally, QDs hold great promise for use in developing sensing schemes based on single-particle FRET (spFRET), notably because of their high photobleaching threshold and strong resistance to degradation. Indeed, spFRET was recently applied to detect DNA hybridization at the single molecule level using QD–DNA conjugates. Two DNA sequences were designed as complementary sequences to different regions of the target DNA; the first sequence carried an acceptor dye, and the second a biotin group that allowed conjugation to commercially available streptavidin-coated QDs. The presence of target molecules initiated a simultaneous and parallel hybridization with both the probe and reporter sequences, which were subsequently conjugated to the QDs (see Figure 8.4) [55]. Single diffusing QD conjugates were then detected using confocal microscopy. The presence of the target sequence was characterized by measuring the QD-donor-dye-acceptor emission ratios, and their dependence on the experimental conditions. Several targets were able to bind to the same QD, which increased the overall FRET efficiency and resulted in a high local concentration of targets, facilitating their detection. Furthermore, as QDs could be excited far from the acceptor absorption spectrum, unbound reporter probes produced extremely low background levels. This resulted in a significantly increased sensitivity of the QD nanosensor compared to a conventional organic dye molecular beacon.

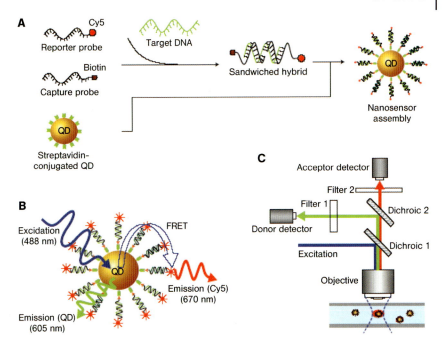

Fig. 8.4 Schematic of single-QD-based DNA nanosensors. (A) Conceptual scheme showing the formation of a sensing assembly in the presence of targets. (B) Fluorescence emission from Cy5 on illumination of QD caused by FRET between QD donor and the Cy5 acceptors in the assembly. (C) Experimental set-up. (Reproduced from Ref. [55].)

The main limiting factor in the performance of QD FRET donors lies with their size. Because FRET efficiency depends on the center-to-center separation between donor and acceptor, the use of thick polymer coating and/or multilayer conjugation strategies (e.g., based on streptavidin and biotin) often yields low FRET efficiencies. Successful QD FRET designs most often include very thin solubilization layers and direct attachment of bioreceptors to the QD surface [48–53].

FRET with QDs using the reverse format with organic dye and QD acceptors seems to be of limited value [56]. Moreover, strong QD direct excitation often overwhelms the contribution resulting from FRET. However, efficient energy transfer from bio- or chemi-luminescent biomolecule donors to QD acceptors has been reported in two instances [57, 58].

In the first study, QDs were conjugated to bioluminescent proteins *Renilla reniformis* luciferases. In the presence of its substrate coelenterazine, the protein experiences a transition into an excited state, and then transfers its energy nonradiatively to the proximal QD. The QD then relaxes to the ground state and emits a PL signal [57]. These findings open new possibilities for cellular and *in-vivo* imaging, as they combine the sensitivity of bioluminescence (absence of tissue autofluorescence) with the multiplex capability of QDs.

In the second study, horseradish peroxidase (HRP) conjugated with an anti-bovine serum albumin (BSA) antibody was added to QD–BSA conjugates [58]. Luminol, the chemiluminescent substrate of HRP, in the presence of the conjugates was able to transfer part of its excitation to the QDs. This energy transfer does not need an excitation light source, but the HRP-luminol reaction requires the presence of an oxidant, H_2O_2, which may represent a limitation for cell and tissue imaging.

8.2
Methods

8.2.1
Synthesis, Characterization, and Capping Strategies

During the early 1990s, it was shown by the Bawendi group [59] – and confirmed by others [60] – that CdSe QDs with a narrow size distribution (∼8–10% as made) can be prepared using the high-temperature reaction of *organometallic* precursors. Distributions can be further improved during post-reaction processing. This preparation followed on the early advances using micelles, and yielded materials with highly crystalline NPs and with improved photophysical properties [e.g., fluorescence quantum yield (QY) and characteristics of the emission spectra]. This reaction scheme initially employed dimethylcadmium ($CdMe_2$) and trioctylphosphine selenide (TOP:Se), diluted in trioctylphosphine (TOP) and their rapid injection into a hot (280–300 °C) coordinating solution of trioctylphosphine oxide (TOPO). Monitoring of the growth can be carried out using UV/visible spectroscopy. When the desired color/size (usually determined from the location of the first absorption peak) is reached, growth is arrested.

Subsequently, Peng and co-workers have further improved the reaction scheme and made it less dependent on the purity of the TOPO, and without use of the pyrophoric $CdMe_2$ precursor [61, 62]. In a series of studies, these authors and others have eventually outlined the importance of impurities to the reaction progress, and shown that these impurities can be externally controlled. For this route, high-purity TOPO and controlled amounts of cadmium coordinating ligands [such as hexylphosphonic acid (HPA) or tetradecylphosphonic acid (TDPA)] and cadmium compounds such as cadmium oxide (CdO) or cadmium acetate [$Cd(Ac)_2$] are used in the reaction; TOP:Se remains the selenium precursor.

During the mid-1990s – and borrowing from the concepts of bandgap engineering developed in semiconductors physics – it was demonstrated that passivating the native QD cores with an additional layer made from a wider band gap material(s) (to create core-shell nanocrystals) could improve the surface quality and dramatically enhance the fluorescence quantum yield. The optimal conditions for creating strongly fluorescent core-shell QDs was detailed in the seminal studies of Hines et al. [63], Dabbousi et al. [64], and Peng et al. [65]. At this point it might be added that the reaction schemes using metal-complexes (e.g., zinc acetate) as precursors for overcoating have also been attempted by several groups following Peng's findings on the core growth.

In order to obtain material with a lower size dispersity, the growth of cores or core-shell nanocrystals is followed by size-selective precipitation using "bad" solvents (e.g., methanol), which also removes impurities and precipitated metals from the reaction solution [1, 59]. Repeated precipitations can further reduce the overall size distribution of the QDs.

8.2.2
Water-Solubilization Strategies

Unless initially grown in an aqueous environment (e.g., using inverse micelle growth or co-precipitation for some of the reported CdTe nanoparticles), highly luminescent QDs, prepared using high-temperature routes (capped with TOP/TOPO ligands), are intrinsically hydrophobic in nature. Consequently, any use in biology requires that they are made hydrophilic, and several water-solubilization strategies have been developed with this in mind. The strategies can be divided into two main categories:

- An approach which involves replacing the native TOP/TOPO capping shell with bifunctional ligands terminated with hydrophilic end functions, typically via thiol–metal interaction. Typical examples of such ligands include mercaptoacetic acid (MAA) [31], DHLA and poly(ethylene-glycol)-terminated dihydrolipoic acid (DHLA-PEG) ligands [35, 66].
- A method which uses encapsulation of the native TOP/TOPO-capped QDs within amphiphilic polymer shells or lipid micelles [37, 39]. These polymers usually contain hydrophobic carbon chains that interdigitate with the TOP/TOPO and a hydrophilic block that provides water solubility.

In both methods, hydrophilicity is facilitated by the presence of charged groups (e.g., carboxylic acids) and/or polyethylene glycol (PEG) chains. When used, the advantages of each strategy must be carefully weighed against the drawbacks, as each approach is undoubtedly suitable for one range of uses but not for others [6]; here, the long-term stability and high PL quantum yields should serve as some guiding factors.

8.2.3
Conjugation Strategies

Conjugation strategies of biomolecular receptors to QDs can essentially be divided into three groups:

- Use of the EDC (1-ethyl-3-(3-dimethylaminopropyl) carbodiimide) condensation to react carboxy groups on the QD surface with amines [16].
- Adsorption or noncovalent self-assembly using engineered proteins [66].
- Metal-affinity driven self-assembly using thiolated peptides or polyhistidine (His) residues [49, 51].

As discussed for the water-solubilization above, each conjugation method has its advantages and limitations. For example, EDC condensation, when applied to QDs capped with thiol-alkyl-COOH ligands, often produces intermediate aggregates due to poor QD stability in neutral and acidic buffers. Nonetheless, when applied to QDs encapsulated with polymeric shells bearing functional groups, it can produce functional but large conjugates with less control over the number of biomolecules per QD-bioconjugate. This approach was used by Invitrogen (formerly QD Corp.) to prepare QD–streptavidin conjugates having ~20 proteins. When using these commercially available QD conjugates to carry out additional assays using biotin–avidin interactions, care must be paid to the fact that these QDs will bind all biotinylated proteins indiscriminately. The direct attachment of proteins/peptides to the QD (using dative thiol-bonding between QD surface ions and thiol or cysteine-terminated proteins/DNAs, or metal-affinity interactions using polyhistidine residues) can reduce aggregation, but still requires that the bioreceptor be engineered with the desired functions before use. We, for example, have shown that by using polyhistidine-appended proteins, control over the bioconjugate valence can be exerted through self-assembly [48, 51].

8.3
Future Outlook

The recent progress made for QD use in biology has clearly proved that they can offer advantages not realized using organic dye labeling, notably for applications such as *in-vivo* cellular and tissue imaging. QDs are very well suited to the development of FRET-based (single and multi-channel) assays, and these inorganic fluorophores have, by far, not yet exhausted their potential for improving the present range of assays and expanding the development of intracellular and tissue imaging. Cellular (and tissue) imaging and sensing based on simple one and two-photon fluorescence and FRET represent the main areas where QDs might encounter substantial development and expansion. Indeed, the development of multiplexed (up to 10 colors) assays, both in solution and inside cell cultures, could benefit from QD high photobleaching thresholds and resistance to degradation. Issues associated with toxicity (e.g., metallic, magnetic and semiconducting NPs) will certainly remain a pressing problem that all scientists active in the field of nanomaterials will continue to investigate and try to overcome. Nonetheless, a number of major hurdles must be overcome before progress is achieved, and these can be summarized as follows:

- To improve the surface properties of QDs, both in organic and in buffer solutions. This will entail the expansion of available surface-functionalization techniques, and potentially consolidate them into an easy-to-use scheme to provide hydrophilic QDs that are stable over a wide range of biologically relevant conditions (pH, counterion excess, etc.).
- The conjugation of QDs to biomolecules, which will require a coordinated effort to develop a simple, reproducible conjugation scheme that involves reactive surface groups, to produce compact conjugates that eventually may have multiple biological functions.
- The development of reproducible and effective means to deliver QD cargos inside live cells, coupled with an ability to direct them towards targeted compartments within the cytoplasm.

Acknowledgments

The authors acknowledge NRL and the Office of Naval Research (ONR grant # N0001406WX20097) for support.

References

1 Murray, C. B., Kagan, C. R., Bawendi, M. G. (2000) Synthesis and characterization of monodisperse nanocrystals and close-packed nanocrystal assemblies. *Annu. Rev. Mater. Sci.* **30**, 545–610.

2 Gaponenko, S. V., Woggon, U. (2004) II–VI semiconductor quantum dots – nanocrystals, In: Klingshirn, C. (Ed.), *Optical properties of semiconductor nanostructures.* Springer-Verlag; Vol. 34C2, pp. 284–347.

3 Efros, A. L., Rosen, M. (2000) The electronic structure of semiconductor nanocrystals. *Annu. Rev. Mater. Sci.* **30**, 475–521.

4 Alivisatos, P. (2004) The use of nanocrystals in biological detection. *Nat. Biotech.* **22**, 47–52.

5 Michalet, X., Pinaud, F. F., Bentolila, L. A., Tsay, J. M., Doose, S., Li, J. J., Sundaresan, G., Wu, A. M., Gambhir, S. S., Weiss, S. (2005) Quantum dots for live cells, in vivo imaging, and diagnostics. *Science* **307**, 538–544.

6 Medintz, I. L., Uyeda, H. T., Goldman, E. R., Mattoussi, H. (2005) Quantum dot bioconjugates for imaging, labelling and sensing. *Nat. Mater.* **4**, 435–446.

7 Borrelli, N. F., Hall, D. W., Holland, H. J., Smith, D. W. (1987) Quantum confinement effects of semiconducting microcrystallites in glass. *J. Appl. Phys.* **61**, 5399–5409.

8 Hall, D. W., Borrelli, N. F. (1988) Absorption saturation in commercial and quantum-confined $CdSe_xS_{1-x}$-doped glasses. *J. Opt. Soc. Am. B* **5**, 1650–1654.

9 Schlamp, M. C., Peng, X. G., Alivisatos, A. P. (1997) Improved

efficiencies in light emitting diodes
made with CdSe(CdS) core/shell type
nanocrystals and a semiconducting
polymer. *J. Appl. Phys.* **82**, 5837–5842.

10 Mattoussi, H., Radzilowski, L. H.,
Dabbousi, B. O., Thomas, E. L.,
Bawendi, M. G., Rubner, M. F. (1998)
Electroluminescence from hetero-
structures of poly(phenylene vinylene)
and inorganic cdse nanocrystals. *J.
Appl. Phys.* **83**, 7965–7974.

11 Coe, S., Woo, W. K., Bawendi, M.,
Bulovic, V. (2002) Electroluminescence from single monolayers of
nanocrystals in molecular organic
devices. *Nature* **420**, 800–803.

12 Steckel, J. S., Coe-Sullivan, S.,
Bulovic, V., Bawendi, M. G. (2003)
1.3 μm to 1.55 μm tunable electroluminescence from PbSe quantum
dots embedded within an organic
device. *Adv. Mater.* **15**, 1862–1866.

13 Greenham, N. C., Peng, X. G.,
Alivisatos, A. P. (1996) Charge
separation and transport in
conjugated-polymer/semiconductor-
nanocrystal composites studied by
photoluminescence quenching and
photoconductivity. *Phys. Rev. B* **54**,
17628–17637.

14 Hu, J. T., Li, L. S., Yang, W. D.,
Manna, L., Wang, L. W., Alivisatos,
A. P. (2001) Linearly polarized
emission from colloidal semiconductor quantum rods. *Science* **292**, 2060–
2063.

15 Schrock, E., duManoir, S., Veldman,
T., Schoell, B., Wienberg, J.,
Ferguson-Smith, M. A., Ning, Y.,
Ledbetter, D. H., BarAm, I., Soenksen,
D., Garini, Y., Ried, T. (1996) Multi-
color spectral karyotyping of human
chromosomes. *Science* **273**, 494–497.

16 Hermanson, G. T. (1996) *Bioconjugate
techniques.* Academic Press, London.

17 Roederer, M., DeRosa, S., Gerstein,
R., Anderson, M., Bigos, M., Stovel,
R., Nozaki, T., Parks, D., Herzenberg,
L. (1997) 8 color, 10-parameter flow
cytometry to elucidate complex
leukocyte heterogeneity. *Cytometry* **29**,
328–339.

18 Lobenhofer, E. K., Bushel, P. R.,
Afshari, C. A., Hamadeh, H. K.

(2001) Progress in the application of
DNA microarrays. *Environm. Health
Perspect.* **109**, 881–891.

19 Bruchez, M., Moronne, M., Gin, P.,
Weiss, S., Alivisatos, A. P. (1998)
Semiconductor nanocrystals as
fluorescent biological labels. *Science*
281, 2013–2016.

20 Henglein, A. (1982) Photochemistry
of colloidal cadmium-sulfide. 2.
Effects of adsorbed methyl viologen
and of colloidal platinum. *J. Phys.
Chem.* **86**, 2291–2293.

21 Mikulec, F. V. (1999) *Semiconductor
nanocrystal colloids: Manganese dopes
cadmium selenide, (core) shell composites
for biological labeling, and highly
fluorescent cadmium telluride.* PhD
Dissertation. Massachusetts Institute
of Technology.

22 Hines, M. A., Guyot-Sionnest, P.
(1998) Bright UV-blue luminescent
colloidal ZnSe nanocrystals. *J. Phys.
Chem. B* **102**, 3655–3657.

23 Weller, H., Schmidt, H. M., Koch, U.,
Fojtik, A., Baral, S., Henglein, A.,
Kunath, W., Weiss, K., Dieman, E.
(1986) Photochemistry of semiconductor colloids. 14. Photochemistry of
colloidal semiconductors – onset of
light-absorption as a function of size
of extremely small CdS particles.
Chem. Phys. Lett. **124**, 557–560.

24 Rossetti, R., Nakahara, S., Brus, L. E.
(1983) Quantum size effects in the
redox potentials, resonance Raman-
spectra, and electronic-spectra of CdS
crystallites in aqueous-solution.
J. Chem. Phys. **79**, 1086–1088.

25 Pietryga, J. M., Schaller, R. D.,
Werder, D., Stewart, M. H., Klimov,
V. I., Hollingsworth, J. A. (2004)
Pushing the band gap envelope: Mid-
infrared emitting colloidal PbSe
quantum dots. *J. Am. Chem. Soc.* **126**,
11752–11753.

26 Jaiswal, J. K., Mattoussi, H., Mauro,
J. M., Simon, S. M. (2003) Long-term
multiple color imaging of live cells
using quantum dot bioconjugates.
Nat. Biotech. **21**, 47–51.

27 Lagerholm, B. C., Wang, M., Ernst,
L. A., Ly, D. H., Liu, H., Bruchez,
M. P., Waggoner, A. S. (2004)

Multicolor coding of cells with cationic peptide coated quantum dots. *Nano Lett.* **4**, 2019–2022.

28 Delehanty, J. B., Medintz, I. L., Pons, T., Brunel, F. M., Dawson, P. E., Mattoussi, H. (2006) Self-assembled quantum dot-peptide bioconjugates for selective intracellular delivery. *Bioconj. Chem.* **17**, 920–927.

29 Bharali, D. J., Lucey, D. W., Jayakumar, H., Pudavar, H. E., Prasad, P. N. (2005) Folate-receptor-mediated delivery of InP quantum dots for bioimaging using confocal and two-photon microscopy. *J. Am. Chem. Soc.* **127**, 11364–11371.

30 Lidke, D. S., Nagy, P., Heintzmann, R., Arndt-Jovin, D. J., Post, J. N., Grecco, H. E., Jares-Erijman, E. A., Jovin, T. M. (2004) Quantum dot ligands provide new insights into erbB/HER receptor-mediated signal transduction. *Nat. Biotech.* **22**, 198–203.

31 Chan, W. C. W., Nie, S. M. (1998) Quantum dot bioconjugates for ultrasensitive nonisotopic detection. *Science* **281**, 2016–2018.

32 Derfus, A. M., Chan, W. C. W., Bhatia, S. N. (2004) Intracellular delivery of quantum dots for live cell labeling and organelle tracking. *Adv. Mater.* **16**, 961–966.

33 Voura, E. B., Jaiswal, J. K., Mattoussi, H., Simon, S. M. (2004) Tracking metastatic tumor cell extravasation with quantum dot nanocrystals and fluorescence emission-scanning microscopy. *Nat. Med.* **10**, 993–998.

34 Courty, S., Luccardini, C., Bellaiche, Y., Cappello, G., Dahan, M. (2006) Tracking individual kinesin motors in living cells using single quantum-dot imaging. *Nano Lett.* **6**, 1491–1495.

35 Uyeda, H. T., Medintz, I. L., Jaiswal, J. K., Simon, S. M., Mattoussi, H. (2005) Synthesis of compact multidentate ligands to prepare stable hydrophilic quantum dot fluorophores. *J. Am. Chem. Soc.* **127**, 3870–3878.

36 Dahan, M., Levi, S., Luccardini, C., Rostaing, P., Riveau, B., Triller, A. (2003) Diffusion dynamics of glycine

receptors revealed by single-quantum dot tracking. *Science* **302**, 442–445.

37 Dubertret, B., Skourides, P., Norris, D. J., Noireaux, V., Brivanlou, A. H., Libchaber, A. (2002) In vivo imaging of quantum dots encapsulated in phospholipid micelles. *Science* **298**, 1759–1762.

38 Kim, S., Lim, Y. T., Soltesz, E. G., De Grand, A. M., Lee, J., Nakayama, A., Parker, J. A., Mihaljevic, T., Laurence, R. G., Dor, D. M., Cohn, L. H., Bawendi, M. G., Frangioni, J. V. (2004) Near-infrared fluorescent type II quantum dots for sentinel lymph node mapping. *Nat. Biotech.* **22**, 93–97.

39 Wu, X., Liu, H., Liu, J., Haley, K. N., Treadway, J. A., Larson, J. P., Ge, N., Peale, F., Bruchez, M. P. (2003) Immunofluorescent labeling of cancer marker her2 and other cellular targets with semiconductor quantum dots. *Nat. Biotech.* **21**, 41–46.

40 Chan, P. M., Yuen, T., Ruf, F., Gonzalez-Maeso, J., Sealfon, S. C. (2005) Method for multiplex cellular detection of mRNAs using quantum dot fluorescent in situ hybridization. *Nucleic Acids Res.* **33**, e161.

41 Pathak, S., Choi, S. K., Arnheim, N., Thompson, M. E. (2001) Hydroxylated quantum dots as luminescent probes for in situ hybridization. *J. Am. Chem. Soc.* **123**, 4103–4104.

42 Hardman, R. (2006) A toxicologic review of quantum dots: Toxicity depends on physicochemical and environmental factors. *Environm. Health Perspect.* **114**, 165–172.

43 Nel, A., Xia, T., Madler, L., Li, N. (2006) Toxic potential of materials at the nanolevel. *Science* **311**, 622–627.

44 Goldman, E. R., Anderson, G. P., Tran, P. T., Mattoussi, H., Charles, P. T., Mauro, J. M. (2002) Conjugation of luminescent quantum dots with antibodies using an engineered adaptor protein to provide new reagents for fluoroimmunoassays. *Anal. Chem.* **74**, 841–847.

45 Goldman, E. R., Clapp, A. R., Anderson, G. P., Uyeda, H. T., Mauro, J. M., Medintz, I. L.,

Mattoussi, H. (2004) Multiplexed toxin analysis using four colors of quantum dot fluororeagents. *Anal. Chem.* **76**, 684–688.

46 Crooker, S. A., Hollingsworth, J. A., Tretiak, S., Klimov, V. I. (2002) Spectrally resolved dynamics of energy transfer in quantum-dot assemblies: Towards engineered energy flows in artificial materials. *Phys. Rev. Lett.* **89**, 186802.

47 Kagan, C. R., Murray, C. B., Nirmal, M., Bawendi, M. G. (1996) Electronic energy transfer in CdSe quantum dot solids. *Phys. Rev. Lett.* **76**, 1517–1520.

48 Clapp, A. R., Medintz, I. L., Mauro, J. M., Fisher, B. R., Bawendi, M. G., Mattoussi, H. (2004) Fluorescence resonance energy transfer between quantum dot donors and dye-labeled protein acceptors. *J. Am. Chem. Soc.* **126**, 301–310.

49 Patolsky, F., Gill, R., Weizmann, Y., Mokari, T., Banin, U., Willner, I. (2003) Lighting-up the dynamics of telomerization and DNA replication by CdSe-ZnS quantum dots. *J. Am. Chem. Soc.* **125**, 13918–13919.

50 Medintz, I. L., Clapp, A. R., Brunel, F. M., Tiefenbrunn, T., Uyeda, H. T., Chang, E. L., Deschamps, J. R., Dawson, P. E., Mattoussi, H. (2006) Proteolytic activity monitored by fluorescence resonance energy transfer through quantum-dot–peptide conjugates. *Nat. Mater.* **5**, 581–589.

51 Medintz, I. L., Clapp, A. R., Mattoussi, H., Goldman, E. R., Fisher, B., Mauro, J. M. (2003) Self-assembled nanoscale biosensors based on quantum dot FRET donors. *Nat. Mater.* **2**, 630–638.

52 Goldman, E. R., Medintz, I. L., Whitley, J. L., Hayhurst, A., Clapp, A. R., Uyeda, H. T., Deschamps, J. R., Lassman, M. E., Mattoussi, H. (2005) A hybrid quantum dot-antibody fragment fluorescence resonance energy transfer-based TNT sensor. *J. Am. Chem. Soc.* **127**, 6744–6751.

53 Kim, J. H., Morikis, D., Ozkan, M. (2004) Adaptation of inorganic quantum dots for stable molecular beacons. *Sens. Act. B* **102**, 315–319.

54 Levy, M., Cater, S. F., Ellington, A. D. (2005) Quantum-dot aptamer beacons for the detection of proteins. *Chembiochem* **6**, 2163–2166.

55 Zhang, C., Yeh, H., Kuroki, M. T., Wang, T. (2005) Single-quantum-dot-based DNA nanosensor. *Nat. Mater.* **4**, 826–831.

56 Clapp, A. R., Medintz, I. L., Fisher, B. R., Anderson, G. P., Mattoussi, H. (2005) Can luminescent quantum dots be efficient energy acceptors with organic dye donors? *J. Am. Chem. Soc.* **127**, 1242–1250.

57 So, M.-K., Xu, C., Loening, A. M., Gambhir, S. S., Rao, J. (2006) Self-illuminating quantum dot conjugates for in vivo imaging. *Nat. Biotech.* **29**, 339–343.

58 Huang, H., Li, L., Qian, H., Dong, C., Ren, J. (2006) A resonance energy transfer between chemiluminescent donors and luminescent quantum-dots as acceptors (CRET). *Angew. Chem. Int. Ed.* **45**, 5140–5143.

59 Murray, C. B., Norris, D. J., Bawendi, M. G. (1993) Synthesis and characterization of nearly monodisperse CdE (E = S, Se, Te) semiconductor nanocrystallites. *J. Am. Chem. Soc.* **115**, 8706–8715.

60 Katari, J. E. B., Colvin, V. L., Alivisatos, A. P. (1994) X-ray photoelectron-spectroscopy of CdSe nanocrystals with applications to studies of the nanocrystal surface. *J. Phys. Chem.* **98**, 4109–4117.

61 Peng, Z. A., Peng, X. G. (2001) Formation of high-quality CdTe, CdSe, and CdS nanocrystals using CdO as precursor. *J. Am. Chem. Soc.* **123**, 183–184.

62 Qu, L. H., Peng, Z. A., Peng, X. G. (2001) Alternative routes toward high quality CdSe nanocrystals. *Nano Lett.* **1**, 333–337.

63 Hines, M. A., Guyot-Sionnest, P. (1996) Synthesis and characterization of strongly luminescing ZnS-capped CdSe nanocrystals. *J. Phys. Chem.* **100**, 468–471.

64 Dabbousi, B. O., Rodriguez-Viejo, J., Mikulec, F. V., Heine, J. R., Mattoussi, H., Ober, R., Jensen, K. F., Bawendi, M. G. (1997) (CdSe)ZnS core-shell quantum dots: Synthesis and characterization of a size series of highly luminescent nanocrystal-lites. *J. Phys. Chem. B* **101**, 9463–9475.

65 Peng, X. G., Schlamp, M. C., Kadavanich, A. V., Alivisatos, A. P. (1997) Epitaxial growth of highly luminescent CdSe/CdS core/shell nanocrystals with photostability and electronic accessibility. *J. Am. Chem. Soc.* **119**, 7019–7029.

66 Mattoussi, H., Mauro, J. M., Goldman, E. R., Anderson, G. P., Sundar, V. C., Mikulec, F. V., Bawendi, M. G. (2000) Self-assembly of CdSe-ZnS quantum dot bioconjugates using an engineered recombinant protein. *J. Am. Chem. Soc.* **122**, 12142–12150.

9
Nanoscale Localized Surface Plasmon Resonance Biosensors

Katherine A. Willets, W. Paige Hall, Leif J. Sherry, Xiaoyu Zhang,
Jing Zhao, and Richard P. Van Duyne

9.1
Overview

The interaction of light with metallic nanoparticles (NPs) has aroused significant interest in recent years due to demonstrated applications in nanoscale lithography [1–3], surface-enhanced spectroscopies [4–9], and chemical and biological sensing applications [10–14]. In each of these examples, light can be localized, manipulated, and amplified on the nanometer scale by exciting a collective electron oscillation in the metallic NPs, known as localized surface plasmon resonance (LSPR). By controlling the size, shape, material and local dielectric environment of the NPs, the resonance condition of the plasmon can be tuned throughout the visible and near-infra-red (IR) ranges [15–21]. It is the latter property – namely the local dielectric environment – which forms the basis of LSPR-based biosensing experiments [10, 11, 14, 22–28].

Before discussing LSPR and its applications in detail, the basic physics of plasmons will be briefly reviewed [20]. When a metal surface (either bulk or nanoscale) is irradiated with electromagnetic radiation (light) of the appropriate frequency, a coherent oscillation of the metal's conduction electrons is induced orthogonal to the propagation direction of the light. This oscillation is a "plasmon", and it can be analyzed as fluctuations in the metal's surface electron density or, in other words, as a longitudinal electron density wave. In the case of metallic NPs with sizes less than the wavelength of the incident radiation, this plasmon is localized on the surface of the NP (in contrast to bulk metal, where the plasmon can propagate along the surface plane and evanescently decay perpendicular to the plane). This principle is illustrated in Figure 9.1, where the electron cloud of a metallic NP oscillates in phase with the incident electromagnetic field.

The wavelength at which this resonance occurs depends on a number of factors related to both the NP itself, as well as its local environment. Typically, gold and silver NPs are chosen for most LSPR applications, although other metals such as

Nanobiotechnology II. Edited by Chad A. Mirkin and Christof M. Niemeyer
Copyright © 2007 WILEY-VCH Verlag GmbH & Co. KGaA, Weinheim
ISBN: 978-3-527-31673-1

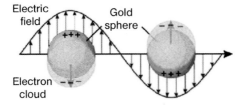

Fig. 9.1 Illustration of the oscillation of conduction band electrons of a gold nanoparticle with an electromagnetic field, resulting in a localized surface plasmon.

aluminum and copper can also support plasmons [29, 30]. The size and shape of these NPs also dictates the resonance condition; for example, a silver cube will have a plasmon resonance red shifted relative to a silver sphere of the same volume [7]. Lastly, the plasmon resonance is affected by the local environment of the NP – either through the bulk (solvent) refractive index or through the adsorption of some species to the NP surface [24, 25, 31–33]. By understanding the contributions from each of these factors, it should be possible to design better LSPR sensors for biosensing applications.

In order to provide a more quantitative description of the relationships described above, Mie theory – which models the extinction of a single sphere of arbitrary material – is often used [34]. Mie theory analytically solves Maxwell's equations with appropriate boundary conditions in spherical coordinates. In the dipole limit, where the dimension of the sphere is much smaller than the incident wavelength ($a \ll \lambda$), the extinction cross-section, $E(\lambda)$, of a sphere can be estimated by the following equation:

$$E(\lambda) = \frac{24\pi^2 a^3 \varepsilon_m^{3/2}}{\lambda \ln(10)} \left[\frac{\varepsilon_i}{\left(\varepsilon_r + 2\varepsilon_m\right)^2 + \varepsilon_i^2} \right] \tag{1}$$

where a is the radius of the sphere, ε_m is the dielectric constant of the medium surrounding the sphere, λ is the wavelength of the absorbing radiation, and ε_r, ε_i are the real and imaginary portions of the sphere's dielectric function, respectively. From this primitive estimation, it can be seen that the extinction of a single sphere depends on its size (a), material ($\varepsilon_r, \varepsilon_i$), and the surrounding environment (ε_m). For non-spherical particles of arbitrary shape, the extinction cross-section also depends on their geometry, and this can be modeled by substituting the term $\chi\varepsilon_m$ for $2\varepsilon_m$ in Eq. (1), where χ contains information about the shape and aspect ratio of a particle of arbitrary geometry [21]. Typically, for these more complicated structures, numerical methods to solve Maxwell's equation are used in order to model the optical properties of objects with arbitrary shapes in various dielectric environments [20]. Two of the most commonly used numerical methods are the discrete dipole approximation (DDA), and the finite-difference time-domain (FDTD) method [35, 36]. In these two approaches, the particle of interest

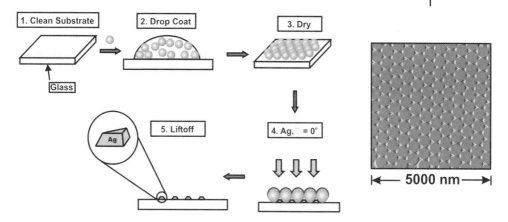

Fig. 9.2 Left: Nanosphere lithographic fabrication of nanoparticle arrays. Right: Atomic force microgram of nanoparticle array fabricated with nanosphere lithography.

is divided into finite elements. In the DDA method, Maxwell's equations are solved in the frequency domain, whereas in the FDTD method they are solved in the time domain. Using these two numerical methods on NPs of various shapes and sizes leads to different predicted resonances.

Because different shapes can tune the LSPR resonance, a great deal of effort has been focused on producing NPs of different geometries. In fact, a large number of protocols has been published for synthesizing both Au and Ag NPs using various methods, often in the presence of stabilizing surfactants [18, 37–43]; as these syntheses become more refined, particles with well-defined shapes and narrow size distributions can be produced. Lithographic techniques, such as electron-beam [44] or nanosphere lithography (NSL) [19], have also been employed in the fabrication of NPs, especially for large-scale arrays. Figure 9.2 (right panel) shows an example of an array of nanoscale triangles produced using NSL. In NSL, a hexagonally close-packed array of polymer spheres is used as a mask through which metal can be deposited (see Figure 9.2). For sensing experiments, such NP arrays are often preferred because they allow a large surface area to be interrogated using a commercially available UV-visible spectrometer; however, single NP LSPR spectroscopy is also possible and can offer zeptomole sensitivity [11].

In addition to designing NPs with specific LSPR resonances, as described above, it is also important for the NPs to provide a large response to changes in their local environment. This response can be modeled using Eq. (2), which relates the shift in the maximum wavelength of the LSPR resonance ($\Delta\lambda_{max}$) to the presence of an adsorbed species [45]:

$$\Delta\lambda_{max} = m\Delta n\left[1 - \exp\left(\frac{-2d}{l_d}\right)\right] \tag{2}$$

Here, m is the refractive index response of the NPs, Δn is the change in refractive index induced by the adsorbate, d is the effective adsorbate layer thickness, and l_d is the characteristic electromagnetic field decay length (approximated as an exponential decay). The ability to sense the binding of an adsorbate to the NP surface through a shift in the LSPR resonance is the key principle behind most biosensing experiments, and thus it is critical to design NP systems that offer both large refractive index responses (m) and short-range electromagnetic field enhancements (small l_d). It is this latter term that provides the biggest advantage in using LSPR versus traditional SPR techniques [28], due to the fact that fields can be highly localized and confined to small volumes around the NPs.

The remainder of this chapter will focus on methods related to LSPR spectroscopy for biosensing. Initially, the different NP geometries for both array and single particle LSPR experiments are discussed, emphasizing the LSPR response to changes in the local environment. Subsequently, an example in which LSPR spectroscopy has been applied to biosensing – specifically, the detection of an Alzheimer's disease biomarker – will also be provided. Finally, the outlook for LSPR spectroscopy for biosensing will be examined.

9.2
Methods

9.2.1
Nanofabrication of Materials for LSPR Spectroscopy and Sensing

As described above, there are a number of factors that determine the LSPR properties of metallic NPs; hence, choosing the correct NP system for biosensing requires an understanding of how NPs of different morphologies respond to controlled changes in their local environment. This is typically characterized by the refractive index sensitivity, m [see Eq. (2)], although a new figure of merit (FOM) has recently been introduced to describe the performance of single NPs as sensors of environmental change [46].

$$FOM = \frac{m(eV\ RIU^{-1})}{FWHM(eV)} \quad (3)$$

By normalizing m to the full-width-half-maximum (FWHM) of the spectral peak, NPs of different shapes and sizes can be directly compared.

Recent studies have highlighted several new strategies and techniques for improvements in the fabrication and characterization of both NP arrays and single NPs for LSPR. For example, the NSL-fabricated triangles described above can be electrochemically oxidized to tune both their size and shape on a length scale ranging from ~1 nm to several tens of nanometers [47]. The particular power of this approach is that the triangles are preferentially oxidized: first at the bottom edges, then at the triangular tips, and finally from the top face, allowing the response of LSPR to changes in morphology to be directly correlated. Alternatively, a novel technique known as atomic layer deposition (ALD) is available for determining the precise distance-dependence of the local field enhancement [i.e., l_d in Eq. (2)] [48]. Atomic layers of Al_2O_3 are deposited on NSL-fabricated arrays, and the shift in the LSPR spectrum is measured as a function of film thickness. Finally, electron beam lithography is available to determine the effect of diffractive coupling on the peak position and linewidth of LSPR spectra. By fabricating one-dimensional (1-D) cylindrical arrays with various interparticle spacing, this coupling can be observed through the emergence of a narrow feature in the extinction spectrum [49]. Because each of these techniques offers precision control over NP structure and local environment, parameters affecting the plasmon resonance can be studied in detail.

Beyond using new methods to better understand the relationships between particle shape, local environment, coupling and the LSPR spectrum, it is also important to develop new LSPR materials for use in biosensing experiments. While this has been a field of active research, with new geometries frequently reported, here we describe three recent systems that have been characterized in detail. In particular, the LSPR response of these systems to changes in the local environment – whether through a bulk refractive index change or interparticle coupling – are described for each system.

9.2.1.1 Film Over Nanowells
Nanohole or nanowell arrays have recently begun to be analyzed as a new plasmonic construct since the discovery of enhanced transmission through sub-wavelength apertures [50–52]. Nanowell structures have been successfully fabricated using reactive ion etching (RIE) through a polystyrene mask (in analogy to the mask used for NSL), followed by thermal vapor deposition of a 50-nm silver film (Figure 9.3) [53]. The thickness of the Ag film (d_m) was selected to be at least 50 nm to allow for efficient reflectance from the surface. When substrates are not optically transparent, the wavelength associated with minimum reflectivity (λ_{min}) provides an alternative measure of the LSPR λ_{max}. This study has revealed that

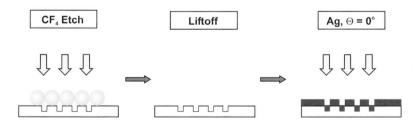

Fig. 9.3 Schematic illustration of the preparation of the film over nanowell surfaces, starting from a sphere mask as shown in Figure 9.2 (steps 1–3).

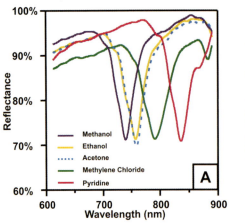
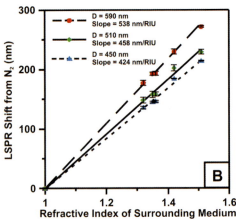

Fig. 9.4 (A) A collection of reflectance spectra of an Ag film over nanowell surface in different solvents (D = 590 nm, d_m = 50 nm). (B) Plots of the [λ_{min}(solvent) − λ_{min}(dry nitrogen)] versus refractive index of the solvent for three nanosphere sizes: D = 450, 510, and 590 nm. Each datum point represents the average value obtained from at least three surfaces; error bars show standard deviations. For all surface preparations, d_m = 50 nm and etching time t_e = 10 min. (Reproduced with permission from Ref. [53]; © 2005 American Chemical Society.)

these nanowells have both extremely narrow plasmon resonances and very strong wavelength sensitivity to the external dielectric constant (~500 nm RIU^{-1}) [53]. The effect of external dielectric media on the plasmon peak was studied by altering the surrounding solvent and monitoring the LSPR changes (see Figure 9.4A) [53]. Figure 9.4B shows plots of [λ_{min}(solvent) − λ_{min}(N$_2$)] for film over nanowell substrates as a function of the refractive index of the surrounding medium [53]. Within this range of refractive index units (RIU), the data points for the surfaces fabricated using the same size polystyrene nanosphere can be fitted well to a linear regression. It is found that the film over nanowell surface using the largest sphere (D = 590 nm) is the most sensitive to changes in the surrounding refractive index, followed by D = 510 nm, then D = 450 nm. For the most sensitive film over nanowell surfaces (D = 590 nm), the linear regression analysis yielded a refractive index sensitivity [i.e., m from Eq. (2)] of 538 nm RIU^{-1}; this means that every change of 0.002 in the refractive index of the solvent will produce a change in the peak position of approximately 1 nm.

9.2.1.2 Solution-Phase NSL-Fabricated Nanotriangles

A novel technique has recently been reported to produce monodisperse solution-phase NPs by releasing NSL-fabricated surface-confined NPs into solution [54]. The fabrication procedure is illustrated schematically in Figure 9.5 (upper panel). Surface-bound NSL-fabricated NPs are incubated in an alkanethiol solution to form a self-assembled monolayer (SAM) on the NPs (steps 1 and 2). Following

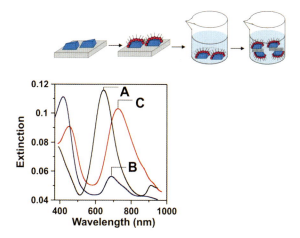

Fig. 9.5 Upper: Schematic illustration of the procedure for releasing NSL-fabricated triangles into solution. Lower: UV-visible extinction spectra of (A) surface-bound, (B) monomeric solution-phase, and (C) dimeric solution phase SAM-functionalized, NSL-fabricated Ag nanoparticles in ethanol. (Reproduced with permission from Ref. [54]; © 2005 American Chemical Society.)

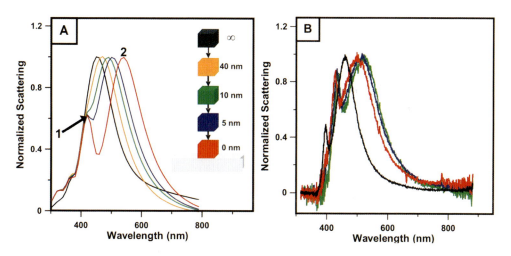

Fig. 9.6 (A) Finite-difference time-domain theory showing the emergence of a second peak as a single nanocube (90 nm diameter) approaches a dielectric substrate. (B) Refractive index sensitivity demonstrated by dark-field scattering spectra in four different dielectric environments [refractive indices: 1.000297 (black); 1.329 (red); 1.3854 (blue); 1.4458 (green)]. (Reproduced with permission from Ref. [46]; © 2005 American Chemical Society.)

SAM functionalization, the Ag NPs are released into solution by sonication in ethanol (step 3). During the releasing process described above, all sides of the released NPs are coated with SAM, except for the bases of the triangles, which were in contact with the glass substrate; this allows the NPs to be asymmetrically functionalized. For example, a dithiol linking agent can be added to the NP solution in order to functionalize the bases, leading to linked NP dimers (step 4). The formation of solution-phase NP monomers and dimers has been verified using transmission electron microscopy (TEM) [54].

The optical properties of the surface-bound and solution-phase SAM-functionalized Ag NPs have been characterized using UV-visible extinction spectroscopy, as shown in Figure 9.5 (lower panel). The spectrum of the released NPs (spectrum B) differs dramatically from that of the particle array on the glass substrate (spectrum A). Two peaks appear in the released NP spectrum – an intense peak at 418 nm and a weak, broad peak centered at 682 nm. Upon forming the linked NP dimers, a further change in the UV-visible spectrum of the NP solution is observed (spectrum C): the high-energy peak shifts from 418 nm to 431 nm and decreases in intensity, while the low-energy peak at 682 nm shifts to 705 nm and increases in intensity and peak width. These observations have been explained using the DDA method, which describes both the origin of the two peaks as well as the shifts upon dimerization [54]. The ability to asymmetrically functionalize these solution-phase NPs to induce formation of specific targets – such as NP dimers – is a significant advantage in certain LSPR biosensing experiments, and will be pursued in future applications.

9.2.1.3 Silver Nanocubes

Silver nanocubes prepared by the polyol synthesis [55, 56] offer a unique plasmonic response when single-particle LSPR spectra are measured on a glass substrate; namely, there are *two* plasmon resonance peaks (as in Figure 9.6) [46]. This is because the dielectric symmetry of their environment is broken when they are placed on a glass substrate. For a NP to yield a new plasmon resonance peak when it is placed on a dielectric surface, it must satisfy two conditions: (i) its near fields must be most intense at the polar regions of the NP; and (ii) it must be thicker than the skin depth of the material (~25 nm for silver) [46].

In order to understand the physical origin of these peaks, FDTD calculations were performed to model the near-field behavior of the cubes [46]. Figure 9.6A shows a series of scattering spectra that were generated from calculations in which the cube is moved progressively closer to a dielectric surface. These spectra show that the dipole mode associated with the solution spectrum shifts into a broad peak at 550 nm when the particle closes to within a few nanometers of the surface. In addition, a blue peak appears at 430 nm that becomes more distinct as the particle approaches the surface. Figure 9.6B shows experimental data from a single nanocube in which a change in the local refractive index is accompanied by a shift in the plasmon resonance of both the high- and low-energy peaks. In particular, the response of the higher energy peak to the change in refractive index proves to be more sensitive than the standard dipole resonances of other NPs due to the narrow linewidth of this peak – that is, the FOM value is 5.4 [46]. This suggests that these particles may offer an advantage over other NP geometries for sensing experiments.

9.2.2
Biosensing

Each of the NP geometries described above – as well as many other NP platforms – can be used for biosensing experiments in which the binding of a biological target can induce a measurable shift in the LSPR spectrum. By appropriately functionalizing NP surfaces, LSPR sensors can be designed to detect a variety of biological and pathogenic molecules, making them a valuable diagnostic tool for the biomedical industry. The sensitivity of LSPR sensors for a number of biologically relevant systems has already been demonstrated [14, 23, 26–28, 57, 58]. In this section, we will describe one of the most important biological applications of the LSPR sensor to date, namely the detection and diagnosis of a possible biomarker for Alzheimer's disease [26, 27].

One hallmark of Alzheimer's disease is the formation of insoluble protein deposits, known as amyloid plaques, in brain tissue. These plaques are formed from the soluble precursor amyloid beta, a small 39- to 43-amino acid protein that is present in elevated levels in the brain and cerebral spinal fluid (CSF) of Alzheimer's disease sufferers [59]. Single units of amyloid beta readily assemble into oligomers of two to 24 units (sometimes referred to as amyloid-β-diffusible ligands, or ADDLs), and these oligomers themselves exhibit significant neuronal toxicity [60]. Though not proven, it is increasingly likely that ADDLs may cause early memory loss in Alzheimer's disease [61, 62]. Currently, the diagnosis of Alzheimer's disease is made based on symptomatic evidence, there being no diagnostic method based on the molecular pathology of the disease currently available.

The LSPR sensor, however, represents a promising step towards the molecular detection of the ADDL biomarker. Antibodies which specifically recognize amyloid beta oligomers [63] were covalently attached to a chemical monolayer on an NSL-fabricated NP surface (Figure 9.7A; step 1). Subsequently, CSF from patients diagnosed with Alzheimer's disease or from healthy, age-matched controls was flowed over the sensor surface (step 2). Any amyloid beta oligomers that remained bound to the antibody-functionalized surface were detected using a second capping antibody (step 3), thus completing the "sandwich" assay. By measuring extinction shifts in response to the addition of capping antibody, it was shown that diseased patients (Figure 9.7C) had significantly higher concentrations of amyloid beta oligomers than did age-matched controls (Figure 9.7B) [27]. The experiment was repeated using soluble post-mortem brain extract from diseased and control patients, with similar results [27].

The results from this study not only confirmed the relationship between elevated amyloid beta levels and Alzheimer's disease, but also suggested that the LSPR sensor could be used as an early detection technique for the condition. Moreover, the sandwich assay described above is broadly generalizable, suggesting that the LSPR sensors might have applications for other biosensing applications.

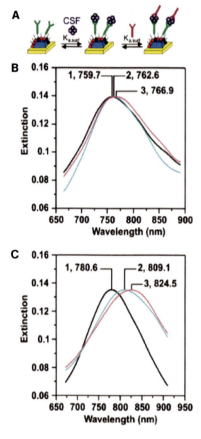

Fig. 9.7 (A) Schematic showing the localized surface plasmon resonance (LSPR) sandwich assay. Antibodies are linked to a SAM-covered nanoparticle (step 1). Amyloid-β-diffusible ligands (ADDLs) are then introduced that bind to the antibodies (step 2). Finally, a second capture antibody binds (step 3), completing the sandwich. (B) LSPR spectra for each step of the assay for an aging patient: (B-1) 100 nM anti-ADDL (λ_{max} = 759.7 nm); (B-2) CSF (λ_{max} = 762.6 nm); and (B-3) 100 nm anti-ADDL (λ_{max} = 766.9 nm). (C) LSPR spectra for each step of the assay for an Alzheimer's patient: (C-1) 100 nm anti-ADDL (λ_{max} = 780.6 nm); (C-2) CSF (λ_{max} = 809.1 nm); and (C-3) 100 nm anti-ADDL (λ_{max} = 824.5 nm). All measurements were made in a nitrogen environment. (Reproduced with permission from Ref. [27]; © 2005 American Chemical Society.)

9.3
Outlook

In this chapter, we have described methods related to the fabrication, characterization, and use of NPs for LSPR biosensing experiments. By using novel techniques such as electrochemistry, atomic layer deposition, and electron-beam lithography of 1-D arrays, the properties of these NP systems can be explored in a controlled manner. Moreover, several architectures for LSPR substrates – including nanowells, released NSL-fabricated nanotriangles and nanocubes – have been described in detail, with a particular emphasis on their ability to sense changes in their local environment. Finally, a sensor for the detection of a biomarker for Alzheimer's disease has been demonstrated using human samples.

In comparison to traditional propagating surface plasmon resonance (SPR) spectroscopy at smooth, thin metal film surfaces, LSPR has been shown to offer comparable speed and sensitivity [28]. However, unlike SPR, LSPR spectroscopy allows very small distances to be probed due to the decay in the electromagnetic field enhancement as one moves away from the NP surface. Moreover, SPR is an inherently "bulk" technique, whereas LSPR spectra can be measured for individual NPs, which may offer improvements in both speed and sensitivity. Several strategies are available to achieve such improvements for both the array and single particle formats. In particular, new NP geometries continue to be explored, with an eye towards narrower LSPR linewidths and better control over lineshapes. Moreover, the development of new amplification strategies in order to maximize the wavelength shift upon analyte binding should reduce the limit of detection into concentration regimes currently inaccessible with this technique. Such strategies include the introduction of secondary labels – such as NPs, enzymes or even resonant molecules – to the binding assays in order to maximize the wavelength shift. Lastly, single-particle LSPR is a promising approach, despite current limitations which mean that the spectra cannot be measured rapidly and in parallel. However, new wavelength-scanning liquid crystal tunable filters may help to overcome this issue. Thus, LSPR promises to remain competitive as a technique for biological sensing applications.

Acknowledgments

The authors gratefully acknowledge support from the Air Force Office of Scientific Research MURI program (grant F49620-02-1-0381), the National Science Foundation (EEC-0118025, DMR-0076097, CHE-0414554, DMR-0520513, BES-0507036), and the National Cancer Institute (1 U54 CA119341-01).

References

1 Srituravanich, W., Fang, N., Sun, C., Luo, Q., Zhang, X. (2004) Plasmonic nanolithography. *Nano Lett.* **4**, 1085–1088.

2 Sundaramurthy, A., Schuck, P. J., Conley, N. R., Fromm, D. P., Kino, G. S., Moerner, W. E. (2006) Toward nanometer-scale optical photolithography: utilizing the near-field of bowtie optical nanoantennas. *Nano Lett.* **6**, 355–360.

3 Kik, P. G., Maier, S. A., Atwater, H. A. (2004) Surface plasmons for nanofabrication. *Proc. SPIE-Int. Soc. Opt. Eng.* **5347**, 215–223.

4 Schatz, G. C., Van Duyne, R. P. (2002) In: Chalmers, J. M., Griffiths, P. R. (Eds.), *Handbook of Vibrational*

Spectroscopy. Wiley, New York, Vol. 1, pp. 759–774.

5 Haynes, C. L., Van Duyne, R. P. (2003) Plasmon-sampled surface-enhanced Raman excitation spectroscopy. *J. Phys. Chem. B* **107**, 7426–7433.

6 Emory, S. R., Nie, S. (1997) Probing single molecules and single nano-particles by surface-enhanced Raman scattering. *Science* **275**, 1102–1106.

7 Haes, A. J., Haynes, C. L., McFarland, A. D., Zou, S., Schatz, G. C., Van Duyne, R. P. (2005) Plasmonic materials for surface-enhanced sensing and spectroscopy. *MRS Bull.* **30**, 368–375.

8 McFarland, A. D., Young, M. A., Dieringer, J. A., Van Duyne, R. P. (2005) Wavelength-scanned surface-enhanced Raman excitation spectroscopy. *J. Phys. Chem. B* **109**, 11279–11285.

9 Haynes, C. L., McFarland, A. D., Van Duyne, R. P. (2005) Surface-enhanced Raman spectroscopy. *Anal. Chem.* **77**, 338A–346A.

10 Haes, A. J., Van Duyne, R. P. (2004) A unified view of propagating and localized surface plasmon resonance biosensors. *Anal. Bioanal. Chem.* **379**, 920–930.

11 McFarland, A. D., Van Duyne, R. P. (2003) Single silver nanoparticles as real-time optical sensors with zeptomole sensitivity. *Nano Lett.* **3**, 1057–1062.

12 Haes, A. J., McFarland, A. D., Van Duyne, R. P. (2003) Nanoparticle optics: sensing with nanoparticle arrays and single nanoparticles. *Proc. SPIE-Int. Soc. Opt. Eng.* **5223**, 197–207.

13 Yonzon, C. R., Stuart, D. A., Zhang, X., McFarland, A. D., Haynes, C. L., Van Duyne, R. P. (2005) Towards advanced chemical and biological nanosensors – An overview. *Talanta* **67**, 438–448.

14 Haes, A. J., Stuart, D. A., Nie, S. M., Van Duyne, R. P. (2004) Using solution-phase nanoparticles, surface-confined nanoparticle arrays and single nanoparticles as biological

sensing platforms. *J. Fluoresc.* **14**, 355–367.

15 Jensen, T. R., Malinsky, M. D., Haynes, C. L., Van Duyne, R. P. (2000) Nanosphere lithography: tunable localized surface plasmon resonance spectra of silver nanoparticles. *J. Phys. Chem. B* **104**, 10549–10556.

16 Jensen, T. R., Duval, M. L., Kelly, L., Lazarides, A., Schatz, G. C., Van Duyne, R. P. (1999) Nanosphere lithography: Effect of the external dielectric medium on the surface plasmon resonance spectrum of a periodic array of silver nanoparticles. *J. Phys. Chem. B* **103**, 9846–9853.

17 Miller, M. M., Lazarides, A. A. (2005) Sensitivity of metal nanoparticle surface plasmon resonance to the dielectric environment. *J. Phys. Chem. B.* **109**, 21556–21565.

18 Xia, Y., Halas, N. J. (2005) Shape-controlled synthesis and surface plasmonic properties of metallic nanostructures. *MRS Bull.* **30**, 338–348.

19 Haynes, C. L., Van Duyne, R. P. (2001) Nanosphere lithography: a versatile nanofabrication tool for studies of size-dependent nanoparticle optics. *J. Phys. Chem. B.* **105**, 5599–5611.

20 Kelly, K. L., Coronado, E., Zhao, L., Schatz, G. C. (2003) The optical properties of metal nanoparticles: the influence of size, shape, and dielectric environment. *J. Phys. Chem. B.* **107**, 668–677.

21 Link, S., El-Sayed, M. A. (1999) Spectral properties and relaxation dynamics of surface plasmon electronic oscillations in gold and silver nano-dots and nano-rods. *J. Phys. Chem. B.* **103**, 8410–8426.

22 Haes, A. J., Van Duyne, R. P. (2002) A nanoscale optical biosensor: sensitivity and selectivity of an approach based on the localized surface plasmon resonance of triangular silver nanoparticles. *J. Am. Chem. Soc.* **124**, 10596–10604.

23 Riboh, J. C., Haes, A. J., McFarland, A. D., Yonzon, C. R., Van Duyne, R. P.

(2003) A nanoscale optical biosensor: Real-time immunoassay in physiological buffer enabled by improved nanoparticle adhesion. *J. Phys. Chem. B* **107**, 1772–1780.

24 Haes, A. J., Zou, S., Schatz, G. C., Van Duyne, R. P. (2004) A nanoscale optical biosensor: the long range distance dependence of the localized surface plasmon resonance of noble metal nanoparticles. *J. Phys. Chem. B.* **108**, 109–116.

25 Haes, A. J., Zou, S., Schatz, G. C., Van Duyne, R. P. (2004) A nanoscale optical biosensor: the short range distance dependence of the localized surface plasmon resonance of silver and gold nanoparticles. *J. Phys. Chem. B.* **108**, 6961–6968.

26 Haes, A. J., Hall, W. P., Chang, L., Klein, W. L., Van Duyne, R. P. (2004) A localized surface plasmon resonance biosensor: First steps toward an assay for Alzheimer's disease. *Nano Lett.* **4**, 1029–1034.

27 Haes, A. J., Chang, L., Klein, W. L., Van Duyne, R. P. (2005) Detection of a biomarker for Alzheimer's disease from synthetic and clinical samples using a nanoscale optical biosensor. *J. Am. Chem. Soc.* **127**, 2264–2271.

28 Yonzon, C. R., Jeoung, E., Zou, S., Schatz, G. C., Mrksich, M., Van Duyne, R. P. (2004) A comparative analysis of localized and propagating surface plasmon resonance sensors: the binding of concanavalin A to a monosaccharide functionalized self-assembled monolayer. *J. Am. Chem. Soc.* **126**, 12669–12676.

29 Zeman, E. J., Schatz, G. C. (1987) An accurate electromagnetic theory study of surface enhancement factors for silver, gold, copper, lithium, sodium, aluminum, gallium, indium, zinc, and cadmium. *J. Phys. Chem.* **91**, 634–643.

30 Athawale, A. A., Katre, P. P., Majumdar, M. B. (2005) Nonaqueous phase synthesis of copper nanoparticles. *J. Nanosci. Nanotechnol.* **5**, 991–993.

31 Malinsky, M. D., Kelly, K. L., Schatz, G. C., Van Duyne, R. P. (2001) Chain length dependence and sensing capabilities of the localized surface plasmon resonance of silver nanoparticles chemically modified with alkanethiol self-assembled monolayers. *J. Am. Chem. Soc.* **123**, 1471–1482.

32 Malinsky, M. D., Kelly, K. L., Schatz, G. C., Van Duyne, R. P. (2001) Nanosphere lithography: effect of substrate on the localized surface plasmon resonance spectrum of silver nanoparticles. *J. Phys. Chem. B* **105**, 2342–2350.

33 El-Sayed, M. A. (2001) Some interesting properties of metals confined in time and nanometer space of different shapes. *Acc. Chem. Res.* **34**, 257–264.

34 Mie, G. (1908) Contributions to the optics of turbid media, especially colloidal metal solutions. *Annalen der Physik (Weinheim, Germany)* **25**, 377–445.

35 Draine, B. T., Flatau, P. J. (1994) Discrete-dipole approximation for scattering calculations. *J. Opt. Soc. Am. A* **11**, 1491–1499.

36 Jensen, T. R., Kelly, K. L., Lazarides, A., Schatz, G. C. (1999) Electrodynamics of noble metal nanoparticles and nanoparticle clusters. *J. Cluster Sci.* **10**, 295–317.

37 Lee, P. C., Meisel, D. (1982) Adsorption and surface-enhanced Raman of dyes on silver and gold sols. *J. Phys. Chem.* **86**, 3391–3395.

38 Sun, Y., Xia, Y. (2002) Shape-controlled synthesis of gold and silver nanoparticles. *Science* **298**, 2176–2179.

39 Jin, R., Cao, Y. C., Hao, E., Metraux, G. S., Schatz, G. C., Mirkin, C. A. (2003) Controlling anisotropic nanoparticle growth through plasmon excitation. *Nature* **425**, 487–490.

40 Daniel, M.-C., Astruc, D. (2004) Gold nanoparticles: assembly, supramolecular chemistry, quantum-size-related properties, and applications toward biology, catalysis, and nanotechnology. *Chem. Rev.* **104**, 293–346.

41 Jin, R., Cao, Y., Mirkin, C. A., Kelly, K. L., Schatz, G. C., Zheng, J. G.

(2001) Photoinduced conversion of silver nanospheres to nanoprisms. *Science* **294**, 1901–1903.

42 Jun, Y.-W., Choi, J.-S., Cheon, J. (2006) Shape control of semiconductor and metal oxide nanocrystals through nonhydrolytic colloidal routes. *Angew. Chem. Int. Ed.* **45**, 3414–3439.

43 Xu, Q., Bao, J., Capasso, F., Whitesides, G. M. (2006) Surface plasmon resonances of free-standing gold nanowires fabricated by nanoskiving. *Angew. Chem. Int. Ed.* **45**, 3631–3635.

44 Haynes, C. L., McFarland, A. D., Zhao, L., Van Duyne, R. P., Schatz, G. C., Gunnarsson, L., Prikulis, J., Kasemo, B., Kaell, M. (2003) Nanoparticle optics: the importance of radiative dipole coupling in two-dimensional nanoparticle arrays. *J. Phys. Chem. B* **107**, 7337–7342.

45 Jung, L. S., Campbell, C. T., Chinowsky, T. M., Mar, M. N., Yee, S. S. (1998) Quantitative interpretation of the response of surface plasmon resonance sensors to adsorbed films. *Langmuir* **14**, 5636–5648.

46 Sherry, L. J., Chang, S.-H., Schatz, G. C., Van Duyne, R. P., Wiley, B. J., Xia, Y. (2005) Localized surface plasmon resonance spectroscopy of single silver nanocubes. *Nano Lett.* **5**, 2034–2038.

47 Zhang, X., Hicks, E. M., Zhao, J., Schatz, G. C., Van Duyne, R. P. (2005) Electrochemical tuning of silver nanoparticles fabricated by nanosphere lithography. *Nano Lett.* **5**, 1503–1507.

48 Whitney, A. V., Elam, J. W., Zou, S., Zinovev, A. V., Stair, P. C., Schatz, G. C., Van Duyne, R. P. (2005) Localized surface plasmon resonance nanosensor: a high-resolution distance-dependence study using atomic layer deposition. *J. Phys. Chem. B.* **109**, 20522–20528.

49 Hicks, E. M., Zou, S., Schatz, G. C., Spears, K. G., Van Duyne, R. P., Gunnarsson, L., Rindzevicius, T., Kasemo, B., Kaell, M. (2005) Controlling plasmon line shapes

through diffractive coupling in linear arrays of cylindrical nanoparticles fabricated by electron beam lithography. *Nano Lett.* **5**, 1065–1070.

50 Kwak, E., Henzie, J., Chang, S., Gray, S. K., Schatz, G. C., Odom, T. W. (2005) Surface plasmon standing waves in large-area sub-wavelength hole arrays. *Nano Lett.* **5**, 1963–1967.

51 Brolo, A. G., Gordon, R., Leathem, B., Kavanagh, K. L. (2004) Surface plasmon sensor based on the enhanced light transmission through arrays of nanoholes in gold films. *Langmuir* **20**, 4813–4815.

52 Ebbesen, T. W., Lezec, H. J., Ghaemi, H. F., Thio, T., Wolff, P. A. (1998) Extraordinary optical transmission through sub-wavelength hole arrays. *Nature* **391**, 667–669.

53 Hicks, E. M., Zhang, X. Y., Zou, S. L., Lyandres, O., Spears, K. G., Schatz, G. C., Van Duyne, R. P. (2005) Plasmonic properties of film over nanowell surface fabricated by nanosphere lithography. *J. Phys. Chem. B* **109**, 22351–22358.

54 Haes, A. J., Zhao, J., Zou, S., Own, C. S., Marks, L. D., Schatz, G. C., Van Duyne, R. P. (2005) Solution-phase, triangular Ag nanotriangles fabricated by nanosphere lithography. *J. Phys. Chem. B* **109**, 11158–11162.

55 Wiley, B., Sun, Y., Mayers, B., Xia, Y. (2005) Shape-controlled synthesis of metal nanostructures: The case of silver. *Chemistry – A European Journal* **11**, 454–463.

56 Im Sang, H., Lee Yun, T., Wiley, B., Xia, Y. (2005) Large-scale synthesis of silver nanocubes: the role of HCl in promoting cube perfection and monodispersity. *Angew. Chem. Int. Ed. Engl.* **44**, 2154–2157.

57 Englebienne, P. (1998) Use of colloidal gold surface plasmon resonance peak shift to infer affinity constants from the interactions between protein antigens and antibodies specific for single or multiple epitopes. *Analyst* **123**, 1599–1603.

58 Raschke, G., Kowarik, S., Franzl, T., Soennichsen, C., Klar, T. A., Feldmann, J., Nichtl, A., Kuerzinger, K. (2003) Biomolecular recognition based on single gold nanoparticle light scattering. *Nano Lett.* **3**, 935–938.

59 Gong, Y., Chang, L., Viola, K. L., Lacor, P. N., Lambert, M. P., Finch, C. E., Krafft, G. A., Klein, W. L. (2003) Alzheimer's disease-affected brain: Presence of oligomeric Ab ligands (ADDLs) suggests a molecular basis for reversible memory loss. *Proc. Natl. Acad. Sci. USA* **100**, 10417–10422.

60 Lambert, M. P., Barlow, A. K., Chromy, B. A., Edwards, C., Freed, R., Liosatos, M., Morgan, T. E., Rozovsky, I., Trommer, B., Viola, K. L., et al. (1998) Diffusible, nonfibrillar ligands derived from Abeta1-42 are potent central nervous system neurotoxins. *Proc. Natl. Acad. Sci. USA* **95**, 6448–6453.

61 Wang, H.-W., Pasternak, J. F., Kuo, H., Ristic, H., Lambert, M. P., Chromy, B., Viola, K. L., Klein, W. L., Stine, W. B., Krafft, G. A., et al. (2002) Soluble oligomers of b-amyloid (1-42) inhibit long-term potentiation but not long-term depression in rat dentate gyrus. *Brain Res.* **924**, 133–140.

62 Walsh, D. M., Selkoe, D. J. (2004) Oligomers in the brain: The emerging role of soluble protein aggregates in neurodegeneration. *Protein Peptide Lett.* **11**, 213–228.

63 Lambert, M. P., Viola, K. L., Chromy, B. A., Chang, L., Morgan, T. E., Yu, J., Venton, D. L., Krafft, G. A., Finch, C. E., Klein, W. L. (2001) Vaccination with soluble A oligomers generates toxicity-neutralizing antibodies. *J. Neurochem.* **79**, 595–605.

10

Cantilever Array Sensors for Bioanalysis and Diagnostics

Hans Peter Lang, Martin Hegner, and Christoph Gerber

10.1
Overview

In recent years there has been an increasing demand for miniaturized, ultrasensitive and fast-responding sensors for applications in biochemistry and medicine. The main requirement is rapidly to obtain reliable qualitative and quantitative data.

Sensors are transducers that respond to a change in a physical parameter and transform one form of energy into another (Figure 10.1). Although many sensors are electrical or electronic devices, some may be electrochemical (pH probe), electromechanical (piezoelectric actuator, quartz, strain gauge), electroacoustic (gramophone pick-up, microphone), photoelectric (photodiode, solar cell), electromagnetic (antenna, photocell), magnetic (Hall-effect sensor, tape or hard-disk head for storage applications), electrostatic (electrometer) or thermoelectric (thermocouple, thermo-resistors). In this chapter we focus only on mechanical sensors, which respond mechanically to changes in an external parameter such as temperature or molecule adsorption, the mechanical response being bending or deflection.

Mechanical sensors consist of a fixed part and also a movable part, such as a thin membrane, a plate or a beam, which is fixed at one or both ends. Here, we will concentrate on cantilevers, which are microfabricated rectangular bar-shaped structures that are supported only at one end. The structures are longer than they are wide, and their thickness is much less than either their length or width. Cantilevers with a sharp tip have been used for almost two decades to image the surface topography of non-conducting samples with subnanometer accuracy in atomic force microscopy (AFM), which was invented during by the mid-1980s by Binnig, Quate and Gerber [1]. Beyond the imaging of surfaces, cantilevers without tips are used as nanomechanical sensors to measure adsorption or desorption processes on their surface.

Nanobiotechnology II. Edited by Chad A. Mirkin and Christof M. Niemeyer
Copyright © 2007 WILEY-VCH Verlag GmbH & Co. KGaA, Weinheim
ISBN: 978-3-527-31673-1

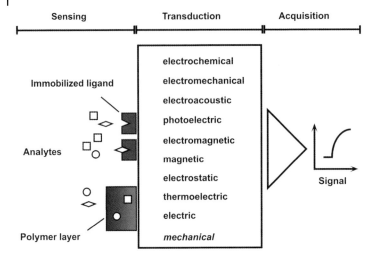

Fig. 10.1 The principles of sensor transduction. Analyte molecules in the environment are recognized by the sensing layer of the sensor. The recognition may either be very specific (i.e., recognition sites on immobilized ligand molecules recognize the analyte molecules), or it may be partially specific (i.e., analyte molecules diffuse into a polymeric layer). Molecule binding may be transduced into a recordable signal by various transduction mechanisms. Subsequently, the acquired signal is further amplified and processed.

10.1.1
Cantilevers as Sensors

The use of beams of silicon as sensors by measuring deflections or changes in resonance frequency was first described during the late 1960s, when Wilfinger et al. [2] investigated large silicon cantilever structures of $50 \times 30 \times 8$ mm for detecting resonances. These authors used localized thermal expansion in diffused resistors (piezoresistors) located near the cantilever support to create a temperature gradient for actuation of the cantilever at its resonance frequency. Similarly, the piezoresistors could also be used to sense mechanical deflection of the cantilever. Later, in 1971, Heng [3] fabricated gold cantilevers that were capacitively coupled to microstrip lines for the mechanical trimming of high-frequency oscillator circuits. In 1979, Petersen [4] constructed cantilever-type micromechanical membrane switches in silicon that should have filled the gap between silicon transistors and mechanical electromagnetic relays, while in 1985 Kolesar [5] suggested the use of cantilever structures as electronic detectors for nerve agents.

The easy availability of microfabricated cantilevers for AFM [1] triggered a number of research investigations into the use of cantilevers as sensors. In 1994, Itoh et al. [6] presented a cantilever coated with a thin film of zinc oxide, and proposed piezoresistive deflection readout as an alternative to optical beam-deflection readout. Later, Cleveland et al. [7] reported the tracking of cantilever resonance frequency to detect nanogram changes in mass loading when small particles are deposited onto AFM probe tips. Thundat et al. [8] showed that the resonance frequency, as well as the static bending of microcantilevers, is influenced by changing ambient conditions (e.g., moisture adsorption), and that the deflection of metal-coated cantilevers could be further influenced by thermal effects (e.g., the bimetallic effect). Gimzewski et al. [9] also showed preliminary chemical sensing applications, in which static cantilever bending revealed chemical reactions with very high sensitivity. Later, Thundat et al. [10] observed changes in the resonance frequency of microcantilevers due to the adsorption of analyte vapor onto exposed surfaces, with the frequency changes occurring due to mass loading or adsorption-induced changes in the cantilever spring constant. However, by coating the cantilever surfaces with hygroscopic materials (e.g., phosphoric acid or gelatin), the cantilever could sense water vapor at picogram mass resolution.

10.1.2
Measurement Principle

The deflection of individual cantilevers can easily be determined using optical beam-deflection electronics, as are common in AFM instrumentation. Unfortunately, single cantilever responses can be prone to artifacts such as thermal drifts or unspecific adsorption, and for this reason the use of passivated reference cantilevers is preferable. The first use of cantilever arrays with sensor and reference cantilevers (see Figure 10.2) was reported in 1998 [11], and this represented significant progress for the understanding of true (difference) cantilever responses, as a possible response of a reference cantilever is subtracted. A scanning electron microscopy image of a microfabricated cantilever array is shown in Figure 10.3.

The cantilever surfaces serve as sensor surfaces, and allow the processes taking place on the surface of the beam [9] to be monitored with unprecedented accuracy, in particular the adsorption of molecules. The formation of molecular layers on the cantilever surface will generate surface stress, and this will eventually result in a bending of the cantilever, provided that the adsorption occurs preferentially on one surface of the cantilever. Adsorption is controlled by coating one surface (typically the upper surface) of a cantilever with a thin layer of a material that

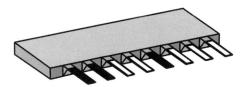

Fig. 10.2 Schematic drawing of an array of cantilever beams without tips for use as cantilever sensors. Cantilevers coated with a sensitive layer to detect the target molecules are called "sensor" cantilevers (white); those coated with a passivation layer inert to target molecules are called "reference" cantilevers (black).

Fig. 10.3 Scanning electron microscopy image of a microfabricated silicon cantilever array. (Illustration courtesy IBM Research GmbH, Zurich Research Laboratory, Rüschlikon, Switzerland.)

exhibits affinity to the molecules to be detected (sensor surface); this surface of the cantilever is referred to as the "functionalized" surface (Figure 10.4). The other surface of the cantilever (typically the lower surface) may be left uncoated or perhaps modified with a passivation layer; this may be a chemical surface that does not exhibit significant affinity for the molecules in the environment to be detected. In order to enable functionalized surfaces to be established, a metal layer is often evaporated onto the surface designed as sensor surface. For this, metal surfaces – such as gold – may be used as a platform to covalently bind a

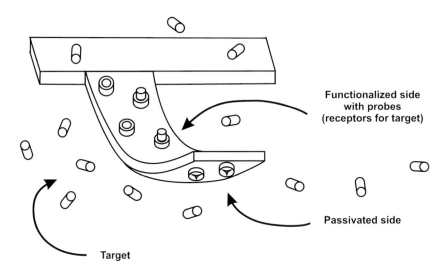

Fig. 10.4 Schematic representation of a cantilever sensor functionalized at its upper side with probe receptors for recognition of target molecules from the environment. The lower side of the cantilever sensor is passivated (i.e., coated with a layer that is inert to target molecules).

monolayer representing the chemical surface that is sensitive to the molecules to be detected. Frequently, a monolayer of thiol molecules covalently bound to a gold surface is applied. The gold layer is also favored for use as a reflection layer if the bending of the cantilever is read out via an optical beam-deflection method.

Given a cantilever coated with gold on its upper surface and left uncoated on its lower surface (consisting of silicon and silicon oxide), the adsorption of thiol molecules will take place on the upper surface of the cantilever, and this will result in a downward bending of the cantilever due to the formation of surface stress. This process is termed "development of compressive surface stress", because the forming self-assembled monolayer produces a downward bending of the cantilever (away from the gold coating). In the opposite situation, when the cantilever bends upwards, we would speak of tensile stress. If both the upper and lower surfaces of the cantilevers are involved in the reaction, then the situation will be much more complex, as a predominant compressive stress formation on the lower cantilever surface might appear as tensile stress on the upper surface. For this reason, it is of utmost importance that the lower cantilever surface is passivated in order that, ideally, no adsorption processes take place on the lower surface of the cantilever.

Single microcantilevers are susceptible to parasitic deflections due to thermal drift or chemical interaction of a cantilever with its environment, especially if the cantilever is operated in a liquid. Then, a baseline drift is often observed during static mode measurements. Moreover, nonspecific physisorption of molecules on the cantilever surface or nonspecific binding to receptor molecules during measurements may contribute to the drift. For these reasons, the use of cantilever arrays with a sensor and a passivated reference cantilever is highly recommended.

10.1.3
Cantilevers: Application Fields

Cantilever array sensors can be operated in a variety of environments, including air, high relative humidity or liquid, for the detection of specific biochemical reactions. Various ways of detecting the bending of the cantilever are possible, such as piezoresistive or piezoelectric detection. However, these detection methods are disadvantaged in that a protection layer (e.g., silicon nitride) must be applied to the cantilever to prevent contact electrode degradation of the read-out device when electrochemically conductive solutions such as salt-containing physiological buffers are used. No degradation problems exist, however, if the cantilever deflection read-out is determined via optical beam deflection. For this reason, many research groups use laser diodes to measure cantilever deflections.

A schematic set-up for cantilever deflection measurement in a liquid is shown in Figure 10.5. An array of eight vertical cavity surface emitting lasers (VCSELs) and a position-sensitive detector (PSD) are used for the optical beam deflection measurement of bending or oscillation of each cantilever separately, in a time-multiplexed manner. The cantilever array is mounted in a liquid cell, and liquids are pumped by a waste syringe from reservoirs and a six-way valve through the measurement chamber.

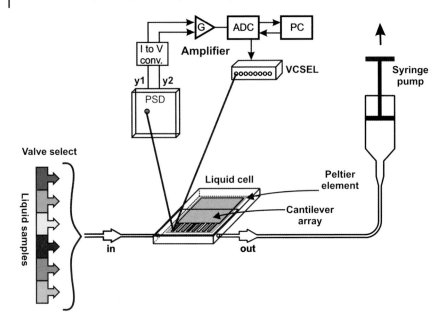

Fig. 10.5 Schematic of the measurement set-up for a liquid environment. The vertical cavity surface emitting laser (VCSEL) light sources are switched on and off in a time-multiplexed manner to facilitate the determination of deflection of each cantilever sensor separately.

Cantilever sensors have been used over a broad range of application areas, and some uses of microcantilever sensors will be described briefly in the following sections. Some of the earliest reported uses involve the adsorption of alkyl thiols onto gold [13, 14], the detection of mercury vapor and relative humidity [15], of dye molecules [16], monoclonal antibodies [17], sugars and proteins [18], solvent vapors [19–22] and fragrance vapors [23], as well as the pH-dependent response of carboxy-terminated alkyl thiols [24], label-free DNA hybridization detection [25, 26], and biomolecular recognition of proteins relevant to cardiovascular diseases [27]. For some more recent reviews, the reader is referred to Refs. [28–31].

10.2
Methods

10.2.1
Measurement Modes

In analogy to AFM, a variety of operating modes for cantilevers have been described. For example, the measurement of static deflection upon the formation

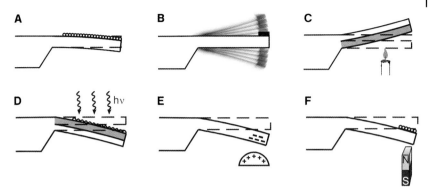

Fig. 10.6 Overview of the various cantilever sensor operation modes.
(A) Static mode; (B) dynamic mode; (C) heat mode; (D) photothermal
spectroscopy; (E) detection of charges; (F) detection of magnetic
forces.

of surface stress during adsorption of a molecular layer is termed "static mode"
(Figure 10.6A). In 1994, Ibach investigated cantilever-like structures to study
adsorbate-induced surface stress [12], while surface-stress induced bending of
cantilevers during the adsorption of alkanethiols on gold was reported by Berger
et al. in 1997 [13]. Cantilever-based sensors can be operated not only in the static
mode but also in a dynamic oscillatory mode, corresponding to non-contact AFM.
In the "dynamic mode" (Figure 10.6B), a cantilever is oscillated at its resonance
frequency and monitored for changes in resonance frequency that are dependent
on molecule adsorption onto or desorption from the cantilever surface. This
technique was described by Cleveland and colleagues [7], who calculated mass
changes from shifts in the cantilever resonance frequency due to tiny tungsten
particle spheres attached to the apex of the cantilever. A third mode of operation
is the so-called "heat mode" (Figure 10.6C), pioneered by Gimzewski et al. [9].
These authors took advantage of the bimetallic effect that results in the bending
of a metal-coated cantilever when heat is produced on its surface. Based on this
effect, a miniaturized calorimeter was constructed with picojoule sensitivity [32,
33].

Further operating modes exploit physical effects such as the production of heat
from the absorption of light by materials deposited on the cantilever (photothermal spectroscopy; Figure 10.6D) [34], or cantilever bending due to electric (Figure
10.6E) or magnetic (Figure 10.6F) forces.

10.2.2
Cantilever Functionalization

It is essential that the surfaces of the cantilever are coated in the correct manner
in order to provide suitable receptor surfaces for the molecules to be detected.
Such coatings should be specific, homogeneous, stable, reproducible, and either

reusable or designed for single use only. For static mode measurements, one side of the cantilever should be passivated for blocking unwanted adsorption. Often, the cantilever's upper side – the sensor side – is coated with a 20 nm-thick layer of gold to provide a platform for the binding of receptor molecules (e.g., via thiol chemistry), whereas the lower side is passivated using silane chemistry for coupling an inert surface such as polyethylene glycol silane (see Figure 10.4).

There are numerous ways to coat a cantilever with molecular layers – some are simple, but others are more advanced. It is very important that the method of choice is fast, reproducible, reliable and allows one or both cantilever surfaces to be coated separately.

The simple methods include thermal or electron-beam-assisted evaporation of material, electrospray, or other standard deposition methods. The disadvantage of these methods is that they are predominantly suitable for coating large areas, but not for individual cantilevers in an array, unless shadow masks are used to protect the areas that are to remain uncoated. The problem is that such masks need to be accurately aligned with the cantilever structures, and this is a very time-consuming process.

By skillful handling, however, tiny particles such as zeolite single crystals can be placed manually onto the cantilever to provide a functional surface [9, 20, 32, 35–37]. A molecule layer on a cantilever may also be formed by directly pipetting solutions of the probe molecules onto the upper surface of the cantilevers [19], or by air-brush spraying the molecules through shadow masks to coat each cantilever separately [20]. All of these methods suffer from a lack from limited reproducibility, and are also very time-consuming if a large number of cantilever arrays has to be coated.

Another strategy for coating cantilevers takes advantage of microfluidic networks (μFN) [38]. These are microfabricated structures of channels and wells, etched several tens to several hundreds of micrometers deep into the silicon wafers. The wells can be easily filled using a laboratory pipette, so that the fluid containing the probe molecules for cantilever coating flows through the channels towards openings at the edge of the μFN. Their pitch matches the distance between individual cantilevers in the array (see Figures 10.7A and 10.8A).

The cantilever array is then inserted into the open channels of the μFN that are filled with a solution of the probe molecules. Incubation of the cantilever array in the channels of the μFN takes from a few seconds (self-assembly of alkanethiol monolayers) to several tens of minutes (coating with protein solutions). To prevent evaporation of the solutions, the channels are covered by a slice of polydimethyl-siloxane (PDMS). In addition, the μFN may be placed in an environment of saturated vapor of the solvent used for the probe molecules to avoid drying out of the solutions.

A modification of this approach is insertion of the cantilever array into an array of dimension-matched disposable glass capillaries. The outer diameter of the glass capillaries is 240 μm, so that they can be placed neatly next to each other to accommodate the pitch of the cantilevers in the array (250 μm). Their inner diameter is 150 μm, allowing sufficient room to insert the cantilevers (width:

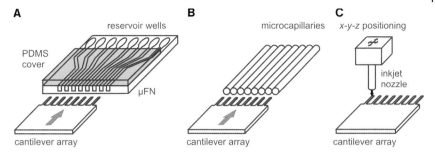

Fig. 10.7 Methods for cantilever sensor functionalization. (A) Insertion of the cantilever array in the channels of a microfluidic network covered with a slice of poly-dimethyl-siloxane (PDMS). (B) Insertion of the cantilever sensor array in the microcapillaries filled with a solution of probe molecules to be adsorbed onto the cantilever sensor surface. (C) Individual coating of each cantilever sensor with the nozzle of an inkjet liquid-dispensing device. μFN = microfluidic network.

100 μm) safely (Figures 10.7B and 10.8B). This method has been successfully applied for the deposition of a variety of materials onto cantilevers, such as polymer solutions [20], self-assembled monolayers [24], thiol-functionalized single-stranded DNA oligonucleotides [26], and proteins [27].

All of the techniques previously described require manual alignment of the cantilever array and functionalization tool, and are therefore not suitable for coating large numbers of cantilever arrays. One method which is appropriate for coating many cantilever sensor arrays in a rapid and reliable manner is inkjet spotting [39, 40] (Figures 10.7C and 10.8C). An *x-y-z* positioning system allows a fine nozzle (capillary diameter: 70 μm) to be positioned with an accuracy of approximately 10 μm over a cantilever. Individual droplets (diameter: 60–80 μm, volume 0.1–0.3 nL) can then be dispensed individually by means of a piezo-driven ejection system in the inkjet nozzle. When the droplets are spotted with a pitch smaller than 0.1 mm, they merge and form continuous films. Thus, by adjusting the number of droplets deposited on the cantilevers the resultant film thickness

Fig. 10.8 Optical microscopy images of cantilever sensor arrays being functionalized using various methods: (A) microfluidic network; (B) microcapillaries; (C) inkjet.

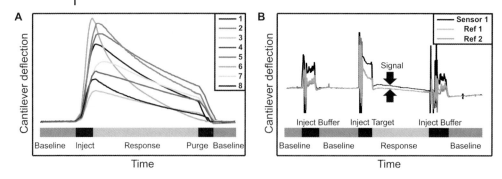

Fig. 10.9 Typical experimental curves for (A) measurement with cantilever array sensors in gas phase and (B) in liquid environment.

can be controlled precisely. The inkjet-spotting technique allows a cantilever to be coated within seconds, and yields very homogeneous, reproducibly deposited layers of well-controlled thickness. Using this approach, successful coating of self-assembled alkanethiol monolayers, polymer solutions, self-assembled DNA single-stranded oligonucleotides [40] and protein layers has been demonstrated. In conclusion, inkjet spotting has proved to be a very efficient and versatile method for functionalization that can even be used to coat arbitrarily-shaped sensors, both reproducibly and reliably [41, 42].

10.2.3
Experimental Procedure

An actual experiment in gas phase proceeds as follows (see Figure 10.9A). A cantilever array functionalized with a different polymer layer on each of the cantilevers is placed in the measurement chamber. First, the measurement cell with the cantilever array is purged with dry nitrogen gas. When a stable baseline has been achieved, dry nitrogen is guided through the headspace of a vial containing the analyte solution to be investigated. The gas stream is saturated with the vapor in the analyte headspace. By mixing the analyte-saturated gas stream with dry nitrogen, the analyte concentration can be adjusted. The gas mixture is then guided into the measurement chamber, where the reaction takes place, resulting in bending of the cantilevers. The numbers in Figure 10.9A refer to the bending profile of each cantilever. When the cantilever response is completed, the measurement chamber is purged with dry nitrogen until a stable baseline is reached, whereupon the set-up is ready again for further experiments. The cantilevers deflect due to diffusion of target molecules into the polymer layers; this produces a swelling of the polymers coating and, in turn, bending of the cantilever.

The procedure of a typical experiment in a liquid environment is shown in Figure 10.9B. A cantilever array functionalized with probe molecules in placed in the measurement chamber. First, the liquid cell containing the cantilever array is filled with buffer. A consecutive injection of buffer produces an injection peak due to pressure changes, but after short time a stable baseline is achieved again. The solution with the target molecules is then injected, whereupon the cantilevers deflect due to the reaction of the target molecules with the probe molecules; this results in the formation of surface stress and bending of the cantilevers. Only the cantilever functionalized with probe molecules matching a recognition site on the target molecules in solution (sensor 1) shows a net bending signal, whereas the passivated reference cantilevers (Ref 1 and Ref 2) show a much smaller response, if at all. When the reaction has taken place, the chamber is purged with buffer until a stable baseline is reached. From the deflection data displayed in Figure 10.9B it is clear that no conclusive result can be deduced from individual cantilever responses only, as both the sensor and reference cantilevers might show a bending response. However, from the *difference* in deflection responses of the probe and reference cantilever, a clear net deflection signal is determined (difference measurement). The conclusion is that it is absolutely mandatory to use at least two cantilevers in an experiment – a reference cantilever and a sensor cantilever – in order to be able to overcome any undesired artifacts such as thermal drift or unspecific adsorption.

Typical measurements in gas phase involve solvent vapor measurements and the use of cantilever arrays as an artificial nose [11, 19, 20, 22, 23], as well as adsorption studies of molecules onto a surface, such as alkanethiols on gold [13, 14], or the detection of humidity [8] or mercury vapor [10]. One particularly promising use is that of molecular recognition phases for the detection of organic vapor mixtures [43].

The potential for measurements in liquids is dominated by biochemical applications, such as the monitoring of hybridization of single-strand DNA molecules with their complements with single base mismatch accuracy at low picomolar concentration range [25, 26], the detection of proteins and biomarkers using highly specific key-lock processes (e.g., for immunoglobulin antibodies) [25], heart attack biomarkers [27], prostate-specific antigen [44], and specific protein conformations of the human estrogen receptor [45]. In the field of antibody detection, the present authors recently demonstrated the biomolecular recognition of single-chain fragment (scFv) antibodies that had been immobilized on the cantilever surface in well-defined orientation with respect to the substrate in order to enhance the accessibility of the receptor. Based on the concentration-dependence data obtained, the estimated sensitivity limit was seen to be as low as 1 nM [46].

In dynamic mode in liquid, we observed coupling of streptavidin-coated latex spheres (diameter 250 nm) to the biotin-functionalized cantilever surface [47]. The detection of 7 ng of latex spheres has been reported for dynamic mode measurement in a physiological buffer. Furthermore, we were able to measure protein interaction with double-stranded DNA oligonucleotides; two different DNA binding proteins – the transcription factors SP1 and NF-κB (which play important roles in protein production and gene expression) were investigated in parallel using cantilever array sensors [48]. Microcantilever arrays can also be used in dynamic mode to monitor the growth of bacteria and fungi by measuring increases in their weight [49, 50]. Advantageously, the cantilever technique yields results much more quickly than observing bacterial growth in a Petri dish, or by using optical absorption (turbidimetric) methods.

In summary, cantilever array sensors have been identified as versatile tools that can be applied to a broad variety of detection problems, notably as the sensitive coating layer can be selected according to the problem encountered.

10.3
Outlook

The applications of cantilever array sensors are manifold, and include gas sensing, quality control monitoring of chemicals, food and air, as well as process monitoring and control, to name a few examples. As an artificial nose, cantilever array sensors can characterize odors and vapors and may be used to assist fragrance design. Due to its extremely high sensitivity, the technique has a large potential to be applied for drugs and explosives detection, as well as for forensic investigations. Moreover, in a liquid environment the major applications are in biochemical analysis and medical diagnosis.

10.3.1
Recent Literature

During the past year the field of cantilever-based sensors has become increasingly diverse, as can be ascertained from the following literature overview. Although this selection of articles is far from complete, it reflects the current trends of research in the field.

For measurements in a gaseous environment, a sensor application in dynamic mode of piezoelectric cantilevers for an ultrasensitive nanobalance has been reported [51]. Elsewhere, micromolded plastic microcantilevers have been proposed for chemical sensing [52], as well as micromachined silicon microcantilevers for gas sensing applications with capacitive read-out [53]. In chemical sensing, ligand-functionalized microcantilevers for characterization of metal ion sensing have been presented [54], whilst an array of flexible microcantilever beams has been used to observe the action of rotaxane-based artificial molecular muscles [55].

The importance of homeland security is discussed in Ref. [56], where electrostatically actuated resonant microcantilever beams in complementary metal-oxide semiconductor (CMOS) technology are applied for the detection of chemical weapons. An integrated sensor platform for homeland defense based on silicon microcantilevers is described in Ref. [57], while in electrochemistry microcantilevers have been used to measure redox-induced surface stress [58], and a differential microcantilever-based system for measuring surface stress changes induced by electrochemical reactions has also been presented [59].

Many recent reports relate to biochemical applications, and several strategies for biochemical detection using microcantilevers are summarized in Figure

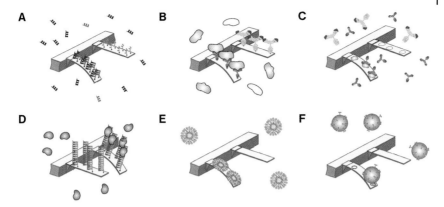

Fig. 10.10 Methods for detecting biomolecules and biosystems using micro-cantilevers. Biochemical recognition or adsorption should result in a change of surface stress and steric effects to be observable, as it is transduced into mechanical motion. The closer to the cantilever surface the surface stress change occurs, the easier it is to observe. (A) DNA hybridization; (B) antigen detection using antibodies immobilized on the cantilever surface; (C) antibody recognition using antigens bound to the surface; (D) detection of transcription factors using double-stranded DNA molecules containing a recognition sequence; (E) adsorption of lipid bilayer vesicles; (F) detection of viruses.

10.10. Recent reports also include a label-free immunosensor array using single-chain antibody fragments [46], and a label-free analysis of transcription factors using microcantilever arrays [48]. Microcantilevers modified by horseradish per-oxidase intercalated nano-assemblies have been used for hydrogen peroxide detection [60], and for the detection of cystamine dihydrochloride and glutaraldehyde [61, 62]; a back-propagation artificial neural network recognition study of analyte species and concentration has also been reported [63]. Cantilever sensors for nanomechanical detection have been used for the observation of specific protein conformation changes [45]. Similarly, in the field of DNA hybridization detection, the chemomechanics of surface stresses induced by DNA hybridization have been studied [64], while the grafting density and binding efficiency of DNA and proteins on gold surfaces has been characterized and optimized [65]. An electro-static microcantilever array biosensor has been applied for DNA detection [66], and microcantilever sensors for DNA hybridization reactions or antibody–antigen interactions without the use of external labels have been tested in dynamic mode [67].

In the realms of diagnostics, an immunoassay of prostate-specific antigen (PSA) which exploits the resonant frequency shift of piezoelectric nanomechani-cal microcantilevers has been reported [68], as well as phospholipid vesicle adsorption measured *in situ* with resonating cantilevers in a liquid cell [69]. Micro-cantilevers have also been used to detect *Bacillus anthracis* [70], while glucose oxidase multilayer-modified microcantilevers have been used to monitor glucose solutions in the nanomolar range [71].

Much effort has been placed in optimizing the cantilever sensor method, and a dimension-dependence study of the thermomechanical noise of microcantilevers has been conducted to determine the minimal detectable force and surface stress [72]. Furthermore, the geometric and flow configurations for enhanced microcantilever detection within a fluidic cell have been investigated [73]. A microcapillary pipette-assisted method to prepare polyethylene glycol-coated microcantilever sensors has been suggested [74], and the role of material microstructures in plate stiffness with relevance to microcantilever sensors has also been investigated [75]. Double-sided surface stress cantilever sensors for more sensitive cantilever surface stress measurement have been proposed [76].

Modified cantilever sensor techniques involve a biosensor based on magneto-strictive microcantilevers [77], piezoelectric self-sensing of adsorption-induced microcantilever bending [78], optical sequential readout of microcantilever arrays for biological detection by scanning the laser beam [79], and cysteine monolayer-modified microcantilevers to monitor flow pulses in a liquid [80]. The photothermal effect has been used to study dynamic elastic bending in microcantilevers [81]. For dynamic mode, the temperature- and pressure-dependence of resonance in multi-layer microcantilevers has been investigated [82], and the inaccuracy in the detection of molecules discussed [83].

The influence of surface stress on the resonance behavior of microcantilevers in higher resonant modes has been studied [84], and an alternative solution has been proposed to improve the sensitivity of resonant microcantilever chemical sensors by measuring in high-order modes and reducing dimensions [85]. A modal analysis of microcantilever sensors with environmental damping has been reported [86]. Theoretical studies have also been conducted on the simulation of adsorption-induced stress of a microcantilever sensor [87], the influence of nanobubbles on the bending of microcantilevers [88], the modeling and simulation of thermal effects in flexural microcantilever resonator dynamics [89], and on surface stress effects related to the resonance properties of cantilever sensors [90]. Furthermore, in a very recent review, the nanotechnologies for biomolecular detection and medical diagnostics have been discussed [91].

10.3.2
Challenges

Although cantilever array sensors might represent one of the answers to increasing demands over recent years for miniaturized, ultrasensitive, and fast-response sensors for use in gas detection, surveillance, and in biochemistry and medicine, there are several shortcomings to this technique. The need to focus a laser beam onto the apex of each cantilever in an array requires the development of pre-alignment strategies, and these will become increasingly difficult as the number of sensors rises. The integration potential of piezoresistive or piezoelectric types of cantilevers is much higher than that of the optical beam-deflection technique, as no alignment is necessary. On the other hand, the operation of electrically actuated piezoresistive or piezoelectric cantilevers in electrolytic liquids (e.g., biochemical salt solutions and buffers) represents a major challenge and requires

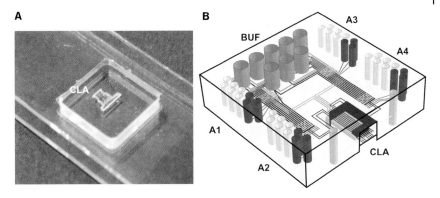

Fig. 10.11 (A) Miniaturized molded PDMS liquid chamber for silicon cantilever sensor arrays. CLA = cantilever array. (Illustration courtesy of A. Bietsch, University of Basel, Switzerland.) (B) Design drawing of a microfluidic environment for cantilever sensor arrays. BUF = buffer reservoirs; A1 to A4 = target molecule solution reservoirs. (Illustration courtesy of J.-P. Ramseyer and H. Breitenstein, University of Basel, Switzerland.)

additional effort to develop a passivation layer such that piezoelectric or piezo-resistive cantilevers can withstand electrochemical etching effects for at least a couple of hours during a biochemical experiment.

Another issue is that while the periphery required for measurements involving cantilever array sensors remains bulky, the cantilever array sensors themselves are tiny, microfabricated structures (see Figure 10.5). The system for liquid sample selection, guiding and flow control is based on standard laboratory equipment-sized instrumentation. In Figure 10.11, two suggestions are illustrated as to how liquid sample management might be optimized using microfluidics and miniaturization. The downscaling problem still persists for the optical readout, as there is no currently available method to shrink the optical beam deflection detection.

Even if the cantilever sensor array is a microfabricated part, its size will most likely eventually limit further miniaturization of the periphery. Thus, in theory the use of nanocantilevers that are a few hundred nanometers long and a few tens of nanometers thick will lead to drastic improvements in their performance [92, 93]. The practical use of these nano-instruments will be very complicated, however, and revolutionary ideas will be needed to obtain efficient read-outs of their microscopic deflections.

Acknowledgments

The authors thank R. McKendry (University College London, London, UK), W. Grange, J. Zhang, A. Bietsch, V. Barwich, M. Ghatkesar, F. Huber, J.-P. Ramseyer, A. Tonin, H.R. Hidber, E. Meyer, and H.-J. Güntherodt (University of Basel,

Basel, Switzerland) for valuable contributions and discussions, as well as U. Drechsler, M. Despont, H. Schmid, E. Delamarche, H. Wolf, R. Stutz, R. Allenspach, and P. F. Seidler (IBM Research, Zurich Research Laboratory, Rüschlikon, Switzerland). They also thank the European Union FP 6 Network of Excellence FRONTIERS for support. This project is funded partially by the National Center of Competence in Research in Nanoscience (Basel, Switzerland), the Swiss National Science Foundation, and the Commission for Technology and Innovation (Bern, Switzerland).

References

1 Binnig, G., Quate, C. F., Gerber, Ch. (1986) Atomic force microscope. *Phys. Rev. Lett.* **56**, 930–933.

2 Wilfinger, R. J., Bardell, P. H., Chhabra, D. S. (1968) Resonistor – A frequency selective device utilizing mechanical resonance of a silicon substrate. *IBM J. Res. Develop.* **12**, January, 113–118.

3 Heng, T. M. S. (1971) Trimming of microstrip circuits utilizing microcantilever air gaps. *IEEE Trans. Microwave Theory Technol.* **19**, 652–654.

4 Petersen, K. E. (1979) Micromechanical membrane switches on silicon. *IBM J. Res. Develop.* **23**, 376–385.

5 Kolesar, E. S. (1983) Electronic nerve agent detector. United States Patent No. 4,549,427, filed September 19, 1983.

6 Itoh, T., Suga, T. (1994) Force sensing microcantilevers using sputtered zinc-oxide thin-film. *Appl. Phys. Lett.* **64**, 37–39.

7 Cleveland, J. P., Manne, S., Bocek, D., Hansma, P. K. (1993) A nondestructive method for determining the spring constant of cantilevers for scanning force microscopy. *Rev. Sci. Instrument.* **64**, 403–405.

8 Thundat, T., Warmack, R. J., Chen, G. Y., Allison, D. P. (1994) Thermal and ambient-induced deflections of scanning force microscope cantilevers. *Appl. Phys. Lett.* **64**, 2894–2896.

9 Gimzewski, J. K., Gerber, Ch., Meyer, E., Schlittler, R. R. (1994) Observation of a chemical-reaction using a micromechanical sensor. *Chem. Phys. Lett.* **217**, 589–594.

10 Thundat, T., Chen, G. Y., Warmack, R. J., Allison, D. P., Wachter, E. A. (1995) Vapor detection using resonating microcantilevers. *Anal. Chem.* **67**, 519–521.

11 Lang, H. P., Berger, R., Andreoli, C., Brugger, J., Despont, M., Vettiger, P., Gerber, Ch., Gimzewski, J. K., Ramseyer, J.-P., Meyer, E., Güntherodt, H.-J. (1998) Sequential position readout from arrays of micromechanical cantilever sensors. *Appl. Phys. Lett.* **72**, 383–385.

12 Ibach, H. (1994) Adsorbate-induced surface stress. *J. Vac. Sci. Technol. A* **12**, 2240–2243.

13 Berger, R., Delamarche, E., Lang, H. P., Gerber, Ch., Gimzewski, J. K., Meyer, E., Güntherodt, H.-J. (1997) Surface stress in the self-assembly of alkanethiols on gold. *Science* **276**, 2021–2024.

14 Berger, R., Delamarche, E., Lang, H. P., Gerber, Ch., Gimzewski, J. K., Meyer, E., Güntherodt, H. J. (1998) Surface stress in the self-assembly of alkanethiols on gold probed by a force microscopy technique. *Appl. Phys. A* **66**, S55–S59.

15 Wachter, E. A., Thundat, T. (1995) Micromechanical sensors for chemical and physical measurements. *Rev. Sci. Instrument.* **66**, 3662–3667.

16 Scandella, L., Binder, G., Mezzacasa, T., Gobrecht, J., Berger, R., Lang, H. P., Gerber, Ch., Gimzewski, J. K., Koegler, J. H., Jansen, J. C. (1998)

Combination of single crystal zeolites and microfabrication: two applications towards zeolite nanodevices. *Microporous Mesoporous Mater.* **21**, 403–409.

17 Raiteri, R., Nelles, G., Butt, H. J., Knoll, W., Skladal, P. (1999) Sensing of biological substances based on the bending of microfabricated cantilevers. *Sens. Act. B* **61**, 213–217.

18 Moulin, A. M., O'Shea, S. J., Welland, M. E. (2000) Microcantilever-based biosensors. *Ultramicroscopy* **82**, 23–31.

19 Lang, H. P., Berger, R., Battiston, F. M., Ramseyer, J.-P., Meyer. E., Andreoli, C., Brugger, J., Vettiger, P., Despont, M., Mezzacasa, T., Scandella, L., Güntherodt, H.-J., Gerber, Ch., Gimzewski, J. K. (1998) A chemical sensor based on a micromechanical cantilever array for the identification of gases and vapors. *Appl. Phys. A* **66**, S61–S64.

20 Baller, M. K., Lang, H. P., Fritz, J., Gerber, Ch., Gimzewski, J. K., Drechsler, U., Rothuizen, H., Despont, M., Vettiger, P., Battiston, F. M., Ramseyer, J.-P., Fornaro, P., Meyer, E., Güntherodt, H.-J. (2000) A cantilever array-based artificial nose. *Ultramicroscopy* **82**, 1–9.

21 Bumbu, G. G., Kircher, G., Wolkenhauer, M., Berger, R., Gutmann, J. S. (2004) Synthesis and characterization of polymer brushes on micromechanical cantilevers. *Macromol. Chem. Phys.* **205**, 1713–1720.

22 Lang, H. P., Baller, M. K., Berger, R., Gerber, Ch., Gimzewski, J. K., Battiston, F. M., Fornaro, P., Ramseyer, J. P., Meyer, E., Güntherodt, H.-J. (1999) An artificial nose based on a micromechanical cantilever array. *Anal. Chim. Acta* **393**, 59–65.

23 Battiston, F. M., Ramseyer, J.-P., Lang, H. P., Baller, M. K., Gerber, Ch., Gimzewski, J. K., Meyer, E., Güntherodt, H.-J. (2001) A chemical sensor based on a microfabricated cantilever array with simultaneous resonance-frequency and bending readout. *Sens. Act. B* **77**, 122–131.

24 Fritz, J., Baller, M. K., Lang, H. P., Strunz, T., Meyer, E., Güntherodt, H.-J., Delamarche, E., Gerber, Ch., Gimzewski, J. K. (2000) Stress at the solid-liquid interface of self-assembled monolayers on gold investigated with a nanomechanical sensor. *Langmuir* **16**, 9694–9696.

25 Fritz, J., Baller, M. K., Lang, H. P., Rothuizen, H., Vettiger, P., Meyer, E., Güntherodt, H.-J., Gerber, Ch., Gimzewski, J. K. (2000) Translating biomolecular recognition into nanomechanics. *Science* **288**, 316–318.

26 McKendry, R., Zhang, J., Arntz, Y., Strunz, T., Hegner, M., Lang, H. P., Baller, M. K., Certa, U., Meyer, E., Güntherodt, H.-J., Gerber, Ch. (2002) Multiple label-free biodetection and quantitative DNA-binding assays on a nanomechanical cantilever array. *Proc. Natl. Acad. Sci. USA* **99**, 9783–9788.

27 Arntz, Y., Seelig, J. D., Lang, H. P., Zhang, J., Hunziker, P., Ramseyer, J.-P., Meyer, E., Hegner, M., Gerber, Ch. (2003) Label-free protein assay based on a nanomechanical cantilever array. *Nanotechnology* **14**, 86–90.

28 Ziegler, C. (2004) Cantilever-based biosensors. *Anal. Bioanal. Chem.* **379**, 946–959.

29 Lavrik, N. V., Sepaniak, M. J., Datskos, P. G. (2004) Cantilever transducers as a platform for chemical and biological sensors. *Rev. Sci. Instrument.* **75**, 2229–2253.

30 Majumdar, A. (2002) Bioassays based in molecular nanomechanics. *Disease Markers* **18**, 167–174.

31 Lang, H. P., Hegner, M., Gerber, Ch. (2005) Cantilever array sensors. *Materials Today* **8**, 30–36.

32 Berger, R., Gerber, Ch., Lang, H. P., Gimzewski, J. K. (1997) Micromechanics: A toolbox for femtoscale science: 'Towards a laboratory on a tip'. *Microelectron. Eng.* **35**, 373–379.

33 Datskos, G., Sepaniak, M. J., Tipple, C. A., Lavrik, N. (2001) Photomechanical chemical microsensors. *Sens. Act. B* **76**, 393–402.

34 Barnes, J. R., Stephenson, R. J., Welland, M. E., Gerber, Ch., Gimzewski, J. K. (1994) Photothermal

spectroscopy with femtojoule sensitivity using a micromechanical device. *Nature* **372**, 79–81.

35 Berger, R., Lang, H. P., Gerber, Ch., Gimzewski, J. K., Fabian, J. H., Scandella, L., Meyer, E., Güntherodt, H.-J. (1998) Micromechanical thermogravimetry. *Chem. Phys. Lett.* **294**, 363–369.

36 Berger, R., Gerber, Ch., Gimzewski, J. K., Meyer, E., Güntherodt, H.-J. (1996) Thermal analysis using a micromechanical calorimeter. *Appl. Phys. Lett.* **69**, 40–42.

37 Scandella, L., Binder, G., Mezzacasa, T., Gobrecht, J., Berger, R., Lang, H. P., Gerber, Ch., Gimzewski, J. K., Koegler, J. H., Jansen, J. C. (1998) Combination of single crystal zeolites and microfabrication: Two applications towards zeolite nanodevices. *Microporous Mesoporous Mater.* **21**, 403–409.

38 Cesaro-Tadic, S., Dernick, G., Juncker, D., Buurman, G., Kropshofer, H., Michel, B., Fattinger, C., Delamarche, E. (2004) High-sensitivity miniaturized immunoassays for tumor necrosis factor a using microfluidic systems. *Lab on a Chip* **4**, 563–569.

39 Bietsch, A., Hegner, M., Lang, H. P., Gerber, Ch. (2004) Inkjet deposition of alkanethiolate monolayers and DNA oligonucleotides on gold: Evaluation of spot uniformity by wet etching. *Langmuir* **20**, 5119–5122.

40 Bietsch, A., Zhang, J., Hegner, M., Lang, H. P., Gerber, Ch. (2004) Rapid functionalization of cantilever array sensors by inkjet printing. *Nanotechnology* **15**, 873–880.

41 Lange, D., Hagleitner, C., Hierlemann, A., Brand, O., Baltes, H. (2002) Complementary metal oxide semiconductor cantilever arrays on a single chip: Mass-sensitive detection of volatile organic compounds. *Anal. Chem.* **74**, 3084–3095.

42 Savran, C. A., Burg, T. P., Fritz, J., Manalis, S. R. (2003) Microfabricated mechanical biosensor with inherently differential readout. *Appl. Phys. Lett.* **83**, 1659–1661.

43 Dutta, P., Senesac, L. R., Lavrik, N. V., Datskos, P. G., Sepaniak, M. J. (2004) Response signatures for nanostructured, optically-probed functionalized microcantilever sensing arrays. *Sensor Lett.* **2**, 238–245.

44 Wu, G., Datar, R. H., Hansen, K. M., Thundat, T., Cote, R. J., Majumdar, A. (2001) Bioassay of prostate-specific antigen (PSA) using microcantilevers. *Nature Biotechnol.* **19**, 856–860.

45 Mukhopadhyay, R., Sumbayev, V. V., Lorentzen, M., Kjems, J., Andreasen, P. A., Besenbacher, F. (2005) Cantilever sensor for nanomechanical detection of specific protein conformations. *Nano Lett.* **5**, 2385–2388.

46 Backmann, N., Zahnd, C., Huber, F., Bietsch, A., Plückthun, A., Lang, H. P., Güntherodt, H.-J., Hegner, M., Gerber, C. (2005) A label-free immunosensor array using single-chain antibody fragments. *Proc. Natl. Acad. Sci. USA* **102**, 14587–14592.

47 Braun, T., Barwich, V., Ghatkesar, M. K., Bredekamp, A. H., Gerber, C., Hegner, M., Lang, H. P. (2005) Micromechanical mass sensors for biomolecular detection in a physiological environment. *Phys. Rev. E* **72**, 0311907.

48 Huber, F., Hegner, M., Gerber, C., Güntherodt, H.-J., Lang, H. P. (2006) Label free analysis of transcription factors using microcantilever arrays. *Biosens. Bioelectron.* **21**, 1599–1605.

49 Nugaeva, N., Gfeller, K. Y., Backmann, N., Lang, H. P., Düggelin, M., Hegner, M. (2005) Micromechanical cantilever array sensors for selective fungal immobilization and fast growth detection *Biosens. Bioelectron.* **21**, 849–856.

50 Gfeller, K. Y., Nugaeva, N., Hegner, M. (2005) Micromechanical oscillators as rapid biosensor for the detection of active growth of *Escherichia coli*. *Biosens. Bioelectron.* **21**, 528–533.

51 Shin, S., Paik, J. K., Lee, N. E., Park, J. S., Park, H. D., Lee, J. (2005) Gas sensor application of piezoelectric cantilever nanobalance; Electrical signal read-out. *Ferroelectrics* **328**, 59–65.

52 McFarland, A. W., Colton, J. S. (2005) Chemical sensing with micromolded plastic microcantilevers. *J. Microelectromech. Syst.* **14**, 1375–1385.

53 Amirola, J., Rodriguez, A., Castaner, L., Santos, J. P., Gutierrez, J., Horrillo, M. C. (2005) Micromachined silicon microcantilevers for gas sensing applications with capacitive read-out. *Sens. Act. B* **111**, 247–253.

54 Dutta, P., Chapman, P. J., Datskos, P. G., Sepaniak, M. J. (2005) Characterization of ligand-functionalized microcantilevers for metal ion sensing. *Anal. Chem.* **77**, 6601–6608.

55 Liu, Y., Flood, A. H., Bonvallett, P. A., Vignon, S. A., Northrop, B. H., Tseng, H. R., Jeppesen, J. O., Huang, T. J., Brough, B., Baller, M., Magonov, S., Solares, S. D., Goddard, W. A., Ho, C. M., Stoddart, J. F. (2005) Linear artificial molecular muscles. *J. Am. Chem. Soc.* **127**, 9745–9759.

56 Voiculescu, I., Zaghloul, M. E., McGill, R. A., Houser, E. J., Fedder, G. K. (2005) Electrostatically actuated resonant microcantilever beam in CMOS technology for the detection of chemical weapons. *IEEE Sensors J.* **5**, 641–647.

57 Pinnaduwage, L. A., Ji, H. F., Thundat, T. (2005) Moore's law in homeland defense: An integrated sensor platform based on silicon microcantilevers. *IEEE Sensors J.* **5**, 774–785.

58 Tabard-Cossa, V., Godin, M., Grutter, P., Burgess, I., Lennox, R. B. (2005) Redox-induced surface stress of polypyrrole-based actuators. *J. Phys. Chem. B* **109**, 17531–17537.

59 Tabard-Cossa, V., Godin, M., Beaulieu, L. Y., Grutter, P. (2005) A differential microcantilever-based system for measuring surface stress changes induced by electrochemical reactions. *Sens. Act. B* **107**, 233–241.

60 Yan, X. D., Shi, X. L., Hill, K., Ji, H. F. (2006) Microcantilevers modified by horseradish peroxidase intercalated nano-assembly for hydrogen peroxide detection. *Anal. Sci.* **22**, 205–208.

61 Yoo, K. A., Na, K. H., Joung, S. R., Nahm, B. H., Kang, C. J., Kim, Y. S. (2006) Microcantilever-based biosensor for detection of various biomolecules. *Jpn. J. Appl. Phys. Pt. 1* **45**, 515–518.

62 Na, K. H., Kim, Y. S., Kang, C. J. (2005) Fabrication of piezoresistive microcantilever using surface micromachining technique for biosensors. *Ultramicroscopy* **105**, 223–227.

63 Senesac, L. R., Dutta, P., Datskos, P. G., Sepaniak, M. J. (2006) Analyte species and concentration identification using differentially functionalized microcantilever arrays and artificial neural networks. *Anal. Chim. Acta* **558**, 94–101.

64 Stachowiak, J. C., Yue, M., Castelino, K., Chakraborty, A., Majumdar, A. (2006) Chemomechanics of surface stresses induced by DNA hybridization. *Langmuir* **22**, 263–268.

65 Castelino, K., Kannan, B., Majumdar, A. (2005) Characterization of grafting density and binding efficiency of DNA and proteins on gold surfaces. *Langmuir* **21**, 1956–1961.

66 Zhang, Z. X., Li, M. Q. (2005) Electrostatic microcantilever array biosensor and its application in DNA detection. *Progr. Biochem. Biophys.* **32**, 314–317.

67 Tian, F., Hansen, K. M., Ferrell, T. L., Thundat, T. (2005) Dynamic microcantilever sensors for discerning biomolecular interactions. *Anal. Chem.* **77**, 1601–1606.

68 Lee, J. H., Hwang, K. S., Park, J., Yoon, K. H., Yoon, D. S., Kim, T. S. (2005) Immunoassay of prostate-specific antigen (PSA) using resonant frequency shift of piezoelectric nanomechanical microcantilever. *Biosens. Bioelectron.* **20**, 2157–2162.

69 Ghatnekar-Nilsson, S., Lindahl, J., Dahlin, A., Stjernholm, T., Jeppesen, S., Hook, F., Montelius, L. (2005) Phospholipid vesicle adsorption measured in situ with resonating cantilevers in a liquid cell. *Nanotechnology* **16**, 1512–1516.

70 Wig, A., Arakawa, E. T., Passian, A., Ferrell, T. L., Thundat, T. (2006)

Photothermal spectroscopy of *Bacillus anthracis* and *Bacillus cereus* with microcantilevers. *Sens. Act. B* **114**, 206–211.

71 Yan, X. D., Xu, X. H. K., Ji, H. F. (2005) Glucose oxidase multilayer modified microcantilevers for glucose measurement. *Anal. Chem.* **77**, 6197–6204.

72 Alvarez, M., Tamayo, J., Plaza, J. A., Zinoviev, K., Dominguez, C., Lechuga, L. M. (2006) Dimension dependence of the thermomechanical noise of microcantilevers. *J. Appl. Phys.* **99**, 024910.

73 Khanafer, K., Vafai, K. (2005) Geometrical and flow configurations for enhanced microcantilever detection within a fluidic cell. *Int. J. Heat Mass Transfer* **48**, 2886–2895.

74 Wright, Y. J., Kar, A. K., Kim, Y. W., Scholz, C., George, M. A. (2005) Study of microcapillary pipette-assisted method to prepare poly-ethylene glycol-coated microcantilever sensors. *Sens. Act. B* **107**, 242–251.

75 McFarland, A. W., Colton, J. S. (2005) Role of material microstructure in plate stiffness with relevance to microcantilever sensors. *J. Micromech. Microeng.* **15**, 1060–1067.

76 Rasmussen, P. A., Grigorov, A. V., Boisen, A. (2005) Double sided surface stress cantilever sensor. *J. Micromech. Microeng.* **15**, 1088–1091.

77 Li, S. Q., Orona, L., Li, Z. M., Cheng, Z. Y. (2006) Biosensor based on magnetostrictive microcantilever. *Appl. Phys. Lett.* **88**, 073507.

78 Adams, J. D., Rogers, B., Manning, L., Hu, Z., Thundat, T., Cavazos, H., Minne, S. C. (2005) Piezoelectric self-sensing of adsorption-induced microcantilever bending. *Sens. Act A* **121**, 457–461.

79 Alvarez, M., Tamayo, J. (2005) Optical sequential readout of microcantilever arrays for biological detection. *Sens. Act. B* **106**, 687–690.

80 Tang, Y. J., Ji, H. F. (2005) Cysteine monolayer modified microcantilevers for impulse monitoring. *Instr. Sci. Technol.* **33**, 131–136.

81 Todorovic, D. M., Bojicic, A. (2005) Photothermal dynamic elastic bending in microcantilever. *Journal de Physique IV* **125**, 459–463.

82 Sandberg, R., Svendsen, W., Molhave, K., Boisen, A. (2005) Temperature and pressure dependence of resonance in multi-layer micro-cantilevers. *J. Micromech. Microeng.* **15**, 1454–1458.

83 Luo, C. (2005) Inaccuracy in the detection of molecules using two microcantilever-based methods. *J. Appl. Mech. – Transactions of the ASME* **72**, 617–619.

84 McFarland, A. W., Poggi, M. A., Doyle, M. J., Bottomley, L. A., Colton, J. S. (2005) Influence of surface stress on the resonance behavior of microcantilevers. *Appl. Phys. Lett.* **87**, 053505.

85 Lochon, F., Dufour, I., Rebiere, D. (2005) An alternative solution to improve sensitivity of resonant microcantilever chemical sensors: comparison between using high-order modes and reducing dimensions. *Sens. Act. B* **108**, 979–985.

86 Dareing, D. W., Thundat, T., Jeon, S. M., Nicholson, M. (2005) Modal analysis of microcantilever sensors with environmental damping. *J. Appl. Phys.* **97**, 084902.

87 Dareing, D. W., Thundat, T. (2005) Simulation of adsorption-induced stress of a microcantilever sensor. *J. Appl. Phys.* **97**, 043526.

88 Jeon, S. M., Desikan, R., Fang, T. A., Thundat, T. (2006) Influence of nano-bubbles on the bending of microcanti-levers. *Appl. Phys. Lett.* **88**, 103118.

89 Jazar, G. N. (2006) Mathematical modeling and simulation of thermal effects in flexural microcantilever resonator dynamics. *J. Vibration Control* **12**, 139–163.

90 Lu, P., Lee, H. P., Lu, C., O'Shea, S. J. (2005) Surface stress effects on the resonance properties of cantilever sensors. *Phys. Rev. B* **72**, 085405.

91 Chang, M. M. C., Cuda, G., Bunimovich, Y. L., Gaspari, M., Heath, J. R., Hill, H. D., Mirkin, C. A., Nijdam, A. J., Terracciano, R.,

Thundat, T., Ferrari, M. (2006)
Nanotechnologies for biomolecular
detection and medical diagnostics.
Curr. Opin. Chem. Biol. **10**, 11–19.

92 Yang, J., Ono, T., Esashi, M. (2000)
Mechanical behavior of ultrathin
microcantilever. *Sens. Act. A* **82**,
102–107.

93 Yang, J. L., Despont, M., Drechsler,
U., Hoogenboom, B. W., Frederix,
P. L. T. M., Martin, S., Engel, A.,
Vettiger, P., Hug, H. J. (2005)
Miniaturized single-crystal silicon
cantilevers for scanning force
microscopy. *Appl. Phys. Lett.* **86**,
134101.

11
Shear-Force-Controlled Scanning Ion Conductance Microscopy

Tilman E. Schäffer, Boris Anczykowski, Matthias Böcker, and Harald Fuchs

11.1
Overview

In multicellular organisms structures such as endothelial or epithelial cell layers form the interface between different fluid compartments, and play an important role for inter- and transcellular processes [1, 2]. Gaining insight into the complex barrier-crossing transport mechanisms is a common interest of cell biology, medicine, and pharmacology, as malfunctioning of these barriers leads to pathological implications. In particular, knowledge about the permeability of barriers for substances such as drugs is highly relevant.

Hence, special electrochemical and microscopic methods are required to study the ion-permeability of barrier-forming cell structures. For example, experimental techniques such as the measurement of transepithelial electrical resistances (TER) provide valuable information about the barrier properties of cell layers [3–5]. In addition to such integrating measurements of the total cell layer impedance, there is a need for complementary microscopic methods which can provide additional information concerning topography and local ion conductance, as well as mechanical, optical, electrical or chemical properties, with spatial resolution down to the nanometer level. Today, these requirements are met by the variety of scanning probe microscopes currently available.

The first scanning tunneling microscope (STM) was constructed by Binnig, Rohrer and coworkers [6] in 1981, and these authors were awarded the Nobel Prize in 1986 for their achievement. This in turn triggered the development of a large family of new microscopes – the so-called scanning probe microscopes (SPM) – all of which are based on a small, locally confined probe that is sensitive to various physical quantities. The most prominent members of the group are the scanning near-field optical microscope (SNOM) [7] and the atomic force microscope (AFM) [8]. One of the main applications of the AFM is to create high-resolution topographic images of surfaces in different environmental conditions (including aqueous buffer solution), which makes it well suited to biological

Nanobiotechnology II. Edited by Chad A. Mirkin and Christof M. Niemeyer
Copyright © 2007 WILEY-VCH Verlag GmbH & Co. KGaA, Weinheim
ISBN: 978-3-527-31673-1

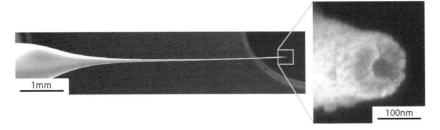

Fig. 11.1 Scanning electron microscopy (SEM) images of a nanopipette. Left: The tapered end of the nanopipette is several millimeters in length. Right: The inner opening diameter at the tip of the nanopipette is typically 30–100 nm. To improve SEM imaging, the nanopipette was sputter-coated with a thin layer of platinum.

applications [9]. Several related SPMs have been developed to date, including the magnetic force microscope (MFM) [10, 11], the electric force microscope [12], and the scanning electrochemical microscope (SECM) [13].

A scanning ion conductance microscope (SICM), which was invented by Hansma et al. in 1989 [14, 15], uses a local probe that is sensitive to ion conductance in an electrolyte solution. Drawn-out glass capillaries ("nanopipettes"), similar to those used in intracellular recording and patch-clamp experiments [16, 17], are well suited for this purpose. A nanopipette puller based on a heated coil or an infrared laser beam locally heats up a glass (e.g., borosilicate) capillary with an initial outer diameter of 1–2 mm that is subsequently drawn apart by force. This results in thin nanopipettes a few millimeters in length (Figure 11.1, left) that have sharp tips with typical opening diameters on the order of 30 to 100 nm (Figure 11.1, right). Opening diameters down to 13 nm were achieved when using quartz capillaries [18]. The use of microfabricated probes has also been reported [19].

The SICM scans a nanopipette over the surface of a sample that is immersed in electrolyte (Figure 11.2a). Two electrodes are placed in the electrolyte: one inside the thick end of the electrolyte-filled nanopipette (the "pipette electrode"), and one outside the nanopipette in the bath over or under the sample (the "bath electrode"). By applying a voltage between both electrodes and recording the ion current through the nanopipette, locally resolved images of ion conductance over the sample surface can be generated. Despite the many possible applications of such a microscope, the SICM is one of the least-developed scanning probe microscopy techniques to date. Only a few set-ups have been described in the literature [14, 15, 20–26].

There is a strong dependence of the measured current on the tip-sample distance (the "current squeezing effect"). The tip-sample distance can therefore be kept constant during scanning by using the current as an input to a feedback loop, thereby generating images of the sample topography. This works particularly well for nonporous samples where the bath electrode is positioned on the same side of the sample as the nanopipette. The pipette and sample do not

A

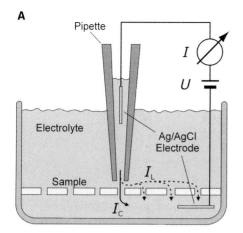

B

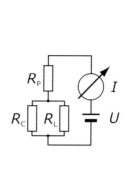

Fig. 11.2 (a) Schematic of the current measurement set-up. When a voltage U is applied between the two electrodes, an ion current flows through the opening of the nanopipette tip. Two alternate paths of the current through the sample are possible: either through a locally defined channel directly below the nanopipette tip (I_C), or through remote pores. As the second path is possible only when current "leaks" through the gap between nanopipette and sample, this current is called leakage current (I_L). The closer the nanopipette tip is to the sample, the more this leakage current is restricted ("squeezed") by the gap. (b) Equivalent circuit. The nanopipette resistance R_P is in series with a parallel configuration of the channel resistance R_C and the leakage resistance R_L. Usually, I_C is the quantity of interest. Therefore, the nanopipette resistance R_P should be as small as possible, and the leakage resistance R_L should be as large as possible, requiring the gap between nanopipette and sample to be small.

come into mechanical contact with each other, and soft and delicate samples can be imaged by using this "non-contact" configuration. Korchev et al., for example, imaged the topography of living cells with SICM [23] in order to monitor dynamic changes in cell volume at a resolution of 2.5×10^{-20} L [27] and to localize single active ion channels on the cell surface [28].

Advanced techniques and a number of different approaches have been conceived for overcoming some limitations and inherent problems of the original SICM concept. One possible option is to employ some type of point-spectroscopy technique, as it is known from other scanning probe methods such as atomic force microscopy [29]. Here, the probe is approached to and then retracted from the sample surface by typically several micrometers for each data point of the scanning area [30–32]. Although the risk of the probe laterally colliding with a protrusion on the surface or it becoming trapped in a depression is reduced, this is achieved at the price of very low scanning speed. Unwanted effects caused by slowly changing direct current (DC) potentials at the electrodes can be reduced for instance by applying either short voltage or current pulses instead of applying a constant voltage between the nanopipette and bath electrodes [31, 32].

Furthermore, modulation techniques have proved to be effective. These share the common idea that the risk of hitting a surface feature during lateral scanning can be effectively reduced by modulating the tip-sample distance. In particular, modulation techniques can help to improve long-term stability by minimizing the influence of DC offsets and drift effects. One such method has been proposed by the groups of Shao [26, 33] and Korchev [25]. These authors keep the voltage between the two electrodes constant, but modulate the z-position of either the sample or the probe with an amplitude of a few tens of nanometers by means of a piezoelectric actuator. If the probe is far away from the sample surface, such a small distance modulation with a typical frequency of a few hundred Hertz does not significantly affect the ion current. However, if the tapered end of the nano-pipette is brought into close proximity to the surface so that the "squeezing effect" sets in, this distance modulation leads to a modulation of the ion current. The amplitude of the resulting alternating current (AC) component in the ion current can then be detected with a lock-in amplifier. The latter allows the recovery of noisy, low-level ion current signals by using the voltage signal which modulates the tip-sample distance as a reference signal. The detected amplitude of the modulated ion current is then used as feedback signal to control the average tip-sample distance – that is, the feedback system tries to maintain a constant amplitude of the ion current while scanning the surface. This method of distance control has advantages over the conventional DC current-based approach because it makes the measurement less sensitive to changes in ionic strength or other DC drift effects.

Such advanced modulation techniques allow gentle scanning of delicate biological surfaces such as living cells. The improved performance of a distance-modulated SICM allows well-resolved images to be obtained of fine surface structures such as microvilli [34] or membrane proteins [18] on living cells. As the introduction of the distance-modulation technique reduces the risk of uncontrolled contact between the probe and sample, and also improves long-term stability, it becomes possible to continuously image specific surface areas for several hours and thereby to study dynamic processes [35, 36]. Gorelik et al., for example, studied the mechanism by which aldosterone activates sodium reabsorption via the epithelial sodium channel, and proposed a new hypothesis that is based on the effect of cell contraction (Figure 11.3) on the interaction of the channel with the F-actin cytoskeleton [37].

So far, we have considered only distance-control methods that are based exclusively on the measurement of ion conductance. Although this works well for samples with a homogeneous distribution of ion conductance, in many cases it is of interest to study inhomogeneities in the local ion conductance, such as those arising due to the presence of channels through a thin porous membrane. In these cases, a mechanism which is independent of the ion conductance is required to keep the tip-sample distance constant. For this purpose, two different techniques have been developed to date: (i) SICM with complementary AFM control [21]; and (ii) SICM with complementary shear-force control [24].

The first technique uses a bent nanopipette that is coated with a reflective metal layer and is scanned over the sample surface. With such a configuration, the sample topography can be imaged using a standard AFM measurement set-up. The

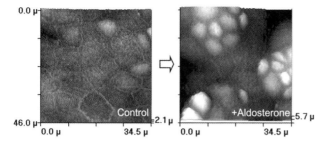

Fig. 11.3 Scanning ion conductance microscopy images of a monolayer of living A6 cells before (left) and 2 h after aldosterone stimulation (right). Prolonged cell morphology changes are observed, similar to changes induced by hypotonic stress. These changes do not occur in every cell, but rather in separate clusters of cells, and are likely the result of cell contraction. (Figure reprinted with permission from Ref. [37]; © 2005 National Academy of Sciences, USA.)

invention of the tapping mode in liquid [38, 39], which simplified the imaging of soft samples in solution, initiated the development of a novel microscope: Tapping-mode AFM combined with SICM [21]. In this design, the bent nanopipette [40, 41] is used both as a force sensor and an ion conductance probe. The bent nanopipette is vibrated perpendicularly to the sample surface with the help of a piezoelectric actuator. The excitation at the nanopipette's base leads to vibration amplitudes at the tip in the range of several nanometers to tens of nanometers. The measured vibration amplitude of the nanopipette serves as an input signal to a feedback loop that controls the nanopipette-sample distance, thereby generating the topography signal when scanning. Simultaneously, the ion current is recorded and used to generate a complementary image of the ion conductance. While such a microscope can also be operated in contact mode (using the DC deflection of the nanopipette), better resolution (both in topography and ion conductance) is generally obtained in tapping mode due to the absence of lateral imaging forces. The tapping mode-based SICM has been applied successfully in the field of biomineralization to investigate details of processes by which living organisms synthesize organic–inorganic composite materials [22].

In the case of shear-force-controlled SICM, a straight-tapered rather than a bent nanopipette is positioned perpendicularly to the sample surface and set into transverse vibrations (Figure 11.4). Arising mechanical shear-forces between the tip and sample provide an independent measure of sample topography. The combination of SICM with such a shear-force-based distance control allows the simultaneous and complementary imaging of sample topography and local ion conductance. Technological aspects as well as application examples of the shear-force based SICM will be discussed below.

The benefits of having two independent information channels are obvious: while the topographic data alone do not permit differentiation between a permeable ion-channel and a closed cavity in the cell layer, the ion current image pro-

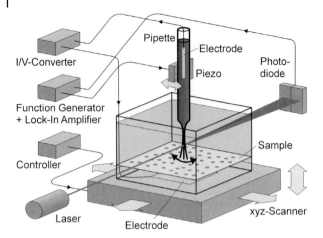

Fig. 11.4 Schematic of a shear-force-controlled scanning ion conductance microscope. A tapered nanopipette is brought into flexural vibrations by a piezoelectric actuator. The vibration amplitude is detected optically by focusing a laser beam onto the thin end of the nanopipette and directing the scattered beam onto a split photodiode. For low-noise amplitude determination, a lock-in amplifier is used whose reference channel is synchronized with the driving signal of the piezoelectric actuator. When approaching the nanopipette to the sample surface, the vibration amplitude decreases due to arising shear forces. The vibration amplitude serves as input to a feedback loop, which keeps the nanopipette at constant distance to the sample surface during xy-scanning. The ion current through the nanopipette is recorded simultaneously, thereby revealing variations in the local ion conductance of the porous sample.

vides this important piece of information. Therefore, the combined set-up is an ideal tool for investigating local variations in the ion conductance of biological specimens. This is of special interest for research into the field of barrier-forming structures. Furthermore, the possibility of recognizing ionic transport channels in hard as well as soft samples opens perspectives for further applications in fields as diverse as biochemistry/pharmacology, corrosion research, quality control of coatings, or artificial membranes.

11.2
Methods

11.2.1
Shear-Force Detection

Shear-force detection is well known from SNOM set-ups, where the tip of a tapered optical fiber is scanned at a constant distance over a sample surface [42, 43]. In a shear-force configuration for SICM, a nanopipette is used instead of the fiber. The nanopipette is mounted perpendicularly to the sample surface (Figure 11.4), and a small piezoelectric actuator that is attached to the nanopipette pro-

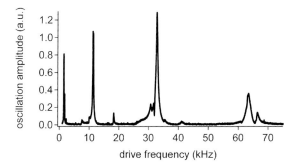

Fig. 11.5 Frequency spectrum of the oscillation of a driven nanopipette that is submerged into liquid electrolyte. Several resonances are typically present. For shear-force imaging, the piezoelectric actuator is driven at one of these resonances.

vides the driving signal that excites flexural mechanical vibrations in the drawn-out end of the nanopipette. The vibration amplitude depends heavily on the frequency of the driving signal, as the resonances in the system typically have a high quality factor (Figure 11.5). Several sharp peaks can be seen that indicate a strong oscillation of the nanopipette tip. When the oscillating nanopipette is approached to the surface, its vibration amplitude decreases sharply due to increasing shear forces at small tip-sample distances (Figure 11.6). The vibration amplitude is therefore well suited as a measure for the sample topography. The technical challenge is to find a method for detecting this vibration amplitude on the nanometer scale. To date, several methods have been established, including optical detection [42] and detection using a piezoelectric tuning fork sensor [44]. Piezoelectric detection methods face some principal problems when they are used for imaging in liquids, as electrical shortcuts may occur in the electrolyte medium. Solutions to

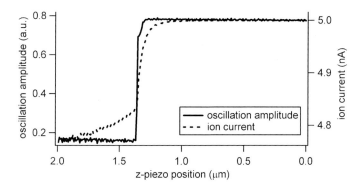

Fig. 11.6 Shear force oscillation amplitude and ion current while approaching a nanopipette to a sample surface. The shear force signal decreases sharply in close vicinity to the surface. The ion current also decreases (though less quickly), but already at larger tip-sample separations than the shear force signal.

this problem include coating the tuning fork with an electrical insulating layer [45], applying a custom piezoelectrical detection [46], or using a diving bell concept, where the water–air interface is forced close to the sample surface [47]. On the other hand, optical detection of shear forces principally works in liquid just as well as in air [48]. In the optical detection method, a laser beam is focused onto the vibrating section of the nanopipette near its tip (Figure 11.4). The incident laser beam is scattered by the vibrating tip, thereby modulating the intensity distribution of the beam. An optoelectronic detector, which is usually based on a split photodiode, detects these intensity modulations of the scattered beam and generates a signal that is proportional to the vibration amplitude of the nanopipette. There are two principal detection modes: detection in transmission and detection in reflection. The latter mode uses the nanopipette as a mirror, and requires the nanopipette to be coated with a reflective layer. It often produces a higher signal-to-noise ratio than the detection in transmission mode. The signal is then fed to a scan controller that keeps the tip-sample separation constant with the help of a feedback loop, thereby generating a topographic image of the sample surface while scanning.

11.2.2
Ion Current Measurement

Typical electrodes for the measurement of ion currents are standard silver/silver chloride (Ag/AgCl) electrodes that are also used as reference electrodes in potentiometry and voltammetry experiments. These are easily fabricated, for example by electrolytic deposition of a layer of silver chloride (AgCl) on a silver (Ag) wire. The Ag/AgCl electrode is described by a reversible redox reaction in which the chloride atoms in the solid silver chloride receive an electron and go into solution as chloride ions, leaving metallic silver. This reaction occurs close to the electrode surface (<1 nm distance). In SICM, two Ag/AgCl electrodes are used as the anode and cathode. This set-up is simpler to that used for voltammetric experiments, where three electrodes (working electrode, auxiliary electrode and reference electrode) are used and controlled by a potentiostat [49]. For the SICM set-up, an external voltage is applied between the two electrodes, thus inducing a Faradaic ion current in the electrolyte that is measured with a low-noise current amplifier. The measured current I, that passes through the nanopipette, depends on the effective ohmic resistance in the current path of the electrolyte. The current can leave the nanopipette opening either through a conductive path in the sample directly below the tip as "channel current" I_C, or it can "escape" sideways through the thin gap between the tip and sample as "leakage current", I_L (see Figure 11.2b). The closer the tip is to the sample surface, the smaller this leakage current is. Considering the equivalent circuit in Figure 11.2b, the effective ohmic resistance in the current path, R_{eff}, can be expressed as:

$$R_{eff} = R_P + \frac{R_C R_L}{R_C + R_L}. \tag{1}$$

This equation demonstrates that, in the case when the channel resistance R_C is of interest, the nanopipette resistance R_P should be small and the leakage resistance R_L should be large compared to R_C. The resistances depend on the particular boundary conditions of the system such as the tip-sample distance: when the tip is far from the sample surface, the leakage resistance is zero and the current through the nanopipette is limited by the nanopipette resistance R_P. However, when the tip approaches the sample surface, the leakage resistance increases as the leakage current is "squeezed" by the narrowing gap [14, 50]. Therefore, the effective resistance increases and the measured current drops. In the ideal case of a perfect seal between nanopipette and surface at zero tip-sample distance, the current becomes zero at contact for a nonporous sample. Nitz et al. [24] constructed a simple analytical model for the distance-dependence of the measured current (in the absence of channels in the sample, i.e. $R_C \rightarrow \infty$). They obtained

$$I(z) \approx I_0 \left(1 + \frac{z_0}{z} \right)^{-1},$$

(2)

where $I_0 = U/R_P$ is the measured current far away from the sample, U is the applied voltage, z the tip-sample separation, and z_0 is approximated by

$$z_0 = \frac{3r_0 r_i}{2L_P} \ln \frac{r_a}{r_i}.$$

(3)

The tapered nanopipette is assumed to be of conical shape with the inner diameter of the thick end r_0, inner diameter of the thin end (at the nanopipette tip) r_i, outer diameter of the thin end r_a, and nanopipette length L_P (Figure 11.7, inset). When the tip approaches the surface, the measured current decreases to zero, with a characteristic length scale z_0. For typical parameters ($r_0 = 0.3$ mm, $L_P = 5$ mm, $r_a = 50$ nm, $r_i = 30$ nm, $\rho = 1.09$ Ωm, $U = 100$ mV), we obtain $I_0 = 519$ pA and $z_0 = 0.60$ nm. In this model, the current begins to drop significantly only at very small tip-sample distances (Figure 11.7). A more detailed method of calculating the distance-dependence of the current that is based on finite element analysis gives qualitatively similar results [51]. For these derivations, it was assumed that other resistances in the system such as resistances in the wires, in the bulk electrolyte and in the solid–liquid junctions, are small. Electrochemically induced potentials were neglected as they can be compensated for by applying a voltage offset. It should be noted that, when fast changes of the current are to be measured, the capacitances in the system need to be considered.

11.2.3
Shear-Force-Controlled Imaging

In order to acquire images of sample conductance that are independent of topography, the nanopipette tip must be held at a constant position over the sample surface. The shear-force distance control serves this purpose, allowing the simul-

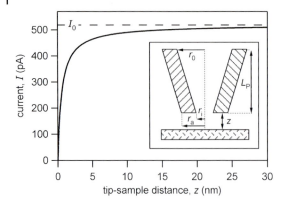

Fig. 11.7 Calculated current versus distance curve, based on a simple analytical model [Eq. (2) and inset]. The parameters used for calculating the curve were $r_0 = 0.3$ mm, $r_i = 30$ nm, $r_a = 50$ nm, $L_P = 5$ mm, $\rho = 1.09$ Ωm and $U = 100$ mV. It is apparent that the current starts to drop only at tip-sample distances well below the nanopipette tip radius.

taneous recording of images of sample topography and ion current [24]. It might appear that the effective vertical spring constant of a vertically oriented nanopipette is relatively large, imposing a possible thread to soft samples. On the other hand, the non-contact nature of shear-force imaging allows the scanning of surfaces with small tip-sample interaction forces. Soft samples as delicate as cells in liquid could be imaged with a shear-force set-up (Figure 11.8), and the cells were well resolved. Furthermore, shear-force-controlled SICM was used to provide further insight into transport mechanisms of cellular membranes. One example is the investigation of the functionality of tight junctions between living cells (Figure 11.9). Features that might be indicative of cell–cell contacts were resolved in

Fig. 11.8 Shear-force topography image of a monolayer of fixed MDCK-II cells on a glass support in buffer solution.

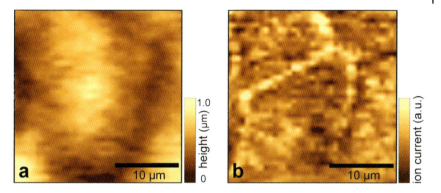

Fig. 11.9 Simultaneous (a) shear-force topography and (b) ion current images of a layer of live MDCK-II cells. Some, but not all, of the cell–cell contacts are visible in topography. The ion current image reveals lines of increased conductance that are independent of local surface topography. This finding suggests increased ion permeability through the tight junctions of the cell–cell contacts. (Sample provided by J. Seebach, experimental data by J. Kamp, University of Münster [51, 70].)

the ion current image as lines of increased conductance, even though the topography image did not reveal such features. This suggests that shear-force-controlled SICM is capable of measuring not only structural, but also functional, data on living cells.

11.3
Outlook

While the combined SICM and shear force microscope provides insight into cellular transport mechanisms and has a broad range of possible applications, further techniques which are based on similar components can be envisioned. The ability of the SICM to image biological samples *in vivo* with nanometer resolution makes it an interesting candidate for combination with other scanning microscopy techniques. For example, whilst adhering to the original SICM idea of using the ion current signal to maintain a constant distance between the scanning probe and the sample, complementary data, such as local optical data, can be recorded simultaneously. Combinations of SICM with scanning near-field optical microscopy (SNOM) or scanning confocal microscopy (SCM) are prominent examples.

The groups of Korchev and Shao modified the original SICM set-up in such a way that the end of the tapered nanopipette also serves as a near-field light source for SNOM [33, 52, 53]. This can be achieved by coupling laser light into the nanopipette via an optical fiber. Coating the outside of the nanopipette with a reflective metal layer helps to confine the laser light to the aperture – that is, the tapered nanopipette's end. Provided that the sample and the substrate are transparent, the SNOM signal can then be collected through an objective and detected by a photomultiplier located underneath the SICM head. Living cells were successfully investigated with such a combined SICM/SNOM set-up. An alternative way to utilize the SICM probe as a confined light source for SNOM was suggested by Bruckbauer et al. [54, 55]. This method is based on fluorescence as it occurs when a calcium indicator, with which the nanopipette is filled, binds with calcium in the sample solution and is illuminated with a laser. The mixing zone where the fluorescent complex forms serves as a localized light source.

SICM has also been successfully combined with SCM [56]. The set-up comprises an inverted light microscope fully configured for SCM, on which the SICM head is placed. During lateral scanning the vertical position of the sample is controlled by a standard SICM, that is, by an ion-current-based feedback loop. As a consequence, the optical confocal volume, which is located just below the end of the nanopipette, follows the topography of the sample. This allows capturing of fluorescence images of a surface simultaneously with topographic data. Interaction of fluorescent nanoparticles (e.g., virus-like particles) with the surface of fixed or living cells can be studied in this way.

Another type of application is to employ the SICM to perform patch-clamp experiments at specific surface sites [34, 36]. In such experiments, the SICM is first used to scan the sample surface, and the spatially resolved images of the surface topography allow the identification of regions or structures of interest. The scan unit is then used to position the tapered SICM probe over a user-selected spot. Finally, the probe is engaged to the sample surface to form a gigaohm seal for subsequent patch-clamp recording. This combination of high-resolution SICM scanning and patch-clamping allows ion channel recordings to be obtained on selective surface spots with lateral position control in the range of a few tens of nanometers. Such high precision is beyond the reach of conventional positioning methods such as those using light microscopy. With the SICM it becomes possible to investigate the activity of single ion channels on nanometer-sized structures on native biological samples such as living cells. For instance, a tapered SICM nanopipette can be positioned deliberately on top of a single microvillus on an epithelial cell and used for subsequent patch-clamp recording. In this way, individual K^+ channels could be identified on the basis of the reversal potential of current-voltage curves [34].

Nanopipettes not only serve as analytical probes but can also be used as powerful tools for the controlled delivery of reagents, and for material deposition. The material flux from a nanopipette to a substrate can either be driven by electric fields, capillary forces, hydrostatic pressure, or by ultrasonic excitation. Such types of nanolithography have been applied successfully to deposit metals [57–60] and chemicals [61, 62], as well as biological substances such as DNA, proteins, or other biomolecules [63–68]. Having the ability to deliver substances in a controlled manner through the nanopipette of a SICM opens the perspective to study the effects of local drug delivery on biological specimens, thereby providing a nanoscopic tool for the drug-induced manipulation of individual cells [69].

Acknowledgments

The authors thank J. Kamp and J. Seebach for contributing to this chapter with their experimental results. They also thank C. Steinem, J. Wegener, P. Heidenreich, S. Schrot, and H.-J. Galla for discussions. This research was supported in part by the German Federal Ministry of Education and Research (BMBF), by the Gemeinnützige Hertie-Stiftung/Stifterverband für die Deutsche Wissenschaft, and by the Deutsche Forschungsgesellschaft (DFG) (SCHA 1264).

References

1 Powell, D. W. (1981) Barrier function of epithelia. *Am. J. Physiol.* **241**, G275–G288.

2 Simionescu, M., Simionescu, N. (1986) Functions of the endothelial cell surface. *Annu. Rev. Physiol.* **48**, 279–293.

3 Diamond, J. M. (1977) The epithelial junction: bridge, gate and fence. *Physiologist* **20**, 10–18.

4 Cereijido, M., Gonzalez-Mariscal, L., Contreras, R. G., Gallardo, J. M., Garcia-Villegas, R., Valdes, J. (1993) The making of a cell junction. *J. Cell Sci. Suppl.* **17**, 127–132.

5 Wegener, J., Sieber, M., Galla, H.-J. (1996) Impedance analysis of epithelial and endothelial cell monolayers cultured on gold surfaces. *J. Biochem. Biophys. Methods* **32**, 151–170.

6 Binnig, G., Rohrer, H. (1982) Surface studies by scanning tunneling microscopy. *Phys. Rev. Lett.* **49**, 57–61.

7 Pohl, D. W., Denk, W., Lanz, M. (1984) Optical stethoscopy: Image recording with resolution $\lambda/20$. *Appl. Phys. Lett.* **44**, 651–653.

8 Binnig, G., Quate, C. F., Gerber, C. (1986) Atomic force microscope. *Phys. Rev. Lett.* **56**, 930–933.

9 Drake, B., Prater, C. B., Weisenhorn, A. L., Gould, S. A., Albrecht, T. R., Quate, C. F., Cannell, D. S., Hansma, H. G., Hansma, P. K. (1989) Imaging crystals, polymers, and processes in water with the atomic force microscope. *Science* **243**, 1586–1589.

10 Martin, Y., Wickramasinge, H. K. (1987) Magnetic imaging by 'force microscopy' with 1000 Å resolution. *Appl. Phys. Lett.* **50**, 1455–1457.

11 Sáenz, J. J., García, N., Grütter, P., Meyer, E., Heinzelmann, H., Wiesendanger, R., Rosenthaler, L., Hidber, H. R., Güntherodt, H. J. (1987) Observation of magnetic forces by the atomic force microscope. *J. Appl. Phys.* **62**, 4293–4295.

12 Martin, Y., Abraham, D. W., Wickramasinghe, H. K. (1988) High-resolution capacitance measurement and potentiometry by force microscopy. *Appl. Phys. Lett.* **52**, 1103–1105.

13 Bard, A. J., Fan, F.-R. F., Kwak, J., Lev, O. (1989) Scanning electrochemical microscopy. Introduction and principles. *Anal. Chem.* **61**, 132–138.

14 Hansma, P. K., Drake, B., Marti, O., Gould, S. A. C., Prater, C. B. (1989) The scanning ion-conductance microscope. *Science* **243**, 641–643.

15 Prater, C. B., Drake, B., Gould, S. A. C., Hansma, H. G., Hansma, P. K. (1990) Scanning ion-conductance microscope and atomic force microscope. *Scanning* **12**, 50–52.

16 Hille, B. (1992) *Ionic Channels of Excitable Membranes.* 2nd edn. Sinauer Associates, Sunderland, Mass.

17 Sakmann, B., Neher, E. (Eds.) (1995) *Single-channel recording.* 2nd edn. Springer, Heidelberg.

18 Shevchuk, A. I., Frolenkov, G. I., Sánchez, D., James, P. S., Freedman, N. Lab, M. J., Jones, R., Klenerman,

D., Korchev, Y. E. (2006) Imaging proteins in membranes of living cells by high-resolution scanning ion conductance micros-copy. *Angew. Chem. Int. Ed.* **45**, 2212–2216.

19 Prater, C. B., Hansma, P. K. (1991) Improved scanning ion-conductance microscope using microfabricated probes. *Rev. Sci. Instrum.* **62**, 2634–2638.

20 Olin, H. (1994) Design of a scanning probe microscope. *Meas. Sci. Technol.* **5**, 976–984.

21 Proksch, R., Lal, R., Hansma, P. K., Morse, D., Stucky, G. (1996) Imaging the internal and external pore structure of membranes in fluid: tapping mode scanning ion conductance microscopy. *Biophys. J.* **71**, 2155–2157.

22 Schäffer, T. E., Ionescu-Zanetti, C., Proksch, R., Fritz, M., Walters, D. A., Almqvist, N., Zaremba, C. M., Belcher, A. M., Smith, B. L., Stucky, G. D., et al. (1997) Does abalone nacre form by heteroepitaxial nucleation or by growth through mineral bridges? *Chem. Mater.* **9**, 1731–1740.

23 Korchev, Y. E., Bashford, C. L., Milovanovic, M., Vodyanoy, I., Lab, M. J. (1997) Scanning ion conductance microscopy of living cells. *Biophys J* **73**, 653–658.

24 Nitz, H., Kamp, J., Fuchs, H., (1998) A combined scanning ion-conductance and shear-force micros-cope. *Probe Microsc.* **1**, 187–200.

25 Shevchuk, A. I., Gorelik, J., Harding, S. E., Lab, M. J., Klenerman, D., Korchev, Y. E., (2001) Simultaneous measurement of Ca^{2+} and cellular dynamics: combined scanning ion conductance and optical microscopy to study contracting cardiac myocytes. *Biophys. J.* **81**, 1759–1764.

26 Pastré, D., Iwamoto, H., Liu, J., Szabo, G., Shao, Z. (2001) Charac-terization of AC mode scanning ion-conductance microscopy. *Ultra-microscopy* **90**, 13–19.

27 Korchev, Y. E., Gorelik, J., Lab, M. J., Sviderskaya, E. V., Johnston, C. L., Coombes, C. R., Vodyanoy, I.,

Edwards, C. R. W. (2000) Cell volume measurement using scanning ion conductance microscopy. *Biophys. J.* **78**, 451–457.

28 Korchev, Y. E., Negulyaev, Y. A., Edwards, C. R. W., Vodyanoy, I., Lab, M. J. (2000) Functional localization of single active ion channels on the surface of a living cell. *Nat. Cell Biol.* **2**, 616–619.

29 Radmacher, M., Cleveland, J. P., Fritz, M., Hansma, H. G., Hansma, P. K. (1994) Mapping interaction forces with the atomic force microscope. *Biophys. J.* **66**, 2159–2165.

30 Gitter, A. H., Bertog, M., Schulzke, J.-D., Fromm, M. (1997) Measure-ment of paracellular epithelial conductivity by conductance scanning. *Eur. J. Physiol.* **434**, 830–840.

31 Mann, S. A., Hoffmann, G., Hengsten-berg, A., Schuhmann, W., Dietzel, I. D. (2002) Pulse-mode scanning ion conductance microscopy: A method to investigate cultured hippocampal cells. *J. Neurosci. Methods* **116**, 113–117.

32 Happel, P., Hoffmann, G., Mann, S. A., Dietzel, I. D. (2003) Monitoring cell movements and volume changes with pulse-mode scanning ion conductance microscopy. *J. Microsc.* **212**, 144–151.

33 Mannelquist, A., Iwamoto, H., Szabo, G., Shao, Z. (2001) Near-field optical microscopy with a vibrating probe in aqueous solution. *Appl. Phys. Lett.* **78**, 2076–2078.

34 Gorelik, J., Yuchun, G., Spohr, H. A., Shevchuk, A. I., Lab, M. J., Harding, S. E., Edwards, C. R. W., Whitaker, M., Moss, G. W. J., Benton, D. C. H., et al. (2002) Ion channels in small cells and subcellular structures can be studied with a smart patch-clamp system. *Biophys. J.* **83**, 3296–3303.

35 Gorelik, J., Shevchuk, A. I., Frolenkov, G. I., Diakonov, I. A., Lab, M. J., Kros, C. J., Richardson, G. P., Vodyanoy, I., Edwards, C. R. W., Klenerman, D., et al. (2003) Dynamic assembly of surface structures in

living cells. *Proc. Natl. Acad. Sci. USA* **100**, 5819–5822.

36 Gorelik, J., Zhang, A., Shevchuk, A., Frolenkov, G. I., Sanchez, D., Lab, M. J., Vodyanoy, I., Edwards, C. R. W., Klenerman, D., Korchev, Y. E. (2002) The use of scanning ion conductance microscopy to image A6 cells. *Mol. Cell. Endocrinol.* **217**, 101–108.

37 Gorelik, J., Zhang, Y., Sánchez, D., Shevchuk, A., Frolenkov, G., Lab, M., Klenerman, D., Edwards, C., Korchev, Y. (2005) Aldosterone acts via an ATP autocrine/paracrine system: The Edelman ATP hypothesis revisited. *Proc. Natl. Acad. Sci. USA* **102**, 15000–15005.

38 Hansma, P. K., Cleveland, J. P., Radmacher, M., Walters, D. A., Hillner, P. E., Bezanilla, M., Fritz, M., Vie, D., Hansma, H. G., Prater, C. B., et al. (1994) Tapping mode atomic force microscopy in liquids. *Appl. Phys. Lett.* **64**, 1738–1740.

39 Putman, C. A. J., Werf, K. O. V. d., Grooth, B. G. D., Hulst, N. F. V., Greve, J. (1994) Tapping mode atomic force microscopy in liquid. *Appl. Phys. Lett.* **64**, 2454–2456.

40 Shalom, S., Lieberman, K., Lewis, A., Cohen, S. R. (1992) A micropipette force probe suitable for near-field scanning optical microscopy. *Rev. Sci. Instrum.* **63**, 4061–4065.

41 Lewis, A., Taha, H., Strinkovski, A., Manevitch, A., Khatchatouriants, A., Dekhter, R., Amman, E. (2003) Near-field optics: from subwavelength illumination to nanometric shadowing. *Nat. Biotechnol.* **21**, 1378–1386.

42 Betzig, E., Trautman, J. K., Harris, T. D., Weiner, J. S., Kostelak, R. L. (1991) Breaking the diffraction barrier: optical microscopy on a nanometric scale. *Science* **5000**, 1468–1470.

43 Toledo-Crow, R., Yang, P. C., Chen, Y., Vaez-Iravani, M. (1992) Near-field differential scanning optical micros-cope with atomic force regulation. *Appl. Phys. Lett.* **60**, 2957–2959.

44 Karraï, K., Grober, R. D. (1995) Piezoelectric tip-sample distance control for near field optical microscopes. *Appl. Phys. Lett.* **14**, 1842–1844.

45 Rensen, W. H. J., van Hulst, N. F., Kämmer, S. B. (2000) Imaging soft samples in liquid with tuning fork based shear force microscopy. *Appl. Phys. Lett.* **77**, 1557–1559.

46 Brunner, R., Hering, O., Marti, O., Hollricher, O. (1997) Piezoelectrical shear-force control on soft biological samples in aqueous solution. *Appl. Phys. Lett.* **71**, 3628–3630.

47 Koopman, M., de Bakker, B. I., Garcia-Parajo, M. F., van Hulst, N. F. (2003) Shear force imaging of soft samples in liquid using a diving bell concept. *Appl. Phys. Lett.* **83**, 5083–5085.

48 Lambelet, P., Pfeffer, M., Sayah, A., Marquis-Weible, F. (1998) Reduction of tip-sample interaction forces for scanning near-field optical micros-copy in a liquid environment. *Ultra-microscopy* **71**, 117–121.

49 Bockris, J. O., Reddy, A. K. N. (2000) *Modern Electrochemistry: Electrodics in Chemistry, Engineering, Biology, and Environmental Science.* 2nd edn. Plenum Publishing Corporation, New York.

50 Bard, A. J., Denuault, G., Lee, C., Mandler, D., Wipf, D. O. (1990) Scanning electrochemical microscopy: a new technique for the characteriza-tion and modification of surfaces. *Acc. Chem. Res.* **23**, 357–363.

51 Schäffer, T. E., Anczykowski, B., Fuchs, H. (2006) Scanning ion conductance microscopy. In: Bhushan, B., Fuchs, H. (Eds.), *Applied Scanning Probe Methods.* Springer-Verlag, Berlin, Heidelberg, New York, Vol. 2, pp. 91–119.

52 Korchev, Y. E., Raval, M., Lab, M. J., Gorelik, J., Edwards, C. R. W., Rayment, T., Klenerman, D. (2000) Hybrid scanning ion conductance and scanning near-field optical microscopy for the study of living cells. *Biophys. J.* **78**, 2675–2679.

53 Mannelquist, A., Iwamoto, H., Szabo, G., Shao, Z. (2002) Near field optical microscopy in aqueous solution:

implementation and characterization of a vibrating probe. *J. Microsc.* **205**, 53–60.

54 Bruckbauer, A., Ying, L., Rothery, A. M., Korchev, Y. E., Klenerman, D. (2002) Characterization of a novel light source for simultaneous optical and scanning ion conductance microscopy. *Anal. Chem.* **74**, 2612–2616.

55 Rothery, A. M., Gorelik, J., Bruckbauer, A., Yu, W., Korchev, Y. E., Klenerman, D. (2003) A novel light source for SICM–SNOM of living cells. *J. Microsc.* **209**, 94–101.

56 Gorelik, J., Shevchuk, A., Ramalho, M., Elliott, M., Lei, C., Higgins, C. F., Lab, M. J., Klenerman, D., Krauzewicz, N., Korchev, Y. (2002) Scanning surface confocal microscopy for simultaneous topographical and fluorescence imaging – Application to single virus-like particle entry into a cell. *Proc. Natl. Acad. Sci. USA* **99**, 16018–16023.

57 Zhang, H., Wu, L., Huang, F. (1999) Electrochemical microprocess by scanning ion-conductance microscopy. *J. Vac. Sci. Technol. B* **17**, 269–272.

58 Lewis, A., Kheifetz, Y., Shambrodt, E., Radko, A., Khatchatryan, E., Sukenik, C. (1999) Fountain pen nanochemistry: Atomic force control of chrome etching. *Appl. Phys. Lett.* **75**, 2689–2691.

59 Müller, A.-D., Müller, F., Hietschold, M. (1998) Electrochemical pattern formation in a scanning near-field optical microscope. *Appl. Phys. A* **66**, S453–S456.

60 Müller, A.-D., Müller, F., Hietschold, M. (2000) Localized electrochemical deposition of metals using micropipettes. *Thin Solid Films* **366**, 32–36.

61 Hong, M.-H., Kim, K. H., Bae, J., Jhe, W. (2000) Scanning nanolithography using a material-filled nanopipette. *Appl. Phys. Lett.* **77**, 2604–2606.

62 Larson, B. J., Gillmor, S. D., Lagally, M. G. (2003) Controlled deposition of picoliter amounts of fluid using an ultrasonically driven micropipette. *Rev. Sci. Instrum.* **75**, 832–836.

63 Ying, L., Bruckbauer, A., Rothery, A. M., Korchev, Y. E., Klenerman, D. (2002) Programmable delivery of DNA through a nanopipet. *Anal. Chem.* **74**, 1380–1385.

64 Bruckbauer, A., Ying, L., Rothery, A. M., Zhou, D., Shevchuk, A. I., Abell, C., Korchev, Y. E., Klenerman, D. (2002) Writing with DNA and protein using a nanopipet for controlled delivery. *J. Am. Chem. Soc.* **124**, 8810–8811.

65 Bruckbauer, A., Zhou, D., Ying, L., Korchev, Y. E., Abell, C., Klenerman, D. (2003) Multicomponent submicron features of biomolecules created by voltage controlled deposition from a nanopipet. *J. Am. Chem. Soc.* **125**, 9834–9839.

66 Taha, H., Marks, R. S., Gheber, L. A., Rousso, I., Newman, J., Sukenik, C., Lewis, A. (2003) Protein printing with an atomic force sensing nanofountainpen. *Appl. Phys. Lett.* **83**, 1041–1043.

67 Rodolfa, K. T., Bruckbauer, A., Zhou, D., Korchev, Y. E., Klenerman, D. (2005) Two-component graded deposition of biomolecules with a double-barreled nanopipette. *Angew. Chem. Int. Ed.* **117**, 7014–7019.

68 Rodolfa, K. T., Bruckbauer, A., Zhou, D., Schevchuk, A. I., Korchev, Y. E., Klenerman, D. (2006) Nanoscale pipetting for controlled chemistry in small arrayed water droplets using a double-barrel pipet. *Nano Lett.* **6**, 252–257.

69 Ying, L., Bruckbauer, A., Zhou, D., Gorelik, J., Shevchuk, A., Lab, M., Korchev, Y., Klenerman, D. (2005) The scanned nanopipette: a new tool for high resolution bioimaging and controlled deposition of biomolecules. *Phys. Chem. Chem. Phys.* **7**, 2859–2866.

70 Seebach, J. (2001) *Einsatz elektrisch leitender Kultursubstrate zur Charakterisierung von Zell-Zell- und Zell-Substrat-Interaktionen.* Dissertation thesis, University of Münster, Münster.

12
Label-Free Nanowire and Nanotube Biomolecular Sensors for In-Vitro Diagnosis of Cancer and other Diseases

James R. Heath

12.1
Overview

Over the past few years, several groups have reported on carbon nanotube (NT) or semiconductor nanowire (NW) chemical [1, 2] and biomolecular sensors [1–3]. The biosensing applications are, in many ways, driven by the emerging concepts of systems biology [4, 5] and the translation of those concepts into the clinic [6]. From both the fundamental (biology) and applied (clinical) perspectives, the technology driver is to develop biomolecular measurement tools that are efficiently produced, sensitive, quantitative, rapid, multiparameter (e.g., mRNAs and proteins), scalable to large numbers of parameters measured, and extendable towards quantitating molecules from ever-smaller amounts of tissues, cells, and serum. In this chapter, we review these sensors, in terms of the needs that are driving their development, in terms of the progress that has been reported in the literature, and in terms of the challenges and questions that remain to be addressed.

12.2
Background

Chip-based methods for the detection of mRNAs [7], cDNAs and (to a lesser extent) proteins [8], are now a part of most experimental biology toolsets. Whilst these tools are not the subject of this chapter, it is instructive to discuss at least one of these methods in some detail, and then to place that within the context of an emerging clinical need. Thus, the point is to illustrate why these emerging biosensor technologies are being developed.

As a general rule, the chip-based methods are predicated upon the optical detection of specific biomolecules through the use of fluorophore labels. The prototypical example for quantitating peptides, proteins, antibodies and hormones is the technique known as the enzyme-linked immunosorbent assay (ELISA) [9].

Nanobiotechnology II. Edited by Chad A. Mirkin and Christof M. Niemeyer
Copyright © 2007 WILEY-VCH Verlag GmbH & Co. KGaA, Weinheim
ISBN: 978-3-527-31673-1

The most common ELISA variant, which achieves an indirect detection of the biomolecule, is often referred to as sandwich assay [10] – that is, the protein or other biomolecule to be detected is sandwiched between a surface-bound and a fluorophore labeled primary and secondary antibody, respectively. It is the binding event of the secondary antibody that is optically detected. For the example of proteins, ELISA assays are typically carried out in multi-well plates, with one type of protein detected per well. Once the well has been prepared with the substrate-bound antibody, incubation of the sample to be tested, followed by incubation with the secondary antibody, takes a few hours, according to standard protocols. As discussed later in this chapter, this long-time scale is not an intrinsic limitation of ELISA methods, but is simply characteristic of the standard protocols. With a primary antibody that has 10^{-9} M affinity, and sufficient incubation time, an ELISA sandwich assay can detect proteins that are present in the low picomolar concentration range.

The limitations of the ELISA are manifold:

- It is designed as a pauciparameter, or single-protein, detection method (although it is extendable by using multi-well plates, given sufficient biological sample).
- The requirement for a fluorophore label means that the number of different molecules that can be detected within a given spatial area is, in principle, diffraction limited and, in practice, typically limited to about one species within a 100 μm diameter area or so (though optically defined chips are extendable to higher densities). (Note: this also applies to mRNA gene chips.) This condition places demands on the amount of tissue needed for a true multiparameter analysis, as well as extending the time scale for sample analysis.
- The range over which the concentration of a specific biomolecule may be quantified is about 10^2, and limited by the minimal detectable signal over background and the tendency for fluorophores to photobleach.
- The method is not real time, although under appropriate conditions it can be reasonably fast.
- The requirement for two complementary antibodies per biomolecule – at least one of which needs to be high affinity – is nontrivial. Significant efforts have been made into improving the ELISA technique, but these will not be discussed here [11–16].

At this point we should perhaps consider the specific challenge of stratifying glioblastoma (brain cancer) patients for therapeutic intervention, based upon the analysis of tissue extracted from a fine-needle biopsy. Brain cancer is the leading cause of cancer deaths in children aged under 15 years, and the second leading cause of cancer deaths among the 15- to 34-year age group. Although brain cancer is less common in adults, it accounts for a disproportionate number of deaths [17], with glioblastoma being the most aggressive of all brain tumors. Hence, this important problem not only represents a scientific and technological challenge but is also typical of the type of clinical diagnosis that can provide an early avenue into personalized medicine.

As with many other cancers, overexpressed or mutated tyrosine kinases can contribute to the development and progression of brain tumors. Various tyrosine kinase inhibitor molecules [18, 19] have been developed and demonstrated as effective therapeutics against brain cancer. However, for any randomly chosen group of glioblastoma patients, the responders to a given inhibitor constitute only a small subset [20] which cannot be distinguished by using traditional pathology practices [21–24]. Clearly, this complicates clinical trials [25] and also makes treating the patients significantly more difficult. If a large piece of the tumor is surgically resected, then combinations of microscopic analysis, mRNA microarray data [26], immunohistochemical staining, Western blotting [20] and other methods can be utilized to detect and/or sort the cancerous cells, interrogate for the presence of oncogenes, characterize the phosphorylation status of kinase pathways, and direct patients towards appropriate therapies. Such an approach truly requires a multi-parameter analysis of cells, genes, and proteins. For example, while the oncogene can be detected at the mRNA level, the proteins within the kinase pathway must be directly measured, as it is the relative abundance of post-translationally modified proteins that is of the greatest interest. A visual inspection of the sorted cancer cells also provides useful information. For example, if it were possible to carry out a full, multi-parameter analysis on a handful of cells extracted from a tumor via a fine-needle biopsy, then not only would the patient likely avoid the surgical procedure, but fundamentally new questions relating to the molecular heterogeneity of the tumor could also be addressed.

The above problem is representative of the types of difficulty that drive the development of new measurement technologies. Although there is clearly a clinical need which drives such developments, these new methods will in time undoubtedly be used to seek new biological answers. One class of these emerging technologies – namely nanowire (NW) and nanotube (NT) nanoelectronic biomolecular sensors – will be discussed in the following section. Although related methods, such as nanomechanical biosensors [27, 28], are currently being developed to meet similar goals, they will not be discussed at this point.

These emerging nanotechnologies face clear challenges, with the need to extend the capabilities of measurement platforms that can sensitively, and in real time, quantitate one or two proteins or mRNAs, to chips that can quantitate large numbers of different biomolecules, being one of the greatest. A related task involves selective control of the nanosensor/biology interface – that is, encoding different sensors so that they each sense a different biomolecule. The standard method of inkjet spotting, for example, cannot achieve the desired densities required for problems such as glioblastoma. A third issue involves the integration of nanoelectronic sensors with microfluidics-based tissue handling and sample delivery systems. This is not only a fabrication challenge; rather, it also involves achieving an understanding of – and control over – the analyte-(surface-bound) antibody-binding kinetics within microenvironments. Finally – and in common with any multiparameter measurement platform – there are issues related to the rapid development of high-affinity protein-capture agents (e.g., antibodies, aptamers, peptides, small molecules) and surface chemistry that allows for proteins, genes, and cells to be detected within a single environment. Some of these challenges are specific to silicon NW sensors, while others are specific to single-wall carbon nanotube (SWNT) sensors. Yet, the emerging body of information relating to both types of devices is beginning to provide guidance on some of these issues. The current state of the art with respect to both NT and NW sensors, as well as some of the key unknowns relating to these devices, will be outlined in the following sections.

12.3
Methods and Current State of the Art

Nanotube sensors were first reported by Dai's group [1] as gas-phase chemical sensors, while Si NW biosensors were first reported by the group of Lieber [3]. A summary of much of the existing literature on NT and NW sensors is provided in Table 12.1, which will serve as a guide for much of the discussion here. Figure 12.1 represents a generic NW or NT sensor, with certain figures of merit and other characteristics indicated.

12.3.1
Mechanisms of Sensing

NT and NW sensors are, in many ways, variants of a traditional class of devices known as chem (field effect transistors) FETs or ion FETs [29–33]. These devices, which function as either pH sensors, or as sensors of specific ions (usually divalent cations), take advantage of the local field generated by the solution-phase species (in this case, ions) to modulate the conductivity of the source-drain channel within a semiconductor FET. Although these traditional (and macroscopic) devices have not yet been demonstrated as good biomolecular sensors, research into related devices for certain types of biosensing remains active [34].

The most commonly reported NT sensor consists of a network of SWNTs grown on-chip [35]. The response of such a type of a sensor is a heterogeneous combination of individual SWNT responses. By contrast, most NW sensors are based upon single NWs, and so physical characteristics such as the NW diameter and doping level will surely all play important roles. No group has yet reported on a systematic investigation of the effect of parameters such as NW diameter and doping level on sensing response, though two groups have shown that p- and n-type Si NW sensors exhibit opposite responses when sensing the same biomolecule [36, 37]. This has been rationalized in terms of the isoelectric point of the sensor–bound capture agent–target biomolecule complex. Si NW DNA sensing

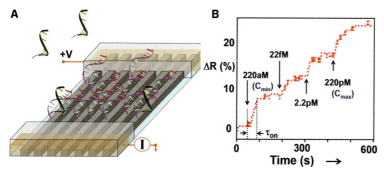

Fig. 12.1 A semiconductor nanowire (NW) sensor and characteristic response metrics. (A) Diagram illustrating some of the important aspects of these devices. The NW is treated as a field-effect transistor, and is electrically contacted (transparent gold regions at either end of the device). A voltage (+V) is applied to the (top) source electrode, and the bottom (drain) electrode is grounded through an ammeter (I). The electrode materials are encased within an insulating dielectric (transparent gray region) to prevent conductivity through the solution, which can interfere with (and dominate) the measured sensor response. The surface of the sensor is coated with an antigen (in this case ssDNA) that exhibits a specific recognition for a particular biomolecule (complementary DNA; cDNA). The environment in which the sensor is utilized can be tissue culture media, serum, desalted serum, water, or even air. A nanotube (NT) sensor is similarly constructed, although it typically will consist of a network of NTs bridging the electrodes. In addition, the electrical transduction of the target–probe binding event can differ between NWs and NTs. (B) The measured sensing response from an n-type Si NW sensor, coated with an ssDNA analyte, to the cDNA antigen (16-mer overlap region) in 0.154 M electrolyte. A change in the nanosensor resistance is detected and correlated a stepwise increasing concentration of the analyte. The nanosensor will exhibit a τ_{on} (i.e., the time between when the sample containing the analyte is introduced to when the response of the nanosensor saturates). τ_{on} is highly dependent upon the on and off target–probe binding constants as well as on how the experiment was conducted (e.g., flowing microfluidics versus a static fluid). The binding constants, in turn, can depend upon the sensing environment, the presence of dications, etc. Some nanosensors also exhibit a τ_{off}, correlating to the time that it takes the nanosensor electrical response to recover to its baseline resistance after the analyte is removed from solution. Nanosensors are often characterized in terms of the range of analyte concentration values over which ΔR responds linearly with respect to $\log[C_{analyte}]$. For this particular sensor, these limits are 220 aM ($= C_{min}$) to 220 pM ($= C_{max}$).

has been carried out in non-electrolyte media by at least two groups [38, 39], each of which has used a different diameter (or width) NW. The results indicate that 50 nm-wide (with a rectangular cross-section) Si NW sensors are significantly less sensitive than 20 nm-diameter Si NW sensors. Variability in doping levels and electrical contacts might also contribute to these reported differences, though it is also likely that the increased sensitivity of the smaller diameter NWs arises in part from the fact that they quasi one-dimensional conductors. This implies that the nontraditional device fabrication methods currently used to prepare NWs [40–42] and NTs [35] will continue to play an important role in their production, as lithographic patterning techniques [43] do not achieve the resolution and regularity required for highly sensitive and reproducible NW sensors.

12.3.2
The Role of the Sensing Environment

The field-gating mechanism that is characteristic of NW sensors means that it is the local change in charge density – and the accompanying change in local chemical potential – that characterizes target–probe binding at the NW sensor surface. Typically, it is a change in the number and nature of the charges that reside at the surface of the NW that is detected (most target–probe binding events are accompanied by such a change). This changing chemical potential affects the source (S) → drain (D) current value, or I_{DS}. However, the change is screened (via Debye screening) from the NW by the solution in which the sensing takes place. Debye screening is a function of electrolyte concentration, and at 0.14 M monovalent electrolyte – which is reasonably representative of most physiological environments such as serum or tissue culture media – the screening length is about 1 nm [44]. In addition, the SiO_2 surface on a Si NW has a low isoelectric point, which means that at pH 7.4 the surface has a high quantity of negative charge [45]. This can potentially exacerbate the Debye screening, as positive ions are recruited to the negatively charged SiO_2 surface, and so the local electrolyte concentration at the surface can be much higher than the ionic strength of the bulk solution [44]. Because of Debye screening, NW sensing experiments have largely been carried out in low ionic strength solutions. Lieber's group, for example, has reported an elegant experiment in which they demonstrated the detection of multiple proteins from serum samples that had been desalted using a microcentrifuge filtration technique [37]. Desalting a tissue or serum sample is challenging when that sample is very small (i.e., a fingerprick or fine-needle biopsy). Thus, the challenge of interrogating the intrinsic biological specimen using NW sensors remains open.

Nanotubes by contrast, sense via a Schottky barrier modulation effect (at the NT contacts [46]), as well as by the chemical gating effect [47]. When the isoelectric point of a protein is close to the pH of the reaction media, the Schottky barrier effect is particularly important. Byon and Choi, for example, were able to increase the sensitivity of SWNT network sensors by increasing the Schottky contact area [48].

For both NWs and NTs, the role that the ionic strength (and the identity of the ions present) in the solution plays has not been systematically investigated. It is unlikely that the high sensitivities to protein detection that have been reported for Si NW sensors [37, 49] in nonionic media would be achieved in 0.14 M electrolyte. Here, the surface-bound biomolecular capture agent plays an important role. For example, extremely high sensitivities for detecting DNA in 0.164 M electrolyte have been reported for Si NW sensors [36]. However, when Si NW sensors with essentially identical diameters and doping levels are utilized to detect the protein tumor necrosis factor-α (TNFα) (using anti-TNFα as the surface-bound capture agent) in the same medium, the sensitivity is very poor (in the few nanomolar limit) [50]. The major difference here is the capture agent. Antibodies are quite large and so hold the captured protein several nanometers – and thus several Debye screening lengths – away from the surface of the NW. ssDNA oligomers, on the other hand, most likely lie down on the NW surface prior to their hybridization with cDNA (at which point they become much stiffer). Thus, DNA-hybridization likely modifies the charge distribution right at the NW surface, while protein detection by antibodies occurs at a larger distance.

It is interesting to note that the electrolyte concentration, in certain situations, can improve the sensitivity of at least NT network sensors. For example, Star and co-workers found that the addition of a 20 mM solution containing Mg^{2+} cations dramatically increased (10^3-fold) the sensitivity of SWNT network sensors towards detecting DNA hybridization events [51]. Mg^{2+} is known to increase the equilibrium binding constant of DNA.

12.3.3
Nanosensor-Measured Target–Probe On/Off Binding Rates

The ionic strength of the solution can, in general, play a significant role in determining on/off target–probe binding rates (see Table 12.1). One commonly utilized label-free method for detecting biomolecules is that of surface plasmon resonance (SPR) [52] which, while not typically used to quantitate protein concentrations, is utilized to quantitate the on/off rates of target–probe binding, and thus to determine binding affinities for antibodies and other capture agents from the equilibrium relationship of $K = k_{on}/k_{off}$. Thus, for a typical antibody binding affinity of 1 nM, $k_{on} = 10^9 \times k_{off}$. NW and NT sensors can also potentially be utilized for measurements of target–probe binding affinities.

Backmann and co-workers have directly compared binding constant measurements determined from nanocantilever-based protein sensors with SPR measurements, and have found quantitative agreement [53]. For the Si NW sensor investigations summarized in Table 12.1, the reports that describe either protein or DNA sensing in low ionic strength media, and that present data from which the on and off times may be estimated (simply by measuring adsorption/desorption times from various sensor plots), exhibit both a rapid sensing response (which is, in general, a function of both τ_{on} and τ_{off}) and a rapid recovery to baseline, which is only dependent upon τ_{off}. τ represents the time to sensor saturation (τ_{on}) or baseline recovery. We utilize τ_{on} and τ_{off} because, for most NW sensor reports, the actual rates are not extractable from the published data. Nevertheless, the label-free, real-time nature of NW-based sensing implies that such devices can be utilized to both extract equilibrium binding constants, or that it can be utilized to discriminate between selective and nonselective binding. Lieber has analyzed his NW sensor data for at least one or two proteins, and has found that the measured on/off rates do reproduce – to at least an order of magnitude – known target–probe binding affinities (C. Lieber, personal communication). As discussed below, when surface-bound assays are integrated into a microfluidics environment, the binding and unbinding kinetics depend heavily upon the flow rate of the solution.

Table 12.1 Summary of nanowire/nanotube sensor reports.

Sensor[a]	Target–probe used	Environment[b]	Sensitivity and dynamic range[c]	Saturation time (concn.)[d]	Reference(s)
SWNT	Gas-phase sensing (dry)		2 ppm	~200 s	1
Polymer-wrapped SWNT rope	Electronic transduction of polymer optical absorption (dry; cryogenic)		10^2–10^3 current amplification of absorbed photons		2
Si NW; D = 20 nm; p-type	Streptavidin/biotin	1 mM PBS; 10 mM NaCl	25 pM; DR = 10^4	τ_{on} = 20 s	3
SWNT net	Streptavidin/biotin	0.01 M PBS	2.5 μM; DR n.a.		79, 82
SWNT net	ssDNA/cDNA	Buffer; ionic strength not provided	100 pm; DR = 10^2	τ_{on} ~ 10 s	81
SWNT	glucose oxidase/glucose	18 MΩ H_2O	10 mM glucose; DR n.a.	τ_{on} ~ 10 s	72
SWNT net	Proteins/antibodies	Demonstrated Schottky effect (i.e., sensing at contacts) as a SWNT (not NW) sensing mechanism			47
Si NW; D = 20 nm; p-type	ssDNA/cDNA	Distilled H_2O	10 fM; DR = 10	τ_{on} < 10 s τ_{off} < 10 s	38
Si NW; D = 50 nm; p,n-type	ssDNA/cDNA	18 MΩ H_2O	25 pM; DR n.a.	τ_{on} ~ 10 s	39
Si NW; D = 20 nm; p-type	Virus particle/antibody	10 μm KCl	1 virus; 50 h^{-1} for 800 virus μL^{-1} 10 h^{-1} for 100 virus μL^{-1}	τ_{on} < 1 min τ_{off} < 1 min	49

Sensor type[a]	Receptor/analyte	Sensing environment[b]	Sensitivity[c]; DR	Response time[d]	Ref.
Si NWs p and n type	Proteins/antibodies	Desalted serum;	0.5 fM; DR > 10^5	τ_{on} ~ 2–3 min, τ_{off} ~ 2–3 min	37
NT net	ssDNA/cDNA	200 mM PBS, 10 mM PBS + 20 mM Mg^{2+}	1 nM; DR = 10^2, 1 pM; DR = 10^5		51
Si NW; 20 nm; p-type	ATP/Abl; Gleevec inhibitor	1 μm electrolyte	100 pM; DR = 200	τ_{on} < 10 s @ 1 nM	80
In_2O_3 and SWNT net	PSA/anti-PSA	Air and PBS; ionic strength not provided	100 pM	τ_{on} (NW) = 20 s; τ_{on} (NT) = 300 s	58
SWNT net	Proteins/antibodies	10 mM PBS	1 pM (nonspecific and specific binding); DR = 10^2	τ_{on} ~ 10 s	48
Si NW; D = 17 nm; n,p-type	ssDNA/cDNA	0.16 M electrolyte	220 aM; DR = 10^6	τ_{on} ~ 60 s	36

[a] Sensor type (NT net = nanotube network); D (sensor width); doping (type and values given when reported for NWs).

[b] Sensing environment (PBS = phosphate buffer solution).

[c] Sensitivity limit to analyte concentration; DR = sensor dynamic range.

[d] τ_{on} = time to sensor saturation at minimum analyte concentration (estimated from plots for many reports); τ_{off} = time for sensor recovery to baseline (estimated from plots from reports; data not provided for many reports). Note that these times do not imply that the slope, or rate or adsorption/desorption, was measured. In most cases, such slopes were not extractable from the published data.

During the writing of this chapter, a systematic investigation regarding protein–antibody binding affinities as a function of ionic strength could not be located. For some very specific cases, reducing the ionic strength to values well below physiological levels ($\ll 0.15$ M electrolyte) can increase k_{on} significantly while impacting k_{off} only slightly, but it can also significantly reduce the selectivity of the assay [54]. For DNA, a lower ionic strength significantly retards DNA hybridization [55], although how k_{on} and k_{off} rates are separately affected is unclear. Again, a systematic study of how ionic strength impacts the device performance metrics of NW sensors is needed. As described below, recent models of target–probe binding in immunoassays highlights the influence of how k_{on} and k_{off} can affect analyte binding rates, especially when the assay is entrained within a flowing microfluidics environment.

12.3.4
The Nanosensor/Microfluidic Environment

Many of the results presented in Table 12.1 were collected from NW or NT sensors entrained within polydimethylsiloxane (PDMS) microfluidics flow channels. Modeling the binding kinetics of surface immunoassays has been recently investigated by two groups [56, 57]. Assuming laminar flow conditions (which may not be appropriate within a microfluidic channel that contains patterned electrodes and static electric fields), Zimmerman and co-workers calculated the various rates for which an ELISA assay could be optically detected as a function of target area, flow rate, analyte concentration, and channel geometry. For a probe that exhibits a nanomolar affinity for the target (typical of most protein–antibody or ssDNA–cDNA molecules utilized in Table 12.1), these authors calculated two limits for the binding kinetics. Under low-flow velocity (the order of 10^{-2} mm s^{-1}), the surface-bound probe is able to exploit a large fraction of the target, but with very slow capture kinetics. Such binding kinetics can be modeled using four partial differential equations that arise from the Navier–Stokes equation, the convection–diffusion equation, and the target–probe k_{on} and k_{off} rates [56]. Higher flow velocities (the order of 1 mm s^{-1}) and small active areas (150 nm^2) are more relevant to the case of nanosensors entrained within microfluidics channels. Under such a limit, the immunoassay binding kinetics reflect the target–probe k_{on} and k_{off} rates:

$$\frac{d\theta_t}{dt} = k_{on} C (\theta_{max} - \theta_t) - k_{off} \theta_t \tag{1}$$

In addition, a characteristic binding time, τ, can be expressed as

$$\tau \approx (k_{on} C + k_{off})^{-1} \tag{2}$$

Under (the quite common) conditions in which $k_{on} C \ll k_{off}$, then $\tau \approx 1/k_{off}$.

Here, θ_t is the surface density of bound target at time t, θ_{max} is the maximum surface density of molecules possible, and C is the target concentration. As seen from Eqs. (1) and (2), at low target concentrations k_{off} plays a more significant role in determining the saturation time for the immunoassay than does k_{on}. In fact, over fairly broad concentration ranges (10^{-16} to 10^{-10} M) of target, τ_{on} (reflecting the time to two-thirds of the sensor saturation limit for a given concentration) can be quite fast (<1 min) for an target–probe binding affinity in the nanomolar range that exhibits values for k_{on} of 10^8 M^{-1} s^{-1} and k_{off} of 10^{-1} s^{-1}. These types of rapid kinetics can also be applied to ELISA assays, and so the previously mentioned slow rate of developing ELISA immunoassays can be substantially improved through microfluidics design and through the optimization of flow parameters. However, if $k_{on} = 10^5$ M^{-1} s^{-1} and k_{off} 10^{-4} s^{-1} (producing the same equilibrium binding affinity), the values for τ_{on} are 10^3-fold longer. This highlights the need for understanding target–probe binding kinetics under different conditions. For example, for cDNA sensing, the ssDNA probe may be immobilized onto the nanosensor surface through electrostatic interactions, or through covalent bonding. It is likely that these two types of immobilization strategies will affect k_{on} and k_{off} values in different ways, as will the ionic strength of the sensing media, or the presence of divalent cations. Modeling studies, capture agent design and optimization are all needed if microfluidic-based nanosensors are to be rationally optimized for real-time, highly sensitive biomolecule detection.

12.3.5
Nanosensor Fabrication

A second challenge involves demonstrating reproducible and high-throughput nanofabrication/assembly methods that can produce near-identical NW or NT sensor circuits time and time again, at low cost, and in a way that allows for many measurements to be executed in parallel on the same chip. Traditionally, the scaling up from single-device demonstrations to robust, integrated technologies has been viewed as an engineering problem. However, for nanotechnologies in which the individual device is non-traditionally fabricated, the development of a manufacturing method itself represents a challenge that is more scientific in nature.

The bulk of reports on NT sensors are based on NT network devices containing many NTs. Some sensitivity is likely sacrificed, as many of the NTs that contribute to the measurement probably do not contribute to the sensing. Although the reproducibility from sensor to sensor is apparently reasonably good [47, 48, 51, 58], no group has yet reported the use of an array of NT sensors, fabricated on the same chip, to detect a panel of biomolecules. Similarly, no group has yet reported on the sensing statistics from a large number of such sensors.

Si NWs for sensing applications have been prepared from three different methods: (i) electron beam lithography (EBL) [39, 59]; (ii) vapor-liquid-solid (VLS) growth [40, 60]; and (iii) superlattice nanowire pattern transfer (SNAP) [42, 61]. Each of these methods has its advantages and disadvantages:

- EBL is a serial patterning method that produces wires exactly where one wants them to be, but it does not scale to the dimensions required for the most sensitive NW devices (see above discussion). However, the wires are readily suspended (a bonus for sensing), doping can be well controlled, and it is straightforward to construct microfluidics channels around EBL NWs.

- The VLS technique relies on using a metal particle (such as Au) to nucleate and direct the growth of a nanowire. This method can produce bulk quantities of semiconductor NWs (both Si and In_2O_3 [58] have been utilized for sensing) of the appropriate dimensions. Those NWs are characterized by a distribution of lengths, and some control over NW diameter and doping has been demonstrated. A batch of n- or p-type 20 ($\pm$3) nm diameter NWs is relatively easy to prepare. The NWs must be assembled into the appropriate device structure [62] (or the device structure must be constructed around the NWs [63]). However, they can be utilized on a variety of sub-strates [64]. For sensing applications, the manner in which the NWs are surface-functionalized with probe, assembled into the appropriate device structure, and entrained within microfluidics has varied, and the integration of these various steps represents ongoing development [37, 49]. VLS nano-wires have been demonstrated as reasonably good FETs, which is a requirement for them also performing as good sensors.

- SNAP NWs (a Si NW sensor array is shown in Figure 12.2) combine components of the EBL and VLS NWs. They are prepared using a templating/stamping approach in which the atomic control achievable in an MBE-grown GaAs/AlGaAs superlattice is translated into the pitch and width of various metal and/or semiconductor nanowires. Si NWs are prepared from a silicon-on-insulator or a bonded substrate that is pre-doped to the appropriate level (both n- and p-type NWs can be simultaneously prepared) within the same circuit. These NWs make excellent FETs [65], and arrays which are a few millimeters long and containing up to 1500 NWs can be prepared. The variation in NW diameter (and pitch) for SNAP NWs is <1 nm. A major limitation with SNAP NW sensors is that each sensor circuit has to be individually prepared, though this is also true of the VLS and EBL NWs. However, recent studies on replicating SNAP NWs using nanoimprinting techniques may pave the way towards high-throughput NW sensor circuit fabrication [66]. These nanowires are sectioned into as many as eight separate 400 NW arrays per chip; from each array several NWs or groups of NWs are electrically contacted. PDMS microfluidics are bonded onto the chip surface so that each separate array is entrained within its own channel.

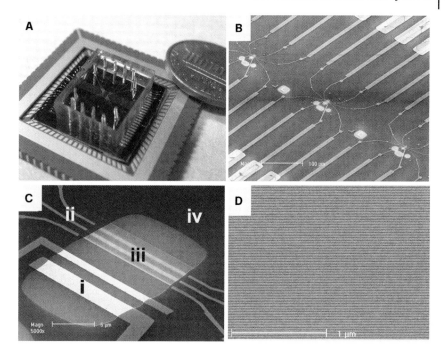

Fig. 12.2 A circuit of Si nanowire (NW) sensors, shown at various levels of resolution. The NWs themselves were fabricated using the superlattice nanowire pattern (SNAP) transfer method, but most of the basic features of this circuit are common to other reported devices. (A) An assembled sensor device, placed onto a chip carrier, and incorporating a polydimethylsiloxane (PDMS) microfluidic layer. A penny is shown for scale. This particular circuit contains six individually addressable NW sensor arrays, with each array containing 400 NWs. For this circuit, only three NW sensor devices are electronically contacted for each array, and each sensing device contains about 10 Si NWs. (B) Expanded view of the NW sensor arrays. The electrical contacts to the NWs extend diagonally to the top right and bottom left of the image. The nanowires themselves are aligned diagonally along the white strip, from bottom right to top left. (C) A single three-element NW sensor array. The individual nanowires are not resolved at this resolution. Various regions are labeled: (i) Gate and counter electrodes (for controlling the reference potential of the NWs and the solution); these electrodes are not a common feature of NW or nanotube (NT) sensor devices, but they do increase sensitivity. (ii) Nanowire electrical contacts. (iii) The bright stripes are groups of contacted Si NWs. (iv) The dark region indicates an insulating Si_2N_3 film deposited on top of the metal contacts. (D) High-resolution electron micrograph showing an array of 18 nm-wide Si NWs used for sensing.

12.3.6
Biofunctionalizing NW and NT Nanosensors

NT and NW sensors are biofunctionalized for sensing in very different ways. While Si NWs can take advantage of the highly developed surface chemistry of the SiO_2 surface (and the lesser-developed surface chemistry of the H–Si surface), NTs are typically biopassivated using noncovalent interactions. Here, the case for NTs is discussed first.

Proteins tend to bind non-selectively to the surfaces of SWNTs [67], and covalent functionalization of SWNTs tends to disrupt their electrical properties. Several groups have investigated the noncovalent functionalization of NTs, using a variety of small molecules, such as pyrene [68], supramolecular structures [69], and polymers [2, 70, 71]. Dai's group has utilized noncovalent interactions to immobilize polyethylene glycol (PEG) side chains [67], which greatly reduces the nonselective binding of proteins. The advantage of these noncovalent methods is that they can be used to attach probes specifically to the SWNT or SWNT network devices without disrupting the covalent bonding within the NTs. This means that the target "sees" probe only within the active sensor region which, according to mathematical models [56], considerably improves the short-time sensitivity of surface-bound immunoassays under certain conditions. For a specific sensing example, Dekker's group has utilized pyrene-labeling to selectively attach glucose oxidase onto SWNTs [72]. For DNA sensing, ssDNA has been demonstrated to interact non-covalently with SWNTs [73], and this has provided a route toward constructing NT network DNA sensors [51].

Most biofunctionalization of NWs has been carried out using traditional methods for coupling biomolecules on SiO_2 surfaces. On a chip that contains multiple individual NW sensors or sensor arrays, the arrays may be isolated from one another using microfluidics channels. If Si NWs are supported on an SiO_2 surface, then the surface surrounding the NWs may also be biofunctionalized, thereby leading to a chemically active surface area that is substantially larger than the actual active area of the NW sensor, and so potentially reducing the overall sensitivity of the assay. If the NWs are biofunctionalized prior to assembly that problem is avoided, but it is not always easy to make ohmic contacts to such devices. Briefly, such nonselective chemistry typically involves introducing an aldehyde group onto the NW surface through the covalent coupling of methoxysilanes. Amide coupling chemistry is then utilized to attach ssDNA and/or antibodies onto the NW surface [49]. DNA may be electrostatically coupled onto a nanowire surface if the surface is prefunctionalized with an amino terminus. The electrical contacts to the NWs are often covered with an insulating dielectric such as Si_3N_4 so that ion-conductance through the solution does not interfere with the sensing measurement.

Most methods for fabricating and/or assembling NW sensor arrays can achieve sensor densities that are substantially higher than can be chemically addressed using either microfluidics channels, or spotting methods for directing surface biofunctionalization. In fact, taking advantage of such high-density NW sensor arrays, by encoding each NW sensor to detect a different biomolecule, constitutes one of the major challenges associated with applying NW sensors for large-scale multi-parameter analysis. The ability to electronically address individual NWs within a dense NW sensory array, is not, however, a particular limitation [61]. Thus, electrochemically encoding a NW sensor array provides a viable approach towards constructing ultra-high-density NW sensor libraries. Mrksich's group has developed chemistry that allows for the electrochemical activation of Au surfaces and enables spatial (or time-dependent) control over biological interactions [74, 75]. Two groups have translated related concepts towards the electrochemically driven biofunctionalization of semiconductor NWs [76–78], although the integration of that chemistry with NW sensors has not yet been reported.

12.4
Outlook

When NT and NW sensors were first reported, a significant amount of activity was apparent involving start-up companies interested in commercializing these devices. However, ultimately most – if not all – of that activity proved to be premature. It is possible that commercial applications in the realm of gas-phase sensors will emerge over the next few years, although in the case of biotech applications this area of research is still young and in a state of rapid evolution. Whilst NW and NT sensors clearly present some unique biotech opportunities, there remain several fundamental scientific challenges that must be met. Most of these challenges have been at least briefly discussed in the preceding sections, and involve all of the different aspects of manufacturability and integration. All NW and/or NT devices and small circuits are currently hand-made and hence are both expensive and time-consuming to construct, there being no firm high-throughput fabrication method available to produce these systems. In addition, a working NW or NT sensor array involves the integration of nanoelectronics, microfluidics, chemistry, and biology. It is doubtful whether there is an existing commercial product that involves a similar level of integration. Nevertheless, progress in this area is moving forward rapidly, and the motivations for developing technologies, as are described in this chapter, will surely increase with time.

Acknowledgments

Some of the studies referred to in this chapter were supported by the National Cancer Institute, a subcontract from the Mitre Corporation, the DARPA MolApps program, and the MARCO Focus Center for Advanced Materials and Devices.

References

1 Kong, J., Franklin, N. R., Zhou, C., Chapline, M. G., Peng, S., Cho, K., Dai, H. (2000) Nanotube molecular wires as chemical sensors. *Science* **287**, 622–625.

2 a. Star, A., Stoddart, J. F., Steuerman, D. W., Diehl, M., Boukai, A., Wong, E. W., Yang, X., Chung, S.-W., Choi, H., Heath, J. R. (2001) Preparation and properties of polymer-wrapped single-walled carbon nanotubes. *Angew. Chem* **40**, 1721–1725. b. Steuermann, D., Star, A., Narizzano, R., Choi, H., Ries, R., Nicolini, C., Stoddart, J., Heath, J. R. (2002) Interactions between conjugated polymers and single-walled carbon nanotubes. *J. Phys. Chem b* **106**, 3124–3130.

3 Cui, Y., Wei, Q., Park, H., Lieber, C. M. (2001) Nanowire nanosensors for highly sensitive and selective detection of biological and chemical species. *Science* **293**, 1289–1292.

4 Davidson, E. H., Rast, J. P., Oliveri, P., Ransick, A., Calestani, C., Yuh, C. H., Minokawa, T., Amore, G., Hinman, V., Arenas-Mena, C., et al. (2003) A genomic regulatory network for development. *Science* **295**, 1669–1678.

5 Kitano, H. (2002) Systems biology: a brief overview. *Science* **295**, 1662–1664.

6 Hood, L., Heath, J. R., Phelps, M. E., Lin, B. (2004) After the Genome Project: Systems biology and new technologies enable predictive and preventive medicine. *Science* **306**, 640–643.

7 Fodor, S. P., Read, J. L., Pirrung, M. C., Stryer, L., Lu, A. T., Solas, D. (1991) Light-directed, spatially addressable parallel chemical synthesis. *Science* **251**, 767–773.

8 MacBeath, G., Schreiber, S. L. (2000) Printing proteins as micro-arrays for high-throughput function determination. *Science* **289**, 1760–1763.

9 Engvall, E., Perlmann, P. O. (1972) Enzyme-linked immunosorbent assay,

ELISA. 3. quantitation of specific antibodies by enzyme-labeled anti-immunoglobulin in antigen-coated tubes. *J. Immunol.* **109**, 129–135.

10 Weller, T. H., Coons, A. H. (1954) Fluorescent antibody studies with agents of varicella and herpes zoster propagated in vitro. *Proc. Soc. Exp. Biol. (New York)* **86**, 789–794.

11 Zhang, H., Cheng, X., Richter, M., Green, M. I. (2006) A sensitive and high-throughput assay to detect low-abundance proteins in serum. *Nature Med.* **12**, 473–477.

12 Schweitzer, B., Wiltshire, S., Lambert, J., O'Malley, S., Kukanskis, K., Zhu, Z. R., Kingsmore, S. F., Lizardi, P. M., Ward, D. C. (2000) Immunoassays with rolling circle DNA amplification: A versatile platform for ultrasensitive antigen detection. *Proc. Natl. Acad. Sci. USA* **97**, 10113–10119.

13 Park, S. J., Taton, T. A., Mirkin, C. A. (2003) Array-based electrical detection of DNA with nanoparticle probes. *Science* **295**, 1503–1506.

14 Nam, J.-M., Thaxton, C. S., Mirkin, C. A. (2003) Nanoparticle-based bio-bar codes for the ultrasensitive detection of proteins. *Science* **301**, 1884–1886.

15 Lasseter, T. L., Cai, W., Hamers, R. J. (2004) Frequency-dependent electrical detection of protein binding events. *Analyst* **129**, 3–8.

16 Demers, L. M., Ginger, D. S., Park, S. J., Li, Z., Chung, S. W., Mirkin, C. A. (2002) Direct patterning of modified oligonucleotides on metals and insulators by dip-pen nanolitho-graphy. *Science* **296**, 1836–1868.

17 Legler, J. M., Ries, L. A. G., Smith, M. A., Warren, J. L., Heineman, E. F., Kaplan, R. S., Linet, M. S. (2000) Brain and other central nervous system cancers: Recent trends in incidence and mortality – Response. *J. Natl. Cancer Inst.* **92**, 77–78.

18 Prados, M., Chang, S., Burton, E., Kapadia, A., Rabbitt, J., Page, M., Federoff, A., Kelly, S., Fyfe, G., et al. (2003) Phase I study of OSI-774 alone

or with temozolomide in patients with malignant glioma. *Proc. Am. Soc. Clin. Oncol.* **22**, 99.

19 Rich, J. N., Reardon, D. A., Peery, T., Dowell, J. M., Quinn, J. A., Penne, K. L., Wikstrand, C. J., Van Duyn, L. B., Dancey, J. E., McLendon, R. E., et al. (2004) Phase II trial of gefitinib in recurrent glioblastoma. *J. Clin. Oncol.* **22**, 133–142.

20 Mellinghoff, I. K., Wang, M. Y., Vivanco, I., Haas-Kogan, D. A., Zhu, S. J., Dia, E. Q., Lu, K. V., Yoshimoto, K., Huang, J. H. Y., Chute, D. J., et al. (2005) Molecular determinants of the response of glioblastomas to EGFR kinase inhibitors. *N. Engl. J. Med.* **353**, 2013–2024.

21 Bailey, P., Cushing, H. (1928) *A Classification of the Tumors of the Glioma Group on a Histogenic Basis with a Correlated Study of Prognosis.* Lippincott, Philadelphia, Pennsylvania.

22 Lu, Q. R., Sun, T., Zhu, Z., M., Ma, N., Garcia, M., Stiles, C. D., Rowitch, D. H. (2002) Common developmental requirement for Olig function indicates a motor neuron/oligodendrocyte connection. *Cell* **109**, 75–86.

23 Doetsch, F. (2003) The glial identity of neural stem cells. *Nature Neurosci.* **6**, 1127–1134.

24 Hemmati, H. D., Nakano, I., Lazareff, J. A., Masterman-Smith, M., Geschwind, D. H., Bronner-Fraser, M., Kornblum, H. I. (2003) Cancerous stem cells can arise from pediatric brain tumors. *Proc. Natl. Acad. Sci. USA* **100**, 15178–15183.

25 Betensky, R. A., Louis, D. N., Cairncross, J. G. (2002) Influence of unrecognized molecular heterogeneity on randomized clinical trials. *J. Clin. Oncol.* **20**, 2495–2499.

26 Mischel, P. S., Cloughesy, T. F., Nelson, S. F. (2004) DNA-microarray analysis of brain cancer: Molecular classification for therapy. *Nature Rev. Neurosci.* **5**, 782–794.

27 Backmann, N., Zahnd, C., Huber, F., Bietsch, A., Plückthun, A., Lang, H.-P., Güntherodt, H.-J., Hegner, M., Gerber, C. (2005) A label-free immunosensor array using single-chain antibody fragments. *Proc. Natl. Acad. Sci. USA* **102**, 14587–14592.

28 Yue, M., Lin, H., Dedrick, D. E., Satyanarayana, S., Majumdar, A., Bedekar, A. S., Jenkins, J. W., Sundaram, S. (2004) A 2-D micro-cantilever array for multiplexed bio-molecular analysis. *J. Microelectromech. Syst.* **13**, 290–299.

29 Lundstroem, I., Shivaraman, S., Svensson, C., Lundkvist, L. (1975) Hydrogen-sensitive MOS field-effect transistor. *J. Appl. Phys.* **26**, 55–57.

30 Kharitonov, A. B., Shipway, A. N., Willner, I. (1999) An Au nanoparticle/bisbipyridinium cyclophane-functionalized ion sensitive field-effect transistor for the sensing of adrenaline. *Anal. Chem.* **71**, 5441–5443.

31 Janata, J. (2001) Centennial retrospective on chemical sensors. *Anal. Chem.* **73**, 150A–153A.

32 Wilson, D. M., Hoyt, S., Janata, J., Booksh, K., Obando, L. (2001) Chemical sensors for portable, handheld field instruments. *IEEE Sensors J.* **1**, 256–274.

33 Covington, J. A., Gardner, J. W., Briand, D., de Rooij, N. F. (2001) A polymer gate FET sensor array for detecting organic vapours. *Sensors Actuators* **B77**, 155–162.

34 Lud, S. Q., Nikolaides, M. G., Haase, I., Fischer, M., Bausch, A. R. (2006) Field effect of screened charges: Electrical detection of peptides and proteins by a thin-film resistor. *Chem. Phys. Chem.* **7**, 379–384.

35 Kong, J., Soh, H. T., Cassell, A. M., Quate, C. F., Dai, H. (1998) Synthesis of individual single-walled carbon nanotubes on patterned silicon wafers. *Nature* **395**, 878–881.

36 Cheng, M. M.-C., Cuda, G., Bunimovich, Y. L., Gaspari, M., Heath, J. R., Hill, H. D, Nijdam, A. J., Mirkin, C. A., Terracciano, R., Thundat, T. G., Ferrari, M. (2006) Nanotechnologies for biomolecular detection and medical diagnostics. *Curr. Opin. Chem. Biol.* **10**, 11–19.

37 Zheng, G., Patolsky, F., Cui, Y., Wang, W. U., Lieber, C. M. (2005)

Multiplexed electrical detection of cancer markers with nanowire sensor arrays. *Nature Biotechnol.* **23**, 1294–1301.

38 Hahm, J., Lieber, C. M. (2004) Direct ultrasensitive electrical detection of DNA and DNA sequence variations using nanowire nanosensors. *Nano Lett.* **4**, 51–54.

39 Li, Z., Chen, Y., Kamins, T. I., Nauka, K., Williams, R. S. (2004) Sequence-specific label-free DNA sensors based on silicon nanowires. *Nano Lett.* **4**, 245–247.

40 Morales, A., Lieber, C. M. (1998) A laser ablation method for the synthesis of crystalline semiconductor nanowires. *Science* **279**, 208–211.

41 Yang, P. D. (2005) The chemistry and physics of semiconductor nanowires. *MRS Bull.* **30**, 85–91.

42 Melosh, N. A., Boukai, A., Diana, F., Geradot, B., Badolato, A., Petroff, P., Heath, J. R. (2003) Ultrahigh-density nanowire lattices and circuits. *Science* **300**, 112–115.

43 Vieu, C., Carcenac, F., Pepin, A., Chen, Y., Mejias, M., Lebib, A., Manin-Ferlazzo, L., Couraud, L., Launois, H. (2000) Electron beam lithography: resolution limits and applications. *Appl. Surf. Sci.* **164**, 111–117.

44 Israelachvili, J. (1985) *Intermolecular and Surface Forces*. Academic Press, London.

45 Hu, K., Fan, F.-R. F., Bard, A. J., Hillier, A. C. (1997) Direct measurement of diffuse double-layer forces at the semiconductor/electrolyte interface using an atomic force microscope. *J. Phys. Chem. B* **101**, 8298–8303.

46 Heinze, S., Tersoff, J., Martel, R., Derycke, V., Appenzeller, J., Avouris, P. (2002) Carbon nanotubes as Schottky barrier transistors. *Phys. Rev. Lett.* **89**, 106801.

47 Chen, R. J., Choi, H. C., Bangsarun-tip, S., Yenilmez, E., Tang, X. W., Wang, Q., Chang, Y. L., Dai, H. J. (2004) An investigation of the mechanisms of electronic sensing of protein adsorption on carbon nanotube

devices. *J. Am. Chem. Soc.* **126**, 1563–1568.

48 Byon, H. R., Choi, H. D. (2006) Network single-walled carbon nanotube-field effect transistors (SWNT-FETs) with increased Schottky contact area for highly sensitive biosensor applications. *J. Am. Chem. Soc.*, **128**, 2188–2189.

49 Patolsky, F., Zheng, G., Hayden, O., Lakadamyali, M., Zhuang, X., Lieber, C. M. (2004) Electrical detection of single viruses. *Proc. Natl. Acad. Sci. USA* **101**, 14017–14022.

50 Bunimovich, Y., Shing, Y. S., Heath, J. R. (2006) Unpublished results.

51 Star, A., Tu, E., Niemann, J., Gabriel, J.-C. P., Joiner, C. S., Valcke, C. (2006) Label-free detection of DNA hybridization using carbon nanotube network field-effect transistors. *Proc. Natl. Acad. Acad. USA* **103**, 921–926.

52 Karlsson, R., Falt, A. (1997) Experimental design for kinetic analysis of protein-protein interactions with surface plasmon resonance. *J. Immunol. Methods* **200**, 121–133.

53 Backmann, N., Zahnd, C., Huber, F., Bietsch, A., Pluckthun, A., Lang, H. P., Guntherodt, H. J., Hegner, M., Gerber, C. (2005) A label-free immunosensor array using single-chain antibody fragments. *Proc. Natl. Acad. Sci. USA* **102**, 14587–14592.

54 Wendt, H., Leder, L., Harma, H., Jelesarov, I., Baici, A., Bosshard, H. R. (1997) Very rapid, ionic strength-dependent association and folding of a heterodimeric leucine zipper. *Biochemistry* **36**, 204–213.

55 Perry-O'Keefe, H., Yao, X.-W., Coull, J. M., Fuch, M., Egholm, M. (1996) Peptide nucleic acid pre-gel hybridization: An alternative to Southern hybridization. *Proc. Natl. Acad. Sci. USA* **93**, 14670–14675.

56 Zimmerman, M., Delamarche, E., Wolf, M., Hunziker, P. (2005) Modeling and optimization of high-sensitivity, low-volume microfluidic-based surface immunoassays. *Biomed. Microdev.* **7**, 99–110.

57 Sheehan, P. E., Whitman, L. J. (2005) Detection limits for nanoscale biosensors. *Nano Lett.* **5**, 803–807.

58 Li, C., Curreli, M., Lin, H., Lei, B., Ishikawa, F. N., Datar, R., Cote, R. J., Thompson, M. E., Zhou, C. (2005) Complementary detection of prostate-specific antigen using ln(2)O(3) nanowires and carbon nanotubes. *J. Am. Chem. Soc.* **127**, 12484–12485.

59 Li, Z., Rajendran, B., Kamins, T. I., Li, X., Chen, Y., Williams, R. S. (2005) Silicon nanowires for sequence-specific DNA sensing: device fabrication and simulation. *Appl. Phys.* **A80**, 1257–1263.

60 Heath, J. R., LeGoues, F. K. (1993) A liquid solution synthesis of single-crystal germanium quantum wires. *Chem. Phys. Lett.* **208**, 263–268.

61 Beckman, R., Johnston-Halperin, E., Luo, Y., Green, J. E., Heath, J. R. (2005) Bridging dimensions: Demultiplexing ultrahigh-density nanowire circuits. *Science* **310**, 465–468.

62 Zhong, Z., Wang, D., Cui, Y., Bockrath, M. W., Lieber, C. M. (2003) Nanowire crossbar arrays as address decoders for integrated nanosystems. *Science* **302**, 1377–1379.

63 Chung, S. W., Yu, J. Y., Heath, J. R. (2000) Silicon nanowire devices. *Appl. Phys. Lett.* **76**, 2068–2070.

64 McAlpine, M. C., Friedman, R. S., Jin, S., Lin, K.-H., Wang, W. U., Lieber, C. M. (2003) High-performance nanowire electronics and photonics on glass and plastic substrates. *Nano Lett.* **3**, 1531–1535.

65 Wang, D., Sheriff, B., Heath, J. R. (2006) Silicon p-FETs from ultrahigh density nanowire arrays. *Nano Lett.* **6**, 1096–1100.

66 Jung, G. Y., Johnston-Halperin, E., Wu, W., Yu, Z. N., Wang, S. Y., Tong, W. M., Li, Z. Y., Green, J. E., Sheriff, B. A., Boukai, A., Bunimovich, Y., Heath, J. R., Williams, R. S. (2006) *Nano Lett.* **6**, 351–354.

67 Chen, R. J., Bangsaruntip, S., Drouvalakis, K. A., Kam, N. W. S.,

Shim, M., Li, Y., Kim, W., Utz, P. J., Dai, H. (2003) Noncovalent functionalization of carbon nanotubes for highly specific electronic biosensors. *Proc. Natl. Acad. Sci. USA* **100**, 4984–4989.

68 Chen, R. J., Zhang, Y., Wang, D., Dai, H. (2001) Noncovalent sidewall functionalization of single-walled carbon nanotubes for protein immobilization. *J. Am. Chem. Soc.* **123**, 3838–3839.

69 Diehl, M. R., Steuerman, D. W., Tseng, H.-R., Vignon, S. A., Star, A., Celester, P. C., Stoddart, J. F., Heath, J. R. (2003) Single-walled carbon nanotube based molecular switch tunnel junctions. *Chem. Phys. Chem.* **4**, 1335–1339.

70 Steuerman, D. W., Star, A., Narizzano, R., Choi, H., Ries, R. S., Nicolini, C., Stoddart, J. R., Heath, J. R. (2002) Interactions between conjugated polymers and single-walled carbon nanotubes. *J. Phys. Chem.* **106**, 3124–3130.

71 O'Connell, M. J., Boul, P., Ericson, L. M., Huffman, C., Wang, Y., Haroz, E., Kuper, C., Tour, J., Ausman, K. D., Smalley, R. E. (2001) Reversible water-solubilization of single-walled carbon nanotubes by polymer wrapping. *Chem. Phys. Lett.* **342**, 265–271.

72 Besteman, K., Lee, J.-O., Wiertz, F. G. M., Heering, H. A., Dekker, C. (2003) Enzyme-coated carbon nanotubes as single-molecule biosensors. *Nano Lett.* **6**, 727–730.

73 Zheng, M., Jagota, A., Strano, M. S., Santos, A. P., Barone, P., Chou, S. G., Diner, B. A., Dresselhaus, M. S., McLean, R. S., Onoa, G. B., et al. (2003) Structure-based carbon nanotube sorting by sequence-dependent DNA assembly. *Science* **302**, 1545–1548.

74 Yeo, W. S., Yousaf, M. N., Mrksich, M. (2003) Dynamic interfaces between cells and surfaces: Electroactive substrates that sequentially release and attach cells. *J. Am. Chem. Soc.* **125**, 14994–14995.

75 Yousaf, M. N., Mrksich, N. (1999) Diels-Alder reaction for the selective immobilization of protein to electro-active self-assembled monolayers. *J. Am. Chem. Soc.* **121**, 4286–4287.

76 Bunimovich, Y., Ge, G., Beverly, K. C., Ries, R. S., Hood, L., Heath, J. R. (2004) Electrochemically programmed, spatially selective biofunctionalization of silicon wires. *Langmuir* **24**, 10630–10638.

77 Curreli, M., Li, C., Sun, Y. H., Lei, B., Gundersen, M. A., Thompson, M. E., Zhou, C. W. (2005) Selective functionalization of In2O3 nanowire mat devices for biosensing applications. *J. Am. Chem. Soc.* **127**, 6922–6923.

78 Rohde, R., Agnew, H., Yeo, W.-S., Bailey, R., Heath, J. R. (2006) A non-oxidative approach towards chemically and electrochemically biofunctionalizing Si(111). *J. Am. Chem. Soc.* **128**, 9518–9525. DOI: 10.1021/ja062012b.

79 Star, A., Gabriel, J. C. P., Bradley, K., Gruner, G. (2003) Electronic detection of specific protein binding using nanotube FET devices. *Nano Lett.* **3**, 459–463.

80 Wang, W. U., Chen, C., Lin, K.-H., Fang, Y., Lieber, C. M. (2005) Label-free detection of small-molecule–protein interactions by using nanowire nanosensors. *Proc. Natl. Acad. Sci. USA* **102**, 3208–3212.

81 Chen, R. J., Bangsaruntip, S., Drouvalikis, K. A., Kam, N. W. S., Shim, M., Li, Y., Kim, W., Utz, P. J., Dai, H. (2003) Noncovalent functionalization of carbon nanotubes for highly specific electronic biosensors. *Proc. Natl. Acad. Sci. USA* **100**, 4984–4989.

82 Star, A., Gabriel, J.-C. P., Bradley, K., Gruner, G. (2003) Electronic detection of specific protein binding using nanotube FET devices. *Nano Lett.* **3**, 459–463.

13
Bionanoarrays

*Rafael A. Vega, Khalid Salaita, Joseph J. Kakkassery,
and Chad A. Mirkin*

13.1
Overview

Patterned arrays of biomolecules, such as DNA, proteins, viruses, and cells, have been utilized as powerful tools in a variety of biological studies. Microarrays, in particular, have led to significant advances in many areas of medical and biological research, opening up avenues for the combinatorial screening and identification of single-nucleotide polymorphisms (SNPs), high-sensitivity expression profiling of proteins, and high-throughput analysis of protein function [1]. With the advent of powerful new nanolithographic methods, such as dip-pen nanolithography (DPN) [2], there is now the possibility of reducing the feature size in such arrays to their physical limit, the size of the structures from which they are made of, and the size of the structures they are intended to interrogate (Figure 13.1) [3]. Such massive miniaturization not only allows one to increase the density of combinatorial libraries, to increase the sensitivity of such structures in the context of a biodiagnostic event, and to reduce the required sample analyte volume, but also to carry out studies that are not possible with the more conventional microarray format. Arrays with features on the nanometer-length scale open up the opportunity to study many biological structures at the single particle level [4]. Such features can be used to immobilize and orient individual virus particles and to study many important processes such as cell infectivity and virus proliferation and transmission. These miniaturized features allow one to contemplate the creation of the equivalent of an entire combinatorial library (e.g., a gene chip or complex protein array) underneath a single cell, thus opening new possibilities for the study of important fundamental, multivalent, processes such as cell-surface recognition, adhesion, differentiation, growth, proliferation, and apoptosis.

In this chapter, methods for preparing biological nanoarrays made from biomaterials – including proteins, oligonucleotides, and viruses – will be high-

Nanobiotechnology II. Edited by Chad A. Mirkin and Christof M. Niemeyer
Copyright © 2007 WILEY-VCH Verlag GmbH & Co. KGaA, Weinheim
ISBN: 978-3-527-31673-1

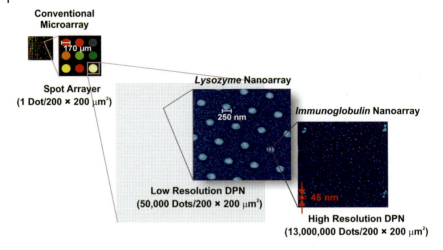

Fig. 13.1 Using low-resolution dip-pen nanolithography (DPN), 55 000 protein spots, each of 250 nm, can be spotted in an area occupied by a single spot (200 µm diameter) in a conventional microarray. Patterns can be further miniaturized using high-resolution DPN to generate a total of 13 000 000 spots in a 200 × 200 µm² area. (Figures reprinted with permission from Ref. [3]; © 2005, American Chemical Society.)

lighted. The focus will be on techniques that allow feature size to be controlled on the sub-500 nm length scale.

13.2
Methods

There is currently a suite of techniques available – with various advantages and disadvantages – for generating nanoscale features of biological molecules. The goal of this section is to highlight some of the more promising methods for generating nanoarrays of biological molecules, and to emphasize how they differ in terms of resolution, multiplexing capabilities, and throughput.

13.2.1
Atomic Force Microscope-Based Methods

Since its inception in 1986, atomic force microscopy (AFM) has been used for elucidating structural details for a variety of surface-bound molecules and materials [5]. Recently, AFM-based lithographies have emerged as powerful tools for patterning biological molecules at the nanometer length scale, most notably through the invention of DPN [4, 6, 7]. In this section, the use of AFM-based methods for generating arrays of biological molecules will be introduced and highlighted.

A

B

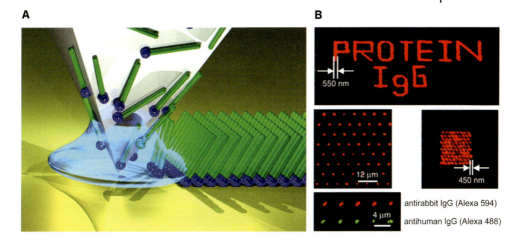

Fig. 13.2 (A) Schematic diagram depicting the dip-pen nanolithography (DPN) process. (B) Fluorescence images of DPN-generated antibody structures on a negatively charged SiO₂ surface. (Part B reprinted with permission from Ref. [13]; © 2003, Wiley-VCH.)

Most AFM-based patterning techniques use a scanning probe tip to selectively eliminate (i.e., etch, electrochemically oxidize, or displace) part of an adsorbed monolayer to generate a template for capturing biomolecules [8–10]. An exception to this method is DPN, where an "ink"-coated AFM tip is used to directly deliver nanoscopic amounts of an adsorbate to a surface on the sub-50 nm to many micrometer-length scale [2, 11] (Figure 13.2). DPN is particularly well-suited for patterning biological and soft organic structures onto surfaces, as it operates under ambient conditions without the need for an electron or ion beam [11]. Also, because DPN is a "constructive" rather than "destructive" patterning tool, the fabrication of high-density combinatorial arrays of different biomolecules with ultrahigh registry is possible. Thus far, DPN has been used to generate arrays of synthetic DNA oligomers, proteins, and peptides [6, 7, 12–19]. In many cases, tip and substrate modification protocols have been developed to maintain the biological fidelity of the deposited structures [12].

The direct deposition of biomolecules using DPN requires the chemical modification of an AFM tip for reproducible tip coating, and the implementation of suitable ink–substrate coupling chemistries for reliable biomolecular transport. For example, the direct deposition of DNA was achieved using hexanethiol-modified oligonucleotides on gold surfaces, and acrylamide-modified oligonucleotides on oxidized silicon wafers that had been pre-modified with 3′-mercaptopropyltrimethoxysilane [7]. The biological activity of the resulting patterns was confirmed by hybridization with fluorophore- or nanoparticle-labeled oligonucleotides (Figure 13.3).

DPN has been used to generate chemical affinity templates for the adsorption of proteins, DNA, and even more recently virus particles from solution [4, 6, 20–

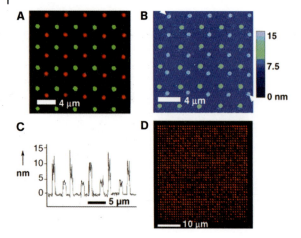

Fig. 13.3 Direct patterning of multiple-DNA molecules by DPN. (A) Combined red-green fluorescence overlay image of two different fluorophore-labeled sequences (Oregon Green 488-X and Texas Red-X) simultaneously hybridized to a two-sequence array deposited on a silanized SiO$_x$ substrate by DPN. (B) Tapping-mode atomic force microscopy (AFM) image of 5-nm (dark) and 13-nm (light) -diameter gold NPs assembled on the same pattern after dehybridization of the fluorophore-labeled DNA. (C) A line plot was taken diagonally through both NP patterns, and the start and finish are indicated by the white arrows in (B). (D) Dark-field optical image showing scattering from densely packed 13 nm-diameter DNA-functionalized NPs hybridized to a 1600-dot DNA array patterned on SiO$_x$. (Parts A–C reprinted with permission from Ref. [7]; © 2003, American Association for the Advancement of Science.)

24]. Indirect patterning approaches are best suited for generating single-component nanoarrays, and can be used without having to optimize a unique set of conditions for the deposition of different classes of biomolecules. How such arrays can be used to study cell-surface recognition is highlighted in Section 13.3.4.

Other AFM-based methods can be used to indirectly pattern DNA and protein nanostructures. These methods rely on a localized tip–surface interaction to eliminate selected areas in an adsorbed monolayer, which in turn allows for another group of molecules to adsorb to these patterned areas. Liu and coworkers [8, 25, 26] have demonstrated that nanografting can be used to generate nanoscale patterns of DNA and proteins, respectively. In nanografting, a scanning probe tip under a high applied load displaces resist molecules within a self-assembled monolayer, which then allows solution species to adsorb to these exposed sites. Lines of DNA, as narrow as 10 nm, have been fabricated using this technique [25].

In a similar approach, Cai and coworkers [9] demonstrated that a conductive AFM tip can be used to selectively eliminate parts of a monolayer on a silicon (Si) substrate. They then used these substrate-exposed areas to immobilize avidin and biotinylated bovine serum albumin (BSA). Matsui and coworkers [10] showed that an AFM tip could be used to "shave" parts of an octadecanthiol monolayer on Au to subsequently fill and immobilize antibodies into these areas. This shaving technique has been used to generate two-component arrays of antibodies (mouse and human IgGs), and the biological activity of such structures was confirmed by reacting them with the corresponding fluorophore-labeled secondary antibodies. The difficulty of this serial process comes from the cross-reactivity and nonspecific binding of the antibodies. The first set of features is always exposed to the solution used to generate the second set of features. Furthermore, feature size-broadening often results from tip damage, and the approach, thus far, has not been amenable to parallelization, which imposes significant throughput limitations.

Overall, AFM-based patterning techniques provide a resolution commensurate with the size of certain individual biomolecules and structures ($\sim$10 nm), offering new opportunities to investigate biological reactions, and to investigate mono- and multivalent processes at the single particle level (for discussions, see Sections 13.3 and 13.5). Such uses cannot be addressed using microarrays, because of a size mismatch between the features in the array and the structures being probed (e.g., viruses, cells, and spores). Although indirect elimination-based techniques – such as nanografting, nanoshaving, or nanoscale oxidation – provide excellent resolution, these techniques are currently limited in throughput and do not offer extended multiplexing capabilities. At present, only DPN is capable of the direct delivery of multiple biomolecules at ultrahigh resolution and registry. Furthermore, recent advances in parallel DPN using linear [27] and large two-dimensional (2-D) arrays of cantilevers [28] have substantially increased the throughput of this method, and this now allows researchers to pattern over square-centimeter areas.

13.2.2
Nanopipet Deposition

Nanopipetting is based on scanning ion-conductance microscopy (SCIM), which was a technique developed to scan soft non-conducting materials by using an electrolyte-filled micropipet as a probe [29]. In SCIM, the ionic-current flowing between an electrode inside the micropipet and a counterelectrode inside the bath is used to control the pipet–substrate distance. By using a pipette with a 100- to 150-nm diameter orifice, research groups have used SCIM to deliver small quantities of DNA and proteins onto various substrates (Figure 13.4) [30]. The number of molecules exiting the tip orifice is dependent on a combination of applied current, electro-osmotic flow, electrophoresis, and dielectrophoresis, depending on the size, charge, and polarizablity of the specific molecules being deposited. It is therefore necessary to characterize the delivery of molecules from the pipet experimentally, as each molecule type has different transport properties that depend upon the magnitude of the electric field.

SCIM deposition of biomolecules occurs in solution, and consequently the feature size depends heavily on the pipet–substrate distance and on the diffusion of

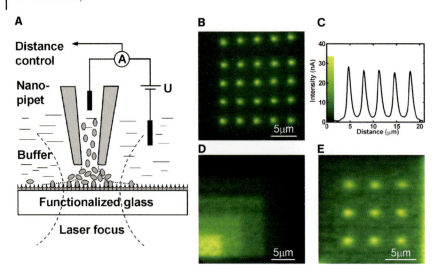

Fig. 13.4 (A) Schematic of a typical nanopipet experiment. Fluorescence images (B) 25 dots of biotinylated DNA deposited with a 10-s hold time each onto a streptavidin-coated glass surface. (C) Line scan of the bottom row in part (B); the FWHM is 830 ± 80 nm. (D) Squares of biotinylated DNA written over each other to create a pattern with increasing intensities. (E) Dots of protein G on a positively charged glass surface. (Reprinted with permission from Refs. [30, 31]; © 2002 and 2003, American Chemical Society.)

biomolecules to the surface. Once the molecules leave the nanopipet tip, there are two contributions to their motion toward the surface: (i) the directed electro-osmotic flow from the pipet; and (ii) subsequent diffusion in solution on the substrate. The highest resolution achieved by nanopipet deposition thus far is approximately ~800 nm. As the deposition occurs in a buffered solution, the biological activity of the deposited molecules is typically preserved [31].

More recently, Klenerman and coworkers [32] reported a "double-barreled nanopipet", where two types of molecules are independently delivered to a surface from the pipet barrels of a single tip. This technique can be carried out in air, which provides higher resolution and eliminates the registry problems involving the deposition of two different species on a surface. This method has been used to deposit spots of DNA or proteins as small as ~500 nm onto a polyethyleneimine (PIE)-coated glass substrate.

Nanopipet-based delivery of biomolecules is among the few techniques capable of directly depositing biomolecules onto surfaces but, at present, it is significantly limited in terms of resolution and throughput.

13.2.3
Beam-Based Methods

Electron-beam lithography (EBL) is a technique widely used to fabricate nano-scale masks in the semiconductor industry [33], and it is capable of producing

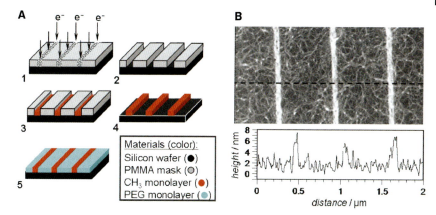

Fig. 13.5 (A) Schematic diagram of the five steps associated with the fabrication of anisotropic chemical nanopatterns. 1, Irradiation of a poly(methyl methacrylate) (PMMA) resist layer, spin-coated on a silicon wafer, with a focused electron beam to create a pattern of parallel lines. 2, Selective dissolution of the irradiated PMMA. 3, Grafting of the methyl-terminated alkylsilanes on bare silicon oxide. 4, Dissolution of the remaining PMMA. 5, Grafting of PEG-alkylsilanes on bare silicon oxide. (B) AFM height image (2×1 μm^2), in air, of a CH_3-50/PEG-550 area after collagen adsorption where the cross-section is taken along the horizontal black-dashed line. (Reprinted with permission from Ref. [37]; © 2005, Wiley-VCH.)

high-quality sub-10 nm features on inorganic substrates. Recently, it has been used to generate nanoscale arrays of biomolecules via an indirect patterning approach. With EBL, patterns are first drawn on a thin layer of photoresist or on a chemisorbed monolayer of molecules. Patterned areas are then chemically modified to capture biomolecules present in a bulk solution, and the surrounding areas are passivated with other molecules that resist the nonspecific adsorption of biomolecules [34–36].

For example, Dupont-Gillain and coworkers [37] generated collagen nanoarrays using EBL. They initially wrote thin lines (line widths = 30–90 nm) in a poly(methyl methacrylate) (PMMA) resist, which was selectively dissolved to allow for the gas-phase grafting of a CH_3-terminated alkylsilane (Figure 13.5A). Collagen molecules can then adsorb onto these CH_3-terminated alkylsilane features. The group then showed that collagen can be aligned and assembled into continuous bundles when the feature size is much smaller than the length of one collagen molecule (~300 nm) (Figure 13.5B). Jonas and coworkers [38] also studied the adsorption of globular proteins onto nanoscopic CH_3-terminated line templates surrounded by chemisorbed oligo(ethylene oxide). The group demonstrated that the orientation of the adsorbed globular proteins can be tailored by reducing the size of the adsorption region below 50 nm, which is believed to reorient the proteins adsorbed at the edges of each stripe. Both of these examples demonstrate that EBL can be used to generate templates that adsorb biomolecules and, in some cases, control their orientation. However, this approach is limited with respect to multiplexing, because only one type of biomolecule can be deposited on a nanostructured array. Although parallel EBL methods are currently be-ing developed, these are prohibitively complex, and thus the throughput of EBL-based patterning is also quite limited [33].

In order to address the issue of low-throughput in EBL, the use of nanoimprint lithography (NIL) was introduced for patterning biomolecules [39, 40]. NIL typi-cally replicates EBL-defined structures in polymer films, and can be used to rap-idly generate extremely small features over large areas. Various proteins such as streptavidin and anti-catalase were immobilized onto aminosilane patterns as small as 75 nm. The biological activities of the immobilized proteins were con-firmed by monitoring their reaction with fluorophore-labeled antibodies. How-ever, NIL is also an indirect approach to patterning biomolecules, and therefore, it is difficult to pattern multiple biomolecules in the context of a single nanoarray. NIL, thus far, can only replicate pre-existing patterns, and therefore each pattern requires the design and fabrication of a different mask, which can be prohibitive if one needs systematically to modify the pitch, size, and/or pattern shape. For certain applications, these parameters are important, and will be discussed in Section 13.3.4.

13.2.4
Contact Printing

Microcontact printing (μCP) techniques have been utilized over the past decade to generate micron-scale biomolecule patterns on a variety of substrates [41]. More recently, nanocontact printing (nCP) techniques have been developed for the di-rect or indirect deposition of a variety of biomolecules on the sub-50 nm length scale [42–44]. μCP and nCP rely on the use of an elastomeric stamp patterned with relief structures to generate patterns and structures with feature sizes rang-ing from 30 nm to 100 μm.

By decreasing the feature size on an elastomeric poly(dimethylsiloxane) (PDMS) stamp to below 100 nm and using dilute protein ink solutions, Delam-arche and coworkers [42] demonstrated that a very small number of green fluo-rescent protein molecules (GFP) can be patterned on a glass substrate. The exact position of the GFP molecules is within $\pm$50 nm of the theoretical center of the dots, although there is a significant fraction ($\sim$60%) of surface sites that remain unoccupied. Huck et al. [43] utilized a combination of "sharp" and "hard" PDMS stamps, with high-molecular weight inks to generate 60-nm multimeric pro-tein lines. The advantage of using nCP is that it is a direct-write technique and therefore a much simpler method for generating nanoscale arrays of biomole-cules compared to indirect methods such as EBL or NIL. Similar to NIL, nCP can be used to pattern over large areas, although pattern registry and pattern size can vary depending on the uniformity of the applied load on the stamp. The ability to deposit multi-component structures in nCP or μCP is quite limited, but DPN has recently been used to ink individual stamp features; it has also been demonstrated that three different DNA strands can be patterned simultaneously [44].

Recently, a novel DNA stamping technique – sometimes referred to as supramolecular nano-stamping (SuNS) – was introduced independently by our group [45], Crooks et al. [46] and Stellacci et al. [47, 48] for the efficient replication and printing of DNA micro- and nanoarrays. SuNS consists of three steps: hybridization; contact; and dehybridization. SuNS has allowed the printing of DNA patterns as small as 50 nm on PMMA substrates, and may prove to be a simple method for rapidly replicating DNA arrays. This method, which shows promise for printing chemically modified DNA, relies on the highly specific and selective hybridization properties of oligonucleotides. The throughput of this technique is very high, although each new pattern requires the use of a new mask.

13.2.5
Assembly-Based Patterning

Periodic nanoarrays of biomolecules have been prepared via three different approaches that use three different classes of materials: colloidal particles [49, 50]; diblock copolymer micelles [51]; and 2-D DNA tiles [52, 53]. In each of these cases, a phase-separated material or prefabricated structure is used to direct the assembly of biomolecules into a preconceived periodic pattern. Directed-assembly techniques are viewed as attractive methods by some researchers because they are often straightforward to implement, inexpensive, and in some cases provide very high resolution. For example, nanosphere lithography (NSL) utilizes the interstitial spaces in a monolayer of close-packed colloidal particles to pattern a variety of materials. NSL provides hexagonally patterned lithographic masks for further surface processing. For example, Ocko et al. [54] have used 300-nm colloidal particles of polystyrene to protect an underlying vinyl monolayer from oxidation and subsequent chemical coupling to polyethylene glycol (PEG)-silane. This approach was used to generate ~120 nm islands that could capture lysozyme proteins from solution, and the biological activity of patterned lysozyme was subsequently confirmed using antibody–antigen binding. Although NSL is limited to hexagonal patterns of single-component structures with modest control over array architectural dimensions (feature size and pitch), this method offers an economical solution for large-scale high-resolution protein patterning. It should be noted that Van Duyne and coworkers have conducted some impressive NSL studies with abiological systems and have dramatically extended the flexibility and capabilities of the approach [55, 56]. It is conceivable that some of their advances could be extended to biological systems [57].

Spatz and coworkers [51] have shown that the assembly of diblock copolymer micelles on glass surfaces results in the formation of a hexagonal, close-packed structure. When gold (Au) nanoparticles (NPs) are enclosed within these micelles, hexagonally packed Au NPs form on glass substrates once the organic polymer is removed by plasma treatment. The distances between the Au NPs can be adjusted through the choice of diblock-copolymer size, and were varied between 28 ± 5, 58 ± 7, 73 ± 8, and 85 ± 9 nm. Such structures were used to study

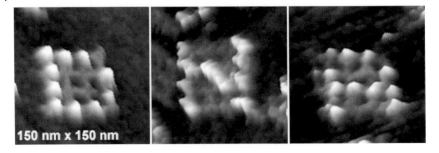

Fig. 13.6 Demonstration of precise programming by assembly. The letters "D, N, and A" were assembled on DNA tile arrays (4 × 4) decorated with protein using the two-step minimal depth strategy. (Reprinted with permission from Ref. [59]; © 2006, Wiley-VCH.)

the activation of integrin function (this is described in more detail in Section 13.3.4). This is an interesting approach towards controlling the spacing between ligands, although the range of inter-particle spacings is limited.

Recently, periodic arrays of assembled DNA tiles or grids have been used as elaborate scaffolds to organize materials at the nanometer-length scale [52, 53, 58, 59]. DNA assemblies have been formed using carefully designed linear oligo-nucleotides with complementary base-pairing segments that form branch-junction motifs. These DNA assemblies were produced with various geometrical structures and functions, ranging from periodic one-dimensional (1-D) materials, to three-dimensional (3-D) polyhedra. The precise positioning of a single strepta-vidin protein by site-selective attachment within finite 2-D DNA grids is shown in Figure 13.6. The synthesis of such nanostructures involves interactions between a large number of short oligonucleotides, and the final yield is highly dependent on the stoichiometry. Although the length scale over which these arrays are regular is small (~1 μm) and the precise position of each DNA assembly cannot be con-trolled or macroscopically addressed, these arrays provide nanometer precision in controlling the spacing between materials.

13.3
Protein Nanoarrays

Protein microarrays have enabled the high-throughput analysis of protein func-tion, protein–protein interactions, catalysis, drug screening, and other biochem-ical reactions [1, 60]. However, biorecognition is inherently a nano- rather than a microscopic phenomenon. Therefore, nanoarrays are beginning to enable scien-tists not only to ask but also answer some detailed questions regarding the inter-actions of biological structures (i.e., cells) with immobilized protein features [51]. Furthermore, nanoarrays of proteins (i.e., antibodies) with well-defined feature size and spacing have been shown to be important and advantageous in developing novel immunoassay schemes for the sensitive and selective detection of macromolecules [20]. In this section, we will describe the immobilization chemistries used to prepare protein nanoarrays, and some of the applications of these arrays.

13.3.1
Strategies for Immobilizing Proteins on Nanopatterns

Because there are many types of proteins with a wide variety of functional groups, different chemical methods have been designed to effectively immobilize the different classes of proteins. Proteins are extremely fragile, contain multiple domain structures, and can have multiple interaction sites. One of the simplest and fastest ways to immobilize proteins is through electrostatic interactions. However, proteins immobilized in this way are often not robust and can be significantly affected by changes in solution pH and ionic strength, thereby allowing for the possibility of protein desorption. Because there is no specific interaction between the desired protein and the surface, the proteins are often randomly attached to such substrates. In the case of microarrays, electrostatic binding, in many cases, results in immobilized proteins with an acceptable level of target binding. This is because the feature size is large enough such that some of the immobilized proteins are in the proper orientation to support analyte binding. Nanoarrays are not likely to perform the same as the feature area is significantly reduced in size [6, 12, 61], and in order to realize a system with feature properties that are similar from spot to spot, immobilization strategies that result in uniform protein immobilization with consistent and optimized bioactivity are needed.

Certain covalent attachment strategies offer researchers the ability to generate immobilized protein structures with substantially higher bioactivities. These techniques have been commonly used in array-type formats to permanently fix proteins on a substrate through a variety of functional groups present on their surface, such as amino-, carboxyl-, hydroxyl-, and thiol-moieties. These moieties can be readily coupled to an amine- or thiol-terminated surface with the appropriate crosslinkers [60]. The covalent approach, usually, produces higher concentrations of proteins on the patterned area than what is produced through electrostatic adsorption. However, the proteins are still immobilized in random forms, which can lead to decreased sensitivities and complete loss of biological activity in surface-based immunoassays compared to immobilization strategies that orient the proteins [62, 63].

A variety of techniques have been developed for the directed immobilization of proteins – specifically antibodies – in active states. The most common method has been the use of antibody-binding proteins A, G, and A/G, which bind the F_c portions of antibodies, leaving their F_{ab} binding sites exposed to solution. Researchers have coupled the covalent approach, described above, for the immobilization of protein A/G to direct the immobilization of antibodies on a surface [64, 65]. However, protein A/G is costly and has varying affinities – and in some cases no affinity – for the different classes of antibodies, and this limits their use in some immunobased assays. Recently, a new approach has been developed in our

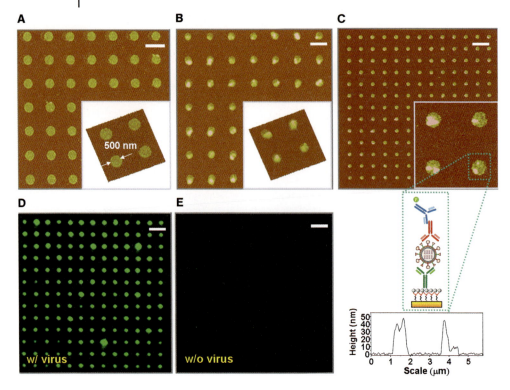

Fig. 13.7 AFM tapping mode image of (A) anti-Udorn (polyclonal goat IgG) immobilized on 500-nm MHA dot arrays pretreated with Zn(NO$_3$)$_2$·6H$_2$O; (B) after the addition and binding of the influenza virus; and (C) followed by the sandwiching of the virus-antibody complex through anti-HA (monoclonal mouse IgG1) and goat anti-mouse (secondary antibody). Fluorescence images of (C) treated with (D) and without (E) influenza virus particles. Scale bars: A, B = 1 μm; C–E = 3 μm for the array and 1 μm for the insert. (Reprinted with permission from Ref. [24]; © 2006, Wiley-VCH.)

group that uses metal-ions as a linker between the surface and the antibody [24]. This approach has the potential to be a low-cost alternative for immobilizing many types of antibodies in active states (Figure 13.7).

A recent study explored the effects of some of the above-described methods of antibody attachment on analyte binding, and found that certain immobilization approaches can increase analyte binding capacity up to 10-fold [66]. Thus, for nanoarray-type applications requiring high sensitivity and low detection limits, such directed protein immobilization techniques are likely to significantly improve their performance in cell-based or immunobased assays.

13.3.2
Bio-Analytical Applications

There can be significant benefits to the miniaturization of chip-based protein assays. These benefits include the use of less capture materials, less analyte, a reduction in the total sample volume and, under certain circumstances, lower limits of detection over conventional immunoassays. Although these nanopatterning techniques are still in their infancy, they represent the next major step in the miniaturization of immunoassays.

A proof-of-concept example of this type of diagnostic application was recently carried out by our group in collaboration with the group of Wolinsky [20], where nanoarrays of monoclonal antibodies for HIV-1 p24 were utilized to detect HIV-1 p24 antigen in blood samples (1 µL) obtained from men with fewer than 50 copies of RNA per mL of plasma (Figure 13.8). The presence of the array-captured p24 was measured by AFM, and the signal was amplified by treating the array

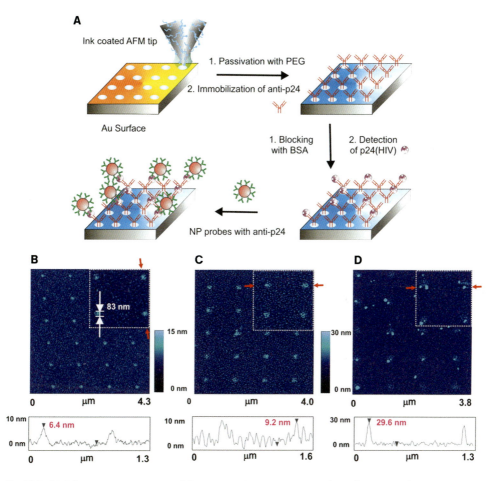

Fig. 13.8 (A) Schematic representation of the detection of HIV-1 p24 through the use of antibody nanoarrays. AFM images and height profiles for (B) an anti-p24 nanoarray (6.4 ± 0.9 nm; n = 10) adsorbed on MHA, (C) after p24 binding (height increase of 2.3 ± 0.6 nm; n = 10), and (D) detection and amplification of p24 through a sandwich complex with anti-p24 coated gold NPs (height increase of 20.3 ± 1.9 nm; n = 10). (Reprinted with permission from Ref. [20]; © 2004, American Chemical Society.)

with captured target with anti-p24-modified Au NPs, allowing for a significant increase in the height of the spots. This nanoarray-based assay exhibited a limit of detection $(0.025$ pg $mL^{-1})$ – three orders of magnitude lower than the conventional enzyme-linked immunosorbent assay (ELISA), with a detection limit of 5 pg mL^{-1}.

13.3.3
Dynamic and Motile Nanoarrays

Chip-based capture and release of target proteins directly from complex mixtures, in the context of nanoarrays, could provide functionality in integrated bioanalytical nanoscale devices in which the transport, separation, and detection of a small number of biomolecules could be performed in one integrated system. Chilkoti, Zauscher, and coworkers [21] have fabricated stimulus-responsive elastin-like polypeptide (ELP) nanostructures that respond to changes in ionic strength and temperature. ELP nanostructures were covalently end-grafted onto 200-nm features of ω-substituted alkane thiolates generated by DPN. A thioredoxin–ELP fusion protein was immobilized onto these ELP nanopatterns above the lower critical solution temperature (LCST). Below the LCST, the thioredoxin could be released back into solution. These experiments demonstrate that "smart," surface-confined biomolecular switches can be built and integrated into nanoscale bioanalytical devices.

Another type of dynamic protein nanoarray has been used in the development of a hybrid nanodevice based on a rotary motor. Soong et al. [67] demonstrated that a multi-subunit protein complex (ATP synthase) with domains that rotate about its membrane-bound axis during the hydrolysis of ATP, could be used to power the rotation of an inorganic Ni rod (nanopropeller) (Figure 13.9). Ni posts, 200 nm in height and 80 nm in diameter, were fabricated by electron-beam lithography. These arrays were subsequently treated with a thermostable form of the F_1-ATPase, with a $10\times$ histidine tag engineered into the β-subunit coding sequence. Finally, biotinylated coated Ni rods were allowed to specifically attach to the lever (streptavidin conjugate) on the γ-subunit of the F_1-ATPase. In the presence of ATP, the nanodevice exhibited an endurance cycle of ~ 2.5 h. Despite the fact that the fabrication yield was low, once optimized, this type of functional dynamic nanoarray might be used to build integrated biomolecular motors with nanoengineered systems to produce nanomechanical devices.

13.3.4
Cell-Surface Interactions

Cellular adhesion is a process mediated by the interaction of cell-surface receptors (i.e., integrin) with ligands from the extracellular matrix (ECM), which results in the clustering of these receptors into focal adhesion complexes. This elaborate and highly regulated process is believed to play a significant role in many essential cellular functions (e.g., motility, proliferation, differentiation, and apoptosis)

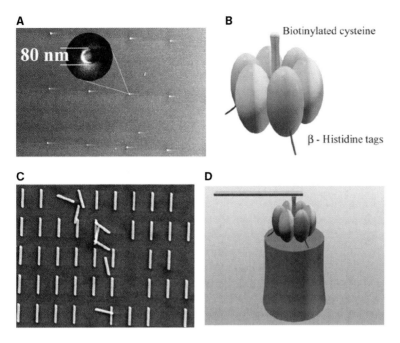

Fig. 13.9 (A) Schematic diagram of the F$_1$-ATPase biomolecular motor-powered nanomechanical device. The device consisted of (A) a Ni post (200 nm height, 80 nm diameter), (B) the F$_1$-ATPase biomolecular motor, and (C) a nanopropeller (length 750– 1400 nm, diameter 150 nm). The device (D) was assembled using sequential additions of individual components and different attachment chemistries. (Reprinted with permission from Ref. [67]; © 2000, American Association for the Advancement of Science.)

[68, 69]. Most studies involving the signaling of focal adhesion complexes with surface-immobilized proteins have involved structures made by conventional microfabrication techniques. These studies have been limited to single-component protein arrays, typically generated by indirect robotic spotting and μCP methods [1, 70, 71]. Recently, researchers have begun to explore how protein arrays fabricated by techniques that offer high resolution can be used to probe surface cellular interactions, and some of these advances are summarized below.

Our group, in collaboration with group of Mrksich, has utilized DPN to generate patterns of retronectin features as small as 200 nm with 700 nm separation distances [6]. Cells were found to adhere and spread, exclusively, on the nanopatterned regions. The cellular morphology was found to be a direct consequence of the sizes and spacings of the patterned protein. These studies were the first to demonstrate that protein structures, well below 1 µm in size, can support cell adhesion.

Since then, others have used novel block copolymer-based strategies to fabricate peptide arrays for probing the feature size dependence in cell adhesion. These experiments were carried out in the context of the RGDfk adhesion peptide (see

Section 13.2.5) [51]. The features in these patterns were designed to be small enough (dot diameter < 8 nm) to capture a single integrin protein per dot. From these studies, Spatz and coworkers [51] have concluded that local dot-to-dot separation, rather than global dot density, is critical for inducing cell adhesion and focal adhesion assembly. A separation of $\geq$ 73 nm between adhesive dots resulted in limited cell attachment and spreading, and dramatically reduced the formation of focal adhesion and actin stress fibers.

Protein nanoarrays have been used to explore how the number, availability, and distribution of fibronectin (an ECM protein) features can dictate the shape and mobility of a cell. Lehnert et al. [72] utilized μCP and nCP techniques to generate patterns of fibronectin where the feature size was varied from the micro- to nano-scale, while keeping the dot-to-dot spacing constant. When the dot-to-dot spacing was less than 5 μm, cells cultured on these substrates could adhere and spread on fibronectin features as small as 0.1 μm^2. However, when the dot spacing was systematically altered over a range of 5 to 25 μm, the cells exhibited binding and adopted unusual shapes, depending on the type of fibronectin pattern (Figure 13.10). Finally, the ability of cells to spread and migrate on these patterns ceased when the dot-to-dot separation exceeded 30 μm. Therefore, the results of these experiments suggest that the extent of cell spreading is directly correlated to the substratum coverage of fibronectin, and not the geometrical pattern.

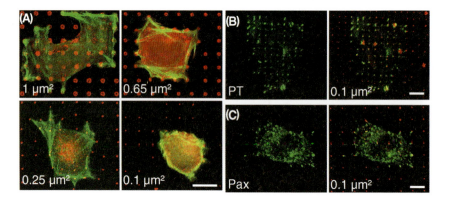

Fig. 13.10 Fibronectin arrays comprising dots of 0.1 μm^2 induce intracellular signaling, but do not support cell spreading at distances >5 μm. (A) B16 cells growing on fibronectin arrays with varying dot sizes (as indicated in lower left corner) and with a constant spacing of 5 μm (center-to-center). Cells spreading and the formation of actin stress fibers could be observed on dots down to 0.25 μm^2. However, dots of 0.1 μm^2 adhered but did not spread. Scale bar = 10 μm. (B) A cell sitting on 0.1 μm^2 dots with 2-μm spacing. Phosphotyrosine (PT, green) accumulates in areas of the cell overlying fibronectin dots (red), indicating that an area of 0.1 μm^2 is sufficient to induce intracellular signaling. (C) Staining for paxilin (Pax, green) of a cell sitting on 0.1 μm^2 fibronectin dots (red) separated by distances of 4 μm. Scale bar = 5 μm. (Reprinted with permission from Ref. [72]; © 2004, The Company of Biologists.)

13.4
DNA Nanoarrays

Although few studies have been conducted to date regarding the development of DNA nanoarrays, such highly miniaturized structures could be quite useful in developing new diagnostic systems that require small sample volumes, and also in controlling the assembly of material building blocks with complementary oligonucleotides. In nanosystems, there is often a need to control the positioning of building blocks at the single particle level in the context of higher-ordered, larger structures. Alternatively, there may be a need to immobilize such building blocks at specific sites in an integrated device in such a way that allows such structures to be addressed macroscopically. Because the chemical recognition properties of DNA can be tailored by virtue of its sequence, DNA represents the ideal building block for creating chemical affinity templates with nanoscopic features that have highly tailorable recognition properties. In this regard, it has been shown that DNA features can be made as small as 50 nm, and the oligonucleotides used in orthogonal assembly approaches, where each feature can be designed to recognize a specific type of building block and immobilize it at a specific location on a substrate [73]. For a 15-mer recognition sequence, there are 4^{15} unique chemical codes. Not all of these are non-overlapping in terms of their recognition properties, but many of them are and so can be selectively used for different assembly applications. Proof-of-concept orthogonal assembly has been demonstrated with DNA nanoarrays and oligonucleotide-modified Au NPs [7, 73]. The assembly strategy has been used in the context of nanoarrays to fabricate unusual single-particle electronic devices [19] and functional bioactive arrays for protein diagnostics [20]. These early advances are briefly discussed in Sections 13.4.2 and 13.4.3.

13.4.1
Strategies for Preparing DNA Nanoarrays

In contrast to the chemistry used to immobilize proteins (see Section 13.3.1), the approach used to immobilize DNA in the context of nanoarrays is almost identical to what has been used with microarrays. The immobilization of DNA on surfaces can be achieved either in a single step or in multiple steps, depending on the nature of the surface functional groups and whether or not the DNA is chemically modified with functional groups that can direct its immobilization. In this section, we will discuss the various immobilization chemistries reported in the literature for generating DNA nanoarrays.

In the simplest approach, DNA can be patterned using electrostatic interactions between a surface and unmodified DNA. At neutral pH, DNA is negatively charged and the phosphate groups in the DNA backbone can bind strongly to a positively charged surface such as a poly-L-lysine-coated glass substrate [74]. Since immobilization occurs nonspecifically, it is often difficult to effect hybridization to complementary targets, and this has been a major drawback with this approach.

One of the attributes of DNA is that it can be chemically modified with functional groups on its 3′ or 5′ ends. Indeed, one of the most widely used immobilization methods utilizes 3′ or 5′ alkylthiol-modified DNA. Such oligonucleotides will readily adsorb onto gold surfaces [7, 48], and this chemistry has been used to prepare nanoarrays by DPN. In this approach, there is a chemical interaction between the sulfur-containing appendage of the DNA and the gold surface to form what many believe to be a gold–thiolate bond [75]. In addition, Michael Addition chemistry can be used to immobilize 5′ acrylamide-modified oligonucleotides [7] on oxide substrates that have been pre-modified with a 3′-mercaptopropyltrimethoxysilane (MPTMS) layer [76]. Recently, Yu and coworkers [47] have generated nanoarrays of DNA on poly(methyl methacrylate) (PMMA) substrates using a covalent immobilization strategy. The PMMA surface was initially activated to generate surface aldehyde groups that react with 5′ amine-modified DNA [77]. In another useful strategy, our group [73] initially patterned 16-mercaptohexadecanoic acid (MHA) on a gold surface and activated the terminal carboxyl moieties with 1-ethyl-3-(3-dimethylaminopropyl) carbodiimide hydrochloride (EDC). 5′-Alkylamine-modified DNA is then coupled to the activated MHA, and this results in the formation of an amide bond [73]. The biotin–streptavidin interaction also has been used as an immobilization modality for DNA. For example, biotinylated DNA has been immobilized onto streptavidin-coated glass surfaces to generate oligonucleotide nanopatterns [30, 32].

13.4.2
DNA-Based Schemes for Biodetection

DNA nanoarrays are just beginning to be explored for their potential in diagnostics. One of the issues pertaining to their utility is the challenge associated with readout. With extraordinarily small features spaced sub-300 nm apart, it is difficult to use conventional optical methods such as fluorescence microscopy to monitor probe-target-array binding events. However, when the feature size of an individual pattern is miniaturized down to the length scale of the DNA structures being probed, the physical properties of that feature – including the height, shape, and hydrophobicity – change substantially upon binding. In addition, an entire array can be built within the field of view of a conventional scanning probe microscope (100 × 100 μM). This allows a variety of scanning probe-based methods to be used, though these are impractical for microarrays because of the large area that must be interrogated, as readout mechanisms for following nanoarray binding events.

We have demonstrated this concept, in part, by preparing nanoarrays of oligonucleotides by DPN [7, 73]. In this case, the feature spacing was not particularly small, and consequently fluorophore-labeled probes and fluorescence could be used, as well as NP probes [78] and AFM to monitor and study the sequence-specific hybridization events (see Figure 13.3). With the advent of massively parallel writing and multi-ink capabilities, DPN has the throughput to generate highly miniaturized DNA chips with extraordinary chemical complexity. Before this capability becomes a practical reality, inking strategies must be developed as well as methods for on-chip, *in-situ* synthesis of the DNA in the context of a DPN experiment.

In addition to the optical and scanning readout of nanoarrays, electrical readout has been preliminarily explored. In this regard, nanoelectrodes have been prepared with gaps selectively functionalized with different oligonucleotides. In this configuration, once the functionalized gaps have hybridized with a target, the area with NP probes can be further developed and the transport properties of the functionalized gap measured [19]. This approach allows the electrical transport properties of particles to be studied down to the single-particle level, and also provides for an on-off readout mechanism for the array in the context of detection.

13.4.3
Applications of Rationally Designed, Self-Assembled 2-D DNA Nanoarrays

DNA tile lattices have been used to construct periodic arrays of well-defined nano-structures from simple oligonucleotide building blocks [79]. Recently, hybridiza-tion strategies have been used to assemble NPs, conjugated to single-stranded DNA molecules, into chain-like structures. Mao et al. generated 1-D arrays of Au NPs by combining DNA-encoded self-assembly with rolling-circle polymerization [80]. These arrays, in certain cases, were several micrometers in length. Yan and coworkers [52, 53] demonstrated the incorporation of a single Au NP into a ratio-nally designed DNA-nanostructured building block, where they used NP-bearing DNA tiles for the directed assembly of 2-D NP arrays with periodic square-like configurations and precisely controlled interparticle spacings. This square ar-rangement of NPs has been considered as a potential tool for constructing novel cellular nonlinear logic networks [81, 82].

The precise control of periodic spacing between proteins has been demon-strated by using DNA nanoarrays generated by programmed self-assembly. Park et al. demonstrated the ability to generate periodic protein arrays through the use of templated self-assembly of streptavidin onto each DNA tile containing a bio-tinylated group terminal to the oligonucleotides (Figure 13.11) [83]. Fully pro-grammable DNA-templated protein nanoarrays, where proteins can be positioned with precision and specificity, might allow functional protein templates with nanometer dimensions to be built for single molecule detection. More recently, LaBean and coworkers [59] developed a fully addressable, finite-sized DNA nano-array displaying a variety of programmed patterns by using a novel stepwise hierarchical assembly technique. In order to properly address the nanoarrays, these authors included loop strands modified with biotin at desired points in the assembly process, and probed the binding of streptavidin at biotinylated sites (see Figure 13.6).

The use of aptamers in self-assembled DNA nanoarrays has been used as a ro-bust platform specifically to attach proteins in periodic arrangements. Aptamers are relatively small RNA or DNA sequences that can be selected through a pro-

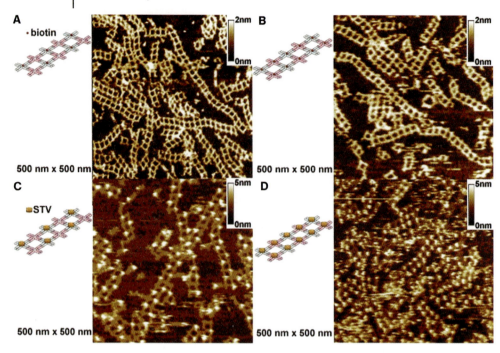

Fig. 13.11 AFM images of the results of programmed assembly of streptavidin on 1-D DNA nanotiles. (A, B) AFM images of bare A*B and A*B* nanotiles before streptavidin binding, respectively. (C, D) AFM images of A*B and A*B* nanotiles after streptavidin binding. (Reprinted with permission from Ref. [83]; © 2005, American Chemical Society.)

cess called SELEX from random pools, based on their ability to bind a specific protein and/or small molecule. Leu et al. first showed how a DNA aptamer could be used in the rational design of DNA nanostructures for the directed assembly of thrombin [84]. Thrombin binding indicates that the thrombin-binding aptamer still functions as the protein-binding moiety upon incorporation into a complex DNA architecture. Aptamer-directed self-assembly of proteins on DNA nanotemplates would allow the relative positions of proteins to be changed in real time, which could enable the study of proximity effects on protein–protein interactions.

In addition to use as directing groups in forming complex NP and protein arrays, self-assembled DNA nanoarrays also have been used as molecular masks in lithographic experiments [85]. In such experiments, the desired DNA nanostructures are generated on a mica substrate, and then used as a mask to control metal deposition. Removal of the DNA mask from the substrate yields a negative replica of the DNA nanostructure. 1-D and 2-D metallic nanopatterns with feature sizes as small as ~10 nm have been generated by this method, though it is possible to create complex structures with more sophisticated DNA motifs [86].

13.5
Virus Nanoarrays

High-resolution nanolithographic techniques allow biological structures to be manipulated at the single-particle level. Although this cannot yet be done routinely with structures as small as proteins, it is possible to manipulate entities such as viruses at this level. The first example of such control involved the use of DPN-generated affinity templates to immobilize tobacco mosaic virus particles (TMV, length = 300 nm, diameter = 18 nm) with excellent control over their orientation and spacing (Figure 13.12) [4]. In these studies, metal ion affinity templates of Zn(II), coordinated to MHA, that bind the natural carboxylate-rich surface moieties of TMV particles were used. Previously, our group [22] and others [23] had generated nanopatterns of NHS-terminated alkanethiols that covalently couple to genetically modified forms of the cow-pea mosaic virus (CPMV) presenting unnatural binding sites (i.e., cystine). However, CPMV particles (diameter = 30 nm) were too small to be isolated and studied at the single-particle level. With the state of the current DPN technology, virus particles 50 nm or larger in diameter are ideal for site-isolation.

Although the above-mentioned methods are very efficient for the immobilization of virus particles with well-known surface chemistries, there are other virus particles of interest that require another strategy. One such strategy has been through the use of antibodies that can recognize and capture surface proteins from a specific virus particle. Our group has reported an antibody immobilization scheme, utilizing metal ion affinity templates that yield active antibody arrays, for

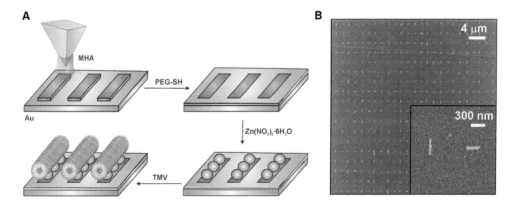

Fig. 13.12 (A) Schematic diagram depicting the process for generating 16-mercaptohexadecanoic acid (MHA) nanotemplates by DPN that were subsequently treated with $Zn(NO_3)_2 \cdot 6H_2O$ for the selective immobilization of single virus particles. (B) AFM tapping mode images of a perpendicular array (40 × 40 μm) of single tobacco mosaic virus (TMV) particles collected at a scan rate of 0.5 Hz. (Reprinted with permission from Ref. [4]; © 2006, Wiley-VCH.)

influenza and simian virus particles [24]. Currently, these surface-immobilized virus particles are being used to study how the number of virus particles, orientation, and presentation (i.e., neutralized) affect cell infectivity at the single-cell and/or single-virus particle level.

13.6
Outlook

Although, at present, we are still in the very early stages of developing ways of making and using bionanoarrays, some very significant advances have been made. Today, it is possible to use techniques such as DPN for the routine fabrication of certain classes of nanoarrays (e.g., one- or two-component structures over larger areas), and we are beginning to learn how to use such arrays to address issues in biology that cannot be addressed with their larger microarray counterparts. The small feature sizes in these arrays are providing research groups with the ability to manipulate biological structures at the single-particle level and to study multivalency in the context of cell-surface interactions. These tools have the potential to revolutionize many aspects of biology, including methods of probing and understanding viral infectivity and proliferation, cellular metabolic and signaling events, and general clinical diagnostics. However, in order to realize their full potential, methods must be developed that allows structures to be built routinely over large areas, with the ability to control feature size, shape, and composition on the sub-100 nm length scale. In particular, these techniques must be developed with biologists in mind, as they are less interested in how a bionanoarray is made but rather how to gain access to it and to acquire control over the structural parameters that make it functional and useful for their studies. Thus, it is imperative that an infrastructure be developed in this field that allows such structures to be built on demand, and in an automated manner. With the advent of high-resolution and massively parallel techniques such as DPN, this goal is achievable in the not-too-distant future.

References

1 Miller, U. R., Nicolau, D. V. (2005) *Microarray Technology and Its Applications.* Springer, New York.

2 Piner, R. D., Zhu, J., Xu, F., Hong, S. H., Mirkin, C. A. (1999) 'Dip-pen' nanolithography. *Science* **283**, 661–663.

3 Rosi, N. L., Mirkin, C. A. (2005) Nanostructures in biodiagnostics. *Chem. Rev.* **105**, 1547–1562.

4 Vega, R. A., Maspoch, D., Salaita, K., Mirkin, C. A. (2005) Nanoarrays of

single virus particles. *Angew. Chem. Int. Ed.* **44**, 6013–6015.

5 Binnig, G., Quate, C. F., Gerber, C. (1986) Atomic Force Microscope. *Phys. Rev. Lett.* **56**, 930–933.

6 Lee, K. B., Park, S. J., Mirkin, C. A., Smith, J. C., Mrksich, M. (2002) Protein nanoarrays generated by dip-pen nanolithography. *Science* **295**, 1702–1705.

7 Demers, L. M., Ginger, D. S., Park, S. J., Li, Z., Chung, S. W., Mirkin,

C. A. (2002) Direct patterning of modified oligonucleotides on metals and insulators by dip-pen nanolithography. *Science* **296**, 1836–1838.

8 Wadu-Mesthrige, K., Xu, S., Amro, N. A., Liu, G. Y. (1999) Fabrication and imaging of nanometer-sized protein patterns. *Langmuir* **15**, 8580–8583.

9 Gu, J. H., Yam, C. M., Li, S., Cai, C. Z. (2004) Nanometric protein arrays on protein-resistant monolayers on silicon surfaces. *J. Am. Chem. Soc.* **126**, 8098–8099.

10 Zhao, Z. Y., Banerjee, P. A., Matsui, H. (2005) Simultaneous targeted immobilization of anti-human IgG-coated nanotubes and anti-mouse IgG-coated nanotubes on the complementary antigen-patterned surfaces via biological molecular recognition. *J. Am. Chem. Soc.* **127**, 8930–8931.

11 Ginger, D. S., Zhang, H., Mirkin, C. A. (2004) The evolution of dip-pen nanolithography. *Angew. Chem. Int. Ed.* **43**, 30–45.

12 Lee, K. B., Lim, J. H., Mirkin, C. A. (2003) Protein nanostructures formed via direct-write dip-pen nanolithography. *J. Am. Chem. Soc.* **125**, 5588–5589.

13 Lim, J. H., Ginger, D. S., Lee, K. B., Heo, J., Nam, J. M., Mirkin, C. A. (2003) Direct-write dip-pen nanolithography of proteins on modified silicon oxide surfaces. *Angew. Chem. Int. Ed.* **42**, 2309–2312.

14 Nam, J. M., Han, S. W., Lee, K. B., Liu, X. G., Ratner, M. A., Mirkin, C. A. (2004) Bioactive protein nanoarrays on nickel oxide surfaces formed by dip-pen nanolithography. *Angew. Chem. Int. Ed.* **43**, 1246–1249.

15 Wilson, D. L., Martin, R., Hong, S., Cronin-Golomb, M., Mirkin, C. A., Kaplan, D. L. (2001) Surface organization and nanopatterning of collagen by dip-pen nanolithography. *Proc. Soc. Natl. Acad. Sci. USA* **98**, 13660–13664.

16 Jiang, H. Z., Stupp, S. I. (2005) Dip-pen patterning and surface assembly

of peptide amphiphiles. *Langmuir* **21**, 5242–5246.

17 Cho, Y., Ivanisevic, A. (2004) SiOx surfaces with lithographic features composed of a TAT peptide. *J. Phys. Chem. B* **108**, 15223–15228.

18 Hyun, J., Kim, J., Craig, S. L., Chilkoti, A. (2004) Enzymatic nanolithography of a self-assembled oligonucleotide monolayer on gold. *J. Am. Chem. Soc.* **126**, 4770–4771.

19 Chung, S. W., Ginger, D. S., Morales, M. W., Zhang, Z. F., Chandrasekhar, V., Ratner, M. A., Mirkin, C. A. (2005) Top-down meets bottom-up: Dip-pen nanolithography and DNA-directed assembly of nanoscale electrical circuits. *Small* **1**, 64–69.

20 Lee, K. B., Kim, E. Y., Mirkin, C. A., Wolinsky, S. M. (2004) The use of nanoarrays for highly sensitive and selective detection of human immunodeficiency virus type 1 in plasma. *Nano Lett.* **4**, 1869–1872.

21 Hyun, J., Lee, W. K., Nath, N., Chilkoti, A., Zauscher, S. (2004) Capture and release of proteins on the nanoscale by stimuli-responsive elastin-like polypeptide 'switches'. *J. Am. Chem. Soc.* **126**, 7330–7335.

22 Smith, J. C., Lee, K. B., Wang, Q., Finn, M. G., Johnson, J. E., Mrksich, M., Mirkin, C. A. (2003) Nanopatterning the chemospecific immobilization of cowpea mosaic virus capsid. *Nano Lett.* **3**, 883–886.

23 Cheung, C. L., Camarero, J. A., Woods, B. W., Lin, T. W., Johnson, J. E., De Yoreo, J. J. (2003) Fabrication of assembled virus nanostructures on templates of chemoselective linkers formed by scanning probe nanolithography. *J. Am. Chem. Soc.* **125**, 6848–6849.

24 Vega, R. A., Maspoch, D., Shen, C. K.-F., Kakkassery, J. J., Chen, B. J., Lamb, R. A., Mirkin, C. A. (2006) Functional antibody arrays through metal ion affinity templates. *ChemBioChem.* **7**, 1653–1657.

25 Liu, M. Z., Amro, N. A., Chow, C. S., Liu, G. Y. (2002) Production of nanostructures of DNA on surfaces. *Nano Lett.* **2**, 863–867.

26 Liu, G. Y., Amro, N. A. (2002)
Positioning protein molecules on
surfaces: A nanoengineering
approach to supramolecular
chemistry. *Proc. Soc. Natl. Acad. Sci.
USA* **99**, 5165–5170.

27 Salaita, K., Lee, S. W., Wang, X. F.,
Huang, L., Dellinger, T. M., Liu, C.,
Mirkin, C. A. (2005) Sub-100 nm,
centimeter-scale, parallel dip-pen
nanolithography. *Small* **1**, 940–945.

28 Salaita, K., Wang, Y., Fragala, J.,
Vega, R. A., Liu, C., Mirkin, C. A.
(2006) Massively parallel dip-pen
nanolithography with 55,000-pen two-
dimensional Arrays. *Angew. Chem.
Int. Ed.* **45**, 7220–7223.

29 Hasnma, P. K., Drake, B., Marti, O.,
Gould, S. A., Prater, C. B. (1989)
The scanning ion-conductance
microscope. *Science* **243**, 641–643.

30 Bruckbauer, A., Ying, L. M., Rothery,
A. M., Zhou, D. J., Shevchuk, A. I.,
Abell, C., Korchev, Y. E., Klenerman,
D. (2002) Writing with DNA and
protein using a nanopipet for
controlled delivery. *J. Am. Chem. Soc.*
124, 8810–8811.

31 Bruckbauer, A., Zhou, D. J., Ying,
L. M., Korchev, Y. E., Abell, C.,
Klenerman, D. (2003) Multicomponent submicron features of biomolecules created by voltage controlled
deposition from a nanopipet. *J. Am.
Chem. Soc.* **125**, 9834–9839.

32 Rodolfa, K. T., Bruckbauer, A., Zhou,
D. J., Korchev, Y. E., Klenerman, D.
(2005) Two-component graded
deposition of biomolecules with a
double-barreled nanopipette. *Angew.
Chem. Int. Ed.* **44**, 6854–6859.

33 Tseng, A. A., Chen, K., Chen, C. D.,
Ma, K. J. (2003) Electron beam
lithography in nanoscale fabrication:
recent development. *IEEE Trans.
Electronics Packaging Manufactur.* **26**,
141–149.

34 Zhang, G. J., Tanii, T., Zako, T.,
Hosaka, T., Miyake, T., Kanari, Y.,
Funatsu, T. W., Ohdomari, I. (2005)
Nanoscale patterning of protein using
electron beam lithography of
organosilane self-assembled
monolayers. *Small* **1**, 833–837.

35 Zhang, G. J., Tanii, T., Funatsu, T.,
Ohdomari, I. (2004) Patterning of
DNA nanostructures on silicon
surface by electron beam lithography
of self-assembled monolayer. *Chem.
Commun.* **15**, 786–787.

36 Hu, W. C., Sarveswaran, K.,
Lieberman, M., Bernstein, G. H.
(2005) High-resolution electron beam
lithography and DNA nano-
patterning for molecular QCA. *IEEE
Trans. Nanotechnol.* **4**, 312–316.

37 Denis, F. A., Pallandre, A., Nysten,
B., Jonas, A. M., Dupont-Gillain, C. C.
(2005) Alignment and assembly of
adsorbed collagen molecules induced
by anisotropic chemical nanopatterns.
Small **1**, 984–991.

38 Pallandre, A., De Meersman, B.,
Blondeau, F., Nysten, B., Jonas, A. M.
(2005) Tuning the orientation of an
antigen by adsorption onto nano-
striped templates. *J. Am. Chem. Soc.*
127, 4320–4325.

39 Falconnet, D., Pasqui, D., Park, S.,
Eckert, R., Schift, H., Gobrecht, J.,
Barbucci, R., Textor, M. (2004) A
novel approach to produce protein
nanopatterns by combining nanoimprint lithography and molecular self-
assembly. *Nano Lett.* **4**, 1909–1914.

40 Hoff, J. D., Cheng, L. J., Meyhofer, E.,
Guo, L. J., Hunt, A. J. (2004)
Nanoscale protein patterning by
imprint lithography. *Nano Lett.* **4**,
853–857.

41 Xia, Y. N., Whitesides, G. M. (1998)
Soft lithography. *Annu. Rev. Mater.
Sci.* **28**, 153–184.

42 Renaultt, J. P., Bernard, A., Bietsch,
A., Michel, B., Bosshard, H. R.,
Delamarche, E., Kreiter, M., Hecht,
B., Wild, U. P. (2003) Fabricating
arrays of single protein molecules on
glass using microcontact printing.
J. Phys. Chem. B **107**, 703–711.

43 Li, H. W., Muir, B. V. O., Fichet, G.,
Huck, W. T. S. (2003) Nanocontact
printing: A route to sub-50-nm-scale
chemical and biological patterning.
Langmuir **19**, 1963–1965.

44 Gerding, J. D., Willard, D. M., Van
Orden, A. (2005) Single-feature
inking and stamping: A versatile

approach to molecular patterning. *J. Am. Chem. Soc.* **127**, 1106–1107.

45 Mirkin, C. A., Demers, L. M., Ginger, D. S. (2003) Patent WO-2003048314.1-76; PCT/US2002/038252.

46 Lin, H. H., Sun, L., Crooks, R. M. (2005) Replication of a DNA microarray. *J. Am. Chem. Soc.* **127**, 11210–11211.

47 Yu, A. A., Savas, T., Cabrini, S., diFabrizio, E., Smith, H. I., Stellacci, F. (2005) High resolution printing of DNA feature on poly(methyl methacrylate) substrates using supramolecular nano-stamping. *J. Am. Chem. Soc.* **127**, 16774–16775.

48 Yu, A. A., Savas, T. A., Taylor, G. S., Guiseppe-Elie, A., Smith, H. I., Stellacci, F. (2005) Supramolecular nanostamping: Using DNA as movable type. *Nano Lett.* **5**, 1061–1064.

49 Denis, F. A., Hanarp, P., Sutherland, D. S., Dufrene, Y. F. (2004) Nanoscale chemical patterns fabricated by using colloidal lithography and self-assembled monolayers. *Langmuir* **20**, 9335–9339.

50 Pammer, P., Schlapak, R., Sonnleitner, M., Ebner, A., Zhu, R., Hinterdorfer, P., Hoglinger, O., Schindler, H., Howorka, S. (2005) Nanopatterning of biomolecules with microscale beads. *ChemPhysChem.* **6**, 900–903.

51 Arnold, M., Cavalcanti-Adam, E. A., Glass, R., Blummel, J., Eck, W., Kantlehner, M., Kessler, H., Spatz, J. P. (2004) Activation of integrin function by nanopatterned adhesive interfaces. *ChemPhysChem.* **5**, 383–388.

52 Sharma, J., Chhabra, R., Liu, Y., Ke, Y. G., Yan, H. (2006) DNA-templated self-assembly of two-dimensional and periodical gold nanoparticle arrays. *Angew. Chem. Int. Ed.* **45**, 730–735.

53 Zhang, J. P., Liu, Y., Ke, Y. G., Yan, H. (2006) Periodic square-like gold nanoparticle arrays templated by self-assembled 2D DNA nanogrids on a surface. *Nano Lett.* **6**, 248–251.

54 Cai, Y. G., Ocko, B. M. (2005) Large-scale fabrication of protein

nanoarrays based on nanosphere lithography. *Langmuir* **21**, 9274–9279.

55 Haynes, C. L., Van Duyne, R. P. (2001) Nanosphere lithography: A versatile nanofabrication tool for studies of size-dependent nanoparticle optics. *J. Phys. Chem. B* **105**, 5599–5611.

56 Hulteen, J. C., Treichel, D. A., Smith, M. T., Duval, M. L., Jensen, T. R., Van Duyne, R. P. (1999) Nanosphere lithography: size-tunable silver nanoparticle and surface cluster arrays. *J. Phys. Chem. B* **103**, 3854–3863.

57 Haes, A. J., Chang, L., Klein, W. L., Van Duyne, R. P. (2005) Detection of a biomarker for Alzheimer's disease from synthetic and clinical samples using a nanoscale optical biosensor. *J. Am. Chem. Soc.* **127**, 2264–2271.

58 Liu, Y., Ke, Y. G., Yan, H. (2005) Self-assembly of symmetric finite-size DNA nanoarrays. *J. Am. Chem. Soc.* **127**, 17140–17141.

59 Park, S. H., Pistol, C., Ahn, S. J., Reif, J. H., Lebeck, A. R., Dwyer, C., LaBean, T. H. (2006) Finite-size, fully addressable DNA tile lattices formed by hierarchical assembly procedures. *Angew. Chem. Int. Ed.* **45**, 735–739.

60 MacBeath, G., Schreiber, S. L. (2000) Printing proteins as microarrays for high-throughput function determination. *Science* **289**, 1760–1763.

61 Chen, S., Liu, L., Zhou, J., Jiang, S. (2003) Controlling antibody orientation on charged self-assembled monolayers. *Langmuir* **19**, 2859–2864.

62 Zull, J. E., Reed-Mundell, J., Lee, Y. W., Vezenov, D., Ziats, N. P., Anderson, J. M., Sukenik, C. N. (1994) Problems and approaches in covalent attachment of peptides and proteins to inorganic surfaces for biosensor applications. *J. Indust. Microbiol. Biotechnol.* **13**, 137–143.

63 Babacam, S., Pivarnik, P., Lecther, S., Rand, A. G. (2000) Evaluation of antibody immobilization methods for piezoelectric biosensor application. *Biosensors Bioelectron.* **15**, 615–621.

64 Lee, S. W., Oh, B.-K., Sanedrin, R. G., Salaita, K., Fujigaya, T., Mirkin, C. A.

(2006) Biologically active protein nanoarrays generated using parallel dip-pen nanolithography. *Adv. Mater.* **18**, 1133–1136.

65 Lynch, M., Mosher, C., Huff, J., Nettikadan, S., Johnson, J., Henderson, E. (2004) Functional protein nanoarrays for biomarker profiling. *Proteomics* **4**, 1695–1702.

66 Peluso, P., Wilson, D. S., Do, D., Tran, H., Venkatasubbaiah, M., Quincy, D., Heidecker, B., Poindexter, K., Tolani, N., Phelan, M. (2003) Optimizing antibody immobilization strategies for the construction of protein microarrays. *Anal. Biochem.* **312**, 113–124.

67 Soong, R. K., Bachand, G. D., Neves, H. P., Olkhovets, A. G., Craighead, H. G., Montemagno, C. D. (2000) Powering an inorganic nanodevice with a biomolecular motor. *Science* **290**, 1555–1558.

68 Blau, H. M., Baltimore, D. (1991) Differentiation requires continuous regulation. *J. Cell Biol.* **112**, 781–783.

69 Ruoslahti, E., Obrink, B. (1996) Common principles in cell adhesion. *Exp. Cell Res.* **227**, 1–11.

70 Chen, C. S., Mrksich, M., Hung, S., Whitesides, G. M., Ingber, D. E. (1997) Geometric control of cell life and death. *Science* **276**, 1425–1428.

71 Shim, J., Bersano-Begey, T. F., Zhu, X. Y., Tkaczyk, A. H., Linderman, J. J., Takayama, S. (2003) Micro- and nanotechnologies for studying cellular function. *Curr. Topics Med. Chem.* **3**, 687–703.

72 Lehnert, D., Wehrle-Haller, B., David, C., Weiland, U., Ballestrem, C., Imhof, B. A., Bastmeyer, M. (2004) Cell behaviour on micropatterned substrata: limits of extracellular matrix geometry for spreading and adhesion. *J. Cell Sci.* **117**, 41–52.

73 Demers, L. M., Park, S. J., Taton, T. A., Li, Z., Mirkin, C. A. (2001) Orthogonal assembly of nanoparticle building blocks on dip-pen nano-lithographically generated templates of DNA. *Angew. Chem. Int. Ed.* **40**, 3071–3073.

74 Tabata, H., Uno, T., Ohtake, T., Kawai, T. (2005) DNA patterning by nano-imprinting technique and its application for bio-chips. *J. Photopolymer Sci. Technol.* **18**, 519–522.

75 Bain, C. D., Troughton, E. B., Tao, Y. T., Evall, J., Whitesides, G. M., Nuzzo, R. G. (1989) Formation of monolayer films by the spontaneous assembly of organic thiols from solution onto gold. *J. Am. Chem. Soc.* **111**, 321–335.

76 Kurth, D. G., Bein, T. (1993) Surface reactions on thin layers of silane coupling agents. *Langmuir* **9**, 2965–2973.

77 Fixe, F., Dufva, M., Telleman, P., Christensen, C. B. V. (2004) Functionalization of poly(methyl methacrylate) (PMMA) as a substrate for DNA microarrays. *Nucleic Acids Res.* **32**, e9.

78 Mirkin, C. A., Letsinger, R. L., Mucic, R. C., Storhoff, J. J. (1996) A DNA-based method for rationally assembling nanoparticles into macroscopic materials. *Nature* **382**, 607–609.

79 Winfree, E., Liu, F. R., Wenzler, L. A., Seeman, N. C. (1998) Design and self-assembly of two-dimensional DNA crystals. *Nature* **394**, 539–544.

80 Deng, Z. X., Tian, Y., Lee, S. H., Ribbe, A. E., Mao, C. D. (2005) DNA-encoded self-assembly of gold nanoparticles into one-dimensional arrays. *Angew. Chem. Int. Ed.* **44**, 3582–3585.

81 Yang, T., Kiehl, R. A., Chua, L. O. (2001) Tunneling phase logic cellular nonlinear networks. *Int. J. Bifurcation Chaos* **11**, 2895–2912.

82 Xiao, S., Liu, F., Rosen, A. E., Hainfeld, J. F., Seeman, N. C., Musier-Forsyth, K., Kiehl, R. A. (2002) Self-assembly of metallic nanoparticle arrays by DNA scaffolding. *J. Nanoparticle Res.* **4**, 313–317.

83 Park, S. H., Yin, P., Liu, Y., Reif, J. H., LaBean, T. H., Yan, H. (2005) Programmable DNA self-assemblies for nanoscale organization of ligands and proteins. *Nano Lett.* **5**, 729–733.

84 Liu, Y., Lin, C. X., Li, H. Y., Yan, H.
(2005) Aptamer-directed self-assembly
of protein arrays on a DNA nanostructure. *Angew. Chem. Int. Ed.* **44**,
4333–4338.

85 Deng, Z. X., Mao, C. D. (2004)
Molecular lithography with DNA

nanostructures. *Angew. Chem. Int. Ed.*
43, 4068–4070.

86 Mao, C. D., LaBean, T. H., Reif, J. H.,
Seeman, N. (2000) Logical computation using algorithmic self-assembly
of DNA triple-crossover molecules
(2000). *Nature* **407**, 750–750.

Part III
Nanostructures for Medicinal Applications

Nanobiotechnology II. Edited by Chad A. Mirkin and Christof M. Niemeyer
Copyright © 2007 WILEY-VCH Verlag GmbH & Co. KGaA, Weinheim
ISBN: 978-3-527-31673-1

14
Biological Barriers to Nanocarrier-Mediated Delivery of Therapeutic and Imaging Agents

Rudy Juliano

14.1
Overview: Nanocarriers for Delivery of Therapeutic and Imaging Agents

The concept of using nanocarriers for the selective delivery of therapeutic and imaging agents in cancer has been reinvigorated by the emergence of exciting new approaches for nanocarrier design and fabrication. In addition to liposomes, which are already widely used for clinical purposes [1, 2], other nanocarriers – including polymeric nanoparticles (NPs), dendrimers, block co-polymer micelles, magnetic NPs and various inorganic NPs – are under intense investigation [3–5]. Each of these technologies has its merits and liabilities, but all are affected by similar biological parameters that govern the abilities of nanocarriers to distribute into tissues and be taken up by cells. Thus, the design of nanocarriers for potential therapeutic or imaging applications should be guided by consideration of the biological barriers that stand between the NP at its point of entry into the body and its ultimate destination. The biological barrier concept has a long history in the drug delivery field [6], and has proven quite useful in this context.

14.2
Basic Characteristics of the Vasculature and Mononuclear Phagocyte System

Although nanocarriers may be used in a variety of local applications (topical, inhalational, etc.), the primary concern here will be with material administered intravenously for systemic use. Thus, the architecture of the vasculature, the composition of the blood, the body's system for particle clearance (the mononuclear phagocyte system), and the permeability characteristics of the vascular endothelium will all be key factors in determining the biodistribution of nanocarriers. Understanding these factors and avoiding the rapid sequestration of administered nanocarriers at undesirable sites represents the first major barrier to the effective use of NP technology for drug delivery.

Nanobiotechnology II. Edited by Chad A. Mirkin and Christof M. Niemeyer
Copyright © 2007 WILEY-VCH Verlag GmbH & Co. KGaA, Weinheim
ISBN: 978-3-527-31673-1

14.2.1
Possible Interactions of Nanocarriers Within the Bloodstream

When injected, nanocarriers will first encounter the cells and plasma proteins of the blood. The binding of plasma proteins to drug carriers has long been of interest to investigators studying liposomes [7, 8], and is of growing interest to NP technologists [9, 10], for very good reasons. Such plasma protein interactions play a key role in determining the state of aggregation of nanocarriers *in vivo*, and their uptake by the phagocytic mononuclear cells that protects the body against foreign pathogens (for further details, see below). In addition, nanocarriers may also interact with circulating blood cells including platelets [11] and leukocytes [12], affecting the functions of these cells, and in turn being affected by the interaction. Thus, careful consideration must be given to controlling the interactions of nanocarriers with the macromolecular and cellular components of blood.

14.2.2
Transendothelial Permeability in Various Tissues and Tumors

If nanocarriers are destined for sites within either normal or tumor tissues, they must first pass across the barrier provided by the vascular endothelium. The blood vessels are lined by endothelial cells, which adhere tightly to the underlying extracellular matrix (ECM) largely via integrins, and form junctions with each other using a variety of cell–cell adhesion molecules including VE-cadherin, junctional adhesion molecule (JAM), occludins, claudins, and platelet and endothelial cell adhesion molecule (PECAM). The integrity of these junctions is regulated by complex signal transduction processes [13].

The permeation of large and small molecules across normal tissue endothelium is usually described in terms of a model involving abundant small pores of about 45 Å diameter and relatively few large pores of about 250 Å [14]. Clearly, a significant component of small pore transport is paracellular, via imperfections in the junctions between the endothelial cells. The identity of the large-pore system has been somewhat controversial. Endothelial cells contain an abundance of small endosomal vesicles derived from caveolae, which are lipid-rich membrane structures containing the protein calveolin. One school of thought suggests that large-pore transport entails "transcytosis" – that is, the uptake of macromolecules into calveolae on one side of the cell, followed by active, energy-requiring movement of the calveolar vesicles across the cell, and then release on the other side. Other investigators believe that the large pores represent rare larger defects in the endothelial junctions, or that multiple calveolar vesicles fuse to form tubes or channels that cross the cell, in either case providing passive pathways for transport (Figure 14.1). One argument for the vesicular transport large-pore model is that while vascular permeability declines dramatically with molecular size up to about 50 Å (the small-pore limit), it is only weakly dependent on molecular size above that. Unfortunately, key experiments testing the energy-dependence of

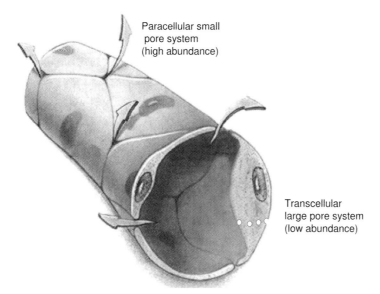

Paracellular small
pore system
(high abundance)

Transcellular
large pore system
(low abundance)

Fig. 14.1 Transcapillary transport of macromolecules and nanoparticles
(NPs). A capillary is depicted lined with tightly abutted endothelial cells.
The transport of most molecules takes place using a paracellular route
that involves very small gaps between the cells. A transcytosis pathway
may be involved in the movement of larger molecules and NPs,
although the magnitude of its contribution is unclear.

large-pore transport, or even the role of calveolin in this process, have provided
contradictory results [15, 16]. In any case, in the normal vasculature of most tissues, egress of entities larger than 250 Å (25 nm) is very limited.

In contrast, in certain specialized tissues the organization of the endothelium
is compatible with the transport of larger moieties. For example, the microvessels
of both the liver and spleen have relatively large "fenestrations" of 100–200 nm
diameter [17]. This, plus the abundance of mononuclear phagocytes in liver and
spleen, strongly contributes to the well-known tendency of injected NPs to accumulate in these organs. It should also be noted that microvascular transport of
water and macromolecules increases during inflammation [18]; thus, sites of inflammation are also potentially accessible to NPs. A further consideration is that
transendothelial permeability is tightly regulated by signaling processes; therefore, there are a number of mediators that increase vascular permeability, including thrombin, histamine, vascular endothelial growth factor (VEGF), tumor
necrosis factor-alpha (TNF-α), and reactive oxygen species (ROS) [13]. Thus, one
could visualize incorporating some of these mediators into NPs in order to promote particle egress from the vasculature.

Tumors are another place where the microvasculature is abnormal in ways that
may potentially enhance delivery via nanocarriers. However, there are complex issues affecting the behavior of nanocarriers within tumors that need to be under-

stood in order to attain tumor targeting. First, the overall architecture of the vasculature differs between tumors and adjacent normal tissue; further, there are often regional differences within the tumor and between the primary tumor and its metastases. Differences include the number, length, degree of branching, and velocity of blood flow [19]. While small tumors may be relatively homogeneous, larger ones usually have several distinct regions including: (i) a necrotic, hypoxic core with almost no blood flow; (ii) a semi-necrotic region with relatively poor flow in unbranched vessels; (iii) a stable region with branched vessels and relatively good flow; and (iv) an advancing front where active angiogenesis is taking place [19, 20]. The transport of macromolecules across the tumor endothelium involves diffusion, convection, and possibly transcytosis. The vascular permeability of tumors is generally higher than that of normal tissue, most likely due to an increased number and size of the "large-pore" component [19]. Some studies have found pores or fenestrations in tumor vessels ranging from 100 to over 700 nm [21]; however, as always, tumors show great heterogeneity with regard to this parameter.

One factor that works against the extravasation of nanocarriers through the larger pores in tumor microvessels is the existence of a high interstitial fluid pressure. This blunts the convective component of transendothelial transport, which relies on the differences in osmotic and hydrostatic pressures between the blood and the fluid of the tissue parenchyma. The existence of the high interstitial pressure in tumors has been ascribed to a paucity of functional lymphatic drainage [19]. Another issue to consider is the movement of a macromolecule or nanocarrier once it leaves the blood. The space between cells is filled with a dense ECM that impedes diffusion. For example, the diffusion of immunoglobulin G (MW $\sim$ 160 000; 6 nm diameter) in tumors has been estimated as 1 hr for a 100 mμ distance [19]; by comparison the intercapillary distance in some tumors has been estimated at about 50 mμ [21]. Thus, IgG would need 15 min to diffuse throughout the tumor parenchyma. Obviously, the diffusion rate through the ECM would be much slower for a typical NP of perhaps 50–100 nm, and therefore the ECM may provide an additional biological barrier to delivery of nanocarriers to tumor cells.

The relatively leaky character of the tumor endothelium, along with poor lymphatic drainage, has combined to provide the so-called "EPR" (enhanced permeability and retention) effect that sometimes allows the selective accumulation of macromolecules or nanocarriers in tumors. The larger pore size provides the opportunity for egress of relatively large entities, and the poor lymph flow means that the extravasated material persists in the tumor longer than it would in normal tissue [21]. Some studies have suggested that it is the retention component due to limited lymphatic drainage that is most important in the EPR effect [22]. This effect of selective retention in tumors is thought to contribute to the therapeutic profile of Doxil®, a liposomal form of the antitumor drug doxorubicin [23]. While the EPR effect may be an important aspect of the successful use of nanocarriers in cancer, the great heterogeneity of tumors should be borne in mind, and thus greater or lesser impacts of the EPR effect with different tumors and different nanocarrier systems should be expected.

14.2.3
Mononuclear Cells and Particle Clearance

The specialized phagocytic cells of the so-called "reticuloendothelial system" or "mononuclear phagocyte system" (MPS) play a key role in the biodistribution of systemically administered NPs. The physiological role of the MPS is primarily to clear the body of invading pathogens, including bacteria and fungi; the mononuclear phagocytes also play a key role in the turnover of apoptotic cells [24].

The membranes of macrophages and other MPS cells display several different classes of receptors that promote the binding of particles and their internalization via several endocytotic pathways [25]. The best-studied pathway involves the receptor for the constant domain of immunoglobulins, the Fc-receptor. There are three forms of this receptor in human macrophages ($Fc\gamma RI$, RIIA, and RIII). These receptors recognize particles that have bound immunoglobulins, anchoring them to the macrophage surface, and then triggering a complex signaling process involving tyrosine kinases, PI-3-kinase, and Rho GTPases, that eventually leads to mobilization of the actinomyosin cytoskeleton, formation of a phagocytic vesicle, and internalization of the particle. Another important class of macrophage receptors are those for complement components. The three complement receptors are comprised of CR1, a single-chain transmembrane protein, and CR3 and 4 that are members of the integrin family of heterodimeric membrane proteins (CR3 = integrin $\alpha m\beta 2$, CR4 = $\alpha x\beta 2$). While CR1 is involved in particle binding, the CR3 and CR4 receptors are primarily responsible for phagosome formation and internalization. Overall, the CR internalization process seems to be less robust than FcR-mediated internalization, and CR phagocytic activity may require addition signals. In addition to FcR and CR, macrophages possess so-called "scavenger" receptors (e.g., SR-A) that recognize conserved motifs found in pathogens, including bacterial lipopolysaccharides, fungal mannans, and many others [26]. Members of the Toll-like family of receptors also recognize similar pathogen motifs, but are involved primarily in triggering inflammatory cytokine responses in macrophages rather than phagocytosis [27]. Macrophages also possess additional integrin receptors that can recognize motifs in other serum proteins such as fibronectin or vitronectin.

Subsequent to engulfment of the particle in a phagosome, fusion with lysosomes takes place leading to enzymatic degradation of the ingested material by proteases and hydrolases that operate efficiently in the low-pH lysosomal environment. The factors controlling the rate and extent of phagosome-lysosome fusion are complex, and may include the number and strength of interactions between the receptors in the phagosome and their ligands on the particle surface. However, the overall process of intracellular trafficking and fusion of intracellular vesicles is also regulated by complex signaling pathways.

Synthetic nanocarriers are recognized by mononuclear phagocytes subsequent to the binding of plasma proteins that are collectively termed "opsonins", and which serve as ligands for the receptors described above. Thus, the binding of immunoglobulins can lead to recognition by the Fc-Receptor and activation of the complement system; binding of the complement components including C3b, iC3b and C1q results in recognition by CR1,3,4, while binding of apolipoproteins can promote interaction with scavenger receptors. Thus, as discussed in detail later, the degree to which the nanocarrier interacts with opsonic proteins in plasma is a key determinant of interactions with the MPS system, and thus of overall biodistribution. In most cases one it would be preferable to avoid extensive MPS uptake, and thus a great deal of attention has been paid to this issue.

Mononuclear phagocytes are found throughout the body, but are particularly prevalent in the liver (Kupffer cells) [28] and spleen (splenic macrophages) [25]. The abundance of MPS cells, along with the relatively open architecture of the hepatic and splenic capillaries, accounts for the fact that much of an administered dose of most NPs accumulates in these two organs. Although several strategies have been devised to avoid the uptake of NPs by mononuclear phagocytes (see below), these strategies have been only partially successful as they work against a very strong biological imperative to direct foreign particles to the liver and spleen phagocytes.

14.3
Cellular Targeting and Subcellular Delivery

Assuming that one can design a nanocarrier that persists in the circulation for an extended time, and is not inappropriately removed by capillary filtration or by the MPS system, then the next set of barriers or challenges is to target the nanocarrier to the correct cell type and have it enter the desired subcellular compartment.

14.3.1
Targeting, Entry, and Trafficking in Cells

The process of targeted nanocarrier-mediated delivery often involves the binding of the nanocarrier to a particular cell-surface receptor, internalization of the nanocarrier into an endosome, and then release of the nanocarrier (or its cargo) into the cytoplasm of the cell. The interplay between cell-surface receptors and the multiple endocytotic pathways found in cells is a complex story, and one that is still not fully understood. What follows is a brief account of current concepts regarding endocytosis of cell-surface receptors.

An important aside concerns the nature of the cell-surface receptors that are the key to cell-type selective targeting. First, it is vital to realize that there is no such thing as a completely cell type-specific receptor. Some receptors are preferentially expressed in certain cell types; well-known examples from cancer biology include: (i) the αvβ3 integrin expressed on capillaries during angiogenesis; (ii) the CD20 antigen expressed on certain types of lymphomas; and (iii) the HER2 receptor tyrosine kinase overexpressed in a subcategory of breast tumors. However, all of these proteins are also expressed to a lesser degree in other cell types, thus leading to possible "off-target" effects. Similar considerations exist concerning receptors prominent in other fields of therapeutic interest. A second consideration is the degree to which a receptor is linked to the endocytotic machinery. In many instances the intent is to deliver a drug into the cell, not simply to the vicinity of a cell. If that is indeed the case, then it is important to choose as a target a receptor that is known to traffic effectively to endosomes. A good example might be the transferrin receptor, which very efficiently enters cells via clathrin-coated vesicles; by contrast, for some integrins only 30% of the receptor on the cell surface is actively involved in internalization [29].

Cells display a surprising variety of endocytotic processes involving different biochemical mechanisms and resulting in differing rates and pathways of internalization (Figure 14.2). It is interesting to note that a good deal of our current knowledge on endocytosis derives from studies of the pathways taken by viruses [30] or by bacterial toxins [31] in entering cells. In addition to phagocytosis (which is largely confined to MPS cells), there are at least four distinct endocytotic pathways [32]:

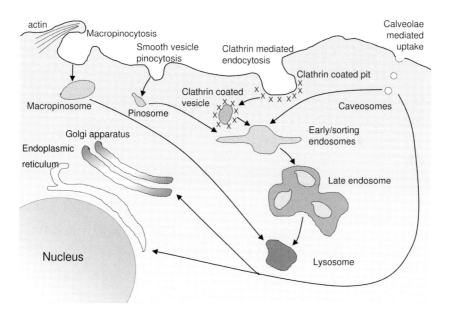

Fig. 14.2 Endocytotic uptake pathways. The diagram shows only those routes leading into the cell; multiple routes of vesicular transport from the inside to the plasma membrane are not shown. Likewise, the many accessory proteins involved in vesicle traffic, such as the RabGTPases, are not depicted.

- The "classic" route is the clathrin-mediated pathway utilized by receptors such as the transferrin receptor and the low-density lipoprotein receptor, as well as many others. These receptors bind to adaptor proteins such as Ap-2 and then engage with clathrin triskelions that form a cage that invaginates the portion of the plasma membrane containing the receptor and its ligand. With the assistance of the dynamin GTPase, a clathrin-coated vesicle buds from the membrane; the clathrin cage is then quickly dis-assembled and the nascent endosome fuses with a sorting endosome of moderately low pH (5.9). Here, the receptor and ligand dissociate, the receptors primarily recycle to the plasma membrane, while the ligand enters late endosomes that eventually fuse with lysosomes, thus degrading the ligands.
- Calveolae are membrane structures rich in sphingolipids and cholesterol, and contain one of the three members of the calveolin protein family. These entities are involved in the internalization of receptors that are glycosylphosphati-dylinositol-anchored proteins; additionally G-protein-coupled receptors primarily transit through calveolae. With the assistance of dynamin, calveolae can internalize to form vesicles called "caveosomes"; these primarily traffic to the Golgi complex and to the endoplasmic reticulum, although in some cases they can also merge with sorting endosomes. Calveolae have been implicated in transcytosis, especially in endothelial cells [13], and some authors believe that this is an important aspect of movement of large molecules across the vascular wall. However, although calveolae have often been associated with internalization of receptors and ligands, there is some evidence that this pathway may not be as active and dynamic as other endocytotic pathways [33].
- Macropinocytosis is a process whereby cells use their actinomyosin contractile machinery to extend membrane protrusions that then pinch off and engulf relatively large volumes of extracellular fluid. The resulting large endosomes can either fuse with lysosomes or recycle to the cell surface. Macropinocytosis is usually triggered by exposure to polypeptide growth factors or other signals.
- There is also a pathway of non-clathrin, non-caveolin-mediated pinocytosis that leads to an early endosomal com-partment and then to the Golgi and endoplasmic reticulum; the mechanistic details of this pathway are poorly understood.

The complex trafficking of vesicles between different cellular endomembrane compartments is highly regulated by a variety of biochemical processes. Thus, different types of endosomes are marked by particular lipid compositions, especially regarding phosphotidylinositols [34], while sphingolipid-rich "rafts" give lateral definition to membrane microdomains. The pattern of vesicle traffic is regulated by GTPases of the Rab and Arf families; for example, Rab 5 is particularly involved in early endocytosis [35]. Finally, the recognition and fusion events involved in merging endomembrane entities and delivery of contents is regulated by multiple proteins including SNARES and sorting nexins [36].

Which of these several pathways is predominant in a particular situation of NP–cell interaction will depend on the surface characteristics of the NP, the cell type, and the ambient physiological conditions. It seems important to determine carefully the route of nanocarrier uptake in order to fully interpret the biological results observed.

14.3.2
Biological and Chemical Reagents for Cell-Specific Targeting

Although the problem of targeting nanocarriers to a specific cell type is challenging, fortunately a number of powerful technologies exist to address this issue. The most obvious approach is to couple antibodies to the nanocarrier; thus, there is a long history of studies with immunoliposomes for targeting and delivery [37, 38], as well as recent examples using immunoliposomes to deliver small interfering RNA (siRNA) [39] or showing that antibody-mediated targeting enhances intracellular uptake and anti-tumor potency of multiple liposomal drugs *in vivo* [40]. Likewise, investigators are now using antibody reagents to target a variety of NPs, including gold nanocages [41] and quantum dots [42, 43]. Recent developments in antibody technology have provided a variety of reagents that may avoid some of the problems associated with use of full-length antibodies that contain Fc domains (and thus are likely to bind and stimulate macrophage Fc receptors). Options include engineered monovalent entities including classic Fab fragments, scFvs that combine the light and heavy chain variable regions in a single polypeptide, and camellid-based heavy chain variable regions that exhibit high affinity. Bi-, tri-, and tetravalent options also exist, including classic Fab'$_2$, minibodies, and other reagents that can recognize one, or two or more, distinct antigenic determinants [44].

A second powerful approach is phage display; here, very large peptide/protein libraries are displayed on the surfaces of bacteriophage and the libraries are screened for binding to specific target molecules [45]. The peptide sequences that are effective in recognizing the intended target are identified by DNA sequencing of the target-bound phage. This potent and flexible technology can be used with libraries comprised of short peptides, sequences from antibodies such as scFvs, or with peptide sequences embedded in protein domain scaffolds that provide structural rigidity. An exciting new thrust for this technology is the emergence of *in-vivo* phage display pioneered by Ruoslahti and Pasqualini [46]. This has allowed the power of phage display to identify determinants that are differentially expressed in the vasculature of different tissues, or that are predominantly expressed in the vasculature of tumors. From the phage libraries small peptides have been obtained that bind preferentially to these determinants and that can be used to target cytotoxic or imaging agents [47, 48].

Another key approach for cellular targeting involves nucleic acid aptamers [49a]. The aptamer concept relies on the fact that single-strand nucleic acids engage in intrastrand base pairing and thus form molecules with a very broad range of three-dimensional shapes that may then bind to complementary "pockets" in proteins. Very complex libraries of RNA or DNA aptamers are screened for binding to protein or cellular targets. The bound aptamers are recovered and then amplified by a polymerase chain reaction process, yielding an enriched population. After several cycles of selection, aptamers with binding affinities and specificities comparable to the best monoclonal antibodies can be obtained. Aptamers can be in the range of 50 bases, and thus can be readily produced on an oligonucleotide synthesizer; this allows the use of chemically modified nucleosides that then provide resistance against nuclease degradation. Aptamers have been used to target clotting factors, cell-surface receptors and intracellular proteins. An exciting recent development is the creation of an RNA aptamer-siRNA chimera that provides both cell type-specific targeting and a powerful biological effect [49b]. Aptamer technology is now also beginning to be applied to the targeting of NPs [50].

14.3.3
Reagents that Promote Cell Entry

In addition to delivering nanocarriers to particular cell types, it is sometimes necessary for the NP or its payload to enter the cytoplasm or nucleus. Delivery of a particle to an endosome does not equate to true intracellular delivery. The inside of an endosome is topologically equivalent to the outside of the cell; a lipid bilayer membrane must still be crossed in order to access the cytosol. Recently, there has been a great deal of interest in a class of reagents that can potentially promote the entry of macromolecules (and even particles) into the cytoplasm; these are collectively termed "cell-penetrating peptides" (CPPs) [51, 52]. The earliest CPPs were polycationic sequences derived from viral or lower eukaryote transcription factors, for example the "TAT" and "Penetratin" (antennepedia) sequences. Other types of CPP have been based on hydrophobic signal sequences or viral fusion domains [53, 54], while additional CPPs are constantly being discovered [55]. CPPs have been used very effectively to promote the intracellular entry of other peptides and of oligonucleotides [56]. There is also a large literature on the use of CPPs to deliver proteins, including some remarkable observations on delivery to the central nervous system (CNS) [57]. However, not all attempts to deliver proteins with CPPs have been successful and there is some controversy about this approach [52, 58, 59]. The precise mechanism(s) of action of the various CPPs is still under investigation; most studies suggest that CPPs assist their "cargos" in escaping from an endosomal compartment. In the case of TAT, recent evidence has suggested that this CPP functions through the macropinocytosis pathway [60]. The use of CPPs for intracellular delivery of nanocarriers has been explored in the contexts of liposomes [61], dendrimers [62], magnetic NPs [63] and others.

14.4
Crafting NPs for Delivery: Lessons from Liposomes

All types of systemically administered particle face the same set of issues in terms of their effective use in the delivery of therapeutic or imaging agents. Liposomes were among the first types of micro/nanoparticles to be developed to the point of clinical applications. Thus, a great deal is known about how various parameters influence the behavior of liposomal delivery systems. Hence, we will use liposomes as a prototype to explore some of these key issues.

14.4.1
Loading

The first major problem that needs to be addressed is that of adequately loading the nanocarrier. Early attempts to load polar drugs into liposomes were not very successful because of the low internal aqueous volumes of these particles. Liposomes would seem an ideal carrier for lipophilic drugs, but these drugs can also rapidly escape. A major step forward came with techniques to load large amounts of amphipathic molecules (e.g., anthracyclines) into liposomes using pH and charge gradients [64]; this can result in the accumulation of sufficient drug to initiate crystallization or the formation of a gel. By using this approach, very high ratios of active drug to inactive carrier (lipid) can be obtained; for example, for Doxil® the doxorubicin:lipid ratio is approximately 1 mg drug to 6 mg lipid carrier (almost 20% active drug in the formulation). It is thought that the drug remains insoluble as the liposome traverses the circulation and is only released as the liposome degrades within or in the vicinity of the tumor cell. In the case of polymer-based NPs, the drug would usually be trapped inside the polymer mesh; the efficiency of entrapment of soluble drugs can be poor (a few percent) as a high mesh density is needed [65]. Another possibility is to couple the drug to the particle surface, but then much of the internal volume (and mass) of the carrier is wasted. A potentially interesting approach is to use drug nanocrystals [66]. Current techniques generate nanocrystals in the 200- to 400-nm range, but further refinements may reduce the size range. These drug-rich particles could then be coated with materials to control release rates and biological interactions. In any case it is clear that a high drug:carrier ratio is an important goal in attaining good pharmacological effects and in minimizing toxicity due to the carrier.

14.4.2
Release Rates

A second key issue is to control the release rate of the drug from the carrier, as well as influencing where in the body the drug will be released. Obviously, the aim would be to minimize release while the nanocarrier–drug complex is traversing the circulation, but then allow effective release at the intended target site, but this is not easy. For various types of liposomal drugs, release rates are influenced by: (i) diffusion across the lipid membrane; (ii) partitioning from the lipid phase; (iii) solubilization of entrapped crystalline drug; and (iv) destabilization of the liposome itself. Liposome destabilization and drug release may be enhanced as liposomes enter the low-pH endosomal compartment, or may even be affected by lipases released from necrotic cells in tumors or inflamed areas [1]. The release of drugs from polymeric particles can take place by diffusion or by degradation/destabilization of the polymer meshwork [67]. One very positive aspect of polymer-based NPs is that there exist numerous approaches for preparing particles that will preferentially release drug at a particular time or anatomic site. For example, NPs that include acid- or esterase-sensitive links may be preferentially degraded in the endosomal-lysosomal system, while particles with S–S bridges will be destabilized in the reducing intracellular environment but be relatively stable in the oxidizing blood environment [68–70]. Nanoparticles containing magnetic or optically active material can potentially be perturbed in particular regions of the body using external sources, thus causing drug release [71, 72]. However, all of these approaches have limitations, and it remains very difficult to design nanocarriers that will release their therapeutic payload at a precisely designated time and place in the body.

14.4.3
Size and Charge

The role of size and charge in determining the rate of particle clearance from the circulation has been known in the case of liposomes for more than three decades [73]. The basic observations that large particles are cleared more rapidly than small, and that anionic particles are cleared more rapidly than neutral ones, have been confirmed many times for liposomes and other particles [74]. However, there are numerous complexities, including favorable effects on liposome clearance by certain types of anionic lipids (for a discussion, see Ref. [1]). Particles larger than 200–300 nm can be removed from the blood by simple filtration in the spleen; additionally, they may activate complement more effectively than smaller particles and thus be removed through macrophage-mediated uptake in liver and spleen [75]. In addition to size, the mechanical properties of particles – such as their degree of rigidity versus deformability – no doubt play a role in their susceptibility to filtration. Positively charged particles are a special case; there are multiple problems associated with their use, including binding to numerous plasma proteins and to cells, that are associated with toxicity. Nonetheless, cationic liposomes (lipoplexes) have been widely used as a platform for the delivery of plasmid DNA and of oligonucleotides, and they may have some advantages in this context [76]. New approaches to toxicity evaluation, including the use of DNA array data [77], may provide important insights into the true degree of the toxicity of cationic carriers.

14.4.4
PEG and the Passivation of Surfaces

One of the most successful approaches to the production of long-circulating NPs is the "passivation" of their surfaces with coating of highly hydrophilic polymers;

this prevents rapid clearance of particles by the macrophage-like cells of the reticuloendothelial system. The premier example of this is the development of "sterically stabilized" or "Stealth®" liposomes via inclusion of a lipid-anchored PEG in the formulation [1]. These entities have circulation lifetimes several-fold greater than those of non-PEGylated liposomes of similar size; this approach has been used in the FDA-approved formulation of liposomal doxorubicin, Doxil®. In addition to PEG, other approaches for "passivation" include the use of dextrans [78] and block polymers such as poloxamines and poloxamers [74]. The precise mechanism underlying the effect of these hydrophilic polymers is not fully understood, but presumably they provide a "brush-like" coating of the particle surface that may exclude large plasma proteins, thus reducing opsonization of the particles [79]. However, this is somewhat controversial as some observations have suggested that PEGylation does not prevent complement activation [80]. Alternative explanations have been suggested, including the unfavorable sequestration of opsonic proteins or binding of "dys-opsonins" that protect against the uptake of opsonized particles.

14.4.5
Decoration with Ligands

Ultimately, the preference would be to have NPs that not only circulate for a protracted time but also have the ability to be targeted to specific sites in the body. This requires the linkage of various targeting ligands, as discussed above. A variety of chemistries are available to link various ligands to liposomes or other NP [6, 81]. An important consideration here is the balance between the effects of the targeting ligand and the need to "passivate" the surface. Early studies on immunoliposomes were hindered by poor placement of the antibody (on the lipid surface rather than external to the PEG layer), the use of full-length antibodies that could interact with Fc receptors on macrophage-like cells, and overconjugation. In this respect, more recent studies have led to substantial improvements, including the availability of PEGs with convenient linker groups, the use of Fab or scFvs that lack the problematic Fc region of antibodies, and the realization that relatively few ligands are enough to provide targeting [23, 81]. A recent study demonstrated the power of these new approaches [40]. Thus, sterically stabilized liposomes containing several types of antitumor drugs were targeted to breast cancer xenografts in mice using an Fab fragment of the anti-EGF-receptor monoclonal cetuximab. Both control sterically stabilized (PEGylated) liposomes and Fab-conjugated liposomes attained a substantial "EPR" effect that resulted in an accumulation in the tumors. However, only the Fab-conjugated liposomes were taken up by the tumor cells, though this provided a substantially enhanced therapeutic effect. A similarly impressive example from the polymeric NP arena is the use of PEGylated polylactide-glycolide NPs containing doxcetaxel and liganded with an RNA aptamer to target prostate tumors expressing the surface prostate-specific membrane antigen (PMSA) [82]. In summary, recent advances in understanding the role of particle size, the need for surface passivation with hydrophilic polymers, and the use of sophisticated targeting ligands are beginning to make a reality of the concept of "smart" nanocarriers.

14.5
Biodistribution of Liposomes, Dendrimers, and NPs

One key aspect of the preclinical development of nanocarriers as therapeutic agents is a full understanding of their pharmacokinetics and biodistribution. Such studies have been extensively pursued for liposomes [83–86], as several liposomal drugs are now available for clinical use. A number of comprehensive studies have also been conducted to determine the biodistribution of dendrimers [particularly of the poly amido amide (PAMAM) variety], as these molecules have evoked interest as image-enhancing agents, and knowledge of their clearance kinetics and tissue distribution is essential in this context [87–89]. By contrast, the extent of formal comprehensive pharmacokinetic/biodistribution studies of organic and inorganic NPs is more limited (an exception is the well-studied Abraxane®, an albumin NP formulation of paclitaxel that is approved for clinical use [90]). Although there is considerable interest in the possible use of carbon nanotubes in drug delivery, most of the investigations conducted thus far have been *in vitro*, and therefore these entities have not been carefully characterized as to biodistribution [91]. In the overall context, undoubtedly the same issues concerning size, charge, the binding of opsonins, and interactions with phagocytic reticuloendothelial cells that affect dendrimers and liposomes will also affect the kinetics and biodistribution of other nanocarrier types.

Although limited in number, several well-conducted studies of the kinetics and biodistribution of several types of NPs have been recently reported, some of which are listed in Table 14.1.

Table 14.1 Typical studies of the kinetics and biodistribution of several types of nanoparticle (NP),

Type of NP	Animal model	Reference
DNA in an emulsifying wax plus detergent oil-in-water emulsion	Biodistribution in mice	92
PEGylated gelatin NPs containing DNA	Functional biodistribution in tumor-bearing mice	93
Polymeric NPs containing indium-oxime	Biodistribution in tumor-bearing mice	94
Chitosan NPs containing doxorubicin	Biodistribution in tumor-bearing mice	95

14.6
The Toxicology of Nanocarriers

One final barrier that must be surmounted prior to the adoption of nanocarriers for use in humans is a rigorous assessment of their potential toxicities. Because of the anticipation that nanotechnology will be a key aspect of the future world economy, there has been widespread discussion of the need to assess the health risks associated with this technology [96]. In this respect, there are two main aspects. The first aspect relates to occupational and environmental exposure to nanomaterials, while the second (which is more germane to this chapter) relates to biomedical applications. Some discussions have been conducted of whether the unique properties of nanomaterials will require new approaches to toxicological evaluation, but this is unlikely to be the case. In terms of environmental or workplace hazards, the most likely portals of entry into the body would be via the airways or the skin. There is actually vast experience with the pulmonary toxicology of very fine particulates, and it seems likely that this will suffice to evaluate NP effects [96–98]. A similar case can likely be made for dermal exposures to nanoparticulates. In the biomedical context it seems clear that nanomaterials should – and indeed inevitably will – be subject to the same level of intense toxicological scrutiny as drugs and biologicals. Overall, it will be difficult (and perhaps inappropriate) to make generalized statements about nanocarrier toxicities. Rather, the toxicological effects will be context-dependent and will be affected by the route of administration, the dose, the chemical nature of the nanocarrier, its surface modifications, and the "payload" that it carries.

Once again, it is the liposome technology that has been most thoroughly evaluated, both in animal models and in humans [99–101]. In general, the toxicities associated with liposomal drugs are due primarily to the drugs, since lipids are natural constituents of the body and are readily metabolized. However, as liposomes are avidly accumulated by liver and splenic macrophages, even "empty" liposomes can cause splenomegaly and hepatotoxicity when administered chronically [102]. It is not clear whether this is an issue in the use of liposomal drugs in patients, but an important lesson to be learned from clinical experience with liposomes is that use of a drug carrier can simultaneously have both desirable and undesirable effects on toxicity. Thus, while one important benefit of liposomal doxorubicin is its reduced cardiac toxicity [100], this comes at the cost of increased dermal toxicity – the so-called "hand–foot syndrome" [103].

The toxicity of dendrimers has also been studied at the preclinical level, and this topic has been thoroughly reviewed recently [104]. Various types of experiments have implicated dendrimers in cytotoxicity to cells in culture, hemolysis, complement activation, cytokine release and generalized in-vivo toxicity. In general, polycationic dendrimers such as unmodified PAMAMs are substantially more toxic than neutral or anionic dendrimers (–COOH- or –OH-terminated PAMAMs, polyesters). The toxicology of dendrimers is still at an early stage, but is likely to expand rapidly as these molecules show so much promise as adjuncts to imaging technologies.

Quantum dots and carbon nanotubes have evoked a great deal of interest for potential use in imaging and drug delivery. Based on very limited studies, claims have been made concerning the lack of toxicity of these materials. These claims are probably very premature as neither of these materials has been subjected to rigorous toxicological analysis. In the case of quantum dots, although their cores are often comprised of highly poisonous substances such as cadmium and selenium, it has been suggested that their surface coatings prevent toxic effects.

More stringent studies of the potential hazards of carbon nanotubes (CNTs) and quantum dots have recently begun to appear in journals dedicated to toxicological assessment. A recent meta-analysis of the observations on quantum dots indicates numerous findings of toxicity in cell culture due to these materials and their constituents [105]. The long persistence of quantum dots in tissues certainly raises a "red flag" regarding possible chronic toxicity, though this has not yet been addressed. In the case of CNTs, one concern is their possible role in the generation of ROS that can cause tissue damage [96]; this can occur either through perturbation of biological processes that regulate cellular ROS levels, or through the generation of reactive species by impurities in the nanotubes. Thus, recent studies have demonstrated substantial pulmonary toxicity of single-walled and multi-walled CNTs when administered to the respiratory system [106, 107]. In summary, the toxicology of truly novel nanomaterials such as CNTs and quantum dots is still at a very early stage and needs to be pursued in a comprehensive and stringent manner.

14.7
Summary

During the past few years a number of new nanomaterials have emerged, the unique physical and chemical properties of which suggest that they may have important applications in biomedicine, particularly as drug carriers. The challenge is now to marry these novel physical characteristics with sophisticated biological characteristics so that effective delivery of therapeutic and imaging agents can be attained *in vivo*. Thus, nanocarriers are at the beginning of a long journey from the laboratory to the clinic; we will surely learn much along the way.

References

1 Drummond, D. C., Meyer, O., Hong, K., Kirpotin, D. B., Papahadjopoulos, D. (1999) Optimizing liposomes for delivery of chemotherapeutic agents to solid tumors. *Pharmacol. Rev.* **51**, 691–743.

2 Allen, T. M., Cullis, P. R. (2003) Drug delivery systems: entering the mainstream. *Science* **303**, 1818–1822.

3 Sahoo, S. K., Labhasetwar, V. (2003) Nanotech approaches to drug delivery and imaging. *Drug Discov. Today* **8**, 1112–1120.

4 LaVan, D. A., McGuire, T., Langer, R. (2003) Small-scale systems for in vivo

drug delivery. *Nat. Biotechnol.* **21**, 1184–1191.

5 Rolland, J. P., Maynor, B. W., Euliss, L. E., Exner, A. E., Denison, G. M., DeSimone, J. M. (2005) Direct fabrication and harvesting of mono-disperse, shape-specific nanobio-materials. *J. Am. Chem. Soc.* **127**, 10096–10100.

6 Poznansky, M. J., Juliano, R. L. (1984) Biological approaches to the controlled delivery of drugs: a critical review. *Pharmacol. Rev.* **36**, 277–336.

7 Bonte, F., Juliano, R. L. (1986) Interactions of liposomes with serum proteins. *Chem. Phys. Lipids* **40**, 359–372.

8 Chonn, A., Semple, S. C., Cullis, P. R. (1992) Association of blood proteins with large unilamellar liposomes in vivo. Relation to circulation lifetimes. *J. Biol. Chem.* **267**, 18759–18765.

9 Luck, M., Paulke, B. R., Schroder, W., Blunk, T., Muller, R. H. (1998) Analysis of plasma protein adsorption on polymeric nanoparticles with different surface characteristics. *J. Biomed. Mater. Res.* **39**, 478–485.

10 Koziara, J. M., Oh, J. J., Akers, W. S., Ferraris, S. P., Mumper, R. J. (2005) Blood compatibility of cetyl alcohol/polysorbate-based nanoparticles. *Pharm. Res.* **22**, 1821–1828.

11 Juliano, R. L., Hsu, M. J., Peterson, D., Regen, S. L., Singh, A. (1983) Interactions of conventional or photopolymerized liposomes with platelets in vitro. *Exp. Cell Res.* **146**, 422–427.

12 Opanasopit, P., Nishikawa, M., Hashida, M. (2002) Factors affecting drug and gene delivery: effects of interaction with blood components. *Crit. Rev. Ther. Drug Carrier Syst.* **19**, 191–233.

13 Mehta, D., Malik, A. B. (2006) Signaling mechanisms regulating endothelial permeability. *Physiol. Rev.* **86**, 279–367.

14 Rippe, B., Rosengren, B. I., Carlsson, O., Venturoli, D. (2002) Transendo-thelial transport: the vesicle con-troversy. *J. Vasc. Res.* **39**, 375–390.

15 Schubert, W., Frank, P. G., Razani, B., Park, D. S., Chow, C. W., Lisanti, M. P. (2001) Caveolae-deficient endothelial cells show defects in the uptake and transport of albumin in vivo. *J. Biol. Chem.* **276**, 48619–48622.

16 Drab, M., Verkade, P., Elger, M., Kasper, M., Lohn, M., Lauterbach, B., Menne, J., Lindschau, C., Mende, F., Luft, F. C., Schedl, A., Haller, H., Kurzchalia, T. V. (2001) Loss of caveolae, vascular dysfunction, and pulmonary defects in caveolin-1 gene-disrupted mice. *Science* **293**, 2449–2452.

17 Scherphof, G. L. (1991) In-vivo behavior of liposomes: interaction with themononuclear phagocyte system and implications for drug targeting. In: Juliano, R. L. (Ed.), *Targeted Drug Delivery Handbook of Experimental Pharmacology*. Vol. 100. Springer-Verlag, Berlin, pp. 285–327.

18 Szekanecz, Z., Koch, A. E. (2004) Vascular endothelium and immune responses: implications for inflam-mation and angiogenesis. *Rheum. Dis. Clin. North Am.* **30**, 97–114.

19 Jain, R. K. (1999) Transport of molecules, particles, and cells in solid tumors. *Annu. Rev. Biomed. Eng.* **1**, 241–263.

20 Rubin, P., Casarett, G. (1966) Microcirculation of tumors. I. Anatomy, function, and necrosis. *Clin. Radiol.* **17**, 220–229.

21 Jang, S. H., Wientjes, M. G., Lu, D., Au, J. L. (2003) Drug delivery and transport to solid tumors. *Pharm. Res.* **20**, 1337–1350.

22 Noguchi, Y., Wu, J., Duncan, R., Strohalm, J., Ulbrich, K., Akaike, T., Maeda, H. (1998) Early phase tumor accumulation of macromolecules: a great difference in clearance rate between tumor and normal tissues. *Jpn. J. Cancer Res.* **89**, 307–314.

23 Park, J. W., Benz, C. C., Martin, F. J. (2004) Future directions of liposome- and immunoliposome-based cancer therapeutics. *Semin. Oncol.* **31**(6 Suppl. 13), 196–205.

24 Aderem, A., Underhill, D. M. (1999) Mechanisms of phagocytosis in

macrophages. *Annu. Rev. Immunol.* **17**, 593–623.

25 Taylor, P. R., Martinez-Pomares, L., Stacey, M., Lin, H. H., Brown, G. D., Gordon, S. (2005) Macrophage receptors and immune recognition. *Annu. Rev. Immunol.* **23**, 901–944.

26 Greaves, D. R., Gordon, S. (2005) Thematic review series: the immune system and atherogenesis. Recent insights into the biology of macrophage scavenger receptors. *J. Lipid Res.* **46**, 11–20.

27 Akira, S., Uematsu, S., Takeuchi, O. (2006) Pathogen recognition and innate immunity. *Cell* **124**, 783–801.

28 Parker, G. A., Picut, C. A. (2005) Liver immunobiology. *Toxicol. Pathol.* **33**, 52–62.

29 Sczekan, M. M., Juliano, R. L. (1990) Internalization of the fibronectin receptor is a constitutive process. *J. Cell Physiol.* **142**, 574–580.

30 Sieczkarski, S. B., Whittaker, G. R. (2005) Viral entry. *Curr. Top. Microbiol. Immunol.* **285**, 1–23.

31 Sandvig, K., van Deurs, B. (2005) Delivery into cells: lessons learned from plant and bacterial toxins. *Gene Ther.* **12**, 865–872.

32 Perret, E., Lakkaraju, A., Deborde, S., Schreiner, R., Rodriguez-Boulan, E. (2005) Evolving endosomes: how many varieties and why? *Curr. Opin. Cell Biol.* **17**, 423–434.

33 Hommelgaard, A. M., Roepstorff, K., Vilhardt, F., Torgersen, M. L., Sandvig, K., van Deurs, B. (2005) Caveolae: stable membrane domains with a potential for internalization. *Traffic* **6**, 720–724.

34 Haucke, V. (2005) Phosphoinositide regulation of clathrin-mediated endocytosis. *Biochem. Soc. Trans.* **33**(Pt. 6), 1285–1289.

35 Stein, M. P., Dong, J., Wandinger-Ness, A. (2003) Rab proteins and endocytic trafficking: potential targets for therapeutic intervention. *Adv. Drug Deliv. Rev.* **55**, 1421–1437.

36 Kavalali, E. T. (2002) SNARE interactions in membrane trafficking: a perspective from mammalian central synapses. *BioEssays* **24**, 926–936.

37 Abra, R. M., Bankert, R. B., Chen, F., Egilmez, N. K., Huang, K., Saville, R., Slater, J. L., Sugano, M., Yokota, S. J. (2002) The next generation of liposome delivery systems: recent experience with tumor-targeted, sterically-stabilized immunoliposomes and active-loading gradients. *J. Liposome Res.* **12**, 1–3.

38 Sapra, P., Tyagi, P., Allen, T. M. (2005) Ligand-targeted liposomes for cancer treatment. *Curr. Drug Deliv.* **2**, 369–381.

39 Pirollo, K. F., Zon, G., Rait, A., Zhou, Q., Yu, W., Hogrefe, R., Chang, E. H. (2006) Tumor-targeting nanoimmuno-liposome complex for short interfering RNA delivery. *Hum. Gene Ther.* **17**, 117–124.

40 Mamot, C., Drummond, D. C., Noble, C. O., Kallab, V., Guo, Z., Hong, K., Kirpotin, D. B., Park, J. W. (2005) Epidermal growth factor receptor-targeted immunoliposomes significantly enhance the efficacy of multiple anticancer drugs in vivo. *Cancer Res.* **65**, 11631–11638.

41 Chen, J., Saeki, F., Wiley, B. J., Cang, H., Cobb, M. J., Li, Z. Y., Au, L., Zhang, H., Kimmey, M. B., Li, X., Xia, Y. (2005) Gold nanocages: bioconjugation and their potential use as optical imaging contrast agents. *Nano Lett.* **5**, 473–477.

42 Gao, X., Cui, Y., Levenson, R. M., Chung, L. W., Nie, S. (2004) In vivo cancer targeting and imaging with semiconductor quantum dots. *Nat. Biotechnol.* **22**, 969–976.

43 Michalet, X., Pinaud, F. F., Bentolila, L. A., Tsay, J. M., Doose, S., Li, J. J., Sundaresan, G., Wu, A. M., Gambhir, S. S., Weiss, S. (2005) Quantum dots for live cells, in vivo imaging, and diagnostics. *Science* **307**, 538–544.

44 Holliger, P., Hudson, P. J. (2005) Engineered antibody fragments and the rise of single domains. *Nat. Biotechnol.* **23**, 1126–1136.

45 Rodi, D. J., Makowski, L., Kay, B. K. (2002) One from column A and two from column B: the benefits of phage display in molecular-recognition studies. *Curr. Opin. Chem. Biol.* **6**, 92–96.

46 Ruoslahti, E. (2004) Vascular zip codes in angiogenesis and metastasis. *Biochem. Soc. Trans.* **32**(Pt. 3), 397–402.

47 Arap, W., Haedicke, W., Bernasconi, M., Kain, R., Rajotte, D., Krajewski, S., Ellerby, H. M., Bredesen, D. E., Pasqualini, R., Ruoslahti, E. (2002) Targeting the prostate for destruction through a vascular address. *Proc. Natl. Acad. Sci. USA* **99**, 1527–1531.

48 Arap, W., Pasqualini, R., Ruoslahti, E. (1998) Cancer treatment by targeted drug delivery to tumor vasculature in a mouse model. *Science* **279**, 377–380.

49 (a) Nimjee, S. M., Rusconi, C. P., Sullenger, B. A. (2005) Aptamers: an emerging class of therapeutics. *Annu. Rev. Med.* **56**, 555–583; (b) McNamara, J. O., II, Andrechek, E. R., Wang, Y., Viles, K. D., Rempel, R. E., Gilboa, E., Sullenger, B. A., Giangrande, P. H. (2006) Cell type-specific delivery of siRNAs with aptamer-siRNA chimeras. *Nat. Biotechnol.* **24**, 1005–1015.

50 Farokhzad, O. C., Jon, S., Khademhosseini, A., Tran, T. N., Lavan, D. A., Langer, R. (2004) Nanoparticle-aptamer bioconjugates: a new approach for targeting prostate cancer cells. *Cancer Res.* **64**, 7668–7672.

51 Zorko, M., Langel, U. (2005) Cell-penetrating peptides: mechanism and kinetics of cargo delivery. *Adv. Drug Deliv. Rev.* **57**, 529–545.

52 Vives, E. (2005) Present and future of cell-penetrating peptide mediated delivery systems: "is the Trojan horse too wild to go only to Troy?" *J. Control. Release* **109**, 77–85.

53 Hawiger, J. (1999) Noninvasive intracellular delivery of functional peptides and proteins. *Curr. Opin. Chem. Biol.* **3**, 89–94.

54 Gros, E., Deshayes, S., Morris, M. C., Aldrian-Herrada, G., Depollier, J., Heitz, F., Divita, G. (2006) A non-covalent peptide-based strategy for protein and peptide nucleic acid transduction. *Biochim. Biophys. Acta* **1758**, 384–393.

55 Rhee, M., Davis, P. (2006) Mechanism of uptake of C105Y, a novel cell-penetrating peptide. *J. Biol. Chem.* **281**, 1233–1240.

56 Juliano, R. L. (2005) Peptide-oligonucleotide conjugates for the delivery of antisense and siRNA. *Curr. Opin. Mol. Ther.* **7**, 132–136.

57 Wadia, J. S., Dowdy, S. F. (2005) Transmembrane delivery of protein and peptide drugs by TAT-mediated transduction in the treatment of cancer. *Adv. Drug Deliv. Rev.* **57**, 579–596.

58 Trehin, R., Merkle, H. P. (2004) Chances and pitfalls of cell penetrating peptides for cellular drug delivery. *Eur. J. Pharm. Biopharm.* **58**, 209–223.

59 Leifert, J. A., Harkins, S., Whitton, J. L. (2002) Full-length proteins attached to the HIV tat protein transduction domain are neither transduced between cells, nor exhibit enhanced immunogenicity. *Gene Ther.* **9**, 1422–1428.

60 Wadia, J. S., Stan, R. V., Dowdy, S. F. (2004) Transducible TAT-HA fusogenic peptide enhances escape of TAT-fusion proteins after lipid raft macropinocytosis. *Nat. Med.* **10**, 310–315.

61 Torchilin, V. P., Rammohan, R., Weissig, V., Levchenko, T. S. (2001) TAT peptide on the surface of liposomes affords their efficient intracellular delivery even at low temperature and in the presence of metabolic inhibitors. *Proc. Natl. Acad. Sci. USA* **98**, 8786–8791.

62 Kang, H., DeLong, R., Fisher, M. H., Juliano, R. L. (2005) Tat-conjugated PAMAM dendrimers as delivery agents for antisense and siRNA oligonucleotides. *Pharm. Res.* **22**, 2099–2106.

63 Zhao, M., Kircher, M. F., Josephson, L., Weissleder, R. (2002) Differential conjugation of tat peptide to superparamagnetic nanoparticles and its effect on cellular uptake. *Bioconj. Chem.* **13**, 840–844.

64 Allen, T. M., Martin, F. J. (2004) Advantages of liposomal delivery systems for anthracyclines. *Semin. Oncol.* **31**(6 Suppl. 13), 5–15.

65 Budhian, A., Siegel, S. J., Winey, K. I. (2005) Production of haloperidol-loaded PLGA nanoparticles for extended controlled drug release of haloperidol. *J. Microencapsul.* **22**, 773–785.

66 Muller, R. H., Keck, C. M. (2004) Challenges and solutions for the delivery of biotech drugs – a review of drug nanocrystal technology and lipid nanoparticles. *J. Biotechnol.* **113**, 151–170.

67 Soppimath, K. S., Aminabhavi, T. M., Kulkarni, A. R., Rudzinski, W. E. (2001) Biodegradable polymeric nanoparticles as drug delivery devices. *J. Control. Release* **70**, 1–20.

68 Shenoy, D., Little, S., Langer, R., Amiji, M. (2005) Poly(ethylene oxide)-modified poly(beta-amino ester) nanoparticles as a pH-sensitive system for tumor-targeted delivery of hydrophobic drugs. 1. In vitro evaluations. *Mol. Pharm.* **2**, 357–366.

69 Carlisle, R. C., Etrych, T., Briggs, S. S., Preece, J. A., Ulbrich, K., Seymour, L. W. (2004) Polymer-coated polyethylenimine/DNA complexes designed for triggered activation by intracellular reduction. *J. Gene Med.* **6**, 337–344.

70 Kommareddy, S., Amiji, M. (2005) Preparation and evaluation of thiol-modified gelatin nanoparticles for intracellular DNA delivery in response to glutathione. *Bioconj. Chem.* **16**, 1423–1432.

71 Edelman, E. R., Langer, R. (1993) Optimization of release from magnetically controlled polymeric drug release devices. *Biomaterials* **14**, 621–626.

72 Hirsch, L. R., Gobin, A. M., Lowery, A. R., Tam, F., Drezek, R. A., Halas, N. J., West, J. L. (2006) Metal nanoshells. *Ann. Biomed. Eng.* **34**, 15–22.

73 Juliano, R. L., Stamp, D. (1975) Effect of particle size and charge on the clearance rates of liposomes and liposome encapsulated drugs. *Biochem. Biophys. Res. Commun.* **63**, 651–658.

74 Moghimi, S. M., Hunter, A. C., Murray, J. C. (2001) Long-circulating and target-specific nanoparticles: theory to practice. *Pharmacol. Rev.* **53**, 283–318.

75 Moghimi, S. M., Hunter, A. C., Murray, J. C. (2005) Nanomedicine: current status and future prospects. *FASEB J.* **19**, 311–330.

76 Chen, W. C., Huang, L. (2005) Non-viral vector as vaccine carrier. *Adv. Genet.* **54**, 315–337.

77 Omidi, Y., Barar, J., Akhtar, S. (2005) Toxicogenomics of cationic lipid-based vectors for gene therapy: impact of microarray technology. *Curr. Drug Deliv.* **2**, 429–441.

78 Moore, A., Marecos, E., Bogdanov, A., Jr., Weissleder, R. (2000) Tumoral distribution of long-circulating dextran-coated iron oxide nanoparticles in a rodent model. *Radiology* **214**, 568–574.

79 Owens, D. E., III, Peppas, N. A. (2006) Opsonization, biodistribution, and pharmacokinetics of polymeric nanoparticles. *Int. J. Pharm.* **307**, 93–102.

80 Moghimi, S. M., Szebeni, J. (2003) Stealth liposomes and long circulating nanoparticles: critical issues in pharmacokinetics, opsonization and protein-binding properties. *Prog. Lipid Res.* **42**, 463–478.

81 Sapra, P., Allen, T. M. (2003) Ligand-targeted liposomal anticancer drugs. *Prog. Lipid Res.* **42**, 439–462.

82 Farokhzad, O. C., Cheng, J., Teply, B. A., Sherifi, I., Jon, S., Kantoff, P. W., Richie, J. P., Langer, R. (2006) Targeted nanoparticle-aptamer bioconjugates for cancer chemotherapy in vivo. *Proc. Natl. Acad. Sci. USA* **103**, 6315–6320.

83 Kamps, J. A., Scherphof, G. L. (2004) Biodistribution and uptake of liposomes in vivo. *Methods Enzymol.* **387**, 257–266.

84 van Etten, E. W., Otte-Lambillion, M., van Vianen, W., ten Kate, M. T.,

Bakker-Woudenbert, A. J. (1995) Biodistribution of liposomal amphotericin B (AmBisome) and amphotericin B-desoxycholate (Fungizone) in uninfected immuno-competent mice and leucopenic mice infected with *Candida albicans. J. Antimicrob. Chemother.* **35**, 509–519.

85 Gabizon, A., Martin, F. (1997) Polyethylene glycol-coated (pegylated) liposomal doxorubicin. Rationale for use in solid tumours. *Drugs* **54**(Suppl. 4), 15–21.

86 Amantea, M. A., Forrest, A., Northfelt, D. W., Mamelok, R. (1997) Population pharmacokinetics and pharmacodynamics of pegylated-liposomal doxorubicin in patients with AIDS-related Kaposi's sarcoma. *Clin. Pharmacol. Ther.* **61**, 301–311.

87 Kobayashi, H., Brechbiel, M. W. (2005) Nano-sized MRI contrast agents with dendrimer cores. *Adv. Drug Deliv. Rev.* **57**, 2271–2286.

88 Lee, C. C., MacKay, J. A., Frechet, J. M., Szoka, F. C. (2005) Designing dendrimers for biological applica-tions. *Nat. Biotechnol.* **23**, 1517–1526.

89 Nigavekar, S. S., Sung, L. Y., Llanes, M., El-Jawahri, A., Lawrence, T. S., Becker, C. W., Balogh, L., Khan, M. K. (2004) 3H dendrimer nanoparticle organ/tumor distribution. *Pharm. Res.* **21**, 476–483.

90 Sparreboom, A., Baker, S. D., Verweij, J. (2005) Paclitaxel repackaged in an albumin-stabilized nanoparticle: handy or just a dandy? *J. Clin. Oncol.* **23**, 7765–7767.

91 Bianco, A., Kostarelos, K., Prato, M. (2005) Applications of carbon nanotubes in drug delivery. *Curr. Opin. Chem. Biol.* **9**, 674–679.

92 Cui, Z., Mumper, R. J. (2002) Plasmid DNA-entrapped nanoparticles engineered from microemulsion precursors: in vitro and in vivo evaluation. *Bioconj. Chem.* **13**, 1319–1327.

93 Kaul, G., Amiji, M. B. (2004) Biodistribution and targeting potential of poly(ethylene glycol)-modified gelatin nanoparticles in subcutaneous murine tumor model. *J. Drug Target.* **12**, 585–591.

94 Shenoy, D., Little, S., Langer, R., Amiji, M. (2005) Poly(ethylene oxide)-modified poly(beta-amino ester) nanoparticles as a pH-sensitive system for tumor-targeted delivery of hydrophobic drugs: part 2. In vivo distribution and tumor localization studies. *Pharm Res.* **22**, 2107–2114.

95 Hyung Park, J., Kwon, S., Lee, M., Chung, H., Kim, J. H., Kim, Y. S., Park, R. W., Kim, I. S., Bong Seo, S., Kwon, I. C., Young Jeong, S. (2006) Self-assembled nanoparticles based on glycol chitosan bearing hydrophobic moieties as carriers for doxorubicin: in vivo biodistribution and anti-tumor activity. *Biomaterials* **27**, 119–126.

96 Nel, A., Xia, T., Madler, L., Li, N. (2006) Toxic potential of materials at the nanolevel. *Science* **311**, 622–627.

97 Oberdorster, G., Oberdorster, E., Oberdorster, J. (2005) Nanotoxicology: an emerging discipline evolving from studies of ultrafine particles. *Environ. Health Perspect.* **113**, 823–839.

98 Tsuji, J. S., Maynard, A. D., Howard, P. C., James, J. T., Lam, C. W., Warheit, D. B., Santamaria, A. B. (2006) Research strategies for safety evaluation of nanomaterials, part IV: risk assessment of nanoparticles. *Toxicol. Sci.* **89**, 42–50.

99 Alberts, D. S., Muggia, F. M., Carmichael, J., Winer, E. P., Jahanzeb, M., Venook, A. P., Skubitz, K. M., Rivera, E., Sparano, J. A., DiBella, N. J., Stewart, S. J., Kavanagh, J. J., Gabizon, A. A. (2004) Efficacy and safety of liposomal anthracyclines in phase I/II clinical trials. *Semin. Oncol.* **31**(6 Suppl. 13), 53–90.

100 Safra, T. (2003) Cardiac safety of liposomal anthracyclines. *Oncologist* **8**(Suppl. 2), 17–24.

101 Tollemar, J., Klingspor, L., Ringden, O. (2001) Liposomal amphotericin B (AmBisome) for fungal infections in immunocompromised adults and children. *Clin. Microbiol. Infect.* **7**(Suppl. 2), 68–79.

102 Allen, T. M., Smuckler, E. A. (1985) Liver pathology accompanying chronic liposome administration in mouse. *Res. Commun. Chem. Pathol. Pharmacol.* **50**, 281–290.

103 Lotem, M., Hubert, A., Lyass, O., Goldenhersh, M. A., Ingber, A., Peretz, T., Gabizon, A. (2000) Skin toxic effects of polyethylene glycol-coated liposomal doxorubicin. *Arch. Dermatol* **136**, 1475–1480.

104 Duncan, R., Izzo, L. (2005) Dendrimer biocompatibility and toxicity. Adv. Drug Deliv. Rev. **57**, 2215–2237.

105 Hardman, R. (2006) A toxicologic review of quantum dots: toxicity depends on physicochemical and environmental factors. Environ. Health Perspect. **114**, 165–172.

106 Lam, C. W., James, J. T., McCluskey, R., Hunter, R. L. (2004) Pulmonary toxicity of single-wall carbon nanotubes in mice 7 and 90 days after intratracheal instillation. *Toxicol. Sci.* **77**, 126–134.

107 Muller, J., Huaux, F., Moreau, N., Misson, P., Heilier, J. F., Delos, M., Arras, M., Fonseca, A., Nagy, J. B., Lison, D. (2005) Respiratory toxicity of multi-wall carbon nanotubes. *Toxicol. Appl. Pharmacol.* **207**, 221–231.

15

Organic Nanoparticles: Adapting Emerging Techniques from the Electronics Industry for the Generation of Shape-Specific, Functionalized Carriers for Applications in Nanomedicine

Larken E. Euliss, Julie A. DuPont, and Joseph M. DeSimone

15.1
Overview

The design and exploitation of materials and structures where at least one dimension is measured in the nanometer range broadly defines the term "nanotechnology" [1]. The umbrella of nanotechnology covers a wide variety of research disciplines, ranging from nanomachines to lithography to the development of nanoparticles (NPs). Nanomedicine is therefore the natural extension of nanotechnology implemented in the medical field. The National Institutes of Health (NIH) have recently coined the term "Nanomedicine" to mean the application of nanotechnology for the treatment, diagnosis, monitoring, and control of biological systems [2]. At the forefront of research in this area is the development of methods to target and deliver pharmaceutically-relevant cargo and diagnostic and imaging agents. One of the current trends in nanomedicine materials development is toward tunable monodisperse nanostructures, and considerable effort has been devoted to the design and fabrication of these materials. However, it is clear that the scientific community has only "scratched the surface" with respect to the ability to control and fabricate complex nanostructures with control over composition, shape, and function.

The use of nanostructures as noninvasive imaging agents – referred to as "nanodiagnostics" – is one current area of research in nanomedicine. Here, "smart-particles" are prepared with the goal of detecting diseases at the earliest stages [3]. As these nanomaterials decrease in size, both their physical and chemical properties are subject to dramatic change, which often makes them amenable diagnostic agents. Examples include colloidal gold [4], iron oxide crystals [5], quantum dots (CdSe/ZnS) [6], and NPs that contain image-active cargos (Gd-based) [7]. For example, Huang and co-workers have described the use of gold nanorods of a specific aspect ratio that absorb and scatter strongly in the near-infrared region. These nanorods are conjugated to anti-epidermal monoclonal antibodies and can be used simultaneously for molecular imaging and photother-

Nanobiotechnology II. Edited by Chad A. Mirkin and Christof M. Niemeyer
Copyright © 2007 WILEY-VCH Verlag GmbH & Co. KGaA, Weinheim
ISBN: 978-3-527-31673-1

mal cancer therapy [8]. Iron oxide NPs cause spin-spin time relaxation changes in nearby water molecules, and this property can be exploited to detect cancers or other diseases [9, 10]. Likewise, Akerman et al. have described the use of quantum dots (CdSe/ZnS) for targeted *in vivo* diagnostic imaging [11]. Here, quantum dots are defined as inorganic nanocrystals less than 10 nm in size and with tunable fluorescent properties. Recently, Whitesides described using the technology of microfluidics to prepare monodisperse micron-sized particles that can be functionalized with magnetic NPs or dyes for use in imaging [12].

In addition to nanoimaging applications, nanomedicine can provide an array of opportunities for targeted drug delivery and controlled release applications. Nanomedicine has both the opportunity and the ability to improve the effectiveness of drug delivery by targeting pharmaceutically-relevant cargo to specific sites, by managing the drug's pharmacokinetics and pharmacodynamics, and improving on its nonspecific toxicity and immunogenicity [3]. The NPs and nanoscopic "vessels" that serve as the most effective drug delivery vehicles are those engineered to be biocompatible, site-specific, that have optimal capability to carry relevant cargo, and can demonstrate controlled release of that cargo. Several areas of research are currently exploring the use of NPs for drug delivery applications; these include liposomes, micelles, and a variety of other polymeric NPs that are composed of organic polymers with specific physical or chemical properties that make them relevant delivery vehicles.

Micelles – and liposomes in particular – are the subject of major interest for drug delivery applications. They possess distinctive advantages over the usual "free" drug administration processes in that they are able to protect the cargo from premature chemical or physical breakdown, while the surface can be modified such that the liposome is targeted to a specific receptor, leading in turn to reduced toxic side effects of the drug. To date, conventional liposomes have been used to carry and transport both antimicrobials [13] and chemotherapeutics [14], as well as DNA and proteins [15, 16].

Unfortunately, there are disadvantages to conventional liposomal carriers, the most prominent being rapid clearance from the blood and chemical and physical instability [17]. Thus, clear advances have been made recently in the development of stabilized liposomes. For example, a polyethylene glycol (PEG) coating can be used to create long-circulating liposomal carriers [18–20], while the placement of targeting ligands on the surface of the PEG-coated liposome to effect site-specific delivery and cargo release is also currently under investigation [21–24]. Although these methods show improvement over conventional liposomes, disadvantages persist for encapsulating non-hydrophilic cargos, and because liposomes are neither shape- nor size-specific, this leads to difficulty in their precise loadability [25].

Additional considerations in the search for the ideal organic NP are the design and synthesis of polymer conjugates. Polymer–drug (or protein) conjugates are hybrid structures that tend to be water-soluble (due to control of the chemical composition of the polymer), to be tumor-specific via the enhanced permeability and retention (EPR) effect and tumor target ligands that decorate the polymer portion of the conjugate, and can be captured via endosomes [26, 27]. Despite these conjugates offering an alternative approach in the field of nanomedicine, they are plagued with various difficulties. For example, they must consist of high-molecular mass, biodegradable polymeric carriers so that they can better exploit the EPR effect. Additionally, they must be able to move away from the heterogeneity of the carrier and to form a more uniform structure. Dendrimers can provide a valuable architecture for these conjugates, but they lack complete control over shape and the ability to load large amounts of therapeutics, targeting moieties, and imaging modalities [28].

By using emerging materials science technologies, synthetic polymer NPs have been developed as a more effective drug delivery method, with considerable research effort having been exerted into the advancement of this architecture for drug transport. For example, recent investigations have led to the production of NPs composed of biodegradable polymer matrices [29, 30]. These have shown great potential for use as delivery agents for drugs, for oligonucleotides in antisense therapy, and for DNA in gene therapy [31, 32] through increased control over release. These degradable polymers can also be signaled to degrade by altering the pH level or temperature [33–36]. Furthermore, the use of charge or surface modification with targeting ligands has led to new ways of targeting cells for uptake [37, 38]. For example, DeSimone's group has demonstrated the cellular uptake of nanogel–DNA complexes by utilizing a positive charge [39].

An ideal therapeutic carrier would be biocompatible, shape- and size-specific, monodisperse, composed of virtually any material, amenable to functionalization, and gentle enough to transport a fragile biological cargo. In this respect, the emerging technology offered by *Particle Replication In Non*wetting *Templates* (PRINT) [40] provides an opportunity to take nanomedicine to the next level by incorporating all of the components of an ideal NP delivery vehicle. PRINT is a decidedly amenable method for the production of monodisperse, organic particles of any size or shape, made from a plethora of biocompatible materials such as poly-L-lactic acid (PLA), PEG, or a degradable polymer. Additionally, by utilizing a general encapsulation technique, these particles can carry chemotherapeutics (doxorubicin) or imaging agents (iron oxide). Further versatility is gained through the ability to readily modify the surface with targeting ligands or proteins.

The long-term strategies for nanomedicine include the ability to encapsulate pertinent cargos such as drugs or imaging agents, to target specific sites such as cells or receptors, and to release these cargos in controlled fashion. Current investigations in this important area are driven by the desire to create a methodology by which to deliver biologically-relevant cargo in an effective manner, to reduce side effects, and to more successfully diagnose and treat disease. Hence, in this chapter we will describe the current trends in nanomedicine whilst exploring future directions that will, ultimately, make nanotechnology a vital aspect in the diagnosis and treatment of disease (Figure 15.1).

Inorganic nanoparticles	**Organic Nanoparticles**	**Microfluidics**
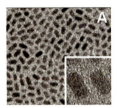	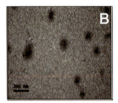	
CdSe/CdS nanocrystals	Inverse micro emulsion PEG hydrogel (200 nm)	Fabrication of micron-sized composite particles using a microfluidic device
Liposomes	**Carbon Nanotubes**	**PRINT**
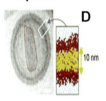	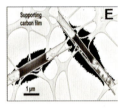	
DOX-loaded nanopolymersome	Carbon Nanotubes filled with magnetic particles	FGMP Shape and size specific, monodisperse, loadable, targetable, comprised of any binding maxtrix, the future is now

Fig. 15.1 Examples of nanocarriers in medicine. (A) CdSe/CdS nanocrystal [79]. (B) Inverse microemulsion PEG hydrogel [39]. (C) Microfluidics used for Lab-on-a-Chip technology and micron-sized composite particle fabrication [12]. (D) DOX-loaded nanopolymersome [80]. (E) Carbon nanotubes filled with magnetic particles [81]. (F) DOX-loaded PRINT particles [82]. (Reprinted with permission from Refs. [12, 39, 80, 81]; © 2002, 2004, 2005, 2006, American Chemical Society; and Wiley Interscience 2005 [79].)

15.2
Methods

The quest for monodisperse, shape-specific nanocarriers is littered with a plethora of challenges. Typically, the process must be gentle enough to be compatible with delicate biologicals, yet amenable to a variety of polymeric materials. The key techniques of bottom-up and top-down fabrication approaches used to achieve these goals are summarized in the following sections.

15.2.1
Bottom-Up Approaches for the Synthesis of Organic Nanoparticles

Both, microscopic and nanoscopic "vessels", such as micelles, vesicles, liposomes, and hollow spheres, have been the subject of much on-going current research regarding their use in applications ranging from gene delivery [41] to waste removal [42]. Here, we will briefly outline the current methods used to synthesize of organic NPs. Micelles and vesicles typically consist of surfactants or block co-polymers, where the intrinsic differences in chemical potentials of the coordinated polymer fragments permit stabilization of the interface between the solvent medium and the final structure [43]. These components have been used to organize a wide array of highly stable and responsive vesicles that can be employed for the transport and delivery of therapeutic agents. The existing approaches available for the synthesis of such organic-based NPs are all based on a "bottom-up" method, for example, grown in a step-wise fashion. The composition of these materials can be easily controlled and modified, such that cargos can be "trapped" into the cores of the structures. Specifically, the use of liposomes in medical applications has received a great deal of attention during recent years. These bilayered membrane structures, which are composed of a phospholipid bilayer surrounding an aqueous or hydrophilic core, show exceptional biocompatibility and thus great potential for clinical use as pharmaceutical carriers, notably in the treatment of cancer. Indeed, several liposome-based drugs (e.g., Doxil®,

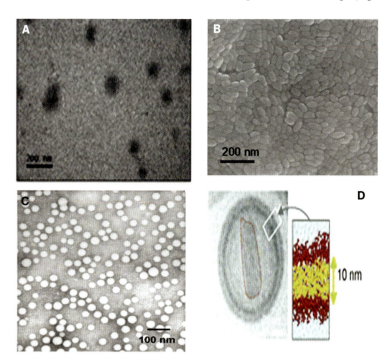

Fig. 15.2 Conventional delivery vectors. (A) 200-nm cationic nano hydrogel delivery vectors produced via inverse microemulsion polymerization [39]. (B) Non-spherical particles synthesized via a "mini-emulsion" technique [61]. (C) Shell crosslinked knedels (SCKs) [59]. (D) DOX-loaded nanopolymersome [80]. (Reprinted with permission from Refs. [39, 59, 80]; © 2002, 2005, 2006, American Chemical Society; and the author Ref. [61], Figure 2B.)

Ambisome®, Daunoxome®) are currently undergoing clinical trials or are already available in the clinical setting.

The original studies conducted by Bangham showed that, when placed in an aqueous system, phospholipids will self-assemble to form bilayer or multilayer structures [44–46], since which time several models have been introduced for the synthesis of liposomes (Figure 15.2). For example, Szoka et al. [47, 48] have described the formation of liposomes via reverse-phase evaporation. Here, the vesicles were formed by introducing an aqueous phase to a mixture of phospholipid and an organic solvent, which is subsequently removed by evaporation. Additionally, Finer has demonstrated a bilayer rearrangement in vesicles by using a sonication procedure; here, the multilamellar particles of egg yolk lecithin were broken by sonication into fragments, which then re-aggregated to form single-shelled vesicles [49]. Further investigations in this area have led to improvements in the synthesis of liposomes, notably in their ability to increase circulation time, loadability, and to tether pendants to the surface. Possible methods for attaining long-circulating liposomes – and therefore increased drug accumulation in the desired target areas – include coating the surface of the liposome with PEG [50], leading in turn to an enhanced retention and permeability (EPR) effect [51]. A recent study conducted by Zalipsky focused on the synthesis of a PEG-coating that could be separated from the liposome at low pH levels, as are found in tumor cells [52, 53]. Furthermore, chemical techniques exist by which liposomes may be modified via surface conjugation with proteins, peptides, or other biologically relevant molecules. The reactions between a carboxyl group and an amino group, for example, lead to the formation of amide bonds on the surface of liposomes, and this allows for surface interaction with proteins [54]. Other surface ligands (e.g., folate) have also been used to modify the liposome surfaces [55]; for example, the anticancer drug doxorubicin, when loaded into folate-modified liposomes, has been delivered into tumor cells via a folate receptor method and subsequently demonstrated higher cytotoxicity [56].

Liposomes – and more broadly, micelles – have been produced by a range of both natural and synthetic amphiphilic polymers, leading to nanoscale structures. Related to the liposomes, shell-crosslinked knedel (SCK) structures have been shown to be biocompatible, stable, and able to carry a pharmaceutical cargo [57, 58]. Wooley and Hawker have reported on the modification of SCKs with functional groups (azido or alkynyl) on the shell or core domains of the micelles and SCKs. The introduction of the Click reactive groups to these nanostructures allows for further interaction with biologically-relevant substrates [59] (Figure 15.2). Likewise, Shen et al. reported on a correlated group of organic NPs, core-surface cross-linked micelles composed of amphiphilic brush copolymers. In this method, the hydrophilic brush polymer backbone is cross-linked with the hydrophobic core, which significantly augments the stability of the micelle. Additionally, the surface properties of these micelles can be readily modified for enhanced targeting for drug delivery of the NPs [60].

The methods described yield liposomes, micelles, and vesicles that permit the encapsulation of various pharmaceutically-relevant cargos and allow modification of the periphery with various targeting ligands. Despite the advantages posed by these fabrication techniques, some significant challenges remain, including a restricted payload size, a lack of robustness, rapid elimination from the blood, and build up in the liver. Additionally, there are limits in the homogeneity of shape (i.e., spherical NPs) as well as limited control over size and dispersity.

While the spherical shape of micelles and liposomes is determined by external forces such as surface tension, several methodologies are currently under investigation for the synthesis of non-spherical polymer NPs. Huck and co-workers have described the synthesis of the first non-spherical liquid-crystalline polymeric NPs utilizing a mini-emulsion technique. Here, a main-chain liquid crystalline polymer (MCLCP) was dissolved in chloroform and mixed with water and a surfactant; this was followed by ultra-sonication to create a mini-emulsion. A suspension of polydisperse NPs ranging from 30 to 150 nm was formed after evaporation of the solvent (Figure 15.2). These particles spontaneously form an ellipsoidal shape with high aspect ratios; the shape can be further controlled by altering the temperature [61].

Inverse microemulsion techniques have been used in the synthesis of polymeric NPs in order to create submicron hydrophilic polymer particles with improved polydispersities for use in drug delivery applications. The group of De-Simone [39] used inverse microemulsion polymerization to synthesize stable, biocompatible polymeric nanogels less than 200 nm in size, for anti-sense and gene delivery to HeLa cells via the exploitation of charge. These spherical particles showed a narrow size distribution, with polydispersity of less than 10% for the nonionic hydrogels [39] (see Figure 15.2).

15.2.2
Top-Down Approaches for the Fabrication of Polymeric Nanoparticles

The ability to control the matrix material, as well as the size and shape of polymeric NPs, is an important goal as materials science approaches the nanometer regime. As noted in Section 15.2.1, polymeric NPs produced via a bottom-up approach have highly diverse matrices as well as an ability to tether pendants to the surface and/or encapsulate materials into the polymeric core, but little control over shape, size, and dispersity. Top-down processes for the fabrication of polymer NP, in addition to controlled composition and loadability, provide the option of imparting shape to the particle. By utilizing "state-of-the-art" engineering processes [62] – for example, microfluidics, molding, embossing, photolithography, and imprint lithography – the materials scientist can produce countless materials for nanomedicine applications that have controlled composition, function, dispersity, and shape. Here, we will outline the various "top-down" methods used to control the shape, size, and matrix material of polymeric materials.

15.2.2.1 Microfluidics
The use of "top-down" fabrication utilizes the ability of an engineering technique to produce devices or templates that are a myriad of shapes and size that are em-

phatically monodisperse on a variety of length scales. By using these techniques, nanostructures can be fabricated and composed of a variety of materials. The promising technique of microfluidics enables one to control the fabrication of non-spherical particles. Microfluidics capitalizes on the changes in physical properties of fluids once they are constrained in the specifically-designed microchannels of the device. Doyle and colleagues [63] have utilized the shearing forces of a photopolymer in a continuous water phase at a specifically-designed microfluidics junction [fabricated by pouring polydimetholxysilane (PDMS) onto a silicon wafer containing positive-relief channels patterned in a SU-8 photoresistor] to produce non-spherical uniform polymer particles on the micron scale. Variation of the speed of the "shearing" liquid provides control over the dimensions and shape of the resulting droplets. The polymeric (PEG based) disc-shaped particles produced are on the order of 16 μm, with a diameter of 40 μm. As the capabilities of lithography grow, it should be possible to scale down the microfluidics device to produce polymeric particles of <10 μm. Additionally, Whitesides et al. [12] have applied a similar method to control the size of monodisperse particles (20 to 1000 μm) by utilizing a microfluidics device, not only to photopolymerize the particles but also to thermally "set" them into a defined shape based on the speed of the shearing material. By using this method, it was also possible to produce multi-component, polymer-based beads with cargos that include copolymers, fluorescent dyes, inorganic NPs (CdSe), liquid crystals, and microporous particles. The loadability of these polymeric particles enables them to be used as carriers of medicines.

Whilst microfluidics offers an attractive method to fabricate polymer particles that can be either photochemically or thermally cured, the alternative engineering techniques of molding and heat embossing may be useful as the quest for nanomaterials is expanded. Yue and co-workers [64] have investigated new materials (e.g., titanium nitride, TiN) for the nanopatterning of silicon via focused ion beam (FIB) lithography. These authors used TiN because it has desirable hydrophobic, chemical and metallurgical stability for multiple mold casting. The group used SYLGARD 184 (rms roughness 10.5 ± 7.0 nm) to cast rubber molds of the TiN-nanopatterned substrate (rms roughness 4 ± 2.0 nm), after which the patterned molds were used to UV-emboss PEG-diacrylate. This was accomplished by pooling a solution of PEG-diacrylate:2,2-dimethoxy-2-phenyl-acetophenone (99.7:0.3, w/w) on the rubber casting of the TiN-modified silicon-patterned substrate. The replicated structures composed were shown to be biocompatible and non-adhesive to protein and cells [65, 66]. Moreover, by slightly altering the method these authors showed that they could fabricate a high aspect ratio (≥ 5) microarray of microcups composed of the same biocompatible material [67].

15.2.2.2 Photolithography

Photolithography constitutes an additional engineering technique that allows for nanofabrication. Customarily, lithography has employed light (photolithography) that is generated by lasers or other sources of various wavelengths (365 nm,

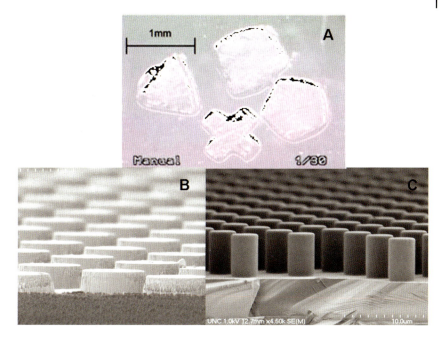

Fig. 15.3 (A) MUFFINS comprised of PEG with encapsulated DNA base pairs [69]. (B and C) Master silicon templates fabricated using reactive ion etching; these master templates are compatible with the PRINT process. (Reprinted with permission from Ref. [69]; © 2004, American Chemical Society.)

248 nm, and 193 nm) to create images or patterns in formulated polymer films known as photoresists [68]. Once the desired pattern is generated (i.e., arbitrarily shaped wells), specifically-designed channels are typically "transferred" into the underlying substrate, using aggressive processing conditions that include (but are not limited to) reactive ion etching (RIE) [68]. Willson and colleagues [69] have utilized this platform of photolithography to engineer discrete shapes (squares, circles, triangles, crosses) on the micron regime (Figure 15.3). The group utilized a modified contact lithographic process to fabricate what they referred to as "MUFFINS" (Mesoscale Underaddressed Functionalized Features Indexed by Shape). This was accomplished by dispensing a PEG-diacrylate pre-polymer solution that contained the desired sensing moiety (various 18-mer target DNA sequences) into a Teflon container, on top of which was placed a photo-mask. The resulting component was subsequently UV-treated. The uncrosslinked polymer portions were washed, leaving isolated shape-specific structures that contained bioactive material. The isolated hydrogel structures served as a biosensor via complementary binding experiments, and were monitored using fluorescence. As photolithography is currently approaching its limit of resolution (90 nm with a

minimum feature size of 37 nm [62]), research groups in this field are now investigating new methods to fabricate materials that will be pertinent to the nanometer regime for nanomedicine applications.

15.2.2.3 Imprint Lithography

In contrast to the traditional methods of lithography, imprint lithography is being applied to the precise fabrication of next-generation NPs, integrated circuits and other electronic and photonic devices (Figure 15.3). Imprint lithography is an alternative to photolithography for the manufacture of integrated circuits and other devices with sub-100-nm features [70–74]. The technique involves a very simple molding process wherein a mask containing shaped cavities is brought into direct contact with curable liquids to create features and patterns on substrates. The performance of perfluoropolyether-based elastomers (PFPEs) in imprint lithography was demonstrated using replica molds generated from master templates created at IBM's Almaden Research Center in California; these had features with a width of 140 nm, a depth of ~50 nm, and a separation of 70 nm [75]. The molds cast using the PFPE-based fluoroelastomer materials (0.4 nm roughness factor) maintained exact preservation of the nanoscale features of the patterned silicon wafer master. The features on the PFPE-based mold, as determined by atomic force microscopy, had an average height of 51 nm, which was in excellent agreement with the measured 54 nm height of the features in the silicon master. In principle, imprint lithography is orders of magnitude less expensive than traditional photolithographic methods used to make features that are smaller than the imaging exposure wavelengths available today. Hence, the technique is attracting great excitement as a replacement for photolithography. However, one ubiquitous aspect of imprint lithography has been the so-called "scum" layer which interconnects all features made using imprint lithography [70–74]. In microelectronics applications, this layer is typically eliminated by using harsh etching processes, such as RIE from an oxygen plasma (O_2-RIE) [68].

Reactive plasmas such as O_2-RIE function by bombarding a surface with an anisotropic stream of high-energy particles that chemically ablate [68] the resist to remove the scum layer uniformly. While these processing methods are well established in the semiconductor industry – where hard, robust, inorganic materials are the norm – they are incompatible with organic materials that are delicate or that would contain biologically-derived moieties. DeSimone and colleagues [75, 76], while investigating new materials to overcome the occurrence of the scum layer, showed that specially-designed, photochemically curable PFPEs [75] could perform accurate nanometer-scale molding when used in lieu of the traditional elastomers presently used in imprint lithography. As mentioned previously, one drawback of imprint lithography is the scum layer that forms when the technique is based on traditional materials such as PDMS [72–74]. PFPE-based molds, on the other hand, are highly fluorinated and have surface energies of 12 dynes cm^{-1}, far less than that of PDMS (20 dynes cm^{-1}) [75] (Figure 15.4). With such non-wetting, non-swelling characteristics, PFPE-based materials enable the gener-

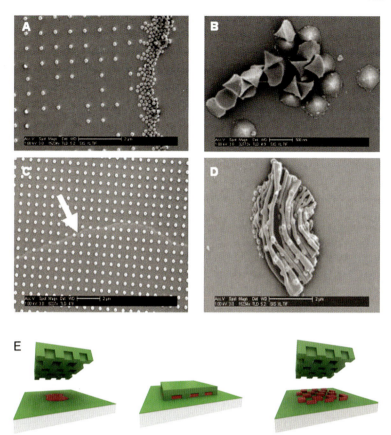

Fig. 15.4 (A) Harvested triacrylate particles using a scalpel. (B) Harvested cones. (C) 200-nm triacrylate particles connected by a scum layer (D) that have been harvested using a doctor's blade. The features are clearly connected by an underlying layer in stark contrast to those seen in (A) which are isolated, discrete objects. (E) In PRINT, the non-wetting nature of fluorinated materials and surfaces (shown in green) confines the liquid precursor inside the features of the mold, allowing for the generation of isolated particles. (Reprinted with permission from Ref. (40); © 2005, American Chemical Society.)

ation of microfluidic devices and harvestable, scum-free objects or particles, by using PRINT.

15.2.2.4 PRINT

PRINT enables the fabrication of monodisperse particles with simultaneous control over structure (i.e., shape, size, composition) and function (i.e., cargo, surface structure). PRINT utilizes the non-wetting properties of the highly fluorinated elastomeric mold and a substrate to produce isolated, harvestable objects with applications ranging from drug delivery agents [40] to photovoltaics [40]. This is accomplished by taking advantage of the reversible seal that is formed between the mold and substrate as slight downward force is applied, and the organic liquid to be molded is either confined within the shaped cavities of the mold or forced out due to the low surface energy of both the mold and the surface. The ability to mold isolated or scum-free objects at the nanoscale is enabled by having fluorinated, low-surface energy materials comprising the internal surfaces exposed to the liquid to be molded (Figure 15.4). Hence, PRINT is the first *general, singular* method capable of forming particles that:

- are monodisperse in size and uniform shape;
- can be molded into any shape;
- can be comprised of essentially any matrix material;
- can be formed under extremely mild conditions (and are therefore compatible with delicate cargos);
- are amenable to post-functionalization chemistry for the bioconjugation of targeting ligands; and
- are in an addressable array (which opens up combinatorial approaches, as the particles can be "bar-coded" using methods similar to DNA array technologies).

As such, PRINT represents a significant scientific and technological breakthrough which will allow the fabrication of previously inaccessible populations of the nanobiomaterials that are poised to revolutionize and accelerate our translational understanding, detection, and treatment of cancer. PRINT is sufficiently delicate and general to be compatible with a wide variety of important nanobiomaterials targeted for advanced understandings and therapies in cancer prevention, detection, diagnosis and treatment. To date, by using PRINT we have fabricated *monodisperse* particles from a wide range of particle matrix materials, including PEG, PLA, poly(pyrrole) (Ppy), and a triacrylate resin.

Monodisperse, shape-specific, and fully bioabsorbable NPs have not previously been fabricated, yet PRINT can be used to make particles from PLA and derivatives thereof such as poly(lactide-*co*-glycolide) (PLGA). It is well known that PLA and PLGA have had a considerable technological impact on the drug delivery and medical device industries, mainly because they are both fully bioabsorbable and nontoxic. Specifically, monodisperse PLA particles using PRINT were fabricated by treating a small amount of the cyclic lactide monomer, $(3S)$-*cis*-3,6-dimethyl-1,4-dioxane-2,5-dione, with the FDA-approved polymerization catalyst, stannous octoate, at 110 °C in a PFPE mold designed to fabricate 200-nm particles. After polymerization had been achieved, the PFPE mold and the flat, non-wetting substrate could be separated to reveal an array of monodisperse 200-nm trapezoidal particles. We have also generated monodisperse, shape-specific 200-nm trapezoidal particles from Ppy; the latter has been used in a variety of applications, ranging from electronic devices and sensors to cell-scaffolds. Ppy particles may be fabricated via a one-step polymerization process, by placing a drop of a 1:1 (v/v) solution of tetrahydrofuran: pyrrole and perchloric acid into the molding apparatus, followed by vacuum evaporation of the solvent. Subsequently, monodisperse 200-nm Ppy trapezoidal particles and 3-μm arrows were fabricated, with good fidelity.

As stated previously, PEG is a material which currently is of tremendous interest to the biotechnology community, due mainly to its commercial availability, nontoxic nature, and biocompatibility. PRINT can be used to produce monodisperse, nanometer-scale PEG particles in a variety of shapes by molding PEG-diacrylate liquid monomer, followed by room-temperature photopolymerization. Because the morphology of the particles is controlled by the master, it is possible to generate complex PEG particles on a variety of length scales.

With the PRINT methodology, particle harvesting is straightforward, and can be accomplished by utilizing several methodologies [77, 78], including the use of a sacrificial adhesive layer composed of poly(cyanoacrylate), which is both biocompatible and non-carcinogenic, as well as being an FDA-approved "medical adhesive". The adhesive is placed between the filled mold and a sheet of glass, such that the embedded particles are left behind when the mold is removed. This sacrificial adhesive is then dissolved in acetone, and removed by centrifugation.

By taking advantage of the delicate nature of PRINT, it is possible to incorporate a myriad of materials into the precursor PRINT polymer solution before particle formation, including image-contrast agents (superparamagnetic iron oxide particles or gadolinium), therapeutics (doxorubicin/paclitaxel/bortezomib), organic dyes (rhodamine), antibodies, proteins, and/or DNA (Figure 15.5). To this end, we have demonstrated the production of PRINT particles composed of a PEG-matrix which contain the chemotherapeutic cargo, doxorubicin, using a general encapsulation technique, and have also demonstrated its release over time into an aqueous phase (Figure 15.5). Additionally, we have shown DNA and proteins to be incorporated into sub-200-nm PEG NPs [40]. In addition to being amenable to carrying pertinent cargo, PRINT can be used to fabricate particles containing relevant chemical handles on their periphery but degradable monomers within the matrix, thereby enhancing both targeting to specific cell types and delivery of the chemotherapeutic cargo.

15.3
Outlook

During recent years, great strides have been made in the use of organic NPs in nanomedicine, such that today, liposomal drugs such as Doxil®, AmBisome®, and DaunoXome® are used in lieu of their free drug counterparts and represent fresh alternatives that are more effective with fewer adverse side effects. The challenges of liposomes, however, include limited payload size, a lack of robustness, difficulties in functionalizing the liposomal material, and problems in their being targeted to specific cell types. One alternative to liposomal delivery, with hopes of increases in payload and greater robustness *in vivo* is to use protein NPs to deliver the drug. A newly available drug, Abraxane™ – an injectable suspension of pacil-

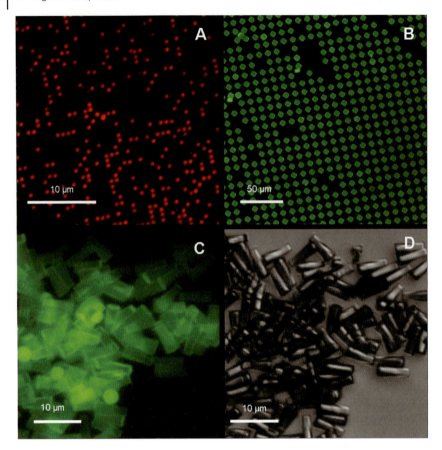

Fig. 15.5 Confocal micrographs of: (A) CY-3 tagged DNA encapsulated in 200-nm trapezoidal PEG PRINT particles [40]. (B) Protein biosensor tagged to the peripheral of 3-μm disc-shaped PEG PRINT particles. (C) Harvested PRINT particles that contain a pertinent chemotherapeutic. (D) Differential interference contrast micrograph showing the same PRINT particles depicted in (C). (Reprinted with permission from Ref. [40]; © 2005, American Chemical Society.)

taxel (a pertinent cancer chemotherapeutic) bound to albumin – has a mean diameter of 130 nm and utilizes this motif. These nanocarriers of medicines offer more advantageous alternatives compared to administration of the free drug, but lack the ability of increased payload, limited control over size and shape, and typically are polydisperse. In contrast, top-down methods for the fabrication of nanocarriers of medicines can produce monodisperse shape-specific materials, but are plagued with limitations of the engineering techniques used as they approach their size limitations. They also display incompatibility between the delicate biologicals and the harsh conditions required to eliminate the residual interconnective layer. Today, new materials are being designed to increase the ability to produce new materials in concert with increasing the size limits of engineering techniques, so that nanocarriers can be fabricated which are robust, functionalizable, loadable, and monodisperse.

The main questions to be addressed when investigating medicines on the nanometer scale include: "What interrelated roles do shape, size and mechanochemico functionality play on the biodistribution and efficacy of nanocarriers *in vivo*?" And, "How can this understanding translate into more efficacious detection, diagnosis, and therapeutic and prevention strategies?" PRINT combines the precision, uniformity and mass production of the nanoscale microelectronics industry with the generation and harvesting of extremely versatile, shape-specific organic carriers that have specific chemical functionality and tailored mechanical properties. The ability to directly fabricate and harvest uniform populations of monodisperse, shape-specific organic nanobiomaterials capable of delivering a variety of therapeutic, detection and imaging agents to specific sites within living organisms will help to answer these questions. Scale-up from the electronics industry should be possible in order to precisely fabricate these "smart" NPs with unprecedented uniformity and capability. Moreover, it should be possible to carry out such mass production in a manner that will allow particles to be customized for use by other research teams worldwide, by establishing a particle "foundry". Indeed, it is hoped that the current Good Manufacturing Practice of Nanomedicine will in the future include materials that are monodisperse, comprised of any matrix, are loadable, shape-specific, tumor- and/or tissue-specific, and utilize imaging techniques to produce medicines that will eliminate the premature death and suffering of patients.

References

1 Brydson, R. M., Hammond, C. (2005) Generic methodologies for nanotechnology: classification and fabrication. In: Kelsall, R. H., Geoghan, M. (Eds.), *Nanoscale Science and Technology.* Wiley, West Sussex, pp. 1–55.

2 Moghimi, S. M., Hunter, A. C., Murray, J. C. (2005) Nanomedicine: current status and future prospects. *FASEB J.* **19**, 311–330.

3 Pison, U., Welte, T., Giersig, M., Groneberg, D. A. (2006) Nanomedicine for respiratory diseases. *Eur. J. Pharmacol.* **533**, 341–350.

4 Sun, Y., Xia, Y. (2002) Shape-controlled synthesis of gold and silver nanoparticles. *Science* **298**, 2176–2179.

5 Euliss, L. E., Grancharov, S. G., O'Brien, S., Deming, T. J., Stuckey, G. D., Murray, C. B., Held, G. A. (2003) Cooperative assembly of magnetic nanoparticles and block copolypeptides in aqueous media. *Nano Lett.* **3**, 1489–1493.

6 Cha, J. N., et al. (2003) Spontaneous formation of nanoparticle vesicles from homopolymer polyelectrolytes. *J. Am. Chem. Soc.* **125**, 8285–8289.

7 Oyewumi, M. O., et al. (2004) Comparison of cell uptake, biodistribution and tumor retention of folate-coated and PEG-coated gadolinium nanoparticles in tumor-bearing mice. *J. Control. Release* **95**, 613–626.

8 Huang, X., et al. (2006) Cancer cell imaging and photothermal therapy in the near-infrared region by using gold

nanorods. *J. Am. Chem. Soc.* **128**, 2115–2120.

9 Dardzinski, B. J. S., Holland, V. J., Boivin, S. K., Imagawa, G. P., Watanabe, T., Lewis, S., Hirsh, J. M. (2001) MR imaging of murine arthritis using ultrasmall super-paramagnetic iron oxide particles. *Magnet. Res. Imaging* **19**, 1209–1216.

10 Dousset, V. B., Delande, C., Coussemacq, M., Canioni, P., Petry, K. G., Caille, J. M. (1999) Comparison of ultrasmall particles of iron oxide-enhanced T2-weighted, conventional T2-weighted, and gadolinium-enhanced T1-weighted MR images in rats with experimental autoimmune encephalomyelitis. *Am. J. Neuroradiol.* **20**, 223–227.

11 Akerman, M. E., et al. (2002) Nano-crystal targeting in vivo. *Proc. Natl. Acad. Sci. USA* **99**, 12617–12621.

12 Xu, S. Q., et al. (2005) Generation of monodisperse particles by using microfluidics: Control over size, shape, and composition. *Angew. Chem. Int. Edn.* **44**, 724–728.

13 Pinto-Alphandary, H., Andremont, A., Couvreur, P. (2000) Targeted delivery of antibiotics using liposomes and nanoparticles: research and applica-tions. *Int. J. Antimicrob. Agents* **13**, 155–168.

14 Drummond, D. C., et al. (1999) Optimizing liposomes for delivery of chemotherapeutic agents to solid tumors. *Pharmacol. Rev.* **51**, 691–743.

15 Nishikawa, M., Hashida, M. (2002) Nonviral approaches satisfying various requirements for effective in vivo gene therapy. *Biol. Pharm. Bull.* **25**, 275–283.

16 Rao, M., Alving, C. R. (2000) Delivery of lipids and liposomal proteins to the cytoplasm and Golgi of antigen-presenting cells. *Adv. Drug Deliv. Rev.* **41**, 171–188.

17 Wang, G. (2005) Liposomes as drug delivery vehicles. In: Wang, B., Siahaan, T., Soltero, R. (Eds.), *Drug Delivery Principles and Applications.* John Wiley & Sons, Hoboken, pp. 411–434.

18 Allen, T. M., Hansen, C. B., Demenezes, D. E. L. (1995) Pharmaco-kinetics of long-circulating liposomes. *Adv. Drug Deliv. Rev.* **16**, 267–284.

19 Gregoriadis, G. (1995) Engineering liposomes for drug delivery: Progress and problems. *Trends Biotechnol.* **13**, 527–537.

20 Torchilin, V. P. (1998) Polymer-coated long-circulating microparticulate pharmaceuticals. *J. Microencapsul.* **15**, 1–19.

21 Park, Y. S. (2002) Tumor-directed targeting of liposomes. *Biosci. Rep.* **22**, 267–281.

22 Maruyama, K. (2002) PEG-immuno-liposome. *Biosci. Rep.* **22**, 251–266.

23 Maruyama, K., et al. (1999) Possibility of active targeting to tumor tissues with liposomes. *Adv. Drug Deliv. Rev.* **40**, 89–102.

24 Papahadjopoulos, D., et al. (1998) Targeting of drugs to solid tumors using anti-HER2 immunoliposomes. *J. Liposome Res.* **8**, 425–442.

25 Torchilin, V. P. (2005) Recent advances with liposomes as pharma-ceutical carriers. *Nat. Rev.* **4**, 145–160.

26 Vicent, M. J., Duncan, R. (2006) Polymer conjugates: nanosized medicines for treating cancer. *Trends Biotechnol.* **24**, 39–47.

27 Malik, N., Evagorou, E. G., Duncan, R. (1999) Dendrimer-platinate: a novel approach to cancer chemo-therapy. *Anti-Cancer Drugs* **10**, 767–776.

28 Lee, C. C., et al. (2005) Designing dendrimers for biological applica-tions. *Nat. Biotechnol.* **23**, 1517–1526.

29 Diwan, M., et al. (2004) Biodegrad-able nanoparticle mediated antigen delivery to human cord blood derived dendritic cells for induction of primary T cell responses. *J. Drug Targeting* **11**, 495–507.

30 Shen, H., et al. (2006) Enhanced and prolonged cross-presentation following endosomal escape of exogenous antigens encapsulated in biodegradable nanoparticles. *Immunology* **117**, 78–88.

31 Langer, R. (2000) Biomaterials in drug delivery and tissue engineering:

One laboratory's experience. *Acc. Chem. Res.* **33**, 94–101.

32 Bogunia-Kubik, K., Sugisaka, M. (2002) From molecular biology to nanotechnology and nanomedicine. *Biosystems* **65**, 123–138.

33 Galindo-Rodriguez, S. A., et al. (2005) Polymeric nanoparticles for oral delivery of drugs and vaccines: A critical evaluation of in vivo studies. *Crit. Rev. Ther. Drug Carrier Systems* **22**, 419–463.

34 Shenoy, D., et al. (2005) Poly(ethylene oxide)-modified poly(beta-amino ester) nanoparticles as a pH-sensitive system for tumor-targeted delivery of hydrophobic drugs: Part 2. In vivo distribution and tumor localization studies. *Pharm. Res.* **22**, 2107–2114.

35 Ramanan, R. M. K., et al. (2006) Development of a temperature-sensitive composite hydrogel for drug delivery applications. *Biotechnol. Prog.* **22**, 118–125.

36 Fernandez, V. V. A., et al. (2006) Thermoresponsive nanostructured poly(N-isopropylacrylamide) hydrogels made via inverse microemulsion polymerization. *Colloid Polymer Sci.* **284**, 387–395.

37 Park, E. K., et al. (2005) Folate-conjugated methoxy poly(ethylene glycol)/poly(epsilon-caprolactone) amphiphilic block copolymeric micelles for tumor-targeted drug delivery. *J. Control. Rel.* **109**, 158–168.

38 Fukumori, Y., Ichikawa, H. (2006) Nanoparticles for cancer therapy and diagnosis. *Adv. Powder Technol.* **17**, 1–28.

39 McAllister, K., et al. (2002) Polymeric nanogels produced via inverse microemulsion polymerization as potential gene and antisense delivery agents. *J. Am. Chem. Soc.* **124**, 15198–15207.

40 Rolland, J. P., et al. (2005) Direct fabrication and harvesting of monodisperse, shape-specific nanobiomaterials. *J. Am. Chem. Soc.* **127**, 10096–10100.

41 Kostarelos, K., Miller, A. D. (2005) Synthetic, self-assembly ABCD nanoparticles; a structural paradigm for viable synthetic non-viral vectors. *Chem. Soc. Rev.* **34**, 970–974.

42 Talens-Alesson, F. I., Salvation, A., Bryce, M. (2006) Removal of phenol by adsorptive micellar flocculation: Multi-stage separation and integration of wastes for pollution minimisation. *Colloids Surfaces, A: Physicochem. Eng. Aspects* **276**, 8–14.

43 Duque, D. (2003) Theory of copolymer micellization. *J. Chem. Phys.* **119**, 5701–5704.

44 Bangham, A. D. (1961) A correlation between surface charge and coagulant action of phospholipids. *Nature* **192**, 1197–1198.

45 Bangham, A. D. (1963) Physical structure and behavior of lipids and lipid enzymes. *Adv. Lipid Res.* **64**, 65–104.

46 Bangham, A. D. (1978) Properties and uses of lipid vesicles: an overview. *Ann. N. Y. Acad. Sci.* **308**, 2–7.

47 Guo, X., Szoka, F. C. (2003) Chemical approaches to triggerable lipid vesicles for drug and gene delivery. *Acc. Chem. Res.* **36**, 335–341.

48 Szoka, F. C., Papahadjopoulos, D. (1978) Procedure for the preparation of liposomes with large internal aqueous space and high capture by reverse phase evaporation. *Proc. Natl. Acad. Sci. USA* **75**, 4194–4198.

49 Finer, E. G., Flook, A. G., Hauser, H. (1971) The use of NMR spectra of sonicated phospholipid dispersions in studies of interactions with the bilayer. *FEBS Lett.* **18**, 331–334.

50 Klibanov, A. L., Maruyama, K., Torchilin, V. P., Huang, L. (1990) Amphipathic polyethyleneglycols effectively prolong the circulation time of liposomes. *FEBS Lett.* **268**, 235–237.

51 Maeda, H., Sawa, T., Konno, T. (2001) Mechanism of tumor-targeted delivery of macromolecular drugs, including the EPR effect in solid tumor and clinical overview of the prototype polymeric drug SMANCS. *J. Control. Release* **74**, 47–61.

52 Kirpotin, D., Hong, K. L., Mullah, N., Papahadjopoulos, D. (1996) Liposomes with detachable polymer coating: Destabilization and fusion of dioleoylphosphatidylethanolamine vesicles triggered by cleavage of surface-grafted poly(ethylene glycol). *FEBS Lett.* **388**, 115–118.

53 Zalipsky, S., Qazen, M., Walker, J. A., Mullah, N., Quinn, Y. P., Huang, S. K. (1999) New detachable poly(ethylene glycol) conjugates: Cysteine-cleavable lipopolymers regenerating natural phospholipid, diacyl phosphatidyl-ethanolamine. *Bioconj. Chem.* **10**, 703–707.

54 Klibanov, A. L., Torchilin, V. P., Zalipsky, S. (2003) In: Torchilin, V. P., Zalipsky, S. (Eds.), *Liposomes, A Practical Approach*. Oxford University Press, Oxford, p. 231–236.

55 Lu, Y., Low, P. S. (2002) Folate-mediated delivery of macromolecular anticancer therapeutic agents. *Adv. Drug Deliv. Rev.* **54**, 675–693.

56 Pan, X. Q., Wang, H. Q., Lee, R. J. (2003) Antitumor activity of folate receptor-targeted liposomal doxorubicin in a KB oral carcinoma murine xenograft model. *Pharm. Res.* **20**, 417–422.

57 Thurmond, K. B., Kowalewski, T., Wooley, K. L. (1996) Water-soluble knedel-like structures: The preparation of shell-cross-linked small particles. *J. Am. Chem. Soc.* **118**, 7239–7240.

58 Thurmond, K. B., Kowalewski, T., Wooley, K. L. (1997) Shell cross-linked knedels: A synthetic study of the factors affecting the dimensions and properties of amphiphilic core-shell nanospheres *J. Am. Chem. Soc.* **119**, 6656–6665.

59 O'Reilly, R. K., Joralemon, M. J., Wooley, K. L., Hawker, C. J. (2005) Functionalization of micelles and shell cross-linked nanoparticles using click chemistry. *Chem. Mater.* **17**, 5976–5988.

60 Xu, P., Tang, H., Li, S., Ren, J., Van Kirk, E., Murdoch, W. J., Radosz, M., Shen, Y. (2004) Enhanced stability of core-surface cross-linked micelles fabricated from amphiphilic brush copolymers. *Biomacromolecules* **5**, 1736–1744.

61 Yang, Z., Huck, W. T. S., Clarke, S. M., Tajbakhsh, A. R., Terentjev, E. M. (2005) Shape-memory nanoparticles from inherently non-spherical polymer colloids. *Nat. Mater.* **4**, 486–490.

62 Gates, B. D., Xu, Q. B., Stewart, M., Ryan, D., Willson, C. G., Whitesides, G. M. (2005) New approaches to nanofabrication: Molding, printing, and other techniques. *Chem. Rev.* **105**, 1171–1196.

63 Dendukuri, D., Tsoi, K., Hatton, T. A., Doyle, P. S. (2005) Controlled synthesis of non-spherical microparticles using microfluidics. *Langmuir* **21**, 2113–2116.

64 Gao, J. X., Chan-Park, M. B., Xie, D. Z., Yan, Y. H., Zhou, W. X., Ngoi, B. K. A., Yue, C. Y. (2004) UV embossing of sub-micrometer patterns on biocompatible polymeric films using a focused ion beam fabricated TiN mold. *Chem. Mater.* **16**, 956–958.

65 Bryant, S. J., Anseth, K. S. (2001) The effects of scaffold thickness on tissue engineered cartilage in photocross-linked poly(ethylene oxide) hydrogels. *Biomaterials* **22**, 619–626.

66 Elisseeff, J., McIntosh, W., Anseth, K., Riley, S., Ragan, P., Langer, R. (2000) Photoencapsulation of chondrocytes in poly(ethylene oxide)-based semi-interpenetrating networks. *J. Biomed. Mater. Res.* **51**, 164–171.

67 Chan-Park, M. B., Yan, Y. H., Neo, W. K., Zhou, W. X., Zhang, J., Yue, C. Y. (2003) Fabrication of high aspect ratio poly(ethylene glycol)-containing microstructures by UV embossing. *Langmuir* **19**, 4371–4380.

68 Levinson, H. J. (2001) *Principles of Lithography*. Bellingham, Washington, SPIE Press, pp. 3–88.

69 Meiring, J. E., Meiring, J. E., Schmid, M. J., Grayson, S. M., Rathsack, B. M., Johnson, D. M., Kirby, R., Kannappan, R., Manthiram, K., Hsia, B., Hogan, Z. L., Ellington, A. D., Pishko, M. V., Willson, C. G. (2004) Hydrogel bio-sensor array platform indexed by shape. *Chem. Mater.* **16**, 5574–5580.

70 Bailey, T., Choi, B. J., Colburn, M., Meissl, M., Shaya, S., Ekerdt, J. G., Sreenivasan, S. V., Willson, C. G. (2000) Step and flash imprint lithography: Template surface treatment and defect analysis. *J. Vacuum Sci. Technol. B* **18**, 3572–3577.

71 Chou, S. Y., Krauss, P. R., Renstrom, P. J. (1996) Imprint lithography with 25-nanometer resolution. *Science* **272**, 85–87.

72 Geissler, M., Xia, Y. N. (2004) Patterning: Principles and some new developments. *Adv. Mater.* **16**, 1249–1269.

73 Xia, Y. N., Whitesides, G. M. (1998) Soft lithography. *Angew. Chem. Int. Ed.* **37**, 551–575.

74 Xia, Y. N., Rogers, J. A., Paul, K. E., Whitesides, G. M. (1999) Unconventional methods for fabricating and patterning nanostructures. *Chem. Rev.* **99**, 1823–1848.

75 Rolland, J. P., Hagberg, E. C., Denison, G. M., Carter, K. R., DeSimone, J. M. (2004) High-resolution soft lithography: enabling materials for nanotechnologies. *Angew. Chem. Int. Ed.* **43**, 5796–5799.

76 Rolland, J. P., Van Dam, R. M., Schorzman, D. A., Quake, S. R., DeSimone, J. M. (2004) Solvent resistant photocurable 'liquid teflon' for microfluidic device fabrication. *J. Am. Chem. Soc.* **126**, 8349.

77 Schaper, C. D., Miahnahri, A. (2004) Polyvinyl alcohol templates for low cost, high resolution, complex printing. *J. Vacuum Sci. Technol. B* **22**, 3323–3326.

78 Linder, V., Gates, B. D., Ryan, D., Parviz, B. A., Whitesides, G. M. (2005) Water-soluble sacrificial layers for surface micromachining. *Small* **1**, 730–736.

79 Talapin, D. V., Mekis, I., Gotzinger, S., Kornowski, A., Benson, O., Weller, H. (2004) CdSe/CdS/ZnS and CdSe/ZnSe/ZnS core-shell-shell nanocrystals. *J. Phys. Chem. B* **108**, 18826–18831.

80 Ahmed, F., Pakunlu, R. I., Srinivas, G., Brannan, A., Bates, F., Klein, M. L., Minko, T., Discher, D. E. (2006) Shrinkage of a rapidly growing tumor by drug-loaded polymersomes: pH-triggered release through copolymer degradation. *Mol. Pharm.* **128**, 340–350.

81 Korneva, G., Ye, H., Gogotsi, Y., Halverson, D., Friedman, G., Bradley, J. C., Kornev, K. G. (2005) Carbon nanotubes loaded with magnetic particles. *Nano Lett.* **5**, 879–884.

82 DeSimone, J. M. (2006) Unpublished results.

16
Poly(amidoamine) Dendrimer-Based Multifunctional Nanoparticles

Thommey P. Thomas, Rameshwer Shukla, Istvan J. Majoros, Andrzej Myc, and James R. Baker, Jr.

16.1
Overview

Conventional cancer chemotherapy using small-molecule drugs is limited by their generally poor solubility, nonspecific toxicity, and drug resistance [1]. Efficient, cell-specific homing of drugs is crucial in cancer therapy in order to completely eradicate all neoplastic cells and thereby prevent the re-growth of cancerous tumors. Targeted therapy using water-soluble macromolecular platforms as carriers of the drug has the potential to decrease the potent systemic toxicity caused by the current chemotherapy and to improve the therapeutic index. This is also based on the notion that the drug delivered through a carrier will have improved solubility and pharmacokinetics [2]. In addition, a targeted drug may have different routes of endocytic pathway from the non-targeted free drug, which could lead to increased cellular retention and action of the drug and decreased drug resistance.

In cancer therapy a "smart" macromolecular drug-carrier should meet several criteria for its *in-vivo* applicability. The ideal macromolecular platform should be: (i) able to carry multiple targeting and drug molecules; (ii) soluble in body fluids; (iii) uniform and monodispersed when dissolved; (iv) diffusible through tissue barriers such as the vasculature and the interstitial fluid; (v) transported into the cancer cell to release the drug into the appropriate cell compartment; (vi) non-toxic before and after biodegradation; and (vii) non-immunogenic. Over the past decade, several drug conjugates using carriers such as liposomes, polymers, and dendrimers which meet many of the above mentioned criteria have been developed [3–6].

Dendrimers have recently emerged as one of the most suitable drug delivery platforms, owing to properties such as their biocompatibility, dimension, and structural architecture which mimic certain biomolecules [3, 4, 7, 8]. Dendrimers are nanosized (2.5 nm to 10 nm) macromolecules that can be synthesized as different "generations" in different molecular weights and sizes. For example, the

Nanobiotechnology II. Edited by Chad A. Mirkin and Christof M. Niemeyer
Copyright © 2007 WILEY-VCH Verlag GmbH & Co. KGaA, Weinheim
ISBN: 978-3-527-31673-1

poly(amidoamine) (PAMAM; Starburst™) dendrimer generations 3, 4, and 5 are 3.6, 4.5, and 5.4 nm in diameter, respectively, and resembles the sizes of the biological molecules insulin (3 nm), cytochrome c (4 nm), and hemoglobin (5 nm). The three-dimensional architecture of the dendrimers is similar to that of some biomacromolecules (e.g., glycogen) and, considering their dimension and biocompatible properties (see below), these dendrimers can be termed as "artificial biomacromolecules". On the basis of these bio-mimicking characteristics, several dendritic molecules have been tested as platforms for the delivery of drugs, either as non-covalent encapsulation complexes, or through carefully engineering the dendrimer to covalently conjugate multiple targeting and drug moieties.

Dendrimers such as PAMAM, poly(propylene imine) (PPI), poly (aryl ether) branches – and those containing core molecules such as carbohydrate or calixarene – have been described as possible drug-delivery agents [8–10]. These dendrimers, when conjugated through a variety of linkages and core groups, are synthesized using different iterative synthetic strategies under controlled conditions. These synthetic strategies can direct the size, shape, and dimension of the dendrimers' interior molecular space and the number of surface functional groups by varying the core molecules and the nature of its branching points [4, 11].

Detailed descriptions on the synthesis and properties of different types of dendrimer have been reviewed recently [4, 8, 11–13]. One of the most extensively studied group of dendrimers for biological applications is the PAMAM ("Starburst") dendrimers, which are also the first dendrimer type to be synthesized and characterized. Studies from the authors' laboratory have shown the potential applicability of the PAMAM dendrimers as a drug-delivery platform [15–17]. The chemical and biological properties and the *in-vivo* applicability of the PAMAM dendrimers as drug carriers are reviewed in this chapter.

16.1.1
PAMAM Dendrimers: Structure and Biological Properties

PAMAM dendrimers resemble globular proteins more than linear-chain polymers (Figure 16.1). They are spherical, highly ordered, multi-branched polymers having positively charged amino groups on the periphery at physiological conditions. These surface amino groups provide useful moieties for functional modification as they allow a variety of reactions to be performed under mild conditions. We and others have recently reported the synthesis and biological properties of conjugates of the PAMAM dendrimer-conjugates with molecules such as folic acid, peptides, and antibodies [18–22].

The small size and hydrophilic nature of the PAMAM dendrimer prevent them from being phagocytosed by the macrophages of the mononuclear phagocytes system (MPS) in the spleen, liver, lungs and bone marrow (the MPS takes up hydrophobic particles larger than 100 nm in diameter). The biocompatibility of the PAMAM dendrimer depends on several factors, such as surface charge and func-

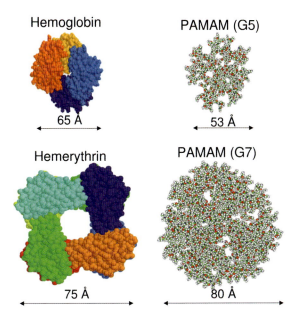

Fig. 16.1 Size comparison between G5 and G7 PAMAM dendrimers and two biomolecules of similar sizes.

tionality, dendrimer generation, and route of administration [23]. Both cationic and anionic PAMAM dendrimers induce cytotoxicity in a concentration- and generation-dependent manner [23–25]. Neutral dendrimers with the surface amino groups "capped" with neutral molecules such as acetamide [14, 16, 26] or lauryl and polyethylene glycol (PEG) chains [25] do not show significant *in-vitro* cytotoxicity. The *in-vivo* application of amine-terminated PAMAM dendrimers can lead to severe toxicity due to its hemolytic activity and nonspecific interaction with cellular membranes [23, 27], whereas acetamide-capped neutral dendrimers do not induce any toxic effect at comparable concentrations [17]. The PAMAM dendrimers show low immunogenicity, and their conjugation with certain molecules such as PEG can make it completely non-immunogenic [28].

The PAMAM dendrimer can serve as a highly suitable platform for tumor delivery of drugs. Higher-generation dendrimers can become entrapped in a tumor tissue through a process termed as "enhanced permeability and retention" (EPR). This is caused by the leaky property of the tumor vasculature (up to 400 nm pore size versus 2–6 nm in normal vasculature) that allows the entry of a macromolecule; the latter is then retained in the tumor due to ineffective lymphatic drainage from the tumor interstitial fluid. Smaller molecules, which easily diffuse out of the tumor, do not accumulate in the lesion by the EPR effect. Although the EPR property has been exploited for macromolecule-mediated "passive" drug targeting, this is an ineffective strategy due to the absence of the EPR effect in all areas of larger tumors, and due to the inability of a carrier to internalize passively into cells in these areas of the tumor. Therefore, it is important to develop specific, targeted delivery of drugs into the tumor.

16.1.2
PAMAM Dendrimers as a Vehicle for Molecular Delivery into Cells

16.1.2.1 PAMAM Dendrimers as Encapsulation Complexes

PAMAM dendrimer has been tested as a "nanocomposite" to carry atoms or molecules into cells. Under this strategy, the dendrimer–drug complex is generated through non-covalent interaction [8, 29–31]. The encapsulation process is expected to yield dispersed small domains of the guest molecules such as metal ions, probably involving H-bonding, ionic and van der Waals forces, and trapping of the molecules within the dendrimer cavity. The metal nanoparticles (NPs) can be prepared by either *in-situ* chemical reaction (e.g., the reduction of metal ions to metal with zero valency) or by physical treatment (e.g., irradiation). Spherical clusters of NPs composed of gold and PAMAM dendrimers with a well-defined size (5–25 nm) have been shown internalize into cells [32]. Similarly, silver-based PAMAM dendrimer nanocomposites have been prepared in which the silver atoms, ions, or silver clusters are dispersed within the dendrimer cavity [33]. Because the dendrimer host is soluble, the silver clusters with extremely high surface areas can slowly diffuse out, making it a useful as a topical antimicrobial agent. As another example, the chemotherapeutic drug cisplatin, upon encapsulation into PAMAM dendrimers, results in an increased chemotherapeutic index as compared to the free drug [34].

Although the above-described "dendrimer-nanocomposites" with drugs and other molecules remain stable during ultrafiltration or dialysis in water, these molecules are released from the dendrimer in the presence of isotonic salt solutions. Therefore, the *in-vivo* applicability of such dendrimer–drug complexes, when entrapped non-covalently in its cargo space, is greatly limited due to the undesirable diffusion of the drugs in the isotonic biological milieu, prior to reaching the target cells [35]. Because of the biological instability of the "nanocomposites", it is important to develop a covalent dendrimer–drug conjugate that is stable in the circulation, but capable of releasing the drug in the cancer cell.

16.1.2.2 Multifunctional Covalent PAMAM Dendrimer Conjugates

Dendrimer-based targeted drug delivery is based on the principle that if a receptor is specifically expressed, or over-expressed, on the surface of the cancer cell, a covalent conjugate of the dendrimer with the drug and a ligand for the receptor travels stably through the circulation and specifically binds, internalizes into the cell, and delivers the drug into the cytosol. The advantage of such covalent conjugates is that additional functions such as a fluorescent sensing agent or an apoptosis-detecting agent can be covalently incorporated onto the dendrimer surface for multifunctional analysis.

For targeted therapy, a tumor-specific cell surface receptor must be initially identified. A variety of cellular receptors have been identified for molecules such as folic acid (FA), peptides, proteins, and antibodies which are relatively more specific for, or overexpressed, in cancer cells, and may serve as potential cell surface-targeting molecules. For example, FA is a small molecule, the receptor (FAR) of which is overexpressed on the surface of a variety of malignancies such as cancer of the ovary and breast. Our studies have shown that KB cells which express high FAR are a suitable model to investigate the binding of FA-conjugated dendrimers (Figure 16.2). Acetamide-capped neutral PAMAM dendrimers with FA as the targeting molecule and methotrexate (MTX) as the chemotherapeutic drug are able to target FAR-expressing KB cells *in vitro*, and in mice xenograft KB tumors *in vivo* [14, 16, 17]. In this conjugate, the FA is conjugated through an amide linkage and MTX through an ester linkage. The ester linkage allows the release of the drug in the endosomes at acidic pH and through hydrolysis by cellular esterases [16]. Other investigators have also used a similar PAMAM dendrimer system to target FAR-expressing tumor cells [18, 36].

Peptides are other small molecules which are currently undergoing extensive testing as cancer-targeting molecules. Targeting using peptides has several advantages over that using the intact protein ligands. Notably, owing to their smaller size, peptides have better pharmacokinetic properties, faster blood clearance, and lower immunogenicity than proteins. Moreover, due to the increased stability of peptides, the conjugation chemistry using peptides can be performed easily. Several identified peptides that bind to receptors such as integrins [37], human endothelial growth factor (EGF) receptor 2 (HER2) [38, 39], and luteinizing hormone-releasing hormone (LHRH) [40] are known to be overexpressed in certain tumors which can serve as the ligands for dendrimer-based targeting. We have conjugated the $\alpha_v\beta_3$ integrin-binding cyclic peptide "RGD4C" containing the binding sequence Arg-Gly-Asp to the PAMAM dendrimer, attaching AlexaFluor as the detecting dye [21]. This conjugate (G5-AF-RGD) binds and internalizes into the $\alpha_v\beta_3$ integrin-expressing human umbilical vein endothelial cells (HUVEC) (Figure 16.3).

As certain antigens and proteins are overexpressed in cancer cells, antibodies have been exploited as targeting agents [41]. Our studies have also demonstrated the applicability of PAMAM dendrimers for antibody-based drug targeting. Acetamide-capped dendrimer linked to fluorescein isothiocyanate (FITC) as the sensing agent and either of the two antibodies, 60bca or J591, which bind to CD14 and prostate-specific membrane antigen (PSMA), respectively, target and internalize into the corresponding antigen-expressing cells *in vitro* [19, 42]. The conjugate G5-PAMAM-Alexa Fluor-Herceptin (G5-AF-HN) binds and internalizes into HER2-expressing MCA207 cells, but not the control MCA207 cells [43]. Studies conducted by others have also demonstrated the applicability of PAMAM dendrimers for the antibody-based delivery of boron in boron neutron capture therapy [44].

After selecting a suitable targeting molecule specific for the tumor to be targeted, the next task is to determine the appropriate apoptosis-inducing chemo-

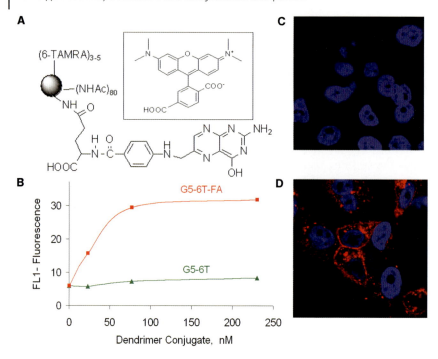

Fig. 16.2 Targeting of G5-6T-FA into KB cells. (A) Chemical structure of G5-6T-FA. The generation 5 (G5) PAMAM was initially partially acetylated (~80–90%) and onto it was conjugated three to five molecules of 6-TAMRA (structure shown in inset) and four to five molecules of folic acid (FA) through amide linkages between the remaining primary amino groups of the dendrimer and the carboxyl groups of the 6-TAMRA or the FA. Control conjugate (G5-6T) was synthesized without performing the third step of conjugating the FA. (B) Dose-dependent binding of G5-6T and G5-6T-FA in FA receptor-expressing KB cells, analyzed by flow cytometry following incubation with different concentrations of the conjugates for 1 h at 37 °C. The data shown are the mean fluorescence of 10 000 cells. (C, D). Confocal microscopic images showing the internalization of G5-6T-FA (D), but not the control conjugate G5-6T (C). The cells were grown on coverslips and incubated with 100 nM each of the conjugates for 1 h at 37 °C. The cells were rinsed, fixed with p-formaldehyde, stained with the nuclear stain DAPI (blue stain), and analyzed by confocal microscopy.

therapeutic drug for conjugation to the dendrimer. Several FDA-approved chemo-therapeutic drugs currently used in the clinic can also be used for conjugation to the PAMAM dendrimer [4]. Hence, we have used the drugs MTX (which inhibits the enzyme dihydrofolate reductase and stops nucleotide and DNA synthesis) [16] and taxol (which promotes microtubule assembly and causes cell cycle arrest) [45] for conjugation to PAMAM dendrimer. In a mouse model developed with FA receptor-expressing xenograft tumors, the MTX-conjugated dendrimer with FA as the targeting agent (G5-FA-MTX) shows increased chemotherapeutic index as compared to free MTX [17]. The increased efficacy of the conjugate is due to FA-

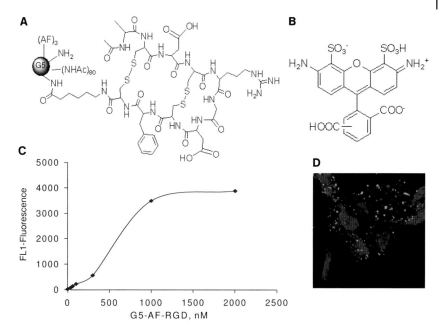

Fig. 16.3 Targeting of G5-AlexaFluor-RGD(G5-AF-FA) into HUVEC cells. (A) Chemical structure of G5-AF-RGD. The G5 PAMAM was initially partially acetylated (~80–90%) and onto it was conjugated an average of three molecules of AF [structure shown in (B)] and two to three molecules of the RGD peptide as described in Ref. [21]. (C) Dose-dependent binding of G5-AF-RGD in FA receptor-expressing HUVEC cells, analyzed by flow cytometry following incubation with different concentrations of the conjugates for 1 h at 37 °C; data shown are mean fluorescence of 10 000 cells. (D) Confocal microscopic images showing internalization of G5-AF-RGD in HUVEC cells. The cells were grown on coverslips and incubated with 100 nM conjugate for 1 h at 37 °C. Cells were rinsed, fixed with p-formaldehyde, stained with the nuclear stain DAPI (blue stain), and analyzed by confocal microscopy.

mediated targeting and entry into the tumor cells, because an equivalent dose of the control conjugate G5-MTX failed to show any effect on tumor regression.

As chemotherapeutics are known to cause cell death through a process called apoptosis or programmed cell death [4, 46], monitoring of the latter can provide useful information on the status of cancer cell death. During apoptosis, several cellular biochemical changes occur, which may be monitored if an intermediate biomolecule of the apoptosis pathway can induce changes in the fluorescent property of a molecule delivered through the dendrimer. Fluorescent apoptosis detection consists of using either a single dye such as rhodamine (which measures mitochondrial membrane permeability), or two dyes in combination utilizing a process termed fluorescence resonance energy transfer (FRET) [4, 47]. The synthesis of a FRET-based apoptosis-detecting PAMAM dendrimer device consists of conjugating a donor-acceptor fluorophore coupled to the dendrimer. For example, a donor and acceptor fluorophore can be coupled though the peptide

DEVD, a substrate of the apoptotic enzyme caspase-3, which will have a reduced fluorescence due to the FRET action between the dyes. The intracellular hydrolysis of this peptide will result in an increase in fluorescence which can be quantified using appropriate techniques.

16.1.2.3 PAMAM Dendrimers as MRI Contrast Agents

Magnetic resonance imaging (MRI) is a noninvasive tissue imaging technique that provides high temporal and spatial imaging resolution of up to 100 μm. Amine-terminated PAMAM dendrimers of varying generations have been tested as Gd-based MRI contrast agents to test the *in-vivo* distribution in mice [22]. Dendrimer–Gd complexes with smaller size (<3 nm) are rapidly extravasated to diffuse throughout the body, whereas complexes between 3 and 6 nm are quickly excreted through the kidney. Dendrimers with sizes of 5–7 nm have a preference for extravasation into the tumor tissue, whereas those of 7–12 nm are largely retained in the circulation [22]. Based on this observed tissue distribution profile, different-sized dendrimers can be selected as suitable MRI contrast agents for different tissues. For example, the higher-generation PAMAM dendrimers are suitable for vascular imaging due to their ability to stay in the blood circulation for extended periods of time. Moreover, these conjugates possess greatly enhanced relaxivities compared to the commonly used contrast agents such as gadolinium chelates, due to the slow rotation of the conjugated paramagnetic ion and increased correlation time [22, 48].

16.1.2.4 Application of Multifunctional Clusters of PAMAM Dendrimer

The conjugation of different types of multiple molecules for targeting, imaging and drug delivery onto the same dendrimer molecule has several limitations. The hydrophobicity of the added functions can result in reduced solubility of the conjugate, leading to low yield during synthesis and reduced solubility in aqueous biological fluids. Also, the physical and chemical interaction between the different molecules conjugated densely on the dendrimer surface and the consequent steric hindrance may reduce the activity of their functions, such as the affinity of a targeting molecule or the fluorescence of a fluorochrome through FRET action. However, these problems can be solved by conjugation of only one function per dendrimer molecule and linking the dendrimer conjugates carrying separate functions to form a "cluster" or "tecto" dendrimer. The versatility of such a cluster dendrimer is the easy combinatorial synthesis of a "custom" conjugate by linking separately synthesized single-function conjugates carrying a specific targeting molecule, a drug, or an apoptosis sensor. We have synthesized dendrimer clusters by linking two dendrimer molecules through a short DNA chain (Figure 16.4). This is easily achieved by the self-assembly (annealing) of two single-stranded complementary DNA molecules covalently conjugated onto two separate dendrimers carrying two different functions. Our recent studies have proven the biological applicability of a synthesized cluster dendrimer with FA and FITC as the two functions on the two dendrimer molecules, linked through DNA [49]. The DNA-linked dendrimer platform would allow the combinatorial

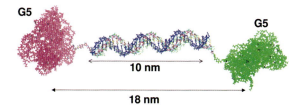

Fig. 16.4 Computer-generated model of two G5-dendrimer molecules linked through a DNA.

synthesis of conjugates containing multiple functions for tumor analysis such as imaging, drug delivery, and cancer cell killing.

16.2
Methods

16.2.1
Synthesis and Characterization of PAMAM Dendrimers

PAMAM dendrimers are commercially available and are synthesized from an ethylene diamine (EDA) or NH_3 initiator core, with exhaustive Michael addition of methyl acrylate (MA) followed by condensation (amidation) reactions of the resulting ester with large excesses of EDA [4, 7, 8]. Repetition of this reaction sequence results in the production of each successive generation (G), resulting in greater molecular weight, size, and functionality (Figure 16.5). Branching occurs at the terminal amine, as two methyl acrylate monomers will be added to each amine. Consequently, each generation of growth doubles the number of termini and approximately doubles the molecular weight, with an increase of about 1 nm in diameter for each added generation.

Dendrimers can be synthesized either through a "divergent" method that allows the growth of the molecule from a central core to its periphery, or through the opposite "convergent" strategy, which begins at the periphery and is directed toward the central core [50–52]. One disadvantage of the divergent synthetic strategy is that, as the end groups branch out in large numbers, there can be structural defects because of an incomplete reaction or the failure of surface groups to form on all the branches. In the convergent synthesis, although the low number of possible side reactions and controlled number of reactive groups allow the synthesis of dendrimers with a higher degree of control, it suffers from steric crowding while anchoring to the central core, especially during the synthesis of higher-generation dendrimers [4].

Dendrimers are characterized by standard polymer characterization techniques such as potentiometric titration, ultraviolet-visible spectroscopy, light scattering, size-exclusion chromatography (SEC), nuclear magnetic resonance (NMR),

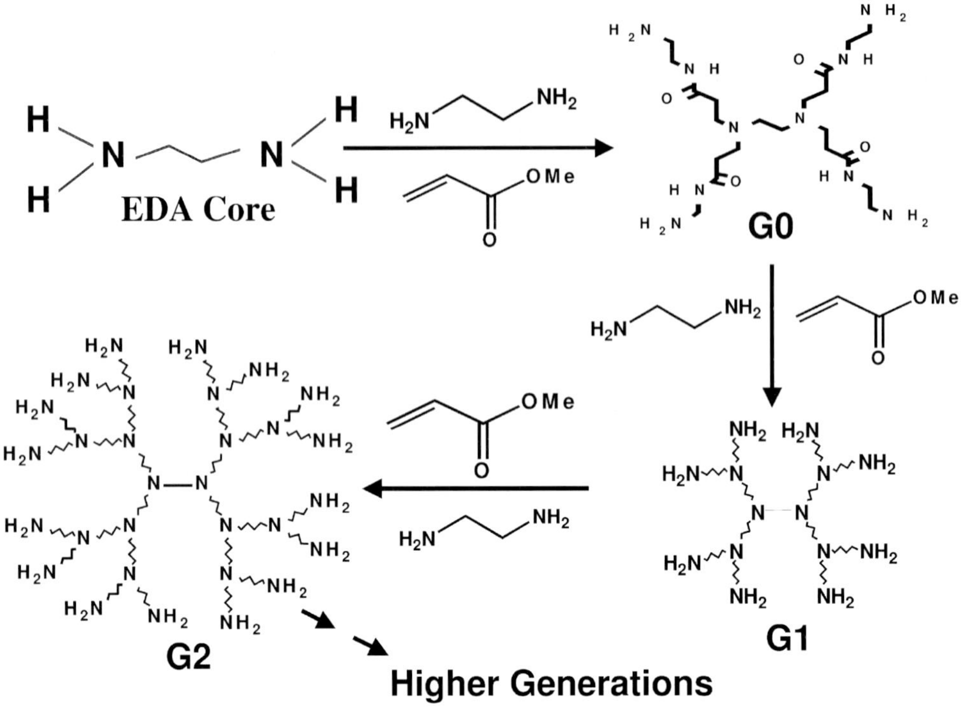

Fig. 16.5 The "divergent" approach for the synthesis of different generations of PAMAM dendrimer. The four branching points of the ethylenediamine core are allowed to react with methacrylate, followed by further amidation with the ethylenediamine to form initially the generation 0 (G0). This iterative reaction cycle is repeated to generate further generations (G1–G9). As shown, the G1 and G2 have eight and 16 functional surface amino groups with molecular weights of 1430 and 3256, respectively. The number of surface amino groups and molecular weights approximately doubles with each higher generation. Because of the controlled synthesis and the well-defined structure of the PAMAM dendrimer, the number of amino groups, molecular weight, etc. can be calculated mathematically. However, there can be minor "defects" in the branching formation during the synthetic procedure which can be determined experimentally (see text for details).

reverse phase-high-pressure liquid chromatography (RP-HPLC), and matrix-assisted, laser desorption ionization-time of flight (MALDI-TOF) mass spectrometry [4, 20, 53]. Due to the regularity of the dendritic structure, NMR (in particular, heteronuclei and ^{13}C) can provide a good deal of information on dendrimer structure. In lower generations, the NMR can reveal any structural defects by showing a multiplicity of peaks. The SEC provides data about molecular weight distribution of dendrimers, whilst with MALDI-TOF it is possible to determine the molecular weight of higher generations. Dendrimers show generation-dependent conformational changes, adopting a globular shape at higher generations. The globular structure gives rise to significantly different solution and bulk properties to dendrimers when compared with their linear analogues, in particular the viscosity and solubility [12].

16.2.2
PAMAM Dendrimer: Determination of Physical Parameters

The theoretical molecular weight (MW), number of terminal groups, etc., can be computed for each generation of dendrimers [54]. This is crucial in determining the defects in branch formation, obtained through comparison of the theoretical and experimentally determined MW. Differences in theoretical and experimental MW for each generation can be attributed to intramolecular coupling (loop formation), bridging, incomplete Michael addition, retro-Michael reactions, and side reactions due to incomplete removal of EDA during the synthesis of each generation. The theoretical MW, degree of polymerization, number of terminal amino groups, etc., can be determined using the following equations:

$$MW = MW_{core} + (MW \text{ of monomers}) \ F_c \left(\frac{F_r^{g+1} - 1}{F_r - 1} \right); \text{ degree of polymerization}$$

$$= F_c \left(\frac{F_r^{g+1} - 1}{F_r - 1} \right); \text{ number of terminal groups: } Z = F_c F_r^g; \text{ number of tertiary}$$

$$\text{amines} = T_c + F_c \left(\frac{2^g - 1}{2 - 1} \right) \text{ (where g is the generation number, } F_c \text{ the number of}$$

functional groups on the core, F_r the multiplicity of the repeating unit MW of monomers, and T_c the number of tertiary amines of the core).

16.2.3
Quantification of Fluorescence of Targeted PAMAM Conjugates

The cellular uptake of a dendrimer–dye conjugate is determined *in vitro* and *in vivo* by various fluorescence detection techniques, such as flow cytometry, confocal microscopy and two-photon optical fiber fluorescence (TPOFF) [54]. Flow cytometry is a powerful tool for determining cellular binding of the conjugate, whilst confocal microscopy shows evidence of tissular, cellular and subcellular localization [55]. Unfortunately, both methods are limited by requiring tissue processing, such as cell isolation or the preparation of cryosections. Also, neither method can provide actual concentrations of the fluorescent material in the cell, and real-time measurements are not possible.

The TPOFF method directly quantifies fluorescence *in vitro* in a cell pellet or in a tissue such as the tumor in anesthetized mice by inserting an optical fiber into the tissue through a 27-gauge needle. By knowing the TPOFF counts of the standard dendrimer–dye conjugate, one can quantify the tissue concentration of the conjugate. Moreover, the TPOFF technique can be used to measure the fluorescence of a small focal volume (microns) of the tissue, and multiple dyes with different emission wavelengths can be measured simultaneously [55].

16.3
Outlook

The PAMAM dendrimers can be synthesized on a large scale under good manufacturing practice (GMP) guidelines, and are biocompatible macromolecules which can be used as suitable carriers of molecules for delivery into tumor cells, both *in vitro* and *in vivo*. Multiple molecules such as tumor-specific targeting agents, chemotherapeutic drugs, and apoptosis-detecting agents can be covalently conjugated onto the dendrimer's surface. Subsequently, the "smart", engineered multifunctional dendrimer nanodevice can mimic a biological molecule and perform multiple biological tasks, such as binding to a cancer cell, releasing a drug to induce apoptosis of the cancer cell, and measuring the extent of cell death. A "cluster" dendrimer may also serve as a useful and versatile nanodevice with separate functions on separate dendrimer molecule, which are linked together through DNA or other linkers. The results of recent *in-vitro* and *in-vivo* studies conducted in the authors' laboratory [14, 16, 17, 19–21, 42, 43] have demonstrated the applicability of the PAMAM dendrimer as a powerful tumor-targeting platform. Using mice tumor models, these studies have shown unequivocally that PAMAM dendrimer-based targeting can achieve increased drug efficacy with significantly less toxicity compared to using the free drug. These results also suggest the possible application of PAMAM-based targeting for a broader range of cancers and other diseases for which current treatment is limited due to the nonspecific toxicity of drugs in cells other than the desired target.

References

1 Luo, Y., Prestwich, G. D. (2002) Cancer-targeted polymeric drugs. *Current Cancer Drug Targets* **2**, 209–226.

2 Nori, A., Kopecek, J. (2005) Intracellular targeting of polymer-bound drugs for cancer chemotherapy. *Adv. Drug Deliv. Rev.* **57**, 609–636.

3 Duncan, R. (2003) The dawning era of polymer therapeutics. *Nat. Rev. Drug Discov.* **2**, 347–360.

4 Majoros, I. J., Thomas, T. P., Baker Jr., J. R. (2006) Molecular engineering in nanotechnology: Engineered drug delivery. In: Rieth, M., Schommers, W. (Eds.), *Molecular Engineering in Nanotechnology: Engineered Drug Delivery. In Handbook of Theoretical and Computational Nanotechnology.* Chapter 8 (in press).

5 Qiu, L. Y., Bae, Y. H. (2006) Polymer architecture and drug delivery. *Pharm. Res.* **23**, 1–30.

6 Vicent, M. J., Duncan, R. (2006) Polymer conjugates: nanosized medicines for treating cancer. *Trends Biotechnol.* **24**, 39–47.

7 Patri, A. K., Majoros, I. J., Baker Jr., J. R. (2002) Dendritic polymer macromolecular carriers for drug delivery. *Curr. Opin. Chem. Biol.* **6**, 466–371.

8 Svenson, S., Tomalia, D. A. (2005) Dendrimers in biomedical applications – reflections on the field. *Adv. Drug Deliv. Rev.* **57**, 2106–2129.

9 Chow, H. F., Mong, T. K., Nongrum, M. F., Wan, C. W. (1998) The synthesis and properties of novel functional dendritic molecules. *Tetrahedron* **54**, 8543–8660.

10 Fischer, M., Vögtle, F. (1999) Dendrimers: from design to application – a progress report. *Angew. Chem. Int. Ed. Engl.* **38**, 884–905.

11 Tomalia, D. A., Majoros, I. J. (2002) Dendrimeric supramolecular and supramacromolecular assemblies. In: Ciferri, A. (Ed.), *Supramolecular Polymers*. Marcel Dekker, New York, pp. 359–435.

12 Bosman, A. W., Janssen, H. M., Meijer, E. W. (1999) About dendrimers: structure, physical properties, and applications. *Chem. Rev.* **99**, 1665–1688.

13 Boas, U., Heegaard, P. M. (2004) Dendrimers in drug research. *Chem. Soc. Rev.* **33**, 43–63.

14 Quintana, A., Raczka, E., Piehler, L., Lee, I., Myc, A., Majoros, I. J., Patri, A. K., Thomas, T. P., Mulé, J., Baker Jr., J. R. (2002) Design and function of a dendrimer-based therapeutic nanodevice targeted to tumor cells through the folate receptor. *Pharm. Res.* **19**, 1310–1316.

15 Majoros, I. J., Keszler, B., Woehler, S., Bull, T., Baker Jr., J. R. (2003) Acetylation of poly(amidoamine) dendrimers. *Macromolecules* **36**, 5526–5529.

16 Thomas, T. P., Majoros, I. J., Kotlyar, A., Kukowska-Latallo, J. K., Bielinska, A. B., Myc, A., Baker Jr., J. R. (2005) Targeting and inhibition of cell growth by an engineered dendritic nanodevice. *J. Med. Chem.* **48**, 3729–3735.

17 Kukowska-Latallo, J. K., Candido, K. A., Cao, Z., Nigavekar, S. S., Majoros, I. J., Thomas, T. P., Balogh, L. P., Khan, M. K., Baker Jr., J. R. (2005) Nanoparticle targeting of anticancer drug improves therapeutic response in animal model of human epithelial cancer. *Cancer Res.* **65**, 5317–5324.

18 Konda, S. D., Aref, M., Wang, S., Brechbiel, M., Wiener, E. C. (2001) Specific targeting of folate-dendrimer MRI contrast agents to the high affinity folate receptor expressed in ovarian tumor xenografts. *MAGMA* **12**, 104–113.

19 Thomas, T. P., Patri, A. K., Myc, A., Myaing, M. T., Ye, J. Y., Norris, T. B., Baker Jr., J. R. (2004) In vitro targeting of synthesized antibody-conjugated dendrimer nanoparticles. *Biomacromolecules* **5**, 2269–2274.

20 Majoros, I. J., Thomas, T. P., Mehta, C. B., Baker Jr., J. R. (2005) Poly(amidoamine) dendrimer-based multifunctional engineered nanodevice for cancer therapy. *J. Med. Chem.* **48**, 5892–5899.

21 Shukla, R., Thomas, T. P., Peters, J., Kotlyar, A., Myc, A., Baker Jr., J. R. (2005) Tumor angiogenic vasculature targeting with PAMAM dendrimer-RGD conjugates. *Chem. Commun. (Camb)*. **46**, 5739–5741.

22 Kobayashi, H., Brechbiel, M. W. (2005) Nano-sized MRI contrast agents with dendrimer cores. *Adv. Drug Deliv. Rev.* **57**, 2271–2286.

23 Duncan, R., Izzo, L. (2005) Dendrimer biocompatibility and toxicity. *Adv. Drug Deliv. Rev.* **57**, 2215–2237.

24 Malik, N., Wiwattanapatapee, R., Klopsch, R., Lorenz, K., Frey, H., Weener, J. W., Meijer, E. W., Paulus, W., Duncan, R. (2000) Dendrimers: relationship between structure and biocompatibility in vitro, and preliminary studies on the biodistribution of ^{125}I-labelled polyamidoamine dendrimers in vivo. *J. Control. Release* **65**, 133–148.

25 Jevprasesphant, R., Penny, J., Jalal, R., Attwood, D., McKeown, N. B., D'Emanuele, A. (2003) The influence of surface modification on the cytotoxicity of PAMAM dendrimers. *Int. J. Pharm.* **252**, 263–266.

26 Hong, S., Bielinska, A. U., Mecke, A., Keszler, B., Beals, J. L., Shi, X., Balogh, L., Orr, B. G., Baker Jr., J. R., Banaszak Holl, M. M. (2004) The interaction of polyamidoamine (PAMAM) dendrimers with supported lipid bilayers and cells: hole formation and relation to transport. *Bioconj. Chem.* **15**, 774–782.

27 Roberts, J. C., Bhalgat, M. K., Zera, R. T. (1996) Preliminary biological evaluation of polyamidoamine (PAMAM) starburst dendrimers. *J. Biomed. Mater. Res.* **30**, 53–65.

28 Kobayashi, H., Kawamoto, S., Saga, T., Sato, N., Hiraga, A., Ishimori, T.,

Konishi, J., Togashi, K., Brechbiel, M. W. (2001) Positive effects of polyethylene glycol conjugation to generation-4 polyamidoamine dendrimers as macromolecular MR contrast agents. *Magn. Reson. Med.* **46**, 781–788.

29 Beck Tan, N., Balogh, L., Trevino, S. (1997) Structure of metallo-organic nano-composites produced from dendrimer complexes. *Proc. ACS PMSE* **77**, 120.

30 Balogh, L., Tomalia, D. A. (1998) Poly (amidoamine) dendrimer-templated nanocomposites. 1. Synthesis of zerovalent copper nanoclusters. *J. Am. Chem. Soc.* **120**, 7355–7356.

31 D'Emanuele, A., Attwood, D. (2005) Dendrimer-drug interactions. *Adv. Drug Deliv. Rev.* **57**, 2147–2162.

32 Bielinska, A., Eichman, J. D., Lee, I., Baker Jr., J. R., Balogh, L. P. (2002) Imaging {Au0-PAMAM} gold-dendrimer nanocomposites in cells. *J. Nanoparticle Res.* **4**, 395–403.

33 Balogh, L. P., Swanson, D. R., Tomalia, D. A., Hagnauer, G. L., McManus, A. T. (2001) Dendrimer-silver complexes and nanocomposites as antimicrobial agents. *Nano Lett.* **1**, 18–21.

34 Malik, N., Evagorou, E. G., Duncan, R. (1999) Dendrimer-platinate: a novel approach to cancer chemo-therapy. *Anticancer Drugs* **10**, 767–776.

35 Patri, A. K., Kukowska-Latallo, J. F., Baker Jr., J. R. (2005) Targeted drug delivery with dendrimers: comparison of the release kinetics of covalently conjugated drug and non-covalent drug inclusion complex. *Adv. Drug Deliv. Rev.* **57**, 2203–2214.

36 Shukla, S., Wu, G., Chatterjee, M., Yang, W., Sekido, M., Diop, L. A., Muller, R., Sudimack, J. J., Lee, R. J., Barth, R. F., Tjarks, W. (2003) Synthesis and biological evaluation of folate receptor-targeted boronated PAMAM dendrimers as potential agents for neutron capture therapy. *Bioconj. Chem.* **14**, 158–167.

37 Arap, W., Pasqualini, R., Ruoslahti, E. (1998) Cancer treatment by targeted drug delivery to tumor vasculature in a mouse model. *Science* **279**, 377–380.

38 Shadidi, M., Sioud, M. (2003) Identifi-cation of novel carrier peptides for the specific delivery of therapeutics into cancer cells. *FASEB J.* **17**, 256–258.

39 Murali, R., Liu, Q., Cheng, X., Berezov, A., Richter, M., Furuchi, K., Greene, M. I., Zhang, H. (2003) Antibody like peptidomimetics as large scale immunodetection probes. *Cell. Mol. Biol. (Noisy-le-grand)* **49**, 209–216.

40 Dharap, S. S., Qiu, B., Williams, G. C., Sinko, P., Stein, S., Minko, T. (2003) Molecular targeting of drug delivery systems to ovarian cancer by BH3 and LHRH peptides. *J. Control. Release* **91**, 61–73.

41 Schrama, D., Reisfeld, R. A., Becker, J. C. (2006) Antibody targeted drugs as cancer therapeutics. *Nature Rev. Drug Disc.* **5**, 147–159.

42 Patri, A. K., Myc, A., Beals, J., Thomas, T. P., Bander, N. H., Baker Jr., J. R. (2003) Synthesis and in Vitro Testing of J591 Antibody-dendrimer conjugates for targeted prostate cancer therapy. *Bioconj. Chem.* **15**, 1174–1181.

43 Shukla, R., Thomas, T. P., Peters, J. L., Kukowska-Latallo, J., Patri, A. K., Kotlyar, A., Baker Jr., J. R. (2006) HER2 specific tumor targeting with dendrimer conjugated anti-HER2 mAb. *Bioconj. Chem.* **17**, 1109–1115.

44 Wu, G., Barth, R. F., Yang, W., Chatterjee, M., Tjarks, W., Ciesielski, M. J., Fenstermaker, R. A. (2004) Site-specific conjugation of boron-containing dendrimers to anti-EGF receptor monoclonal antibody cetuximab (IMC-C225) and its evaluation as a potential delivery agent for neutron capture therapy. *Bioconj. Chem.* **15**, 185–194.

45 Majoros, I. J., Myc, A., Thomas, T., Mehta, C. B., Baker Jr., J. R. (2006) PAMAM dendrimer-based multi-functional conjugate for cancer therapy: synthesis, characterization, and functionality. *Biomacromolecules* **7**, 572–579.

46 Green, D. R., Kroemer, G. (2005) Pharmacological manipulation of cell death: clinical applications in sight? *J. Clin. Invest.* **115**, 2610–2617.

47 Berney, C., Danuser, G. (2003) FRET or no FRET: a quantitative comparison. *Biophys. J.* **84**, 3992–4010.

48 Kobayashi, H., Kawamoto, S., Jo, S. K., Sato, N., Saga, T., Hiraga, A., Konishi, J., Hu, S., Togashi, K., Brechbiel, M. W. (2002) Renal tubular damage detected by dynamic micro-MRI with a dendrimer-based magnetic resonance contrast agent. *Kidney Int.* **61**, 1980–1985.

49 Choi, Y. S., Thomas, T. P., Kotlyar, A., Islam, M. T., Baker Jr., J. R. (2005) Synthesis and functional evaluation of DNA-assembled polyamidoamine dendrimer clusters for cancer cell-specific targeting. *Chem. Biol.* **12**, 35–43.

50 Tomalia, D. A., Baker, H., Dewald, J. R., Hall, M., Kallos, G., Martin, S., Roeck, J., Ryder, J., Smith, P. B. (1985) A new class of polymers: Starburst dendritic molecules. *Polymer J.* **17**, 117–132.

51 Moors, R., Vögtle, F. (1993) Dendrimere polyamine. *Chem. Ber.* **126**, 2133–2135.

52 Hawker, C., Fréchet, J. M. J. (1990) A new convergent approach to monodisperse dendritic macromolecules. *J. Chem. Soc., Chem. Commun.* **1**, 1010–1013.

53 Islam, M., Majoros, I. J., Baker Jr., J. R. (2005) HPLC Analysis of PAMAM dendrimer based multifunctional devices. *J. Chromatogr. B* **822**, 21–26.

54 Majoros, I. J., Mehta, C. B., Baker Jr., J. R. (2004) Mathematical Description of Dendrimer Structure. *J. Comp. Theoret. Nanosci.* **1**, 193–198.

55 Thomas, T. P., Myaing, M. T., Ye, J. Y., Candido, K., Kotlyar, A., Beals, J., Cao, P., Keszler, B., Patri, A. K., Norris, T. B., Baker Jr., J. R. (2004) Detection and analysis of tumor fluorescence using a two-photon optical fiber probe. *Biophys. J.* **86**, 3959–3965.

17
Nanoparticle Contrast Agents for Molecular Magnetic Resonance Imaging

Young-wook Jun, Jae-Hyun Lee, and Jinwoo Cheon

17.1
Introduction

The inner depths of the human body have long been an unexplored "dark world", as visible light is unable to penetrate the tissues. Although observation of the body cavity is possible by the use of surgical procedures, it would be preferable to observe the inner spaces of the body by noninvasive means. The relatively recent developments of imaging techniques such as computed tomography (CT), near-infra-red (IR) fluorescence imaging, positron emission spectroscopy (PET), and magnetic resonance imaging (MRI) have made such exploration possible, and has enabled mankind to make major advances in the field of biomedical diagnosis and therapy.

As most biological processes and diseases are related to molecular and cellular events, the precise observation of these detailed biological processes is important [1, 2]. However, classical imaging techniques are generally unsatisfactory for such "molecular imaging" applications, and the development of high-performance imaging systems capable of identifying detailed biological processes at the molecular and subcellular levels is important.

Inorganic nanoparticles (NPs) as probes have the potential to revolutionize conventional imaging systems [3, 4], and the enhanced optical and magnetic properties of inorganic NPs arising from nanoscale quantum properties have afforded major enhancements in imaging sensitivity and resolution [9–11]. The small size of NPs, comparable to that of biological functional units (e.g., proteins), makes them ideally suited to observing and tracking molecular events [12]. Today, as a variety of molecular markers for specific biological events are available through rapid advances in molecular biology, the conjugation of inorganic NP probes with bioactive molecules permits the tracking of molecular events [5–8]. Whilst in the past, both quantum dots and magnetic NPs have served as NP molecular imaging agents for fluorescent optical imaging and for molecular MRI, respectively. this chapter will focus on magnetic NP-based molecular MR imaging.

Nanobiotechnology II. Edited by Chad A. Mirkin and Christof M. Niemeyer
Copyright © 2007 WILEY-VCH Verlag GmbH & Co. KGaA, Weinheim
ISBN: 978-3-527-31673-1

17.2
NP-Assisted MRI

Currently, MRI is one of the most powerful medical diagnostic tools available, due mainly to its noninvasive nature and multi-dimensional tomographic capabilities, coupled with high spatial resolution [13]. Although in terms of sensitivity MRI lags behind other tools [2], any such weakness can be significantly improved by using magnetic NP contrast agents [10, 11]. Under an applied magnetic field, NPs are magnetized and generate an induced magnetic field, which perturbs the magnetic relaxation processes of the protons in the water molecules surrounding the magnetic NP. In turn, this leads to a shortening of the spin-spin relaxation time (T2) of the proton, and a consequent darkening of the MR images (Figure 17.1).

According to the outer sphere spin-spin relaxation formula of solvent protons by solute magnetic particles, the spin-spin relaxation time (T2) of the proton is

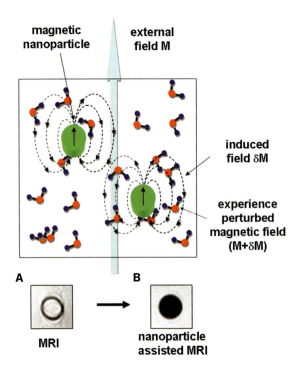

Fig. 17.1 MR contrast effects of magnetic NPs. Under an applied magnetic field (M), magnetic NPs are magnetized and generate an induced magnetic field (δM), which perturbs the magnetic relaxation processes of the proton in water molecules, which is reflected as dark MR contrast. Axially sliced MR images of (A) water and (B) magnetic NP solution in microcentrifuge tubes. The circular spots in the MR images arise from the wall of microcentrifuge tubes.

$$\frac{1}{T2} = \left(\frac{32\pi N_A[M]}{405000rD}\right)\gamma_I{}^2\mu^2\{6.5j_2(\omega_s, \tau) + 1.5j_1(\omega_I, \tau) + j_1(0, \tau)\},$$

where, γ_I is the gyromagnetic ratio of protons in water, M is the molarity of magnetic NPs, r is their radius, N_A is Avogadro's number, μ is the magnetic moment of the NP, ω_s and ω_I are the respective Larmor angular precession frequencies of the solute electronic and water proton magnetic moments, the functions $j_n(\omega, \tau)$ are spectral density functions, and τ $(= r^2/D)$ is the time scale of fluctuations in the particle-water proton magnetic dipolar interaction arising from the relative diffusive motion (D) of a particle and water molecules [14]. Therefore, the shortening of T2 is achieved by increasing the magnetic moment of the NPs. Recently, Cheon, Suh, and co-workers demonstrated such magnetic moment effects on T2 by elucidating the correlated nanoscale effects of iron oxide NPs between size, magnetism, and T2 relaxivity [15]. The transmission electron microscopy (TEM) images of highly monodispersed iron oxide NPs of size 4, 6, 9, and 12 nm, respectively, are shown in Figure 17.2a. These magnetic NPs exhibit size-dependent magnetic moments and, as the NP size is increased from 4 to 6, 9, and to 12 nm, the mass magnetization value at 1.5 T changes from 25 to 43, 80, and 102 emu (g^{-1} Fe) (Figure 17.2e). Such a trend is clearly reflected in the T2-weighted MR images. The 1/T2 relaxivity gradually increases from 56 to 106, 130, and to 190 L mol^{-1} s^{-1}, which is imaged by the gradual change of the MR contrast from white to black through gray (Figure 17.2).

Such effect of magnetic NPs on MR contrast provides them with an ability to identify a variety of biological events. For example, magnetic NPs larger than 30 nm have been used for phagocytosis imaging [16, 17]. Magnetic NPs taken up by phagocytes are imaged as dark contrast, but tumor cells without phagocytic ability appear as white contrast, and liver metastases [18, 19], spleen [18] and lymph nodes [20] have been detected in this way.

In contrast, smaller NPs (e.g., 10 nm) are able to escape from phagocytes, such that magnetic NPs, upon conjugation with a target-specific biomolecule, can be used to detect target tissues through molecular interactions between NP–biomolecule conjugates and molecular markers expressed by target tissues [21, 22]. A variety of clinically benign iron oxide-based magnetic NPs [e.g., superparamagnetic iron oxide (SPIO)] have been investigated, and the imaging of infarction [23–25], angiogenesis [26], apoptosis [27], gene expression [28, 29], and cancer [15, 30–33] have been reported. Unfortunately however, the MR signal sensitivity and the specificity of NP probes to target tissues remain unsatisfactory for clinical applications, and further efforts are required for their improvement. Recently developed biocompatible magnetic NPs and their use in molecular MR imaging are briefly reviewed in the following sections.

17.2.1
Magnetic NP Contrast Agents

In order to be used successfully as molecular MR contrast agents, magnetic NP must possess the following properties:

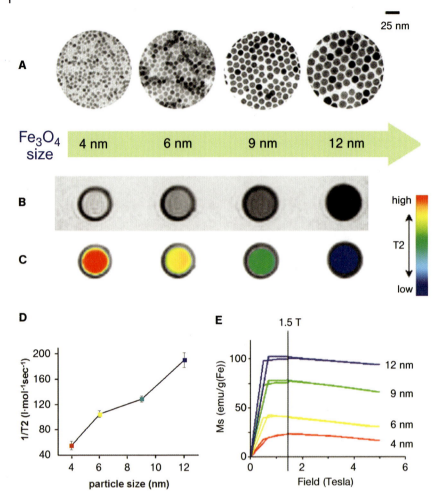

Fig. 17.2 Nanoscale-size effects of iron oxide NPs on magnetism and induced magnetic resonance (MR) signals. (A) TEM images of Fe_3O_4 nanocrystals of 4 to 6, 9, and 12 nm. (B) Size-dependent T2-weighted MR images of iron oxide NPs in aqueous solution at 1.5 Tesla. (C) Size-dependent changes from red to blue in color-coded MR images based on T2 values. (D) Graph of 1/T2 relaxivity value versus iron oxide NP size. (E) Magnetization of iron oxide NPs measured by a SQUID magnetometer. (Reproduced with permission, from Ref. [15].)

- a uniform and high superparamagnetic moment;
- a high colloidal stability under physiological conditions (e.g., high salt concentration and pH changes);
- the ability to escape from the reticuloendothelial system (RES);
- low toxicity and high biocompatibility; and

- a functionality to be linked to biologically active species (e.g.,
  nucleic acid, proteins).

As these properties are largely related to the size, stoichiometry and surface structures of the NPs, various types of iron oxide NPs have been developed for these purposes.

17.2.1.1 Silica- or Dextran-Coated Iron Oxide Contrast Agents

For conventional MR contrast agents, iron oxide NPs are synthesized by the precipitation of iron oxide in an aqueous solution containing ferrous salt by adding an alkaline solution [34]. These iron oxide NPs are usually insoluble as prepared, and consequently they must be coated to render them soluble in aqueous media. Early attempts to achieve such water solubility involved the use of silica as a coating material [35]. The size of the core magnetic iron oxide can range from 4 to 10 nm, and the total particle size from 10 nm to 1 µm, including the coating materials. As these NPs have a broad size distribution, further size-sorting procedures (including differential centrifugation and dialysis) are required. One such silica-coated iron oxide contrast agent is AMI-121 (generic name Ferumoxsil), and this is commercially available as Lumirem® (Guerbet) and Gastromark® (Advance Magnetics). The core, which is composed of polycrystalline iron oxide, is approximately ~10 nm in size, and the hydrodynamic size is approximately ~300 nm. This agent is delivered orally and used for MR imaging of the abdomen [36].

Although silica-coated iron oxide NPs are reasonably stable in aqueous media, they tend to aggregate in blood and therefore cannot be injected into the bloodstream. In order to enhance the colloidal stability of iron oxide NPs, another type of coating agent – dextran or carbodextran – has been utilized [37]. Dextran possesses high colloidal stability under harsh physiological conditions, and hence dextran-coated iron oxide NPs should be very stable. Dextran-coated iron oxide NPs are prepared by co-precipitation from aqueous solution containing ferrous salt and dextran by adding an alkaline solution [34]. Currently, three representative dextran-coated iron oxide NPs are available, namely AMI-25 (Feridex®; Berlex Laboratories and Endorem®; Guerbet), SHU 555A (Resovist®; Schering), and AMI-227 (Combidex®; Advanced Magnetics and Sinerem®; Guerbet) (see Table 17.1).

- AMI-25 is composed of 5–6 nm iron oxide core and dextran-coating materials; the total size ranges from 80 to 150 nm, with a T2 relaxivity of ~98.3 L mmol^{-1} s^{-1} [38].
- SHU 555A has a ~4.2 nm iron oxide core coated with carbodextran, and a total size of ~62 nm; Resovist has a higher T2 relaxivity value of 151.0 L mmol^{-1} s^{-1} [39], and has no known adverse side effects after rapid intravenous injection [40]. These magnetic contrast agents are generally trapped and accumulated by the reticuloendothelial cells in the liver, with a short blood half-life of less than 10 min; thus, they are used for liver imaging [17].

Table 17.1 Currently available silica- or dextran-coated iron oxide contrast agents.

Agent (Tradename) [Reference]	Parameter					
	Iron oxide core size [nm]	Total size [nm]	Coating material	Magnetization at 1.5 T [emu g^{-1}]	T2 relaxivity [L mol^{-1} s^{-1}]	Half-life in blood
AMI-121[a] (Lumirem; Gastromark) [36]	~10	~300	Silica	NA	72	<5 min
AMI-25[a] (Feridex; Endorem) [38]	5~6	80~150	Dextran	78	98	~6 min
SHU 555A[a] (Resovist) [39]	~4.2	~62	Carbo-dextran	NA	151	3 min
AMI-227[a] (Combidex; Sinerem) [41]	4~6	20~40	Dextran	69.8	53	>24 h
MION; CLIO [42–44]	~2.8	10~30	Dextran	60~68	~69	~10 h

[a] Commercially available.

- Compared to these two iron oxide contrast agents, AMI-227 has a similar iron oxide core size (4–6 nm) but a smaller overall size of 20–40 nm. Although AMI-227 has lower T2 contrast effects (T2 relaxivity of ~53 L mmol^{-1} s^{-1}), its smaller size provides a much higher blood half-life of ~24 h, which enables MR angiography and lymph node detection [41].

Smaller dextran-coated iron oxide NPs, including monocrystalline iron oxide (MION) and its derivative, cross-linked iron oxide (CLIO), are composed of a ~2.8 nm core iron oxide and dextran shell with a total size of 10–30 nm [42–44]. Since these NPs are relatively small in size and have long blood half-lives and their surface can be readily linked with biologically active molecules, consequently they are valuable for *in-vivo* molecular MR imaging of biological targets [42] (see Section 17.2.1).

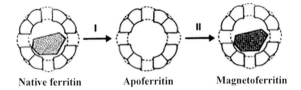

Native ferritin   Apoferritin   **Magnetoferritin**

Fig. 17.3 Synthetic scheme of magnetoferritin. Removal of ferrihydrite from native ferritin produces apoferritin, and subsequent formation of magnetite NPs inside apoferritin results in the formation of magnetoferritin. (Reproduced with permission, from Ref. [46].)

17.2.1.2 Magnetoferritin

Ferritin is a well-known iron storage protein used to sequester and store iron in the body, and is composed of a $\sim$6 nm hydrated iron oxide, a ferrihydrite $(5Fe_2O_3 \cdot 9H_2O)$ core, and a polypeptide apoferritin shell [45]. Ferritin has been used as an efficient synthesizer for other magnetic materials [46–48]. For example, magnetically less useful ferrihydrite can be replaced by iron sulfide or magnetite NPs (Figure 17.3) [46]. Magnetoferritin has also been used as a contrast agent for MR imaging, and possesses a reasonably high T2 relaxation value of 157 L mmol^{-1} s^{-1} [49]. Although magnetoferritin is expected to have high biocompatibility and colloidal stability in the blood (considering that it mimics naturally occurring ferritin), the results obtained have been contradictory. *In vivo*, the magnetoferritin particles are cleared rapidly from the blood (half-life < 10 min) by the reticuloendothelial system in the liver, spleen, and lymph nodes [49]. Thus, magnetoferritins are suitable only for liver, spleen, and lymph node detection rather than for molecular imaging.

17.2.1.3 Magnetodendrimers and Magnetoliposomes

The unique pore structures and multiple functional end-groups of dendrimers make them useful as host materials in drug and gene delivery. Similarly, dendrimers can efficiently deliver magnetic NPs to cells. Bulte, Frank, and co-workers have demonstrated the use of carboxy-terminated dendrimer (G = 4.5) -coated iron oxide contrast agents [50, 51]. Typically, magnetodendrimers are synthesized via pH-controlled reaction of a ferrous salt and a trimethylamine oxide oxidant in a methanol/water mixture containing polyamidoamine dendrimers (Figure 17.4a). The core size of the magnetodendrimers produced is 7–8 nm, and they tend to aggregate to oligomers with a size of 20–30 nm (Figure 17.4b). Magnetodendrimers show enhanced magnetic properties (saturation magnetism $\sim$94 emu g^{-1} Fe) and a high T2 relaxivity of 200 to 406 L mmol^{-1} s^{-1} [50], compared to those of dextran-coated MION. As dendrimers can be efficiently transfected to cells without the need for a transfection agent, these magnetodendrimers can be used as labelers for cellular MR imaging and trafficking [50].

Similarly, liposomes which are also widely used for drug and gene delivery can serve as good coating materials to solubilize iron oxide NPs. Liposomes have a

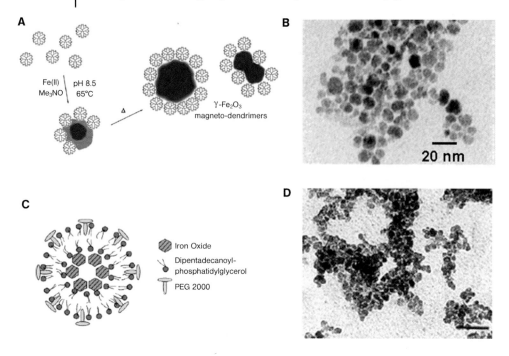

Fig. 17.4 Schematics and TEM images of (A,B) magnetodendrimers and (C,D) magnetoliposomes. (Reproduced with permission, from Refs. [51, 52].)

bilayer assembly of surfactant molecules with a hydrophlic head and a hydrophobic tail. As shown in Figure 17.4c, the hydrophilic ends of the inner layer surfactants encapsulate the iron oxide NPs, and the hydrophilic heads of the outer layer surfactant make them soluble in water. Bulte, Frank, and co-workers have reported that such magnetoliposomes can be used as bone marrow MR contrast agents [52]. The iron oxide core size of the magnetoliposomes is ~16 nm and the entire size is ~40 nm (Figure 17.4d) with a T2 relaxivity of ~240 L mmol^{-1} s^{-1}.

17.2.1.4 Non-Hydrolytically Synthesized High-Quality Iron Oxide NPs: A New Type of Contrast Agent

With the exception of MION and CLIO, previously developed iron oxide MR contrast agents undergo rapid uptake by the RES, and thus are effective for liver, spleen, and lymph node detection. On the other hand, difficulties have been encountered when these agents are used for molecular MR imaging [16–20]. In order for molecular imaging to be successful, it is necessary to have magnetic NP systems which exhibit excellent magnetic properties, are able to escape from the RES, and possess an active functionality that can be linked with biologically active molecules [10]. As the magnetic properties of NPs depend heavily on the material's properties – such as size, shape, stoichiometry, and crystallinity [15, 53, 54] – it is critical that such properties can be optimized. Unfortunately, conventional water-phase protocols – which have been widely used for superparamagnetic iron oxide (SPIO) contrast agents – generally lack precise size-controllability and monodispersity, and provide relatively poor crystallinity and wide stoichiometric composition [34]. In contrast, non-hydrolytic, high-temperature growth methods allow one to achieve size-controllability, and to produce high single crystallinity and good stoichiometry [15, 54, 56, 57]. For example, the NP size can easily be controlled from 4 nm to ~20 nm with a very narrow size distribution ($\sigma < 8\%$) by varying the growth conditions [15, 54]. One difficulty that must be overcome prior to the use of these NPs as MR contrast agents is to achieve water-solubility, as non-hydrolytically synthesized iron oxide NPs are soluble only in organic media. Various surface-modification methods have been developed, including bifunctional ligand [15, 33, 58], micellular [59, 60], polymer [61–63], and siloxane-linking procedures [64, 65]. For example, non-hydrolytically synthesized NPs can be transferred to aqueous media by overcoating the NPs with polyethylene glycol-(PEG)-ylated phospholipid micelles. Such a micellular coating strategy has been demonstrated in the case of quantum dots [66], and Bao and co-workers have successfully extended this strategy to produce water-soluble iron oxide NPs (Figure 17.5a) [60]. The PEGylated NPs can be further linked to cellular transfection Tat peptides and used for MR cellular labeling [60].

Bawendi and co-workers have proposed another approach to transfer iron oxide NPs from organic to aqueous media by coating them with polymeric phosphine oxide ligands [67] that bind tightly to the iron oxide NP surface through multi-dendate bondings (Figure 17.5b,c).

It is well known that the siloxane linkage to a metal oxide surface is efficient and strong, and Zhang and colleagues have successfully applied this strategy for the synthesis of water-soluble iron oxide NPs [68]. Refluxing toluene solution containing triethoxysilyl-terminated PEG ligands and non-hydrolytically synthesized iron oxide NPs provides iron oxide NPs with high colloidal stability in aqueous media (Figure 17.5d).

The major advantage of these non-hydrolytic-synthesized iron oxide NPs, as mentioned above, is the precise size control with high monodispersity. Cheon, Suh, and colleagues have demonstrated such advantages for the synthesis of iron oxide MR contrast agents [15, 33, 58]. As shown in TEM images (Figure 17.6a), the NPs obtained are ~9 nm, with a narrow size distribution ($\sigma < 8\%$). High-resolution TEM and X-ray analyses have shown the NPs to be single crystalline stoichiometric Fe_3O_4. Water-soluble iron oxide (WSIO) NPs are then obtained by introducing the 2,3-dimercaptosuccinic acid (DMSA) ligand onto the NP surface. This ligand endows the particles with high water-phase stability through: (i) carboxylate chelate bonding to iron; and (ii) disulfide cross-linkages between the ligands (Figure 17.6b) [15]. Furthermore, the remaining free thiol group of the ligand can be used for the attachment of target-specific biomolecules. The Fe_3O_4 nanocrystals obtained with the DMSA ligand are fairly stable in water

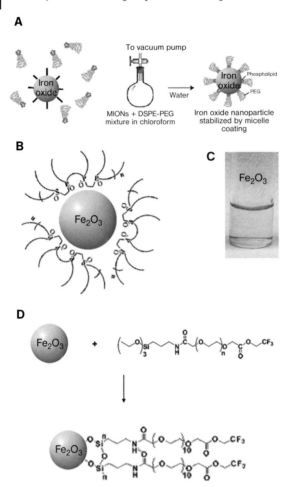

Fig. 17.5 (A) Synthetic scheme of polyethylene glycol (PEG)-ylated iron oxide NPs. (Reproduced with permission, from Ref. [60].) (B) Multi-dentate phosphine oxide ligand approach for the synthesis of iron oxide NPs. The phosphine oxide functional groups bind to the surface of iron oxide, and exposed PEG groups make them water-soluble. (C) Iron oxide NPs dissolved in water. (Reproduced with permission, from Ref. [67].) (D) Siloxane-polyethylene glycol (PEG)-coated iron oxide NPs. Silanization of the terminal ethoxysilane group of the PEG ligand on top of iron oxide NPs induces the formation of PEG-coated iron oxide NPs. (Reproduced with permission, from Ref. [68].)

(Figure 17.6c) and phosphate-buffered saline (PBS) up to a NaCl concentration of 250 mM, without aggregation. These NPs are used as MR probes, upon conjugation with cancer-targeting antibodies, not only for the *in-vitro* detection of cancer cells but also for *in-vivo* imaging of cancer implanted in mice [15, 33] (see Section 17.2.2.5).

A

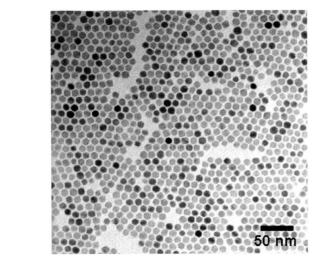

B

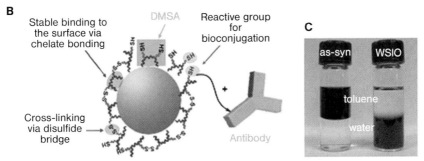

Fig. 17.6 (A) TEM image of ∼9 nm Fe₃O₄ NPs. (B) Schematic of 2,3-dimercaptosuccinic acid (DMSA)-coated iron oxide NPs. The carboxylic ends of DMSA bind to the surface iron oxide NPs, and are further stabilized through interligand disulfide cross-linkages. Remaining free thiol can be used for further conjugation for biomolecules such as antibodies. (C) Solubility test of as-synthesized and DMSA-coated iron oxide NPs. (Reproduced with permission, from Ref. [15].) WSIO = water-soluble iron oxide.

17.2.2
Iron Oxide NPs in Molecular MR Imaging

When iron oxide NPs are conjugated with biologically active materials (e.g., antibodies), the resulting iron oxide–biomolecule conjugates possess dual functionalities of both the MR contrast enhancers and the molecular recognition capability. These conjugates act as molecular imaging probes which can efficiently report on various molecular/biological events occurring in region-of-interest targets (Figure 17.7). Molecular MR studies utilizing such iron oxide–biomolecule conjugates include the imaging of inflammation [69, 70], infarction [23–25], angiogenesis

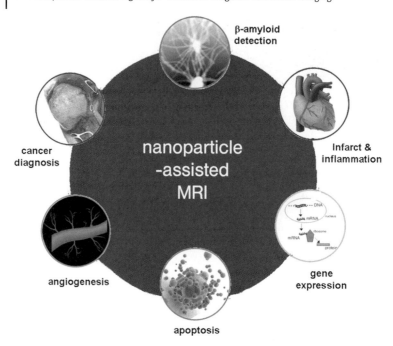

Fig. 17.7 Nanoparticle-assisted molecular MR imaging of biological systems.

[26], apoptosis [27], gene expression [28, 29], β-amyloid plaques [71], and cancer [15, 30–33].

17.2.2.1 Infarction and Inflammation

For the imaging of infarctions and inflammations, MION NPs are conjugated with specific antibodies through electrostatic interactions, or via covalent linkages by the reaction of potassium periodate-activated surface hydroxyl groups with lysine residues of the antibodies. For example, $R_{11}D_{10}$ antimyosin Fab was conjugated electrostatically to hydroxyl groups on the MION surfaces for cardiac infarct imaging [24]. As infarcted cardiac cells possess increased membrane porosity compared with normal cells, iron oxide–antimyosin Fab conjugates can efficiently be transported into damaged cells and recognize myosin. Figure 17.8a,b shows T2-weighted MR images of a mouse with a cardiac infarction after the injection of MION-$R_{11}D_{10}$ antimyosin Fab conjugates. The infarcted region is clearly observed as dark MR images, but no contrast effect was seen when unconjugated MIONs were administered. Such a targeting effect of MION-$R_{11}D_{10}$ antimyosin Fab conjugates was evaluated through *ex-vivo* immunohistological analyses using Prussian blue staining. Weissleder and co-workers further extended this strategy for the detection of inflammation by conjugating MIONs with polyclonal human immunoglobulin G. MION–IgG conjugates consistently detected the area of inflammation in T2-weighted spin-echo MR images (Figure 17.8c,d), and this was further confirmed histologically by using Prussian blue staining [69].

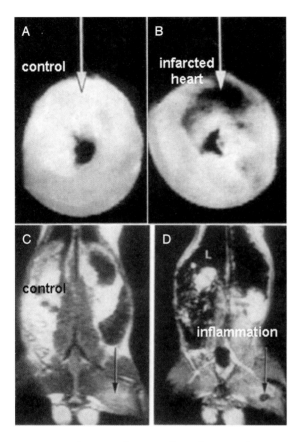

Fig. 17.8 MR detection of (A,B) cardiac infarction and (C,D) inflammation in a mouse. Following the intravenous injection of iron oxide-antimyosin antibody, a dark MR contrast is observed in the infarcted area. Similarly, inflammation is not imaged in the control mouse, but the inflammation site is clearly shown as dark contrast. (From Ref. [24] & [69], with permission)

17.2.2.2 Angiogenesis

Angiogenesis is the fundamental growth process of new blood vessels for development, reproduction, and wound repair. This process is also related to the progression of tumor growth. Therefore, the development of anti-angiogenic agents might represent a potential pathway to efficient cancer treatment. The imaging of angiogenesis is also related to cancer diagnosis and the evaluation of anti-cancer agents. Several molecular markers are involved in angiogenesis, including vascular endothelial growth factor (VEGF), fibroblast growth factors (FGF), platelet-derived endothelial cell growth factor (PD-EDGF), Tie-2 receptor, integrin, and E-selectin [72]. Among the various angiogenesis markers, VEGF, Tie 2 receptor, integrin, and E-selectin have been extensively studied [73–75]. Bogdanov and col-

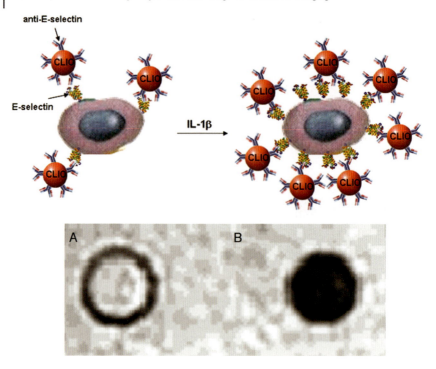

Fig. 17.9 *In-vitro* MR detection of E-selectin stimulated by interleukin-1β (IL-1β). (A) Without IL-1b, a white MR image is obtained from human umbilical vein endothelial cells (HUVEC) treated only with CLIO-anti E-selectin antibody. (B) In contrast, after sequential dosing of IL-1β and CLIO-anti E-selectin to HUVEC, E-selectin expression is clearly imaged as a dark MR region. (Reproduced with permission, from Ref. [26].)

leagues reported that E-selectin expression in human endothelial cells can be imaged by using a CLIO–monoclonal anti-human E-selectin antibody conjugate MR contrast agent [26]. When only CLIO–antibody conjugates are applied to endothelial cells with a low E-selectin expression level, the MR contrast effect is hardly observed (Figure 17.9a). In contrast, when interleukin-1β (which stimulates the expression of E-selectin) is applied to the cells and the NP–antibody is administered, a significant MR contrast effect is apparent (Figure 17.9b).

17.2.2.3 Apoptosis

Apoptosis is an active process for the programmed self-destruction of cells. During the early stages of apoptosis, there is a redistribution of phosphatidylserine in the cell membrane, and the detection of such processes can serve as indicators of apoptosis. Representative binding proteins to phosphatidylserine include annexin V and synaptotagmin I. Although the imaging of apoptosis using these antibodies has been carried out previously using radio-isotopic techniques [76], the spatial resolution is only ~1–3 mm and further improvement is needed. Brindle and col-

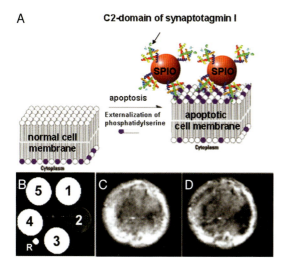

Fig. 17.10 (A) MR Imaging of apoptosis using SPIO-C2 domain of synaptotagmin I. (B) T2-weighted MR images of: (B, 1) water; (B, 2) SPIO-C2 conjugate treated apoptotic cells; (B, 3) SPIO-BSA control conjugate treated apoptotic cells; (B, 4) SPIO-only treated apoptotic cells; and (B, 5) SPIO-C2 conjugate-treated normal cells. *In-vivo* MR images of tumors implanted in a mouse before (C) and after (D) drug treatment. (Reproduced with permission, from Ref. [27].)

leagues have shown that conjugates of the SPIO and C2 domain of synaptotagmin I (C2-SPIO) can detect apoptotic cells through MRI with ~0.1 mm resolution (Figure 17.10a) [27]. Whilst there are no MR signals for nonspecific SPIO (BSA-SPIO) (Figure 17.10b(3)), SPIO-only treated apoptotic cells (Figure 17.10b(4)), and C2-SPIO-treated normal cells (Figure 17.10b(5)), C2-SPIO-treated apoptotic cells (Figure 17.10b(2)) clearly show dark MR contrast with a significant change in T2 values ($\Delta T2 = \sim 90\%$). Further extension of this strategy to an *in-vivo* animal study was also successful. When C2-SPIO was injected intravenously into drug-treated, tumor-bearing mice, the NP conjugates were able to detect apoptotic regions with a significant MR signal change (Figure 17.10c,d).

17.2.2.4 Gene Expression

Gene expression is an emerging field in the biomedical sciences, and the imaging of such processes is very important. Although several approaches to detect *in-vivo* gene expression have been performed through optical [77, 78] and radioisotopic imaging techniques [79], a number of limitations have been apparent, including the low-penetration depth of light for optical imaging and the poor spatial resolution of radioisotopic imaging. Weissleder and colleagues showed that the MR detection of transgene expression of engineered transferrin receptor (ETR) in tumors was possible by using MION-transferrin (MION-Tf) contrast agents (Figure 17.11a) [29]. When MION-Tf is applied to cells with various ETR expression levels, a gradual decrease in T2 was observed as the ETR expression levels of cells

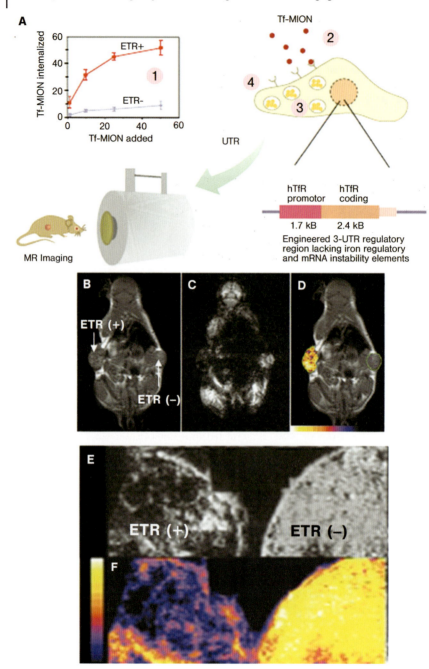

Fig. 17.11 *In-vivo* and *ex-vivo* imaging of engineered transferrin receptor (ETR) expression in tumors. (A) Schematic of MR imaging; (B–D) *in-vivo* imaging of ETR(+) and ETR(−) tumor-implanted mouse before (B) and after (C,D) the injection of MION-transferrin. The ETR(+) tumor is color-coded based on T2 (D). *Ex-vivo* imaging of excised ETR(+) and ETR(−) tumors (E) and their color maps based on T2 (F). (Reproduced with permission, from Ref. [29].)

was increased due to proportional binding of MION-Tf conjugates to the expressed ETR. These authors also determined whether ETR expression could be detected *in vivo* in live mice with ETR-positive and ETR-negative tumors. The results showed that the MR contrast effect was observed only for ETR-positive tumors (Figure 17.11b–d). *Ex-vivo* MR imaging of excised tumors showed more dramatic differences between the two tumors (Figure 17.11e,f).

17.2.2.5 Cancer Imaging

The noninvasive detection of cancer in its early stages is of great interest, as it can significantly increase the survival rate of patients. With conventional MRI techniques, detectable size of cancer is approximately ∼1 cm^3. However, if NP contrast agents can specifically recognize cancer cells through molecular interaction, then selective enhancement of the MR signal of cancer cells might provide a means of clearly distinguishing smaller sized cancer from normal tissues. Tiefenauer and co-workers suggested that the detection of cancers might be possible through such molecular recognition of SPIO-NP–antibody conjugates [32]. Conjugation of poly(glutamic acid-lysine-tyrosine)-coated NPs with anti-carcinoembryonic antigen (CEA) antibody is performed through a sulfo-MBS cross-linking method. In the T2-weighted MR images, dark contrast is imaged in the CEA-expressed tumors, but the contrast difference was not highly pronounced (Figure 17.12a,b). Artemov and colleagues then used another approach to detect cancer cells [31]. As avidin–biotin interaction is known to be very strong, avidin-conjugated SPIO can efficiently detect biotinylated cancer-specific antibodies which bind to the cancer cells (Figure 17.12c). The *in-vitro* fluorescence-assisted cell-sorting analyses and MR imaging conducted by these authors confirmed cancer detection (Figure 17.12d). Here, Au-565 cells with a high expression of HER2/neu cancer markers were imaged as dark MR contrast, whereas no contrast effect was obtained from MDA-MB-231 cells with low HER2/neu expression.

Recently, Cheon, Suh, and co-workers showed that highly sensitive cancer targeting could be achieved by using high-quality, small-sized water-soluble iron oxide (WSIO) NP–antibody conjugates [33]. The WSIO NPs have a high magnetic momentum (∼100 emu g^{-1} Fe) and small hydrodynamic size (∼9 nm) – factors which are advantageous for both *in-vitro* and *in-vivo* cancer imaging. When these NPs are conjugated with herceptin, they were able successfully to detect cancer cells (SK-BR-3) as dark MR images (Figure 17.13b) through molecular interaction between NP surface-bound herceptin and HER2/neu cancer markers, compared to non-treated (Figure 17.13a) and WSIO-irrelevant conjugate-treated cells (Figure 17.13c). This MRI result was confirmed by an optical technique where vivid green fluorescence from the fluorescein (FITC) was clearly observed only for FITC-WSIO-herceptin probe conjugate-treated cells (Figure 17.13d,e). Furthermore, WSIO NPs enabled the detection of various cell lines with different levels of HER2/neu cancer marker expression: Bx-PC-3, MDA-MB-231, BT-474, and NIH3T6.7 cell lines, which are arranged in the order of increasing HER2/*neu* expression level. T2-weighted MR signals of the cell lines treated with WSIO-herceptin probe conjugates become darker as the expression level of the HER2/

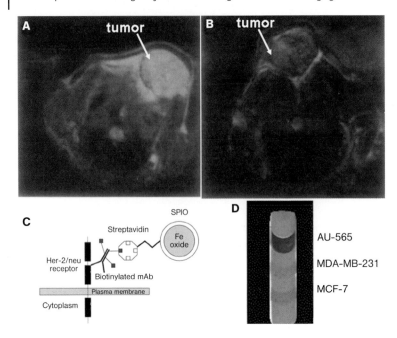

Fig. 17.12 (A,B) MR detection of carcinoembryonic antigen (CEA) overexpressed tumors by using iron oxide-CEA antibody conjugates. MR images made before (A) and after (B) injection of SPIO-CEA antibody. (Reproduced with permission, from Ref. [32].) (C,D) *In-vitro* MR detection of HER2/neu overexpressed cancer cells by using avidin-coated SPIO and biotinylated herceptin. (C) Schematic and (D) MR images of SPIO-avidin conjugate-treated cells (AU-565, MDA-MB-231, MCF-7). (Reproduced with permission, from Ref. [31].)

neu receptors increased (Figure 17.13f). It is noteworthy that the MR signal intensities of the cell lines treated with WSIO-herceptin probe conjugates showed a marked difference from that of control conjugates, indicating excellent specific binding efficiency of the probe conjugates.

These magnetic probes have been extended to the *in-vivo* detection of cancer cells implanted in mice [33]. When these WSIO-herceptin conjugates are injected intravenously in mice, they successfully reach and recognize HER2/neu receptors overexpressed from cancer cells, and this results in a significant MR contrast effect in the tumor sites, with a ~20% decrease in T2 value compared to the control experiments (Figure 17.14a–c). In high-resolution MR images of WSIO-herceptin conjugate-treated mice measured at 9.4 T MRI, a dark MR image initially appeared near the bottom region of the tumor and then gradually grew and spread to the central and upper regions of the tumor as time elapsed (Figure 17.14d). This time-dependent MR signal change was useful to reveal the heterogeneous pattern of the intratumoral vasculature, where the lower side of the tumor had well-developed vascular structures.

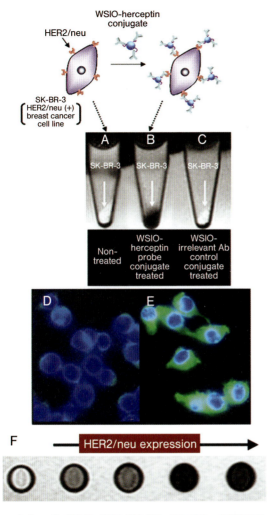

Fig. 17.13 *In-vitro* cancer detection using water-soluble iron oxide (WSIO)-herceptin conjugates. MR images of: (A) non-treated; (B) WSIO-herceptin-treated; (C) WSIO-irrelevant antibody-treated breast cancer cells (SK-BR-3). (D,E) Fluorescence images of (D) FITC-WSIO-irrelevant antibody-treated and (E) FITC-WSIO-herceptin-treated SK-BR-3 cell lines. (F) MR images of WSIO-herceptin conjugate-treated cell lines with increasing expression levels of HER2/*neu* receptors: Bx-PC-3, MDA-MB-231, BT-474, and NIH3T6.7 cell lines. Control conjugates were applied to Bx-PC-3 cell lines. (Reproduced with permission, from Ref. [33].)

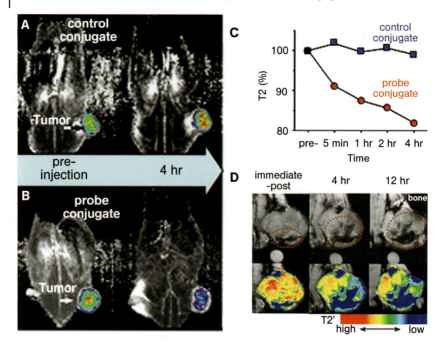

Fig. 17.14 *In-vivo* MRI of cancer targeting events of WSIO-herceptin conjugates. Color maps of T2-weighted MR images of cancer cell-implanted (NIH3T6.7) mice at different temporal points (pre-injection and 4 h) after the intravenous injection of: (A) WSIO-irrelevant antibody control conjugates; and (B) WSIO-herceptin probe conjugates. (C) Plot of T2 values versus time after injection of WSIO-antibody conjugates in (A) and (B) samples. (D) T2*-weighted MR images of cancer cell-implanted (NIH3T6.7) mouse at 9.4 Tesla and their color maps at different temporal points after probe conjugate injection. A dark MR image initially appears near the bottom region of the tumor and then gradually grows and spreads to the central and upper regions of the tumor as time elapses (red circles).

17.3
Outlook

Although much progress has been made in the development of magnetic NP contrast agents for molecular MR imaging during the past few years, their successful use is limited to *in-vitro* systems, except for a few *in-vivo* cases. Two main difficulties lie in the poor MR contrast effects and limited stability and biocompatibility under *in-vivo* conditions. The MR signal-enhancing effect of conventional iron oxide-based NPs is unsatisfactory compared to other diagnostic tools such as fluorescence and PET, and there is a clear need for improvement. Thus, it is important to develop new types of magnetic NP contrast agents which can significantly improve contrast effects. As NPs with higher magnetization values provide stronger MR contrast effects, the development of novel NPs with superior magnetism is the first prerequisite. Concurrently, as the MR contrast effects of

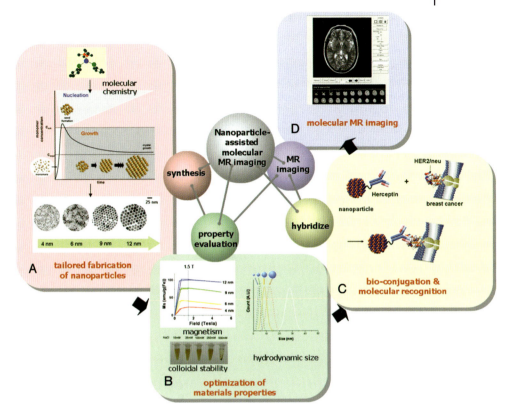

Fig. 17.15 NP-assisted molecular MR imaging. (A) The modern molecular chemistry approach enables the tailored synthesis of high-quality magnetic NPs with desired sizes and monodispersity. (B) Subsequent evaluation and optimization of the material's properties, such as magnetism, hydrodynamic size, and colloidal and biostabilities, are important. (C) When these NPs are conjugated to biomolecules, the resulting NP-biomolecule conjugates possess the capabilities of both MR contrast effects and molecular recognition of target biosystems. (D) This will enable molecular MR imaging which can identify biological events such as pinpointing cancer diagnosis, cell migration, cell signaling, and genetic developments, with high sensitivity and specificity.

NPs are strongly correlated to the materials' characteristics in terms of their size, shape, composition, single crystallinity, and magnetism, it is important to develop a reasonable NP model system which can clearly describe the relationship between nanoscale materials characteristics and MR contrast effects (Figure 17.15a).

The next major step is to impart high colloidal stability and biocompatibility to the magnetic NPs. Although a variety of coating materials has been developed for such applications, it remains necessary to develop general and more reliable protocols for tailoring NP surfaces with desired coating materials. As a smaller overall size is advantageous for escaping the RES, the coating materials should be as small as possible whilst possessing high colloidal stability, without any aggregation under physiological conditions (Figure 17.15b).

The toxicity of magnetic NPs is also a vital issue that must be resolved in prior to clinical utilization. Although iron oxide NPs have in the past been regarded as clinically benign materials, the potential cytotoxicity arising from their size, shape, and coating materials should be examined, along with systematic guidelines for the nanotoxicity evaluation of newly developed NPs.

In time, when novel NPs with highly enhanced magnetism, small size, and high colloidal and biostability have been developed, significant improvements in MR detection sensitivity and target specificity are expected (Figure 17.15c,d), and this will undoubtedly bring about major advances in cancer diagnosis and biomedical imaging. For example, the highly enhanced MR contrast of a biological target through the molecular recognition of such NP contrast agents should provide an *in-vivo* diagnosis of early-stage cancer, with submillimeter resolution. In addition, many presently unrevealed biological processes – for example, *in-vivo* pathways of cell evolution, cell differentiation, cell-to-cell interactions, and molecular signaling pathways – will most likely be monitored with precision, at the molecular level, by using next-generation NP contrast agents.

References

1 Weissleder, R., Mahmood, U. (2001) Molecular imaging. *Radiology* **219**, 316–333.

2 Massoud, T. F., Gambhir, S. S. (2003) Molecular imaging in living subjects: seeing fundamental biological processes in a new light. *Gene Dev.* **17**, 545–580.

3 Michalet, X., Pinaud, F. F., Bentolila, L. A., Tsay, J. M., Doose, S., Li, J. J., Sundaresan, G., Wu, A. M. S., Gambhir, S., Weiss, S. (2005) Quantum dots for live cells, in vivo imaging, and diagnostics. *Science* **307**, 538–544.

4 LaVan, D. A., Lynn, D. M., Langer, R. (2002) Moving smaller in drug discovery and delivery. *Nat. Rev. Drug. Discov.* **1**, 77–84.

5 Whaley, S. R., English, D. S., Hu, E. L., Barbara, P. F., Belcher, A. M. (2000) Selection of peptides with semiconductor binding specificity for directed nanocrystal assembly. *Nature* **405**, 665–668.

6 Bruchez, Jr., M., Moronne, M., Gin, P., Weiss, S., Alivisatos, A. P. (1998) Semiconductor nanocrystals as fluorescent biological labels. *Science* **281**, 2013–2016.

7 Chan, W. C. W., Nie, S. (1998) Quantum dot bioconjugates for ultrasensitive nonisotopic detection. *Science* **281**, 2016–2018.

8 LaConte, L., Nitin, N., Bao, G. (2005) Magnetic nanoparticle probes. *Nanotoday* **8**, 32–38.

9 Wu, X., Liu, H., Liu, J., Haley, K. N., Treadway, J. A., Larson, J. P., Ge, N., Peale, F., Bruchez, M. P. (2003) Immunofluorescent labeling of cancer marker Her2 and other cellular targets with semiconductor quantum dots. *Nat. Biotechnol.* **21**, 41–46.

10 Bulte, J. W. M., Kraitchman, D. L. (2004) Iron oxide MR contrast agents for molecular and cellular imaging. *NMR Biomed.* **17**, 484–499.

11 Wang, Y.-X. J., Hussain, S. M., Krestin, G. P. (2001) Superparamagnetic iron oxide contrast agents: physicochemical characteristics and applications in MR imaging. *Eur. Radiol.* **11**, 2319–2331.

12 Han, M., Gao, X., Su, J. Z., Nie, S. (2001) Quantum-dot-tagged

microbeads for multiplexed optical coding of biomolecules. *Nat. Biotechnol.* **19**, 631–635.

13 Mitchell, D. G. (1999) *MRI Principles.* W.B. Saunders Company, Philadelphia.

14 Koenig, S. H., Keller, K. E. (1995) Theory of 1/T1 and 1/T2 NMRD profiles of solutions of magnetic nanoparticles. *Magn. Reson. Med.* **34**, 227–233.

15 Jun, Y.-W., Huh, Y.-M., Choi, J.-S., Lee, J.-H., Song, H.-T., Kim, S. J., Yoon, S., Kim, K.-S., Shin, J.-S., Suh, J.-S., Cheon, J. (2005) Nanoscale size effect of magnetic nanocrystals and their utilization for cancer diagnosis via magnetic resonance imaging. *J. Am. Chem. Soc.* **127**, 5732–5733.

16 Hamm, B., Staks, T., Taupitz, M., Maibauer, R., Speidel, A., Huppertz, A., Frenzel, T., Lawaczeck, R., Wolf, K. J., Lange, L. (1994) Contrast-enhanced MR imaging of liver and spleen: First experience in humans with a new superparamagnetic iron oxide. *J. Magn. Reson. Imaging* **4**, 659–668.

17 Reimer, P., Tombach, B. (1998) Hepatic MRI with SPIO: detection and characterization of focal liver lesions. *Eur. Radiol.* **8**, 1198–1204.

18 McLachlan, S. J., Morris, M. R., Lucas, M. A., Fisco, R. A., Eakins, M. N., Fowler, D. R., Scheetz, R. B., Olukotun, A. Y. (1994) Phase I clinical evaluation of a new iron oxide MR contrast agent. *J. Magn. Reson. Imaging* **4**, 301–307.

19 Weissleder, R. (1994) Liver MR imaging with iron oxides: toward consensus and clinical practice. *Radiology* **193**, 593–595.

20 Bengele, H. H., Palmacci, S., Rogers, J., Jung, C. W., Crenshaw, J., Josphson, L. (1994) Biodistribution of an ultrasmall superparamagnetic iron oxide colloid, BMS 180549, by different routes of administration. *Magn. Reson. Imaging* **12**, 433–442.

21 Suwa, T., Ozawa, S., Ueda, M., Ando, N., Kitajima, M. (1998) Magnetic resonance imaging of esophageal squamous cell carcinoma using magnetite particles coated with anti-epidermal growth factor receptor antibody. *Int. J. Cancer* **75**, 626–634.

22 Kresse, M., Wagner, S., Pfefferer, D., Lawaczeck, R., Elste, V., Semmler, W. (1998) Targeting of ultrasmall super-paramagnetic iron oxide (USPIO) particles to tumor cells in vivo by using transferrin receptor pathways. *Magn. Reson. Med.* **40**, 236–242.

23 Krieg, F. M., Andres, R. Y., Winter-halter, K. H. (1995) Superpara-magnetically labelled neutrophils as potential abscess-specific contrast agent for MRI. *Magn. Reson. Imaging* **13**, 393–400.

24 Weissleder, R., Lee, A. S., Khaw, B. A., Shen, T., Brady, T. J. (1992) Antimyosin-labeled monocrystalline iron oxide allows detection of myocardial infarct: MR antibody imaging. *Radiology* **182**, 381–385.

25 Kraitchman, D. L., Heldman, A. W., Atalar, E., Amado, L. C., Martin, B. J., Pittenger, M. F., Hare, J. M., Bulte, J. W. M. (2003) In vivo magnetic resonance imaging of mesenchymal stem cells in myocardial infarction. *Circulation* **107**, 2290–2293.

26 Kang, H. W., Josephson, L., Petrovsky, A., Weissleder, R., Bogdanov, Jr., A. (2002) Magnetic resonance imaging of inducible e-selectin expression in human endothelial cell culture. *Bioconj. Chem.* **13**, 122–127.

27 Zhao, M., Beauregard, D. A., Loizou, L., Davletov, B., Brindle, K. M. (2001) Non-invasive detection of apoptosis using magnetic resonance imaging and a targeted contrast agent. *Nat. Med.* **7**, 1241–1244.

28 Högemann, D., Josephson, L., Weissleder, R., Basilion, J. P. (2000) Improvement of MRI probes to allow efficient detection of gene expression. *Bioconj. Chem.* **11**, 941–946.

29 Weissleder, R., Moore, A., Mahmood, U., Bhorade, R., Benveniste, H., Chiocca, E. A., Basilion, J. P. (2000) In vivo magnetic resonance imaging of transgene expression. *Nat. Med.* **6**, 351–355.

30 Tiefenauer, L. X., Kühne, G., Andres, R. Y. (1993) Antibody-magnetite

nanoparticles: In vitro characterization of a potential tumor-specific contrast agent for magnetic resonance imaging. *Bioconj. Chem.* **4**, 347–352.

31 Artemov, D., Mori, N., Okollie, B., Bhujwalla, Z. M. (2003) MR molecular imaging of the Her-2/neu receptor in breast cancer cells using targeted iron oxide nanoparticles. *Magn. Reson. Med.* **49**, 403–408.

32 Tiefenauer, L. X., Tschirky, A., Iwhne, G., Andres, R. Y. (1996) In vivo evaluation of magnetite nanoparticles for use as a tumor contrast agent in MRI. *Magn. Reson. Imaging* **14**, 391–402.

33 Huh, Y.-M., Jun, Y.-W., Song, H.-T., Kim, S. J., Choi, J.-S., Lee, J.-H., Yoon, S., Kim, K.-S., Shin, J.-S., Suh, J.-S., Cheon, J. (2005) In vivo magnetic resonance detection of cancer by using multifunctional magnetic nanocrystals. *J. Am. Chem. Soc.* **127**, 12387–12391.

34 Sjögren, C. E., Johansson, C., Naevestad, A., Sontum, P. C., Briley-Saebo, K., Fahlvik, A. K. (1997) Crystal size and properties of superparamagnetic iron oxide (SPIO) particles. *Magn. Reson. Imaging* **15**, 55–67.

35 Hahn, P. F., Stark, D. D., Lewis, J. M., Saini, S., Elizondo, G., Weissleder, R., Fretz, C. J., Ferrucci, J. T. (1990) First clinical trial of a new superparamagnetic iron oxide for use as an oral gastrointestinal contrast agent in MR imaging. *Radiology* **175**, 695–700.

36 Bach-Gansmo, T. (1993) Ferrimagnetic susceptibility contrast agents. *Acta. Radiol. Suppl.* **387**, 1–30.

37 Weissleder, R., Elizondo, G., Josephson, L., Compton, C. C., Fretz, C. J., Stark, D. D., Ferrucci, J. T. (1989) Superparamagnetic iron oxide-enhanced MR imaging: pulse sequence optimization for detection of liver cancer. *Radiology* **171**, 835–839.

38 Weissleder, R., Stark, D. D., Engelastad, B. L., Bacon, B. R., Compton, C. C., White, D. L., Jacobs, P., Lewis, J. (1989) Superpara-

magnetic iron oxide: pharmacokinetics and toxicity. *Am. J. Roentgenol.* **152**, 167–173.

39 Reimer, P., Rummeny, E. J., Daldrup, H. E., Balzer, T., Tombach, B., Berns, T., Peters, P. E. (1995) Clinical results with Resovist: a phase 2 clinical trial. *Radiology* **195**, 489–496.

40 Grubnic, S., Padhani, A. R., Revell, P. B., Husband, J. E. (1999) Comparative efficacy of and sequence choice for two oral contrast agents used during MR imaging. *Am. J. Roentgenol.* **173**, 173–178.

41 Bartolozzi, C., Lencioni, R., Donati, F., Cioni, D. (1999) Abdominal MR: liver and pancreas. *Eur. Radiol.* **9**, 1496–1512.

42 Moore, A., Marecos, E., Bogdanov, A., Weissleder, R. (2000) Tumoral distribution of long-circulating dextran-coated iron oxide nanoparticles in a rodent model. *Radiology* **214**, 568–574.

43 Palmacci, S., Josephson, L. (1993) Synthesis of polysaccharide covered superparamagnetic oxide colloids. *U. S. Patent*, 5,262,176.

44 Shen, T., Weissleder, R., Papisov, M., Bogdanov, A., Brady, T. J. (1993) Monocrystalline iron oxide nanocompounds (MION): Physicochemical properties. *Magn. Reson. Med.* **29**, 599–604.

45 Ford, G. C., Harrison, P. M., Rice, D. W., Smith, J. M. A., Treffry, A., White, J. L., Yariv, J. (1984) Ferritin: Design and formation of an iron-storage molecule. *Philos. Trans. R. London Ser. B*, **304**, 551–565.

46 Meldrum, F. C., Heywood, B. R., Mann, S. (1992) Magnetoferritin: in vitro synthesis of a novel magnetic protein. *Science* **257**, 522–523.

47 Bulte, J. W. M., Douglas, T., Mann, S., Frankel, R. B., Moskowitz, B. M., Brooks, R. A., Baumgarner, C. D., Vymazal, J., Strub, M.-P., Frank, J. A. (1994) Magnetoferritin: Characterization of a novel superparamagnetic MR contrast agent. *J. Magn. Reson. Imaging* **4**, 497–505.

48 Meldrum, F. C., Wade, V. J., Nimmo, D. L., Heywood, B. R., Mann, S.

(1991) Synthesis of inorganic nano-phase materials in supramolecular protein cages. *Nature* **349**, 684–687.

49 Bulte, J. W. M., Douglas, T., Mann, S., Vymazal, J., Laughlin, P. G., Frank, J. A. (1995) Initial assessment of magnetoferritin biokinetics and proton relaxation enhancement in rats. *Acad. Radiol.* **2**, 871–878.

50 Bulte, J. W. M., Douglas, T., Witwer, B., Zhang, S.-C., Strable, E., Lewis, B. K., Zywicke, H., Miller, B., Gelderen, P., Moskowitz, B. M., Duncan, I. D., Frank, J. A. (2001) Magnetodendrimers allow endosomal magnetic labeling and in vivo tracking of stem cells. *Nat. Biotechnol.* **19**, 1141–1147.

51 Strable, E., Bulte, J. W. M., Moskowitz, B., Vivekanandan, K., Allen, M., Douglas, T. (2001) Synthesis and characterization of soluble iron oxide-dendrimer composites. *Chem. Mater.* **13**, 2201–2209.

52 Bulte, J. W. M., De Cuyper, M., Despres, D., Frank, J. A. (1999) Short- vs. long-circulating magnetolipo-somes as bone marrow-seeking MR contrast agents. *J. Magn. Reson. Imaging* **9**, 329–335.

53 Cheon, J., Kang, N.-J., Lee, S.-M., Lee, J.-H., Yoon, J.-H., Oh, S. J. (2004) Shape evolution of single-crystalline iron oxide nanocrystals. *J. Am. Chem. Soc.* **126**, 1950–1951.

54 Sun, S., Zeng, H. (2002) Size-controlled synthesis of magnetite nanoparticles. *J. Am. Chem. Soc.* **124**, 8204–8205.

55 Redl, F. X., Black, C. T., Papaefthymiou, G. C., Sandstrom, R. L., Yin, M., Zeng, H., Murray, C. B., O'Brien, S. P. (2004) Magnetic, electronic, and structural characteriza-tion of non-stoichiometric iron oxides at the nanoscale. *J. Am. Chem. Soc.* **126**, 14583–14599.

56 Park, J., An, K., Hwang, Y., Park, J.-G., Noh, H.-J., Kim, J.-Y., Park, J.-H., Hwang, N.-M., Hyeon, T. (2004) Ultra-large-scale syntheses of monodisperse nanocrystals. *Nat. Mater.* **3**, 891–895.

57 Jana, N. R., Chen, Y., Peng, X. (2004) Size- and shape-controlled magnetic (Cr, Mn, Fe, Co, Ni) oxide nano-crystals via a simple and general approach. *Chem. Mater.* **16**, 3931–3935.

58 Song, H.-T., Choi, J.-s., Huh, Y.-M., Kim, S., Jun, Y.-w., Suh, J.-S., Cheon, J. (2005) Surface modulation of magnetic nanocrystals in the development of highly efficient magnetic resonance probes for intracellular labeling. *J. Am. Chem. Soc.* **127**, 9992–9993.

59 Pileni, M.-P. (2003) The role of soft colloidal templates in controlling the size and shape of inorganic nanocrystals. *Nat. Mater.* **2**, 145–150.

60 Nitin, N., LaConte, L. E. W., Zurkiya, O., Hu, X., Bao, G. (2004) Functionalization and peptide-based delivery of magnetic nanoparticles as an intracellular MRI contrast agent. *J. Biol. Inorg. Chem.* **9**, 706–712.

61 Yee, C., Kataby, G., Ulman, A., Prozorov, T., White, H., King, A., Rafailovich, M., Sokolov, J., Gedanken, A. (1999) Self-assembled monolayers of alkanesulfonic and –phosphonic acids on amorphous iron oxide nanoparticles. *Langmuir* **15**, 7111–7115.

62 Harris, L. A., Goff, J. D., Carmichael, A. Y., Riffle, J. S., Harburn, J. J., St. Pierre, T. G., Saunders, M. (2003) Magnetite nanoparticle dispersions stabilized with triblock copolymers. *Chem. Mater.* **15**, 1367–1377.

63 Burke, N. A. D., Stover, H. D. H., Dawson, F. P. (2002) Magnetic nanocomposites: Preparation and characterization of polymer-coated iron nanoparticles. *Chem. Mater.* **14**, 4752–4761.

64 Lu, Y., Yin, Y., Mayers, B. T., Xia, Y. (2002) modifying the surface properties of superparamagnetic iron oxide nanoparticles through a sol-gel approach. *Nano Lett.* **2**, 183–186.

65 Santra, S., Tapec, R., Theodoropoulou, N., Dobson, J., Hebard, A., Tan, W. (2001) Synthesis and characterization of silica-coated

iron oxide nanoparticles in micro-emulsion: The effect of nonionic surfactants. *Langmuir* **17**, 2900–2906.

66 Dubertret, B., Skourides, P., Norris, D. J., Noireaux, V., Brivanlou, A. H., Libchaber, A. (2002) In vivo imaging of quantum dots encapsulated in phospholipid micelles. *Science* **298**, 1759–1762.

67 Kim, S.-W., Kim, S., Tracy, J. B., Jasanoff, A., Bawendi, M. G. (2005) Phosphine oxide polymer for water-soluble nanoparticles. *J. Am. Chem. Soc.* **127**, 4556–4557.

68 Kohler, N., Fryxell, G. E., Zhang, M. (2004) A bifunctional poly(ethylene glycol) silane immobilized on metallic oxide-based nanoparticles for conjugation with cell targeting agents. *J. Am. Chem. Soc.* **126**, 7206–7211.

69 Weissleder, R., Lee, A. S., Fischman, A. J., Reimer, P., Shen, T., Wilkinson, R., Callahan, R., Brady, T. J. (1991) Polyclonal human immunoglobulin G labeled with polymeric iron oxide: antibody MR imaging. *Radiology* **181**, 245–249.

70 Schmitz, S. A., Taupitz, M., Wagner, S., Wolf, K.-J., Beyersdorff, D., Hamm, B. (2001) Magnetic resonance imaging of atherosclerotic plaques using superparamagnetic iron oxide particles. *J. Magn. Reson. Imaging* **14**, 355–361.

71 Wadghiri, Y. Z., Sigurdsson, E. M., Sadowski, M., Elliott, J. I., Li, Y., Scholtzova, H., Tang, C. Y., Aguinaldo, G., Pappolla, M., Duff, K., Wisniewski, T., Turnbull, D. H. (2003) Detection of Alzheimer's amyloid in transgenic mice using

magnetic resonance microimaging. *Magn. Reson. Med.* **50**, 293–302.

72 Risau, W. (1997) Mechanisms of angiogenesis. *Nature* **386**, 671–674.

73 Veikkola, T., Karkkainen, M., Claesson-Welsh, L., Alitalo, K. (2000) Regulation of angiogenesis via vascular endothelial growth factor receptors. *Cancer Res.* **60**, 203–212.

74 Kim, I., Kim, H. G., Si, J. N., Kim, J. H., Kwak, H. J., Koh, G. Y. (2000) Angiopoietin-1 regulates endothelial cell survival through the phosphati-dylinositol 3'-Kinase/Akt signal transduction pathway. *Circ. Res.* **86**, 24–29.

75 Saeed, M., Wendland, M. F., Engelbrecht, M., Sakuna, H., Higgins, C. B. (1998) Value of blood pool contrast agents in magnetic resonance angiography of the pelvis and lower extremities. *Eur. Radiol.* **8**, 1047–1053.

76 Lahorte, C. M., Vanderheyden, J. L., Steinmetz, N., Van de Wiele, C., Dierckx, R. A., Slegers, G. (2004) Apoptosis-detecting radioligands: current state of the art and future perspectives. *Eur. J. Nucl. Med. Mol. Imaging* **31**, 887–919.

77 Shah, K., Weissleder, R. (2005) Molecular optical imaging: applications leading to the development of present day therapeutics. *NeuroRx.* **2**, 215–225.

78 Gross, S., Piwnica-Worms, D. (2005) Spying on cancer: molecular imaging in vivo with genetically encoded reporters. *Cancer Cell* **7**, 5–15.

79 Haberkorn, U., Altmann, A. (2003) Functional genomics and radio-isotope-based imaging procedures. *Ann. Med.* **35**, 370–379.

18

Micro- and Nanoscale Control of Cellular Environment for Tissue Engineering

Ali Khademhosseini*, Yibo Ling*, Jeffrey M. Karp, and Robert Langer

18.1
Overview

Tissue engineering is a potentially powerful approach for restoring organ functionality and overcoming the shortage of transplantable organs. In tissue engineering, the principles of engineering and life sciences are used to develop biological substitutes, typically composed of biological and synthetic components that restore, maintain, or improve tissue function [1]. Although relatively simple tissues such as cartilage and skin have already been successfully engineered, many basic challenges persist in the engineering of more complex tissues. These challenges – which include the generation of vascularized tissues and complex geometries – can be traced to our limited abilities to control the cellular environment at micro- and nanoscale resolution. Cells in the body reside in an environment that is regulated by cell–cell, cell–extracellular matrix (ECM) and cell–soluble factor interactions presented in a spatially and temporally dependent manner (Figure 18.1). In order for tissue engineering to succeed, it is critical to reproduce these *in-vivo* factors outside the body. In this chapter we analyze the use of micro- and nanoscale engineering techniques for controlling and studying cell–cell, cell–substrate and cell–soluble factor interactions, as well as for fabricating organs with controlled architecture and resolution.

18.1.1
Cell–Substrate Interactions

Decisions such as cell growth, migration, and differentiation can all be affected by a cell's interaction with the surrounding surfaces. Numerous micro- and nano-engineering approaches have been used to control cell–substrate interactions *in vitro* through presenting specific molecules to cells. These molecules, which

---

*) These authors contributed equally to this chapter.

Nanobiotechnology II. Edited by Chad A. Mirkin and Christof M. Niemeyer
Copyright © 2007 WILEY-VCH Verlag GmbH & Co. KGaA, Weinheim
ISBN: 978-3-527-31673-1

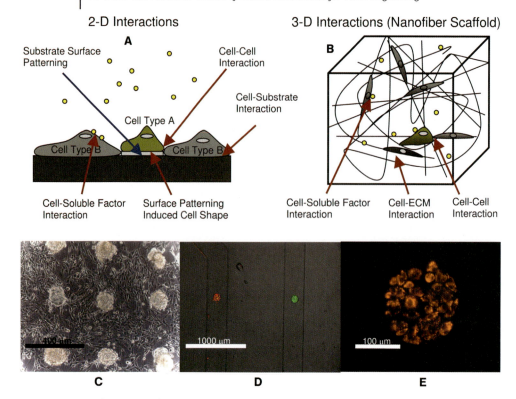

Fig. 18.1 Cell–microenvironment interactions for tissue engineering. The upper row depicts a schematic of interactions in (left) two dimensions (2-D) and (right) three dimensions (3-D). The lower row provides examples of: (c) patterned co-cultures; (d) cells within microfluidic channels; and (e) cells that are micropatterned on a substrate.

range from adsorbed protein, to engineered peptides, to non-adhesive polymers, can be used to engineer many cellular functions *in vitro*. For example, self-assembled monolayers of alkanethiols have been shown to affect the action potential generation capacity of neurons [2], as well as the differentiation and proliferation of myoblasts [3]. Self-assembled multi-layer structures with cholesteryl moieties have also been shown to improve fibroblast adhesion and spreading [4]. Alternatively, surfaces can be engineered to prevent protein adsorption and cell adhesion on a variety of surfaces [5].

Many biological processes such as cell migration, axon extension and angiogenesis are regulated by spatially dependent signals, including surface gradients of molecules. Traditional macroscale techniques have been limited in their ability to form spatially regulated patterns of molecules. Micro- and nanoengineering approaches can be used to generate *in-vitro* gradients of molecules on substrates to mimic natural environments. For example, surface gradients of laminin have been generated using a microfluidics system and have been used to study the extension of axons under controlled conditions [6]. Alternatively, microscale gradients of surface molecules [7] were engineered into planar surfaces by merging the gradient generation capacity of microfluidics and photopolymerization chemistry. Using this technique, gradients of the adhesive peptide RGDs (Arg-Gly-Asp-Ser) were formed and could be used to study cell adhesion [8]. The gradients of hydrogels with varying polymer concentrations can also be used to study the response of cells to surface stiffness.

Micro- and nano-textured substrates have been shown to significantly influence cell adhesion, gene expression [9–11], and migration [12]. Nanoengineered topographies can be incorporated into tissue engineering scaffolds to provide functional cues to cells. In fact, surface texture has already proven to be an important parameter in current orthopedic replacement/augmentation applications [13]. It is generally believed that implantable orthopedic materials must be hospitable for osteoblasts which deposit new bone matrix directly adjacent to the material. As such, many investigations in the field of orthopedics have focused on the fabrication of biomaterials which maximize cellular adhesion, and indeed it has been found that nanotextured materials which mimic the nanoscale features of bone surfaces are especially suitable for these applications [13]. It is likely that the superior adhesive properties of textured materials are due to the increased particle boundaries and increased surface area available for osteoblast adhesion. Such nanoscale features, which are typically less than 100 nm, may be created through a number of techniques, such as chemical etching and anodization. These techniques have proven to be especially efficient for generating surface roughness on metals such as titanium, and it has been shown that nanotextured surfaces prepared by chemical etching [14] induce significantly higher levels of metabolic activity.

Another approach for generating nanotopography is to embed nanoscale objects within biomaterials. For example, carbon nanofiber-embedded composites have been generated with increased osteoblast adhesion and decreased adhesion for other cell types [15]. The decreased adhesion for other cell types is especially advantageous in that it reduces the formation of disruptive fibrous tissue around the implant. Another alternative technique that is widely applicable to a range of materials (e.g., metals, ceramics, carbon fibers, composites) is the use of nanoparticles (NPs) as composite materials [16–19]. A detailed review of such technologies is beyond the scope of this chapter, but for the interested reader more comprehensive reviews of nanotextured materials [13] and cell–surface interactions [20] can be found elsewhere.

One potential advantage of micro- and nanoscale technologies is that they miniaturize experiments and can be used to perform high-throughput analysis. For example, robotic spotters can be used to perform high-throughput analysis of cell–substrate interactions. In these studies, stem cells were interfaced with thousands of different polymeric materials that were patterned using microarray technology. The effects were observed for human embryonic stem (ES) cells [21] and human mesenchymal stem cells (hMSCs) [22], and resulted in the identification of some surprising cell–material interactions. Although the specific biological mechanisms underlying these interactions are not known, this method can be used to identify novel and potentially useful biomaterials, and for monitoring unexpected cell responses. A variation of this approach has also been used to analyze cell differentiation in response to combinations of ECM molecules [23].

18.1.2
Cell Shape

The effect of cell shape on various cell fate decisions has been studied by immobilizing cells onto micropatterned substrates. Differences in the size and shape of the adhesive region cause cytoskeletal rearrangements, which have been shown to effect proliferation and apoptosis [5]. For example, smooth muscle cells and endothelial cells [24–26], when micropatterned on poly(lactic-co-glycolic acid) (PLGA) surfaces, and have been shown to provide a better maintenance of cell function and morphology. In addition, cell shape has been shown to direct stem cell differentiation. Human MSCs that were patterned on small and large fibronectin patterns differentiated differently based on the size of the patterns. Typically, small islands induced the cells to form spheres, while large islands induced them to flatten and to adhere to the surface. Staining for differentiation markers indicated that the spherical cells differentiated into adipocytes, while flattened cells became osteoblasts [27].

Within the body, cells are exposed to dynamic environments which may alter their shapes in response to changing mechanical forces. Typical patterning techniques generate fixed patterns which cannot be changed after the deposition of adhesive molecules. However, in order to study the dynamics of cell shape change there is a need for substrates that can be dynamically altered to regulate the adhesive cell environments. Towards this end, photoinitiated gels [28, 29] and thermally responsive polymer surfaces [30] have been developed. In addition, other techniques such as the electrochemical modulation of self-assembled monolayers [31, 32] have enabled the reversible switching of surface properties. An alternative dynamic patterning approach uses surface cracks to reversibly modulate the adhesive properties of the substrate at the scale of adhesion complexes [33]. By using this approach, adhesive signals can be patterned within nanocracks of a poly(dimethylsiloxane) (PDMS) substrate, and the availability of such signals can be reversibly modulated multiple times by varying the strain applied to the substrate. However, this technique is only capable of generating parallel nanoscale cracks, and is ineffective for the generation of more complicated nonlinear patterns. Clearly, many further studies must be carried out to develop approaches capable of generating dynamic patterns at the nanoscale with more flexibility and precision.

18.1.3
Cell–Cell Interactions

One of the biggest challenges in tissue engineering is to reproduce *in vitro* the specific arrangement of cells found *in vivo*. The co-culture of different cell types is one approach to artificially recreate these arrangements. For example, hepatocytes in co-culture with endothelial cells better maintain their differentiated phenotype [34]. Similarly, homotypic hepatocyte cell–cell interactions, such as hepatocyte spheroids, can be used to better maintain a differentiated phenotype [35]. However, the degree of cell–cell interactions in these co-cultures cannot easily be controlled without the use of microscale technologies. In an effort to control the degree of heterotypic and homotypic cell–cell interactions, micropatterned co-cultures have been used; these are patterned through the use of micropatterned adhesive substrates which selectively position various cell types relative to each other. Patterned co-cultures were initially used to study the cell–cell interactions between hepatocytes and non-parenchymal fibroblasts [36], and have led to important findings about the nature of hepatocyte and fibroblast interactions [37–39].

Other methods for generating patterned co-cultures utilize such techniques as thermally reversible polymers [40, 41], layer-by-layer deposition of ionic polymers [42], microfluidic deposition [43], and micromolding of hydrogels [44]. Although cells have recently been patterned [45] onto three-dimensional (3-D) scaffolds using replica printing techniques, further investigations will need to be carried out on the incorporation of such fundamentally two-dimensional (2-D) techniques into 3-D tissue engineering constructs.

Beyond patterning surfaces to generate co-cultures, another approach might be to physically confine cells and cell aggregates within defined spaces. It is known that characteristics such as cellular phenotypic expression and stem cell fate decisions may be affected both by physical interaction with surfaces and by the diffusion limitations introduced by confining cells and cell aggregates within spaces. The recent use of microwells [46] for generating arrays of many different cell types is a potentially useful tool for facilitating and studying the effect of culturing and co-culturing cell types within confined spaces. Beyond its applications for generating high-throughput multicellular arrays, this approach may also be amenable for generating co-cultures of many cell types.

18.1.4
Cell-Soluble Factor Interactions

Soluble factors such as signaling proteins and nutrients are also important components of a cell's microenvironment. Both, micro- and nanofluidic technology can be used to control the spatial and temporal presentation of soluble factors to cells *in vitro*. For example, laminarly flowing fluids within microchannels can be used to pattern cells and their substrates [47, 48]. By using laminar flows, a cell may be simultaneously exposed to multiple spatially segregated soluble factors [49]. Such a technique may be useful for a wide range of studies in which the delivery of spatially segregated factors or conditions are relevant and allows for characterizations of intracellular molecular kinetics. Microfluidic patterning can also be used to study the effects of soluble factors on cells. For example, temperature gradients have been used to control the development of different sides of an embryo [50]. Microfluidic gradient generators have also been used to study the effects of molecular gradients on cells [51, 52]. Using this technique, neutrophils were exposed to gradients of interleukin (IL)-8 to generate a novel understanding of the migration behavior of cells in response to nonlinear gradients [53]. Microfluidic gradients have also been shown to effect neural stem cell differentiation [54] and axon extension [55, 56].

18.1.5
3-D Scaffolds

Traditionally, "engineered tissues" have been fabricated by seeding cells within porous 3-D scaffolds, with such scaffolds serving as an environment within which nutrient and oxygen transport, as well as mechanical support [57], are provided for cell growth and proliferation, leading to the formation of 3-D tissues. Under ideal conditions, the scaffold would gradually degrade and become replaced by the ECM molecules deposited by cells, eventually leading to 3-D structures which resemble native tissue architecture. Porous scaffolds are currently generated through a variety of processes, including solvent casting and particulate leaching [58]. Although these technologies can be potentially used to generate nanoscale structures, such processes may not be able to engineer properties such as pore size, geometry, interconnectivity, and spatial distribution of the scaffolds. Furthermore, the diffusion limitations imposed by the scaffolds generated using current techniques has prevented the engineering of larger pieces of (more than a few hundred microns) viable tissue [59, 60].

Emerging techniques which have been used to generate scaffolds with microscale resolution include 3-D printing, microsyringe deposition, and tissue spin casting. 3-D printing (3DP) is typically used to generate ceramic [61] scaffolds in orthopedic tissue engineering applications. Similarly, in organ printing the cells and matrix material may be printed and built up [62]. Although 3DP confers a great deal of control over the macrostructure of the scaffold, like solvent casting and particulate leaching it suffers from process-derived limitations on pore properties. Alternatively, microsyringe deposition which has been used to generate layered PLGA scaffolds [63] confers a much higher degree of control, with fairly precise resolutions over the macrostructure of the scaffold, albeit at the expense of rapidity. One can imagine a combination of 3DP with microsyringe deposition which might enable the generation of much higher-resolution scaffolds. Also, the layering or combination of multiple "pieces" of engineered tissue is generally applicable to other materials such as polyurethane [64], and recently has been used to build larger structures using microfluidics and spin-coating approaches combined with particulate leaching [63]. Variations on these approaches are extensive, however, and the interested reader will find a more comprehensive review of such technologies elsewhere [65–67].

The aforementioned techniques have, despite their applicability to nanoscale features, only been applied to the generation of microscale features. One current approach for creating nanoscale scaffolds builds these structures from polymer nanofibers which are typically around a few hundred nanometers in diameter. Indeed, many natural biomaterials such as collagen and chitin are composed of fibrous structures. Techniques for fabricating nanofibers include electrospinning, melt-blowing, phase separation, self-assembly, and template synthesis.

Electrospinning is a popular approach as it is relatively inexpensive and capable of producing nanofibers from a variety of polymers and other materials. A number of polymers – including but not limited to PLGA, collagen, polycaprolactone (PCL), poly(L-lactic acid) (PLLA) and so forth – have been used to generate scaffolds for orthopedic [68], cartilage [69], and cardiac tissue engineering [70] applications. The resultant structures may be extremely porous and have very high surface area to volume ratios. Specific nanoscale features play a prominent role both in promoting cell proliferation and in guiding cell growth and general tissue architecture. Towards this end, recent investigations have shown that nanofibers aligned in desired directions and patterns [71] can induce growth and proliferation of cardiac cells into biologically relevant contractile spindle structures. In addition, as nanofibers can be generated from well-characterized polymers such as PLGA, it is possible to take advantage of properties such as biodegradability and surface functionalization. In fact, the size of biodegradable fibers can be used to modulate the degradation rate of the material.

Self-assembled amphiphilic peptides (Figure 18.2) were recently formed into hydrogels for tissue engineering [72] which resembled bone matrix alignments [73]. Peptide groups may be customized to direct cell behavior and polymerized directly into the hydrogel. For example, it was shown that directed differentiation of neural stem cells could be modulated using such a hydrogel functionalized

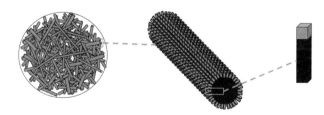

Fig. 18.2 Structure of self-assembled amphiphilic peptide hydrogels. (Left) A macroscopic hydrogel is composed of (center) fibers (which form in solution) that are in turn composed of (right) amphiphilic peptides. The hydrophobic hydrocarbon tails (black) of the peptides are buried within the fiber interior, while the hydrophilic peptide headgroups (gray) are exposed [73].

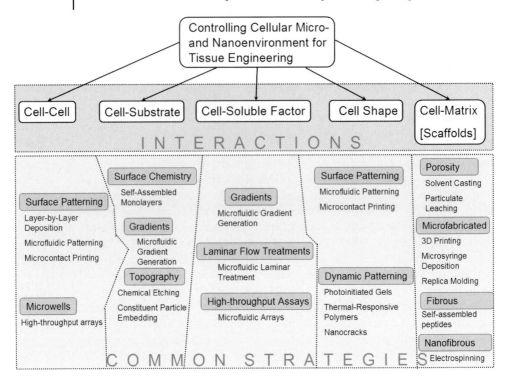

Fig. 18.3 An overview of cell–microenvironment interactions and relevant techniques.

with isoleucine-lysine-valine-alanine-valine (IKVAV, a laminin-derived sequence) without the use of additional biochemical factors [74].

18.2
Methods

The general methods and themes related to the techniques mentioned in Section 18.1 are described in the following sections (see Figure 18.3).

18.2.1
Soft Lithography

Traditional lithography, as developed by the semiconductor industry, has inspired the recent emergence of soft lithography and other fabrication techniques in the generation of micro- and nanoscale features for tissue engineering applications [75]. Soft lithography can be used to fabricate small-scale features without the use of expensive clean rooms by utilizing elastomeric molds made from patterned silicon wafers (Figure 18.4). A number of techniques for controlling the

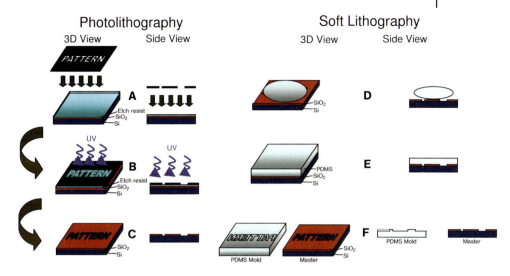

Fig. 18.4 A schematic overview of photolithography and soft lithography. In photolithography (a), a mask is placed on a silicon wafer that is coated with a thin layer of photoresist (b). The mask and wafer are exposed to ultraviolet (UV) light, which generates a patterned silicon wafer (c). In soft lithography (d), a liquid prepolymer (typically PDMS) is molded on a patterned silicon wafer/master (e). The polymer is then cured and the PDMS mold is subsequently peeled from the stamp (f).

micro- and nanoscale cellular environment rely on this approach. In particular, many micropatterning techniques for facilitating cell–cell interaction and controlling cell shape utilize microfabrication techniques. For example, techniques such as capillary force lithography, micromolding, microcontact printing, and microfluidic patterning for generating specific patterns of molecules on planar surfaces often rely on microfabrication approaches [76]. In capillary force lithography and micromolding, molds placed on a layer of prepolymer solution displace the polymer solution into the void regions of the mold before polymerization is induced [77]. In microcontact printing, molds may be "inked" with molecules and placed on top of surfaces in order to print them in the shape of the mold pattern [78]. In microfluidic patterning, molecules may be flowed through microfluidics channels produced by layering a mold on top of a flat surface [75]. Recently, the gradient generation capacity of microfluidics techniques has been coupled with photopolymerization chemistry to generate hydrogels with spatial control of the gel properties [7, 79]. In these studies, gradients of photocrosslinkable monomers were generated within microfluidics channels and then photopolymerized via exposure to ultraviolet light. In addition to creating crosslinking density gradients across a material, gradients of conjugated signaling or adhesive molecules were also generated for directing and inducing cell migration, adhesion, and differentiation.

Nanocrack structures that enable dynamic cell patterning are also made from PDMS substrates generated by soft lithography [33]. Such substrates are exposed to oxygen plasma to "harden" or to create a silica-like film on the surface rendered resistant to protein adsorption via the deposition of tridecafluoro-(1,1,2,2-tetrahydrooctyl)-1-trichlorosilane. The substrates are then subjected to uniaxial mechanical strain (a combination of strains can be applied at once and in different directions) and, due to differences in surface and bulk properties, parallel arrays of nanocracks are generated. The bulk material exposed by cracking is not resistant to protein adsorption, and so the exposure of this bulk material allows for the adsorption of ECM proteins that in turn facilitates cell attachment along these parallel lines. The crack widths can be modulated by varying the strain applied, and the crack can be closed by reducing the strain, thereby reversing the availability of adhesive portions of substrate. This technique can be applied repeatedly in different directions to dynamically direct the growth and retraction of cells along linear nanocracks.

Microfabrication approaches may further be used to generate features within 3-D scaffolds, and indeed finer structures such as microvasculature have been generated in tissue scaffolds using microfabrication approaches [80, 81]. Initially, standard photolithography techniques were used to create vascular network patterns directly upon silicon and Pyrex surfaces. Cells were seeded onto the surface and allowed to grow to confluence, and then peeled off as a monolayer and "rolled" together into a rough approximation of tissue. Later studies relied on soft lithographic replica molding of the patterned networks generated using these standard micromachining technologies with biocompatible materials such as PDMS, PLGA [82] and poly(glycerol sebacate) (PGS) [63]. Generally, replica molding is performed by the pouring of a mixture of prepolymer solution and crosslinking agent onto a (typically silicon) patterned surface, followed by subsequent photo- or thermally catalyzed polymerization of the mixture. Replica molding, coupled with particulate leaching, can further generate biodegradable porous microfluidics structures. For example, artificial capillary networks were generated through layer-by-layer stacking. Finally, layer-by-layer techniques may also be used with microfluidic patterning to generate cell-seeded scaffolds [83–85]. Here, cell-prepolymer solutions are repeatedly flowed, deposited, and polymerized within regions of microfluidics channels in a sequential manner. The sequential deposition of different cell-prepolymer solutions allowed for the controlled generation of layers of different cells within scaffolds.

18.2.2
Self-Assembled Monolayers

Self-assembled monolayers of alkanethiols used for cell culture may be formed on gold surfaces, either through direct or vapor deposition of alkanethiols to the surface. In this technique, a silicon or glass substrate is typically first coated through vacuum evaporation with a thin layer (e.g., 1 nm) of titanium to facilitate gold adhesion, followed by coating with gold. Upon vapor deposition of the alkanethiol,

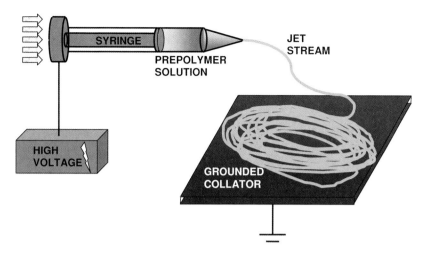

Fig. 18.5 A schematic diagram of polymer nanofiber electrospinning. A high voltage is applied to a syringe in order to charge the prepolymer solution contained within. The prepolymer solution is then ejected from the syringe in a controlled manner, and a jet stream of the charged prepolymer is drawn towards a grounded (earthed) collator. As the jet stream moves through the air towards the collator, the polymer dries to form nanofibers. The charge of the polymer slowly dissipates after collecting on the collator.

the molecules' sulfur groups will orient themselves to contact with the gold surface, while the alkyl chains are packed away from the surface.

18.2.3
Electrospinning

Electrospinning – a technique derived from electrostatic spraying – is the most common method of producing nanofiber scaffolds as it is capable of producing fibers of various sizes (from tens of nanometers up to microns) from various materials (Figure 18.5). In brief, a charged material solution (e.g., a polymer such as PLA) dissolved in a solvent is discharged from a reservoir through a tip that ranges in size from few hundred microns to a few millimeters. A grounded (earthed) collecting surface (collator) below the tip that is 10–30 cm away will draw the "whipping jet" of ejected solution towards it. As the jet travels through the air towards its target, the solvent dissolves into a nanofiber, which collects upon the grounded surface. With time, the nanofiber eventually loses its charge and the result is a mat of fiber which can be used as the basis for scaffold material. In order to achieve alignment in particular directions or dimensions, the collecting surface can translated relative to the dispensing tip. Fiber dimensions and characteristics may be controlled through modulation of material concentration in solvent. Layers of woven or non-woven mats of nanofibers may be stacked upon one another to build larger scaffold structures.

18.2.4
Nanotopography Generation

In order to generate nanoscale surface roughness, chemical etching techniques are often used. For example, a machined surface may be treated with a solution of H_2SO_4 or other corrosive solution to generate nanotextures. Alternatively, nanotextured surfaces may be generated through powder metallurgy (or the embedding of constituent NPs), which can be applied to materials such as pure titanium, Ti6A14V and Co28Cr6Mo. In this approach, powders are typically loaded into a steel-tool die and compacted at high uniaxial pressures (on the order of many GPa) using hydraulic presses. In this spirit, carbon nanotubes can also be pressed into materials at high pressures to generate surface roughness.

18.2.5
Layer-by-Layer Deposition

Sequential deposition is a recurrent theme for the generation of patterned co-cultures for controlling cell–cell interactions. One variation on "sequential deposition" is to use thermally responsive polymers [40] whereby electron beam irradiation can be used to pattern a thin layer of thermally responsive acrylamide onto surfaces. Cells can then be cultured to confluence at 37 °C. The temperature is subsequently reduced, to induce the thermally responsive regions to become non-adhesive and force cell detachment. Once the detached cells have been removed, the temperature is increased to enable the adhesion of a second cell type on the exposed substrate. Layer-by-layer deposition of ionic polymers has also been used to generate patterned co-cultures. In this approach, a surface that is patterned with an ionic polymer is treated with an adhesive protein, resulting in preferential attachment of the protein to non-patterned regions, such that cell patterns can then be formed on the adhesive regions. Next, the non-adhesive surfaces can be treated with another layer of cell-adhesive ionic polymer. The subsequent deposition of a secondary cell type results in the formation of patterned co-cultures [86].

18.2.6
3D Printing

Traditionally, 3DP – a solid freeform fabrication technique – has been applied to orthopedic tissue engineering applications to generate channels and networks within scaffolds [61]. For ceramic scaffolds, printing relies on the ability to bond wetted regions of powdered base material, but generally a bed of powder may be printed with a particular binder solution which bonds the wetted regions of powder in the shape of the wetted regions into a monolithic mold. The subsequent removal of non-bonded powder produces the final structure. The technique may be performed sequentially in a layer-by-layer manner to produce complicated 3-D structures. Microsyringe deposition is similar to 3DP in that it relies on controlled deposition, and the shape of the resulting structure is a direct function of deposition patterns. However, instead of rastering a printhead across the surface and depositing a "glue", the deposited material is typically a polymer which polymerizes to form the scaffold directly. Here, the resolution of the deposited lines of polymer is a function of a number of parameters such as syringe pressure and dimensions, as well as the solution viscosity. Complex structures are also generated through layer-by-layer application of this technique.

18.3
Outlook

The merger of novel biomaterials and nanofabrication strategies has led to dramatic enhancements in the complexity and biomimicry of today's tissue engineering constructs. Emerging tools for manipulating the micro- and nanoscale cellular environment have provided much insight into the fundamental biology of how cells interact with the surrounding components such as cell–cell, cell–soluble factors, and cell–ECM molecules. This knowledge can be used to direct cell fates, and can be incorporated into tissue engineering scaffolds. As our understanding of the relevant parameters increases, new nanomaterials and technologies that provide proper signals and environmental cues to cells provide exciting opportunities for the generation of clinically viable tissues. Significant challenges remain to be addressed, however, including the lack of suitable materials with the desired degradation rates, and the mechanical properties for the desired tissue. Another challenge is the optimization of scaffold architecture, including pore size, morphology, surface topography, and bioactivity. Also, new and optimized processing methods must be developed to address issues related to cell seeding, vascularization and scale up into 3-D structures. In addition, research is required to test and validate the *in-vivo* functionality of micro- and nanofabricated constructs, and to assess the performance of these constructs against existing clinically applied technologies.

References

1 Langer, R., Vacanti, J. P. (1993) Tissue engineering. *Science* **260**, 920–926.

2 Romanova, E. V., Oxley, S. P., Rubakhin, S. S., Bohn, P. W., Sweedler, J. V. (2006) Self-assembled monolayers of alkanethiols on gold modulate electrophysiological parameters and cellular morphology of cultured neurons. *Biomaterials* **27**, 1665–1669.

3 Lan, M. A., Gersbach, C. A., Michael, K. E., Keselowsky, B. G., Garcia, A. J. (2005) Myoblast proliferation and differentiation on fibronectin-coated self assembled monolayers presenting different surface chemistries. *Biomaterials* **26**, 4523–4531.

4 Hwang, J. J., Iyer, S. N., Li, L. S., Claussen, R., Harrington, D. A., Stupp, S. I. (2002) Self-assembling biomaterials: liquid crystal phases of

cholesteryl oligo(L-lactic acid) and their interactions with cells. *Proc. Natl. Acad. Sci. USA* **99**, 9662–9667.

5 Chen, C. S., Mrksich, M., Huang, S., Whitesides, G. M., Ingber, D. E. (1997) Geometric control of cell life and death. *Science* **276**, 1425–1428.

6 Dertinger, S. K., Jiang, X., Li, Z., Murthy, V. N., Whitesides, G. M. (2002) Gradients of substrate-bound laminin orient axonal specification of neurons. *Proc. Natl. Acad. Sci. USA* **99**, 12542–12547.

7 Burdick, J. A., Khademhosseini, A., Langer, R. (2004) Fabrication of gradient hydrogels using a microfluidics/photopolymerization process. *Langmuir* **20**, 5153–5156.

8 Zaari, N., Rajagopalan, S. K., Kim, S. K., Engler, A. J., Wong, J. Y. (2004) Photopolymerization in microfluidic gradient generators: microscale control of substrate compliance to manipulate cell response. *Adv. Mater.* **16**, 2133–2137.

9 den Braber, E. T., de Ruijter, J. E., Ginsel, L. A., von Recum, A. F., Jansen, J. A. (1998) Orientation of ECM protein deposition, fibroblast cytoskeleton, and attachment complex components on silicone microgrooved surfaces. *J. Biomed. Mater. Res.* **40**, 291–300.

10 Walboomers, X. F., Croes, H. J., Ginsel, L. A., Jansen, J. A. (1999) Contact guidance of rat fibroblasts on various implant materials. *J. Biomed. Mater. Res.* **47**, 204–212.

11 van Kooten, T. G., Whitesides, J. F., von Recum, A. (1998) Influence of silicone (PDMS) surface texture on human skin fibroblast proliferation as determined by cell cycle analysis. *J. Biomed. Mater. Res.* **43**, 1–14.

12 Teixeira, A. I., Abrams, G. A., Bertics, P. J., Murphy, C. J., Nealey, P. F. (2003) Epithelial contact guidance on well-defined micro- and nano-structured substrates. *J. Cell Sci.* **116**, 1881–1892.

13 Sato, M., Webster, T. J. (2004) Nanobiotechnology: implications for the future of nanotechnology in orthopedic applications. *Expert Rev. Med. Devices* **1**, 105–114.

14 de Oliveira, P. T., Nanci, A. (2004) Nanotexturing of titanium-based surfaces upregulates expression of bone sialoprotein and osteopontin by cultured osteogenic cells. *Biomaterials* **25**, 403–413.

15 Price, R. L., Waid, M. C., Haberstroh, K. M., Webster, T. J. (2003) Selective bone cell adhesion on formulations containing carbon nanofibers. *Biomaterials* **24**, 1877–1887.

16 Webster, T. J., Siegel, R. W., Bizios, R. (1999) Osteoblast adhesion on nanophase ceramics. *Biomaterials* **20**, 1221–1227.

17 Webster, T. J. (2001) Nanophase ceramics: The future of orthopedic and dental implant material. In: Ying, J. Y. (Ed.), *Nanostructured Materials*. Academy Press, New York, pp. 126–166.

18 Webster, T. J., Ergun, C., Doremus, R. H., Siegel, R. W., Bizios, R. (2000) Enhanced functions of osteoblasts on nanophase ceramics. *Biomaterials* **21**, 1803–1810.

19 Webster, T. J., Ergun, C., Doremus, R. H., Siegel, R. W., Bizios, R. (2001) Enhanced osteoclast-like cell functions on nanophase ceramics. *Biomaterials* **22**, 1327–1333.

20 Stevens, M. M., George, J. H. (2005) Exploring and engineering the cell surface interface. *Science* **310**, 1135–1138.

21 Anderson, D. G., Levenberg, S., Langer, R. (2004) Nanoliter-scale synthesis of arrayed biomaterials and application to human embryonic stem cells. *Nat. Biotechnol.* **22**, 863–866.

22 Anderson, D. G., Putnam, D., Lavik, E. B., Mahmood, T. A., Langer, R. (2005) Biomaterial microarrays: rapid, microscale screening of polymer-cell interaction. *Biomaterials* **26**, 4892–4897.

23 Flaim, C. J., Chien, S., Bhatia, S. N. (2005) An extracellular matrix microarray for probing cellular differentiation. *Nat. Methods* **2**, 119–125.

24 Thakar, R. G., Ho, F., Huang, N. F., Liepmann, D., Li, S. (2003) Regulation of vascular smooth muscle cells by micropatterning. *Biochem. Biophys. Res. Commun.* **307**, 883–890.

25 Miller, D. C., Thapa, A., Haberstroh, K. M., Webster, T. J. (2004) Endothelial and vascular smooth muscle cell function on poly(lactic-*co*-glycolic acid) with nano-structured surface features. *Biomaterials* **25**, 53–61.

26 Thapa, A., Webster, T. J., Haberstroh, K. M. (2003) Polymers with nano-dimensional surface features enhance bladder smooth muscle cell adhesion. *J. Biomed. Mater. Res. A* **67**, 1374–1383.

27 McBeath, R., Pirone, D. M., Nelson, C. M., Bhadriraju, K., Chen, C. S. (2004) Cell shape, cytoskeletal tension, and RhoA regulate stem cell lineage commitment. *Dev. Cell* **6**, 483–495.

28 Elbert, D. L., Hubbell, J. A. (2001) Conjugate addition reactions combined with free-radical cross-linking for the design of materials for tissue engineering. *Biomacromolecules* **2**, 430–441.

29 Schutt, M., Krupka, S. S., Milbradt, A. G., Deindl, S., Sinner, E. K., Oesterhelt, D., Renner, C., Moroder, L. (2003) Photocontrol of cell adhesion processes: model studies with cyclic azobenzene-RGD peptides. *Chem. Biol.* **10**, 487–490.

30 Okano, T., Yamada, N., Okuhara, M., Sakai, H., Sakurai, Y. (1995) Mechanism of cell detachment from temperature-modulated, hydrophilic-hydrophobic polymer surfaces. *Biomaterials* **16**, 297–303.

31 Lahann, J., Mitragotri, S., Tran, T. N., Kaido, H., Sundaram, J., Choi, I. S., Hoffer, S., Somorjai, G. A., Langer, R. (2003) A reversibly switching surface. *Science* **299**, 371–374.

32 Yeo, W. S., Yousaf, M. N., Mrksich, M. (2003) Dynamic interfaces between cells and surfaces: electro-active substrates that sequentially release and attach cells. *J. Am. Chem. Soc.* **125**, 14994–14995.

33 Zhu, X., Mills, K. L., Peters, P. R., Bahng, J. H., Liu, E. H., Shim, J., Naruse, K., Csete, M. E., Thouless, M. D., Takayama, S. (2005) Fabrication of reconfigurable protein matrices by cracking. *Nat. Mater.* **4**, 403–406.

34 Morin, O., Normand, C. (1986) Long-term maintenance of hepatocyte functional activity in co-culture: requirements for sinusoidal endothelial cells and dexamethasone. *J. Cell Physiol.* **129**, 103–110.

35 Fukuda, J., Okamura, K., Ishihara, K., Mizumoto, H., Nakazawa, K., Ijima, H., Kajiwara, T., Funatsu, K. (2005) Differentiation effects by the combination of spheroid formation and sodium butyrate treatment in human hepatoblastoma cell line (Hep G2): a possible cell source for hybrid artificial liver. *Cell Transplant.* **14**, 819–827.

36 Bhatia, S. N., Balis, U. J., Yarmush, M. L., Toner, M. (1999) Effect of cell-cell interactions in preservation of cellular phenotype: cocultivation of hepatocytes and nonparenchymal cells. *FASEB J.* **13**, 1883–1900.

37 Bhatia, S. N., Balis, U. J., Yarmush, M. L., Toner, M. (1998) Microfabrication of hepatocyte/fibroblast co-cultures: role of homotypic cell interactions. *Biotechnol. Prog.* **14**, 378–387.

38 Bhatia, S. N., Balis, U. J., Yarmush, M. L., Toner, M. (1998) Probing heterotypic cell interactions: hepatocyte function in micro-fabricated co-cultures. *J. Biomater. Sci. Polym. Ed.* **9**, 1137–1160.

39 Bhatia, S. N., Yarmush, M. L., Toner, M. (1997) Controlling cell interactions by micropatterning in co-cultures: hepatocytes and 3T3 fibroblasts. *J. Biomed. Mater. Res.* **34**, 189–199.

40 Yamato, M., Konno, C., Utsumi, M., Kikuchi, A., Okano, T. (2002) Thermally responsive polymer-grafted surfaces facilitate patterned cell seeding and co-culture. *Biomaterials* **23**, 561–567.

41 Hirose, M., Yamato, M., Kwon, O. H., Harimoto, M., Kushida, A., Shimizu,

T., Kikuchi, A., Okano, T. (2000) Temperature-responsive surface for novel co-culture systems of hepatocytes with endothelial cells: 2-D patterned and double layered co-cultures. *Yonsei Med. J.* **41**, 803–813.

42 Khademhosseini, A., Suh, K. Y., Yang, J. M., Eng, G., Yeh, J., Levenberg, S., Langer, R. (2004) Layer-by-layer deposition of hyaluronic acid and poly-l-lysine for patterned cell co-cultures. *Biomaterials* **25**, 3583–3592.

43 Chiu, D. T., Jeon, N. L., Huang, S., Kane, R. S., Wargo, C. J., Choi, I. S., Ingber, D. E., Whitesides, G. M. (2000) Patterned deposition of cells and proteins onto surfaces by using three-dimensional microfluidic systems. *Proc. Natl. Acad. Sci. USA* **97**, 2408–2413.

44 Tang, M. D., Golden, A. P., Tien, J. (2003) Molding of three-dimensional microstructures of gels. *J. Am. Chem. Soc.* **125**, 12988–12989.

45 Stevens, M. M., Mayer, M., Anderson, D. G., Weibel, D. B., Whitesides, G. M., Langer, R. (2005) Direct patterning of mammalian cells onto porous tissue engineering substrates using agarose stamps. *Biomaterials* **26**, 7636–7641.

46 Khademhosseini, A., Yeh, J., Eng, G., Karp, J., Kaji, H., Borenstein, J., Farokhzad, O. C., Langer, R. (2005) Cell docking inside microwells within reversibly sealed microfluidic channels for fabricating multiphenotype cell arrays. *Lab on a Chip* **5**, 1380–1386.

47 Takayama, S., Ostuni, E., Qian, X. P., McDonald, J. C., Jiang, X. Y., LeDuc, P., Wu, M. H., Ingber, D. E., Whitesides, G. M. (2001) Topographical micropatterning of poly(dimethylsiloxane) using laminar flows of liquids in capillaries. *Adv. Mater.* **13**, 570–574.

48 Takayama, S., McDonald, J. C., Ostuni, E., Liang, M. N., Kenis, P. J. A., Ismagilov, R. F., Whitesides, G. M. (1999) Patterning cells and their environments using multiple laminar fluid flows in capillary

networks. *Proc. Natl. Acad. Sci. USA* **96**, 5545–5548.

49 Takayama, S., Ostuni, E., LeDuc, P., Naruse, K., Ingber, D. E., Whitesides, G. M. (2001) Subcellular positioning of small molecules. *Nature* **411**, 1016.

50 Lucchetta, E. M., Lee, J. H., Fu, L. A., Patel, N. H., Ismagilov, R. F. (2005) Dynamics of *Drosophila* embryonic patterning network perturbed in space and time using microfluidics. *Nature* **434**, 1134–1138.

51 Jeon, N. L., Dertinger, S. K. W., Chiu, D. T., Choi, I. S., Stroock, A. D., Whitesides, G. M. (2000) Generation of solution and surface gradients using microfluidic systems. *Langmuir* **16**, 8311–8316.

52 Dertinger, S. K. W., Chiu, D. T., Jeon, N. L., Whitesides, G. M. (2001) Generation of gradients having complex shapes using microfluidic networks. *Anal. Chem.* **73**, 1240–1246.

53 Jeon, N. L., Baskaran, H., Dertinger, S. K. W., Whitesides, G. M., Van de Water, L., Toner, M. (2002) Neutrophil chemotaxis in linear and complex gradients of interleukin-8 formed in a microfabricated device. *Nat. Biotechnol.* **20**, 826–830.

54 Chung, B. G., Flanagan, L. A., Rhee, S. W., Schwartz, P. H., Lee, A. P., Monuki, E. S., Jeon, N. L. (2005) Human neural stem cell growth and differentiation in a gradient-generating microfluidic device. *Lab on a Chip* **5**, 401–406.

55 Dertinger, S. K. W., Jiang, X. Y., Li, Z. Y., Murthy, V. N., Whitesides, G. M. (2002) Gradients of substrate-bound laminin orient axonal specification of neurons. *Proc. Natl. Acad. Sci. USA* **99**, 12542–12547.

56 Taylor, A. M., Blurton-Jones, M., Rhee, S. W., Cribbs, D. H., Cotman, C. W., Jeon, N. L. (2005) A microfluidic culture platform for CNS axonal injury, regeneration and transport. *Nat. Methods* **2**, 599–605.

57 Cohen, S., Bano, M. C., Cima, L. G., Allcock, H. R., Vacanti, J. P., Vacanti, C. A., Langer, R. (1993) Design of synthetic polymeric structures for

cell transplantation and tissue engineering. *Clin. Mater.* **13**, 3–10.

58 Karp, J., Dalton, P., Shoichet, M. (2003) Scaffolds for tissue engineering. *MRS Bull.* **28**, 301–306.

59 Folkman, J., Hochberg, M. (1973) Self-regulation of growth in three dimensions. *J. Exp. Med.* **138**, 745–753.

60 Hutmacher, D. W., Sittinger, M., Risbud, M. V. (2004) Scaffold-based tissue engineering: rationale for computer-aided design and solid free-form fabrication systems. *Trends Biotechnol.* **22**, 354–362.

61 Seitz, H., Rieder, W., Irsen, S., Leukers, B., Tille, C. (2005) Three-dimensional printing of porous ceramic scaffolds for bone tissue engineering. *J. Biomed. Mater. Res. B Appl. Biomater.* **74**, 782–788.

62 Mironov, V., Boland, T., Trusk, T., Forgacs, G., Markwald, R. R. (2003) Organ printing: computer-aided jet-based 3D tissue engineering. *Trends Biotechnol.* **21**, 157–161.

63 Vozzi, G., Flaim, C., Ahluwalia, A., Bhatia, S. (2003) Fabrication of PLGA scaffolds using soft lithography and microsyringe deposition. *Biomaterials* **24**, 2533–2540.

64 Folch, A., Mezzour, S., During, M., Hurtado, O., Toner, M., Muller, R. (2000) Stacks of microfabricated structures as scaffolds for cell culture and tissue engineering. *Biomed. Microdevices* **2**, 207–214.

65 Yeong, W. Y., Chua, C. K., Leong, K. F., Chandrasekaran, M. (2004) Rapid prototyping in tissue engineering: challenges and potential. *Trends Biotechnol.* **22**, 643–652.

66 Bhatia, S. N., Chen, S. C. (1999) Tissue engineering at the micro-scale. *Biomed. Microdevices* **2**, 131–144.

67 Tsang, V. L., Bhatia, S. N. (2004) Three-dimensional tissue fabrication. *Adv. Drug Deliv. Rev.* **56**, 1635–1647.

68 Yoshimoto, H., Shin, Y. M., Terai, H., Vacanti, J. P. (2003) A biodegradable nanofiber scaffold by electrospinning and its potential for bone tissue engineering. *Biomaterials* **24**, 2077–2082.

69 Fertala, A., Han, W. B., Ko, F. K. (2001) Mapping critical sites in collagen II for rational design of gene-engineered proteins for cell-supporting materials. *J. Biomed. Mater. Res.* **57**, 48–58.

70 Zong, X., Bien, H., Chung, C. Y., Yin, L., Fang, D., Hsiao, B. S., Chu, B., Entcheva, E. (2005) Electrospun fine-textured scaffolds for heart tissue constructs. *Biomaterials* **26**, 5330–5338.

71 Xu, C. Y., Inai, R., Kotaki, M., Ramakrishna, S. (2004) Aligned biodegradable nanofibrous structure: a potential scaffold for blood vessel engineering. *Biomaterials* **25**, 877–886.

72 Zhang, S. (2003) Fabrication of novel biomaterials through molecular self-assembly. *Nat. Biotechnol.* **21**, 1171–1178.

73 Hartgerink, J. D., Beniash, E., Stupp, S. I. (2001) Self-assembly and mineralization of peptide-amphiphile nanofibers. *Science* **294**, 1684–1688.

74 Silva, G. A., Czeisler, C., Niece, K. L., Beniash, E., Harrington, D. A., Kessler, J. A., Stupp, S. I. (2004) Selective differentiation of neural progenitor cells by high-epitope density nanofibers. *Science* **303**, 1352–1355.

75 Whitesides, G. M., Ostuni, E., Takayama, S., Jiang, X. Y., Ingber, D. E. (2001) Soft lithography in biology and biochemistry. *Annu. Rev. Biomed. Eng.* **3**, 335–373.

76 Folch, A., Toner, M. (2000) Micro-engineering of cellular interactions. *Annu. Rev. Biomed. Eng.* **2**, 227–256.

77 Suh, K. Y., Seong, J., Khademhos-seini, A., Laibinis, P. E., Langer, R. (2004) A simple soft lithographic route to fabrication of poly (ethylene glycol) microstructures for protein and cell patterning. *Biomaterials* **15**, 557–563.

78 Mrksich, M., Dike, L. E., Tien, J., Ingber, D. E., Whitesides, G. M. (1997) Using microcontact printing to pattern the attachment of mammalian cells to self-assembled monolayers of alkanethiolates on transparent films of gold and silver. *Exp. Cell. Res.* **235**, 305–313.

79 Koh, W. G., Revzin, A., Pishko, M. V. (2002) Poly(ethylene glycol) hydrogel microstructures encapsulating living cells. *Langmuir* **18**, 2459–2462.

80 Borenstein, J. T., Terai, H., King, K. R., Weinberg, E. J., Kaazempur-Mofrad, M. R., Vacanti, J. P. (2002) Microfabrication technology for vascularized tissue engineering. *Biomed. Microdevices* **4**, 167–175.

81 Kaihara, S., Borenstein, J., Koka, R., Lalan, S., Ochoa, E. R., Ravens, M., Pien, H., Cunningham, B., Vacanti, J. P. (2000) Silicon micromachining to tissue engineer branched vascular channels for liver fabrication. *Tissue Eng.* **6**, 105–117.

82 King, K., Wang, C., Kaazempur-Mofrad, M., Vacanti, J., Borenstein, J. (2004) Biodegradable microfluidics. *Adv. Mater.* **16**, 2007–2012.

83 Tan, W., Desai, T. A. (2003) Microfluidic patterning of cells in extracellular matrix biopolymers: effects of channel size, cell type, and matrix composition on pattern integrity. *Tissue Eng.* **9**, 255–267.

84 Tan, W., Desai, T. A. (2004) Layer-by-layer microfluidics for biomimetic three-dimensional structures. *Biomaterials* **25**, 1355–1364.

85 Tan, W., Desai, T. A. (2005) Microscale multilayer cocultures for biomimetic blood vessels. *J. Biomed. Mater. Res. A* **72**, 146–160.

86 Fukuda, J., Khademhosseini, A., Yeh, J., Eng, G., Cheng, J., Farokhzad, O. C., Langer, R. (2006) Micropatterned cell co-cultures using layer-by-layer deposition of extracellular matrix components. *Biomaterials* **27**, 1479–1486.

19
Diagnostic and Therapeutic Targeted Perfluorocarbon Nanoparticles

Patrick M. Winter, Shelton D. Caruthers, Gregory M. Lanza, and Samuel A. Wickline

19.1
Overview

Traditional clinical imaging techniques provide gross anatomical characterization of diseases at very late stages of progression. Treatment options at these stages tend to have limited efficacy, and the clinical prognosis is poor. A new medical imaging approach is to detect disease in its earliest, most easily treatable, pre-symptomatic phase. Molecular imaging with targeted contrast agents aims to detect and characterize the biochemical signatures of disease before anatomical abnormalities become apparent [1–3]. In conjunction with early detection, targeted therapeutics are being pursued as a means to treat presymptomatic diseases with very low doses of highly potent agents [4–6]. By targeting the therapy directly to the site of pathology, high drug concentrations are achieved at the desired location without accumulating in normal tissues, leading to decreased incidence of unwanted side effects. This chapter will outline the use of perfluorocarbon nanoparticles (NPs) for molecular imaging with magnetic resonance imaging (MRI) and targeted drug delivery, including:

- Diagnostic imaging;
- Targeted therapeutics; and
- Other imaging modalities

Perfluorocarbon (PFC) NPs consist of a heterogeneous mixture of at least one liquid PFC intimately dispersed in water, forming droplets that are 200–300 nm in diameter. The chemical structure of PFCs is similar to that of hydrocarbons, with all the hydrogen atoms replaced with fluorine. PFCs are clear, colorless, odorless, volatile, biologically inert, chemically stable, nontoxic, and do not become degraded in the body through metabolism [7, 8]. The NPs are stabilized for use *in-vivo* with surfactants, the most common being phospholipids, which isolate the PFC core from the surroundings. The phospholipid surface provides ideal chemistry for the incorporation of specialized targeting ligands, imaging

Nanobiotechnology II. Edited by Chad A. Mirkin and Christof M. Niemeyer
Copyright © 2007 WILEY-VCH Verlag GmbH & Co. KGaA, Weinheim
ISBN: 978-3-527-31673-1

agents or therapeutic drugs. Targeting and imaging agents are designed to adopt only one orientation on the NP surface: the hydrophilic end (imaging chelate or targeting molecule) points out to facilitate interactions with the biological environment, while the hydrophobic end intercalates into the surface of the particle. The incorporation of hydrophobic drugs into the particle surface assures limited dissociation from the NP until binding onto a target cell membrane.

NPs provide a "platform" for the incorporation and transportation of large numbers of imaging agents and targeting vectors, such as antibodies, peptides, polysaccharides, and aptamers [1, 3, 4]. Incorporating multiple targeting ligands can increase the avidity of these structures. In addition, by carrying tens to hundreds of thousands of imaging beacons, NPs allow imaging modalities which traditionally are considered to have low sensitivity (e.g., MRI or computed tomography; CT) to be able to detect targets at very low concentrations [9, 10]. PFC NPs are particularly well suited for the roles of molecular imaging and targeted drug delivery due to their long circulating half-life, sensitive and selective binding to the epitope of interest, prominent contrast to noise enhancement, acceptable toxicity, ease of clinical use, and applicability with standard commercially available imaging systems [11–14].

NPs can be formulated to carry vast numbers of imaging agents, such as gadolinium chelates, thereby amplifying the signal enhancement properties of the agent [5, 15]. MRI is sensitive to the relaxation properties of water in the body. In some cases, the native differences in relaxation between two tissues are not sufficient to provide a clinical diagnosis, however, and contrast agents are used to increase the detectability of an abnormal anatomy. The most common forms of MRI contrast agent utilize gadolinium, which increases the relaxation rate of the MRI signal and increases the signal intensity of tissues in contact with the contrast agent. As expected, the attachment of multiple gadolinium ions onto the surface of a NP imparts a linear increase in the signal enhancement delivered to each binding site. In addition to this simple multiplicative effect, the physical properties of the NPs instill an additional increase in the efficiency of the interactions between the gadolinium agent and the water molecules. These two factors allow NPs carrying 74 000 gadolinium ions to be 620 000-fold more effective than traditional small-molecule gadolinium contrast agents [15].

In addition to carrying thousands of gadolinium chelates, NPs can express many copies of the targeting ligand, which in turn increases the sensitivity for the biomarker of interest. Incorporating multiple targeting and imaging agents allows the reliable detection of even very sparse epitopes associated with cardiovascular disease and cancer. One biomarker that has gained widespread interest in several diseases is the $\alpha_v\beta_3$-integrin, a cellular receptor associated with angiogenesis [16, 17]. This integrin plays a critical part in smooth muscle cell migration and cellular adhesion [18, 19], both of which are required for the formation of new blood vessels. Other biomarkers associated primarily with cardiovascular disease include fibrin (an early marker of ruptured atherosclerotic plaques, which lead to myocardial ischemia and stroke), tissue factor (a receptor expressed on vascular smooth muscle cells that is involved with repair of damaged vessels), and collagen III [a component of the extracellular matrix (ECM) that is exposed following angioplasty procedures]. The incorporation of targeting ligands that bind to these biomarkers onto the NP surface provides a means to selectively guide particles directly to pathological tissues, even at very early stages of disease progression.

As well as imaging biomarkers of disease, PFC NPs are capable of specifically and locally delivering therapeutic drugs through a novel process called "contact facilitated drug delivery" [20]. Lipophilic drugs can be incorporated into the outer lipid shell of targeted NPs. Upon binding to the appropriate cell type, as directed by the targeting ligand, the drug is directly transferred from the NP's surfactant monolayer to the targeted cell membrane [21]. This lipid–drug mixing process is very slow and inefficient for free unbound particles, but binding to the cell surface substantially increases the frequency and duration of these interactions. Combining drug delivery and molecular imaging spatially delineates delivery and permits the local therapeutic concentration to be estimated (i.e., rational drug dosing). Therefore, a single targeted NP formulation can provide not only the diagnosis of a disease but also site-specific drug delivery and monitoring of therapeutic efficacy [4, 6, 22].

19.2
Methods

Through the use of various targeting ligands, NPs can transport imaging agents directly to the biochemical epitopes associated with several diseases, allowing detection as well as characterization of pathology on the molecular scale. As targeted NPs specifically bind to the disease sites, they can also be used for the precise delivery of drugs. Selected applications of molecular imaging and targeted drug delivery with PFC NPs for the diagnosis and treatment of cardiovascular disease and cancer are outlined in the following sections.

19.2.1
Diagnostic Imaging

Fibrin is an abundant component of thrombi, and an early marker of ruptured atherosclerotic plaques. Plaque rupture is the proximate cause of myocardial ischemia and stroke, which represent the most prevalent causes of death in America today [23]. Fibrin-targeted NPs have been developed for *in-vivo* molecular imaging of clot formation and demonstrated as effective in experimental animals [24]. Anti-fibrin antibodies were covalently coupled to the NP surface through the use of a reactive phospholipid included in the surfactant, N-4-(p-maleimidophenyl)butyramide 1,2-dipalmitoyl-sn-glycero-3-phosphoethanolamine (MPB-PE). An imaging agent, gadolinium-diethylenetriamine pentaacetic acid-phosphatidylethanolamine (Gd-DTPA-PE), was also included in the phospholipid mixture to provide MRI enhancement.

The PFC (perfluorooctylbromide), surfactants (lecithin, cholesterol, MPB-PE and Gd-DTPA-PE) and water were emulsified for 4 min in a microfluidizer. This process drives the components through microchannels in an interaction chamber at very high pressure (20 000 lb/in^2; 1400 bar). The high shear rates force the material to assemble into small, stable droplets of PFC covered in a surfactant monolayer, thus segregating them from the surrounding water. Purified monoclonal anti-fibrin antibodies [25] were coupled to MPB-PE on the NPs overnight, followed by dialysis to remove unbound antibodies. This procedure created 250 nm-diameter NPs that expressed multiple copies of the gadolinium chelate and anti-fibrin antibodies on the particle surface for imaging and targeting, respectively.

Clots were treated with fibrin-targeted NPs and displayed distinct contrast enhancement in images collected with a clinical 1.5 T scanner [24]. The high payload of gadolinium on the NPs, combined with the abundant availability of fibrin, results in contrast-to-noise levels greater than 100 with *in-vivo* MRI. This study established the feasibility of molecular imaging of fibrin with routine MRI techniques, allowing the detection and assessment of ruptured atherosclerotic plaques. This technology could aid in directing clinical interventions to the site of plaque rupture or in the identification of microthrombi before they produce significant hemodynamic stenoses.

While fibrin is an abundant target corresponding to late-stage atherosclerosis, early biomarkers of vascular disease, such as those associated with angiogenesis, may be much more difficult to image because they are expressed at much lower concentrations. Angiogenic capillaries proliferate throughout the vessel wall to meet the high metabolic demands of growing plaques [26] and contribute to subsequent lesson rupture [27, 28]. Neovascular proliferation is closely associated with the "culprit" lesions that cause clinical symptoms, including unstable angina, myocardial infarction and stroke [29–31]. The ability to characterize the extent and severity of angiogenesis within the arterial wall could provide a surrogate marker for atherosclerotic burden in clinical patients.

Molecular imaging of angiogenesis associated with atherosclerosis was demonstrated using NPs targeted to the $\alpha_v\beta_3$-integrin (Figure 19.1) [32]. NPs were targeted to the $\alpha_v\beta_3$-integrin by incorporating a peptidomimetic vitronectin antagonist (i.e., $\alpha_v\beta_3$-integrin antagonist) (US patent No. 6,130,231) onto the particle surface. The peptidomimetic contained a thiol group which was coupled to a phospholipid included in the surfactant mixture, N-[{ω-[4-(p-maleimidophenyl)butanoyl]amino} poly(ethylene glycol)$_{2000}$] 1,2-distearoyl-sn-glycero-3-phosphoethanolamine (MPB-PEG$_{2000}$-PE). Gadolinium-diethylenetriamine pentaacetic acid-bis-oleate (Gd-DTPA-BOA) was also included in the surfactant to allow MRI detection of the NPs. The surfactant mixture (consisting of lecithin, cholesterol, dipalmitoyl-phosphatidylethanolamine, Gd-DTPA-BOA and lipid-conjugated $\alpha_v\beta_3$-integrin antagonist) was added to perfluorooctylbromide and water for emulsification at a pressure of 1400 bar. Cholesterol-fed rabbits were imaged on a clinical MRI system, and targeted particles produced 47 ± 5% signal enhancement averaged across the abdominal aorta. Angiogenesis displayed a heterogeneous distribution,

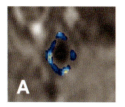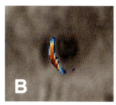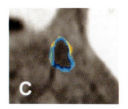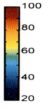

Fig. 19.1 Molecular imaging of angiogenesis in the abdominal aorta of a cholesterol-fed rabbit with $\alpha_v\beta_3$-targeted nanoparticles. False-colored maps of percentage MRI enhancement 2 h post-injection from individual aortic segments at three different anatomic levels: (A) renal artery; (B) mid-aorta; (C) diaphragm. (Reprinted with permission from Ref. [32].)

with individual slices enhancing by 80% and individual pixels enhancing by more than 100%.

In addition to playing a central role in the progression of atherosclerosis, angiogenesis is essential for the growth of primary tumors and the proliferation of malignancies. The detection of very early oncologic signatures, such as angiogenesis, could foster timely diagnosis when treatment options are plentiful and ample interventional opportunities are available [33]. Targeted NPs were produced by emulsifying water, perfluorooctylbromide and a surfactant mixture that included the lipid-conjugated peptidomimetic vitronectin antagonist and Gd-DTPA-BOA. NPs were administered intravenously to mice bearing very small melanomas [34] that were extremely difficult to recognize prior to injection (Figure 19.2a). The tumors displayed a clear signal enhancement within 30 min after injection of $\alpha_v\beta_3$-targeted NPs, and the image intensity continued to increase up to 2 h post-injection (Figures 19.2b and 19.3).

Vascular agents, such as NPs, are only exposed to $\alpha_v\beta_3$-integrins expressed on angiogenic vasculature. Rupture of the vessel wall, however, allows NPs to interact with smooth muscle cells, which also express the $\alpha_v\beta_3$-integrin. Other epitopes, including collagen III in the ECM, could also become available for NP binding following vessel wall injury. Perfluorooctylbromide NPs were targeted to the $\alpha_v\beta_3$-integrin by including the peptidomimetic vitronectin antagonist in the surfactant prior to emulsification. Collagen III NPs were produced by coupling anti-collagen antibodies to MPB-PEG$_{2000}$-PE on the surface after particle formation [35]. Antibodies are easily denatured and lose their biological activity under the high shear forces used to create the NP, but the peptidomimetic antagonist is much more resistant to these harsh procedures. Therefore, peptidomimetic targeting ligands can be included in the surfactant mixture and emulsified with the particles, while antibody ligands must be coupled to the NP in a separate step following emulsification. Both targeted NP formulations included Gd-DTPA-BOA for visualization under MRI.

NPs targeted to either $\alpha_v\beta_3$-integrin or collagen III were used to visualize vascular damage in the carotid arteries of pigs following balloon injury [35]. Three-

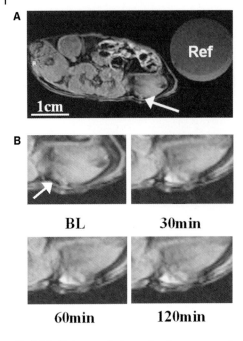

Fig. 19.2 (A) Image of a tumor-bearing mouse collected before injection of $\alpha_v\beta_3$-targeted nanoparticles. The arrow indicates the location of a small tumor (~2 mm diameter), which is difficult to distinguish from normal tissue. Ref: Test tube phantom for signal normalization. (B) Magnified images showing signal enhancement in the tumor before (BL) and up to 2 h after injection of $\alpha_v\beta_3$-targeted nanoparticles. (Reprinted with permission from Ref. [34].)

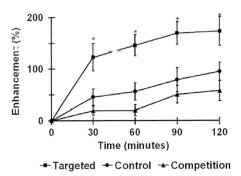

Fig. 19.3 Temporal evolution of mouse tumor enhancement with $\alpha_v\beta_3$-targeted contrast agent (■), non-targeted contrast agent (●), or a competition procedure (▲). MRI signal enhancement in the targeted group was almost twofold higher than in non-targeted animals (*, $p < 0.05$). Competitive blockade of $\alpha_v\beta_3$-integrin sites greatly diminished MRI contrast compared to that of targeted particles ($p < 0.05$), confirming the specificity of the targeted nanoparticles for the $\alpha_v\beta_3$-integrin. (Reprinted with permission from Ref. [34].)

dimensional reconstruction of the MRI enhancement pattern was similar for both NP formulations. The contrast-to-noise ratio achieved with $\alpha_v\beta_3$-integrin targeted NPs was much higher (13.8 ± 5.2) compared to collagen III-targeted NPs (3.3 ± 0.3), most likely reflecting the relative density of each biomarker and/or the relative avidity of each targeting probe. The low molecular weight ($\sim$1000 Da) of the $\alpha_v\beta_3$-integrin-targeting molecule allows 250–300 homing ligands to be incorporated onto the surface of each NP, whereas only 25–50 of the larger (50 000 Da) collagen III antibody fragments can be attached to each particle. Imaging the spatial distribution of targeted NPs allowed morphological analysis of the vessel injury, revealing an injury length of $\sim$31 mm, which exceeded the actual balloon length (20 mm) by 50%.

Another potential target for molecular imaging applications in cardiovascular disease is tissue factor, a transmembrane glycoprotein that is expressed on vascular smooth muscle cells (VSMC). Tissue factor is involved in a variety of normal and pathological processes, including thrombosis, hemostasis, angiogenesis, cell signaling and mitogenesis. Tissue factor expression on cultured VSMC was imaged using ligand-targeted NPs on a clinical 1.5 T MRI scanner [9].

NPs were targeted to tissue factor using a convenient three-step binding approach. Perfluorooctylbromide NPs were produced with Gd-DTPA-BOA, for MRI enhancement, and biotinylated phosphatidylethanolamine, for subsequent targeting, in the surfactant mixture. The NPs were 273 nm in diameter, with a polydispersity index of 0.15. The concentration of gadolinium was 6.17 mmol L^{-1} emulsion, while the particle concentration was 59 nM, corresponding to 94 200 gadolinium ions on the surface of each NP. Cell cultures were serially exposed to biotinylated anti-tissue factor antibodies, avidin, and biotinylated NPs, with thorough washing between each step. Quantitative MRI showed that tissue factor-targeted NPs bound to VSMC at a concentration of 468 ± 30 pM, producing a contrast-to-noise ratio of 17.7 with respect to untreated cells. The quantitative results were verified with destructive chemical analysis (gas chromatography) of the perfluorocarbon content.

19.2.2
Targeted Therapeutics

The use of NPs for combined imaging and targeted drug delivery offers the potential to individualize therapy based on the identification and distribution of disease biomarkers [1, 4, 6]. Patients can be segmented and prescribed treatment regiments designed specifically to intercede in the biochemical pathways present. The ability to directly image the drug delivery vehicle assures that the drug is reaching the intended target. Follow-up assessment of target expression can verify the early response to therapy and can indicate non-responders that may require secondary treatments.

The concept of targeted drug delivery was demonstrated with tissue factor-targeted particles loaded with anti-proliferative drugs (paclitaxel or doxorubicin) and applied to cultured VSMC [20]. Biotinylated phosphatidylethanolamine was included in the NP surfactant for three-step targeting to tissue factor expressed on the cell surface. The surfactant also included doxorubicin or paclitaxel – lipophilic drugs that remain in the phospholipid surface layer because they are not soluble in either the water outside of the particle or the perfluorocarbon inside. The addition of free drug into the culture medium did not retard cellular proliferation, which suggested that the overall drug concentrations were below the therapeutic threshold. Incorporating the drugs into tissue factor-targeted NPs, however, significantly reduced cell growth, showing that targeted therapeutics can effectively concentrate the drugs onto the target cell.

Subsequent targeted drug delivery experiments have focused on establishing the feasibility of these methods for *in-vivo* applications. Since $\alpha_v\beta_3$-targeted NPs accumulate in angiogenic vasculature [32], the incorporation of anti-angiogenic compounds could permit combined diagnosis and treatment in both cardiovascular disease and cancer. Under the category of cardiovascular disease, $\alpha_v\beta_3$-targeted NPs containing fumagillin, a lipophilic anti-angiogenic agent [36], were injected into atherosclerotic rabbits [37]. Perfluorooctylbromide NPs were formulated with the lipid-conjugated peptidomimetic vitronectin antagonist (targeting ligand), Gd-DTPA-BOA (imaging agent), fumagillin (therapeutic drug), lecithin and phosphatidylethanolamine in the surfactant mixture. MRI signal enhancement averaged over the abdominal aorta was identical at 4 h after injection for animals receiving therapeutic (fumagillin) or control (no drug) NPs (Figure 19.4). One week later, $\alpha_v\beta_3$-targeted NPs (no drug) were administered to reassess angiogenesis in the aorta. The MR signal enhancement was distinctly lower in fumagillin-treated

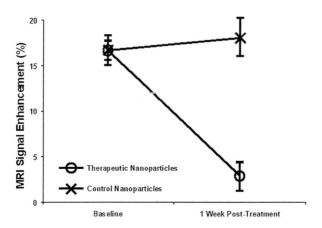

Fig. 19.4 Aortic MRI signal enhancement in rabbits treated with $\alpha_v\beta_3$-targeted nanoparticles with (O) and without (×) fumagillin. At the time of treatment, both nanoparticle formulations produced identical enhancement. Rabbits receiving $\alpha_v\beta_3$-targeted fumagillin nanoparticles had lower angiogenesis at 1 week post-treatment compared to rabbits injected with nanoparticles lacking fumagillin (*, p < 0.05).

rabbits: $2.9 \pm 1.6\%$ ($p < 0.05$), whereas control animals showed no change: $18.1 \pm 2.1\%$.

Other studies have utilized agents similar to fumagillin, and demonstrated anti-angiogenic effects in animal models of atherosclerosis [38]. In these studies, however, the drug was not specifically targeted and required multiple administrations over the course of several weeks. Without targeting, the dose required was more than 50 000-fold greater than the dose given with targeted NPs. These high dosages can lead to adverse side effects, including central nervous system toxicity [39].

Targeted cancer therapy can be accomplished by emulsifying water, perfluorooctylbromide, the MPB-PEG$_{2000}$-PE conjugated peptidomimetic vitronectin antagonist, Gd-DTPA-BOA, lecithin and paclitaxel [20], a commonly prescribed drug that arrests cellular proliferation. The effectiveness of these NPs was evaluated in rabbits bearing small V×2 tumors (<1 cm across) in the hind limb. Rabbits received an intravenous injection of therapeutic NPs, and MRI signal enhancement was measured. The successful delivery of targeted NPs was confirmed by a significant increase in MRI intensity. At seven days post-treatment, the tumors treated with paclitaxel NPs were significantly smaller (1.1 ± 0.2 cm^3) than untreated tumors (2.7 ± 0.6 cm^3), suggesting that $\alpha_v\beta_3$-targeted paclitaxel NPs may suppress tumor cell proliferation and inhibit tumor development. These methods may prove useful in the treatment of early primary or metastatic tumors alone, or in conjunction with adjuvant therapy. Such findings highlight the role of MRI for nanomedicine systems that combine the power of imaging with targeted therapy.

19.2.3
Other Imaging Modalities

PFC NPs are also amenable for use with several other common clinical imaging modalities. For instance, NPs can be detected with ultrasound imaging because the speed of sound in the PFC core is slower than in biological tissues (670 versus 1540 m s^{-1}) [40], providing inherent echogenicity. Therefore, NPs do not have to be formulated with gadolinium ions or radioisotopes, as is normally required for detection by MRI or nuclear imaging. Typically, differences in the speed of sound are not sufficient to visualize NPs when dispersed in the bloodstream. Upon binding to a tissue, however, the NPs induce a local acoustic impedance mismatch, leading to a strong ultrasound reflection. Thus, ultrasound imaging is sensitive to NPs bound to the target, but very little background signal is generated from circulating NPs. With targeted NPs, ultrasound has been used for molecular imaging of fibrin, tissue factor and angiogenesis [12, 40–42].

Targeted NPs and intravascular ultrasound have been utilized to study the expression of tissue factor following balloon angioplasty in an animal model [41, 43]. Balloon inflation produced over-stretch injury in the vessel wall, but did not intrinsically alter the ultrasound image intensity. The application of tissue factor-targeted NPs increased the image intensity in the vessel wall by 140%, clearly illustrating the locations of tissue damage. The pattern of NP enhancement was consistent with the severity of vascular injury determined by histological analysis. Non-targeted NPs, on the other hand, produced no change in ultrasound signal [43].

Ultrasound has also been extensively studied as a stand-alone therapeutic tool by applying high-power ultrasound pulses to induce heating or cavitation [44]. However, lower levels of ultrasound energy can be utilized in conjunction with PFC NPs to increase drug delivery. Insonification at normal clinical imaging levels can increase the contact frequency and likelihood of fusion between NPs and the target cell membrane [21]. The application of 2 MHz ultrasound with a mechanical index of 1.9 for 5 min leads to a 10-fold greater lipid delivery from targeted NPs to the targeted cell membrane. This method may represent a simple, low-cost approach to directing drug delivery and increasing therapeutic efficacy with targeted NPs.

One other commonly used medical imaging technique is CT, which can be used to diagnose and characterize advanced atherosclerotic disease using angiography and calcium scoring. Multi-slice CT offers the ability to obtain high signal-to-noise images with excellent spatial and temporal resolutions [45]. PFC compounds are naturally radio-opaque, and have been utilized for CT angiography [46]. For molecular imaging applications, however, the PFC in the NP core does not provide adequate X-ray attenuation for reliable visualization. One major advantage of the NP platform is the flexibility to exchange the core material to suit the imaging method. The incorporation of a more radio-opaque compound, such as iodinated oil, into the particles can improve the detection of pathology. For example, fibrin-targeted PFC core NPs provided only modest contrast enhancement on treated clots, whereas targeted iodinated NPs produced a contrast-to-noise ratio greater than 25 [10].

Targeted NPs also can be designed specifically for detection with nuclear imaging systems [47]. The particles can be labeled with radioisotopes, such as indium and technetium, and detected with conventional scintillation or single photon emission computed tomography (SPECT) imaging systems. The chelates that bind radioisotopes onto the particle surface are very similar to those utilized for attaching gadolinium for MRI applications. Accordingly, NPs can carry a large payload of radioisotopes, allowing very sparse epitopes and very minute areas of pathology to be sensitively detected. The ability to formulate NPs for each of the various imaging techniques provides the opportunity to design multi-modal contrast agents that are applicable to hybrid imaging systems, such as MRI/ultrasound or SPECT/CT.

19.3
Outlook

The costs of treating chronic diseases, such as atherosclerosis and cancer, are increasing at unsustainable rates, leading medical research efforts to focus on early detection and preventive strategies. The ability to detect the earliest markers of pathology with molecular imaging is opening avenues for the non-invasive diagnosing of presymptomatic disease. The biochemical information could be employed to segregate patient populations and to apply personalized therapies. Serial evaluation of these biomarkers would then provide an early detection of therapeutic response, and alternative, more aggressive, therapies could be pursued if traditional methods fail. Finally, combining molecular imaging with targeted drug delivery enables the linking of diagnosis and therapy, where the detection of disease, characterization of pathological signatures, selective delivery of potent chemotherapeutic agents, quantification of drug delivered and evaluation of disease regression could be achieved simultaneously with a single agent.

The future for molecular imaging technologies is limited only by our understanding of normal and pathological biology. Potential targets are being rapidly proposed and classified through the work of genetic and proteomic researchers. The complex interplay of inhibitory and stimulatory cellular signaling is being mapped on the laboratory bench, yielding new clinical biomarkers to detect and even predict the onset of disease. Likewise, targeted drug delivery could lead to the rapid application of investigational drugs to the clinic. Site-specific deposition of potentially harmful drugs could provide effective therapy while minimizing adverse side effects. A whole range of previously intolerable drugs may soon be re-examined with these technologies, and in the future drug candidates may not need to be abandoned due to unacceptable biodistribution or clearance kinetics. The ability to quantitate the distribution and retention of targeted delivery vehicles non-invasively enables the confirmation of drug deposition and prediction of the subsequent therapeutic response.

One of the greatest strengths of NP agents is their inherent flexibility to carry and deliver a wide range of therapeutic agents. For example, NPs can be designed to deliver conventional drugs, such as cytotoxins and antibiotics, or to deliver unorthodox treatments, such as gene therapy constructs. In addition to using NPs to carry a drug payload, the therapeutic device itself may be labeled with NPs to allow *in-vivo* monitoring and tracking. In this way, NPs may be employed in clinical applications of cellular therapies [48].

Developments in molecular imaging and targeted drug delivery are proceeding very rapidly in both academic and commercial laboratories. The application of these agents to the clinic, however, has been relatively slow because they are considerably more complex than traditional drug candidates. The use of multiple active pharmaceutical ingredients can make high-volume manufacturing very challenging, and can lead to difficulties in assuring the stability and safety of the agent. A variety of different platform technologies, including PFC NPs, is emerging with broad applications in the detection, treatment and monitoring of nascent disease, and several candidates are likely soon to enter clinical trials. It is clear that targeted contrast agents have been shown as feasible in numerous animal model studies by several independent investigators [49–52]. Hence, the question is no longer whether agents based on nanotechnology will make it to the clinic, but only which agents will be approved – and when such tools will be available in the marketplace.

The ultimate clinical application of these molecular-imaging techniques will most likely require the deliberate development of task-specific imaging hardware, software and analysis products. For MRI, specialized coils may be needed to provide robust imaging performance for clinical decision-making. For example, some molecular-imaging applications may require intravascular coils that are deployed on catheter systems. The imaging of fibrin deposition on ruptured atherosclerotic plaques, or the imaging of vascular epitopes exposed by angioplasty procedures, may greatly benefit from the increased sensitivity associated with intravascular coils [53, 54]. These coils can be placed precisely near the area of interest during interventional procedures, obviating the need for large field-of-view imaging. In addition, specialized image reconstruction or analysis packages may need to be developed for routine diagnostic evaluation with MRI. Automated image subtraction or thresholding procedures may provide a better detection of signal enhancement due to targeted contrast agent binding, and aid the clinical efficacy of these methods. Other imaging modalities may also require specialized hardware or software development. Ultrasound detection of targeted NPs may benefit from high-frequency probes that offer higher resolution than is currently available with traditional clinical systems. Furthermore, the development of hybrid imaging systems, such as SPECT/CT or SPECT/MRI, may greatly aid the application of targeted NPs in the clinic. These imaging systems can combine the high resolution of MRI and CT imaging with the high sensitivity of nuclear-imaging techniques.

In conclusion, the goals of molecular imaging and targeted drug delivery aim to switch current clinical practice from the evaluation of symptoms and systemic administration of potentially hazardous drugs towards detecting the early molecular signatures of disease, and pinpoint application of therapy. Clearly, the accomplishment of these aims could significantly reduce the mounting burden of chronic debilitating diseases in today's society.

Acknowledgments

These studies were supported by grants from the National Institutes of Health (HL-78631, HL-73646, CO-37007, CA-119342 and EB-01704), the American Heart Association, and Philips Medical Systems.

References

1 Cyrus, T., Winter, P. M., Caruthers, S. D., Wickline, S. A., Lanza, G. M. (2005) Magnetic resonance nanoparticles for cardiovascular molecular imaging and therapy. *Expert Rev. Cardiovasc. Ther.* **3**, 705–715.

2 Lanza, G. M., Winter, P. M., Caruthers, S. D., Morawski, A. M., Schmieder, A. H., Crowder, K. C., Wickline, S. A. (2004) Magnetic resonance molecular imaging with nanoparticles. *J. Nucl. Cardiol.* **11**, 733–743.

3 Winter, P. M., Caruthers, S. D., Wickline, S. A., Lanza, G. M. (2006) Molecular imaging by MRI. *Curr. Cardiol. Rep.* **8**, 65–69.

4 Lanza, G. M., Wickline, S. A. (2003) Targeted ultrasonic contrast agents for molecular imaging and therapy. *Curr. Probl. Cardiol.* **28**, 625–653.

5 Lanza, G. M., Winter, P., Caruthers, S., Schmeider, A., Crowder, K., Morawski, A., Zhang, H., Scott, M. J., Wickline, S. A. (2004) Novel paramagnetic contrast agents for molecular imaging and targeted drug delivery. *Curr. Pharm. Biotechnol.* **5**, 495–507.

6 Wickline, S. A., Lanza, G. M. (2003) Nanotechnology for molecular imaging and targeted therapy. *Circulation* **107**, 1092–1095.

7 Flaim, S. F. (1994) Pharmacokinetics and side effects of perfluorocarbon-based blood substitutes. *Artif. Cells Blood Substit. Immobil. Biotechnol.* **22**, 1043–1054.

8 McGoron, A. J., Pratt, R., Zhang, J., Shiferaw, Y., Thomas, S., Millard, R. (1994) Perfluorocarbon distribution to liver, lung and spleen of emulsions of perfluorotributylamine (FTBA) in pigs and rats and perfluorooctyl bromide (PFOB) in rats and dogs by 19F NMR spectroscopy. *Artif. Cells Blood Substit. Immobil. Biotechnol.* **22**, 1243–1250.

9 Morawski, A. M., Winter, P. M., Crowder, K. C., Caruthers, S. D., Fuhrhop, R. W., Scott, M. J., Robertson, J. D., Abendschein, D. R., Lanza, G. M., Wickline, S. A. (2004) Targeted nanoparticles for quantita-tive imaging of sparse molecular epitopes with MRI. *Magn. Reson. Med.* **51**, 480–486.

10 Winter, P. M., Shukla, H. P., Caruthers, S. D., Scott, M. J., Fuhrhop, R. W., Robertson, J. D., Gaffney, P. J., Wickline, S. A., Lanza, G. M. (2005) Molecular imaging of human thrombus with computed tomography. *Acad. Radiol.* **12** (Suppl. 1), S9–S13.

11 Lanza, G. M., Lorenz, C. H., Fischer, S. E., Scott, M. J., Cacheris, W. P., Kaufmann, R. J., Gaffney, P. J., Wickline, S. A. (1998) Enhanced detection of thrombi with a novel fibrin-targeted magnetic resonance imaging agent. *Acad. Radiol.* **5** (Suppl. 1), S173–S176; discussion S183–S174.

12 Lanza, G. M., Wallace, K. D., Scott, M. J., Cacheris, W. P., Abendschein, D. R., Christy, D. H., Sharkey, A. M., Miller, J. G., Gaffney, P. J., Wickline, S. A. (1996) A novel site-targeted ultrasonic contrast agent with broad biomedical application. *Circulation* **94**, 3334–3340.

13 Lanza, G. M., Winter, P. M., Neubauer, A. M., Caruthers, S. D., Hockett, F. D., Wickline, S. A. (2005) (1)H/(19)F magnetic resonance molecular imaging with perfluorocarbon nanoparticles. *Curr. Top. Dev. Biol.* **70**, 57–76.

14 Yu, X., Song, S. K., Chen, J., Scott, M. J., Fuhrhop, R. J., Hall, C. S., Gaffney, P. J., Wickline, S. A., Lanza, G. M. (2000) High-resolution MRI characterization of human thrombus using a novel fibrin-targeted para-magnetic nanoparticle contrast agent. *Magn. Reson. Med.* **44**, 867–872.

15 Winter, P. M., Caruthers, S. D., Yu, X., Song, S. K., Chen, J., Miller, B., Bulte, J. W., Robertson, J. D., Gaffney, P. J., Wickline, S. A., et al. (2003) Improved molecular imaging contrast agent for detection of human thrombus. *Magn. Reson. Med.* **50**, 411–416.

16 Brooks, P. C., Stromblad, S., Klemke, R., Visscher, D., Sarkar, F. H., Cheresh, D. A. (1995) Antiintegrin alpha v beta 3 blocks human breast cancer growth and angiogenesis in human skin. *J. Clin. Invest.* **96**, 1815–1822.

17 Kerr, J. S., Mousa, S. A., Slee, A. M. (2001) Alpha(v)beta(3) integrin in angiogenesis and restenosis. *Drug News Perspect.* **14**, 143–150.

18 Bishop, G. G., McPherson, J. A., Sanders, J. M., Hesselbacher, S. E., Feldman, M. J., McNamara, C. A., Gimple, L. W., Powers, E. R., Mousa, S. A., Sarembock, I. J. (2001)

Selective alpha(v)beta(3)-receptor blockade reduces macrophage infiltration and restenosis after balloon angioplasty in the atherosclerotic rabbit. *Circulation* **103**, 1906–1911.

19 Corjay, M. H., Diamond, S. M., Schlingmann, K. L., Gibbs, S. K., Stoltenborg, J. K., Racanelli, A. L. (1999) alphavbeta3, alphavbeta5, and osteopontin are coordinately upregulated at early time points in a rabbit model of neointima formation. *J. Cell. Biochem.* **75**, 492–504.

20 Lanza, G. M., Yu, X., Winter, P. M., Abendschein, D. R., Karukstis, K. K., Scott, M. J., Chinen, L. K., Fuhrhop, R. W., Scherrer, D. E., Wickline, S. A. (2002) Targeted antiproliferative drug delivery to vascular smooth muscle cells with a magnetic resonance imaging nanoparticle contrast agent: implications for rational therapy of restenosis. *Circulation* **106**, 2842–2847.

21 Crowder, K. C., Hughes, M. S., Marsh, J. N., Barbieri, A. M., Fuhrhop, R. W., Lanza, G. M., Wickline, S. A. (2005) Sonic activation of molecularly-targeted nanoparticles accelerates transmembrane lipid delivery to cancer cells through contact-mediated mechanisms: implications for enhanced local drug delivery. *Ultrasound Med. Biol.* **31**, 1693–1700.

22 Wickline, S. A., Neubauer, A. M., Winter, P., Caruthers, S., Lanza, G. (2006) Applications of nanotechnology to atherosclerosis, thrombosis, and vascular biology. *Arterioscler. Thromb. Vasc. Biol.* **26**, 435–441.

23 Thom, T., Haase, N., Rosamond, W., Howard, V. J., Rumsfeld, J., Manolio, T., Zheng, Z. J., Flegal, K., O'Donnell, C., Kittner, S., et al. (2006) Heart disease and stroke statistics – 2006 update: a report from the American Heart Association Statistics Committee and Stroke Statistics Subcommittee. *Circulation* **113**, e85–e151.

24 Flacke, S., Fischer, S., Scott, M. J., Fuhrhop, R. J., Allen, J. S., McLean, M., Winter, P., Sicard, G. A., Gaffney,

P. J., Wickline, S. A., et al. (2001) Novel MRI contrast agent for molecular imaging of fibrin: implications for detecting vulnerable plaques. *Circulation* **104**, 1280–1285.

25 Raut, S., Gaffney, P. J. (1996) Evaluation of the fibrin binding profile of two anti-fibrin monoclonal antibodies. *Thromb. Haemost.* **76**, 56–64.

26 Gossl, M., Rosol, M., Malyar, N. M., Fitzpatrick, L. A., Beighley, P. E., Zamir, M., Ritman, E. L. (2003) Functional anatomy and hemodynamic characteristics of vasa vasorum in the walls of porcine coronary arteries. *Anat. Rec. A Discov. Mol. Cell. Evol. Biol.* **272**, 526–537.

27 Moulton, K. S. (2002) Plaque angiogenesis: its functions and regulation. *Cold Spring Harbor Symp. Quant. Biol.* **67**, 471–482.

28 O'Brien, E. R., Garvin, M. R., Dev, R., Stewart, D. K., Hinohara, T., Simpson, J. B., Schwartz, S. M. (1994) Angiogenesis in human coronary atherosclerotic plaques. *Am. J. Pathol.* **145**, 883–894.

29 Khurana, R., Zhuang, Z., Bhardwaj, S., Murakami, M., De Muinck, E., Yla-Herttuala, S., Ferrara, N., Martin, J. F., Zachary, I., Simons, M. (2004) Angiogenesis-dependent and independent phases of intimal hyperplasia. *Circulation* **110**, 2436–2443.

30 Moreno, P. R., Purushothaman, K. R., Fuster, V., Echeverri, D., Truszczynska, H., Sharma, S. K., Badimon, J. J., O'Connor, W. N. (2004) Plaque neovascularization is increased in ruptured atherosclerotic lesions of human aorta: implications for plaque vulnerability. *Circulation* **110**, 2032–2038.

31 Tenaglia, A. N., Peters, K. G., Sketch, M. H., Jr., Annex, B. H. (1998) Neovascularization in atherectomy specimens from patients with unstable angina: implications for pathogenesis of unstable angina. *Am. Heart J.* **135**, 10–14.

32 Winter, P. M., Morawski, A. M., Caruthers, S. D., Fuhrhop, R. W.,

Zhang, H., Williams, T. A., Allen, J. S., Lacy, E. K., Robertson, J. D., Lanza, G. M., et al. (2003) Molecular imaging of angiogenesis in early-stage atherosclerosis with alpha(v)-beta3-integrin-targeted nanoparticles. *Circulation* **108**, 2270–2274.

33 Miller, J. C., Pien, H. H., Sahani, D., Sorensen, A. G., Thrall, J. H. (2005) Imaging angiogenesis: applications and potential for drug development. *J. Natl. Cancer Inst.* **97**, 172–187.

34 Schmieder, A. H., Winter, P. M., Caruthers, S. D., Harris, T. D., Williams, T. A., Allen, J. S., Lacy, E. K., Zhang, H., Scott, M. J., Hu, G., et al. (2005) Molecular MR imaging of melanoma angiogenesis with alphanubeta3-targeted paramagnetic nanoparticles. *Magn. Reson. Med.* **53**, 621–627.

35 Cyrus, T., Abendschein, D. R., Caruthers, S. D., Harris, T. D., Glattauer, V., Werkmeister, J. A., Ramshaw, J. A., Wickline, S. A., Lanza, G. M. (2006) MR three-dimensional molecular imaging of intramural biomarkers with targeted nanoparticles. *J. Cardiovasc. Magn. Reson.* **8**, 535–541.

36 Ingber, D., Fujita, T., Kishimoto, S., Sudo, K., Kanamaru, T., Brem, H., Folkman, J. (1990) Synthetic analogues of fumagillin that inhibit angiogenesis and suppress tumour growth. *Nature* **348**, 555–557.

37 Winter, P. M., Neubauer, A. M., Caruthers, S. D., Harris, T. D., Robertson, J. D., Williams, T. A., Schmieder, A. H., Hu, G., Allen, J. S., Lacy, E. K., et al. (2006) Endothelial alpha(v)beta3 integrin-targeted fumagillin nanoparticles inhibit angiogenesis in atherosclerosis. *Arterioscler. Thromb. Vasc. Biol.* **26**, 2103–2109.

38 Moulton, K. S., Heller, E., Konerding, M. A., Flynn, E., Palinski, W., Folkman, J. (1999) Angiogenesis inhibitors endostatin or TNP-470 reduce intimal neovascularization and plaque growth in apolipoprotein E-deficient mice. *Circulation* **99**, 1726–1732.

39 Herbst, R. S., Madden, T. L., Tran, H. T., Blumenschein, G. R., Jr., Meyers, C. A., Seabrooke, L. F., Khuri, F. R., Puduvalli, V. K., Allgood, V., Fritsche, H. A., Jr., et al. (2002) Safety and pharmacokinetic effects of TNP-470, an angiogenesis inhibitor, combined with paclitaxel in patients with solid tumors: evidence for activity in non-small-cell lung cancer. *J. Clin. Oncol.* **20**, 4440–4447.

40 Lanza, G. M., Trousil, R. L., Wallace, K. D., Rose, J. H., Hall, C. S., Scott, M. J., Miller, J. G., Eisenberg, P. R., Gaffney, P. J., Wickline, S. A. (1998) In vitro characterization of a novel, tissue-targeted ultrasonic contrast system with acoustic microscopy. *J. Acoust. Soc. Am.* **104**, 3665–3672.

41 Lanza, G. M., Abendschein, D. R., Hall, C. S., Marsh, J. N., Scott, M. J., Scherrer, D. E., Wickline, S. A. (2000) Molecular imaging of stretch-induced tissue factor expression in carotid arteries with intravascular ultrasound. *Invest. Radiol.* **35**, 227–234.

42 Morawski, A. M., Lanza, G. A., Wickline, S. A. (2005) Targeted contrast agents for magnetic resonance imaging and ultrasound. *Curr. Opin. Biotechnol.* **16**, 89–92.

43 Lanza, G. M., Abendschein, D. R., Hall, C. S., Scott, M. J., Scherrer, D. E., Houseman, A., Miller, J. G., Wickline, S. A. (2000) In vivo molecular imaging of stretch-induced tissue factor in carotid arteries with ligand-targeted nanoparticles. *J. Am. Soc. Echocardiogr.* **13**, 608–614.

44 Kennedy, J. E. (2005) High-intensity focused ultrasound in the treatment of solid tumours. *Nat. Rev. Cancer* **5**, 321–327.

45 Morgan-Hughes, G. J., Marshall, A. J., Roobottom, C. A. (2002) Multislice computed tomography cardiac imaging: current status. *Clin. Radiol.* **57**, 872–882.

46 Mattrey, R. F. (1994) The potential role of perfluorochemicals (PFCs) in diagnostic imaging. *Artif. Cells Blood Substit. Immobil. Biotechnol.* **22**, 295–313.

47 Wickline, S. A., Lanza, G. M. (2002) Molecular imaging, targeted therapeutics, and nanoscience. *J. Cell. Biochem. Suppl.* **39**, 90–97.

48 Ahrens, E. T., Flores, R., Xu, H., Morel, P. A. (2005) In vivo imaging platform for tracking immuno-therapeutic cells. *Nat. Biotechnol.* **23**, 983–987.

49 Artemov, D., Bhujwalla, Z. M., Bulte, J. W. (2004) Magnetic resonance imaging of cell surface receptors using targeted contrast agents. *Curr. Pharm. Biotechnol.* **5**, 485–494.

50 Atri, M. (2006) New technologies and directed agents for applications of cancer imaging. *J. Clin. Oncol.* **24**, 3299–3308.

51 Delikatny, E. J., Poptani, H. (2005) MR techniques for in vivo molecular and cellular imaging. *Radiol. Clin. North Am.* **43**, 205–220.

52 Persigehl, T., Heindel, W., Bremer, C. (2005) MR and optical approaches to molecular imaging. *Abdom. Imaging* **30**, 342–354.

53 Choi, C. J., Kramer, C. M. (2002) MR imaging of atherosclerotic plaque. *Radiol. Clin. North Am.* **40**, 887–898.

54 Larose, E., Yeghiazarians, Y., Libby, P., Yucel, E. K., Aikawa, M., Kacher, D. F., Aikawa, E., Kinlay, S., Schoen, F. J., Selwyn, A. P., et al. (2005) Characterization of human athero-sclerotic plaques by intravascular magnetic resonance imaging. *Circulation* **112**, 2324–2331.

Part IV
Nanomotors

Nanobiotechnology II. Edited by Chad A. Mirkin and Christof M. Niemeyer
Copyright © 2007 WILEY-VCH Verlag GmbH & Co. KGaA, Weinheim
ISBN: 978-3-527-31673-1

20
Biological Nanomotors

Manfred Schliwa

20.1
Overview

The increase in cell size that characterizes eukaryotic cells was accompanied by the elaboration of molecular machineries that stabilize cell shape, power cell movement, secure segregation of the genetic material, and deliver goods to specific destinations within the cell. These tasks are accomplished by a special class of machines termed "molecular motors", which use polymers of two classes of cytoskeletal fiber as tracks on which to move: (i) microfilaments composed of actin subunits; and (ii) microtubules made from tubulin dimers. Whereas relatives of these cytoskeletal polymers already form part of the prokaryotic make-up, motors apparently are novel inventions of the eukaryotic cell. Three classes of these linear molecular motors are known to date myosins, which use actin filaments as tracks; and kinesins and dyneins, which move on microtubules. For almost a century, myosin from skeletal muscle was the only protein known to be involved in force generation and movement [1], but it was joined in 1965 by dynein, an ATPase present in flagella and cilia [2]. Many biologists at the time probably were quite happy with the view of one motor (myosin) being responsible for cytoplasmic movements, and a second (dynein) for ciliary and flagellar beating. However, many cellular movements could not clearly be associated with either myosin or dynein, and this eventually led to the discovery of a new type of cytoplasmic motor, kinesin, in 1985 [3, 4]. With respect to different motor categories, this seemed to be the end of the line, but subsequently further complexity arose within each group. A combination of biochemical, molecular genetic and genomic approaches revealed that each of the three motor classes comprises superfamilies of motors of strikingly varied make-up and function. Today, we can distinguish at least 24 different classes of myosins [5], 14 different families of kinesins [6], and two groups of dyneins (axonemal and cytoplasmic) [7]. Schematic overviews of the domain organization and molecular architecture of representative motor families are shown in Figure 20.1.

Nanobiotechnology II. Edited by Chad A. Mirkin and Christof M. Niemeyer
Copyright © 2007 WILEY-VCH Verlag GmbH & Co. KGaA, Weinheim
ISBN: 978-3-527-31673-1

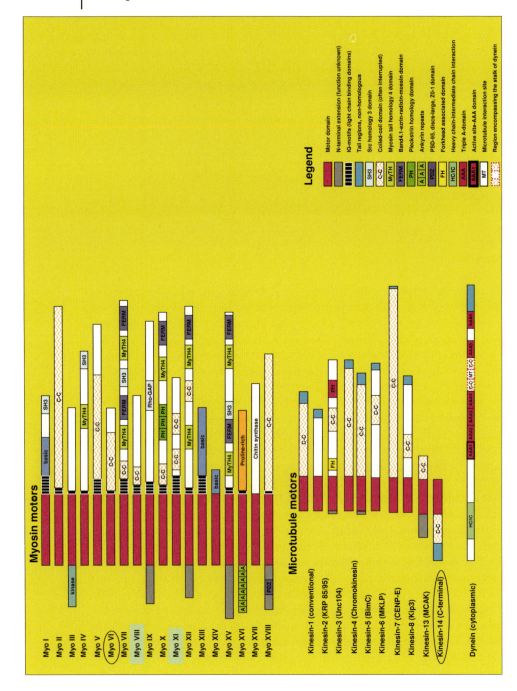

Myosin motors

Myo I
Myo II
Myo III
Myo IV
Myo V
Myo VI
Myo VII
Myo VIII
Myo IX
Myo X
Myo XI
Myo XII
Myo XIII
Myo XIV
Myo XV
Myo XVI
Myo XVII
Myo XVIII

Microtubule motors

Kinesin-1 (conventional)
Kinesin-2 (KRP 85/95)
Kinesin-3 (Unc104)
Kinesin-4 (Chromokinesin)
Kinesin-5 (BimC)
Kinesin-6 (MKLP)
Kinesin-7 (CENP-E)
Kinesin-8 (Kip3)
Kinesin-13 (MCAK)
Kinesin-14 (C-terminal)

Dynein (cytoplasmic)

Legend

Motor domain
N-terminal extension (function unknown)
IQ-motifs (light chain binding domains)
Tail regions, non-homologous
SH3   Src homology 3 domain
C-C   Coiled-coil domain (often interrupted)
MyTH   Myosin tail homology 4 domain
FERM   Band4.1-ezrin–radixin–moesin domain
PH   Pleckstrin homology domain
A|A|A Ankyrin repeats
PDZ   PSD-95, discs-large, ZO-1 domain
FH   Forkhead associated domain
HC/IC Heavy chain–intermediate chain interaction
AAA   Triple A-domain
AAA   Active site-AAA domain
MT   Microtubule interaction site
       Region encompassing the stalk of dynein

A mammalian organism may harbor the genes for over 100 different motors, and a given cell may express well over 50 of these – a number that probably can be multiplied several-fold due to alternative splicing, post-translational modifications, or the make-up of associated proteins. Many of these motors have not yet been characterized, and clear functions are assigned only to a subset. Nevertheless, despite large gaps in our knowledge, amazing insights into the workings of these machines have been gained during the past few years, and these form the basis of this chapter.

The three classes of motors share several features:

- They require a polar track along which to move – actin filaments in the case of myosin, and microtubules for dyneins and kinesins. The information on polarity is provided by the uniform molecular orientation of the subunit proteins actin and tubulin, respectively, that make up the track. Intermediate filaments – the third major structural component of eukaryotic cells – are non-polar and are not known to support oriented movement.

- All motors use the energy derived from nucleoside triphosphate hydrolysis (ATP in most cases) to undergo conformational changes that result in directional movement along the track. The basic mechanisms of these conformational changes appear to be similar for kinesins and myosins, but different from dyneins.

- The binding sites for both ATP and the track are located in a globular catalytic (or motor) domain, also referred to as the "head". In the process of movement, the two sites communicate with each other and with sites that amplify the conformational change generated upon ATP hydrolysis.

- The non-motor domains include secondary structure elements important for motor function, such as coiled-coil segments for dimerization and domains or associated proteins involved in regulation and cargo binding (Figures 20.1 and 20.2).

- All motors studied so far in some detail can generate a force that is sufficient to move even large objects through viscous cytoplasm.

These features are discussed in more detail in the following sections.

◄──

Fig. 20.1 Overview of the domain organization of cytoskeletal motors. The motor domain is shown in magenta; other domains are labeled as outlined in the legend. The two plant-specific myosin families are boxed in green, motor families with members that move towards the minus-end are circled. For the kinesin motors, the family name preceding the new nomenclature of Lawrence et al. [6] is shown in parentheses.

20.2
The Architecture of the Motor Domain

The motor domain is the center of action in force generation and movement. Its size differs markedly between the three types of motors, being surprisingly small in kinesins (45 kDa), about twice that size in myosins (100 kDa), and exceptionally large in dyneins (500 kDa). As structural information at atomic resolution is the key to an understanding of molecular function, considerable efforts were directed at obtaining the crystal structures of the motor domain. The structure of the myosin II motor domain was solved by X-ray crystallography in 1993 [8], followed in 1996 by the structures of conventional kinesin and the kinesin-like protein ncd [9, 10] (Figure 20.2). As yet, no high-resolution structure is available for any of the dyneins.

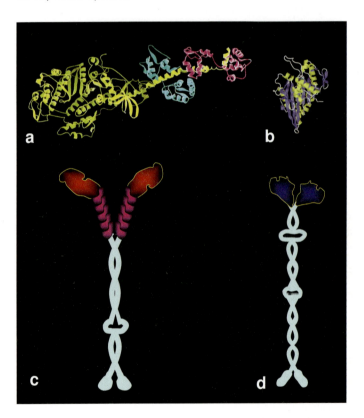

Fig. 20.2 Structure of myosin and kinesin motors. (a) Crystal structure of the motor domain of myosin II; the heavy chain is shown in yellow; the two light chains associated with the neck are shown in blue and magenta. (b) Crystal structure of the motor domain of Kinesin-1; beta-sheets and alpha-helices are shown in blue and yellow, respectively. (c) Schematic overview of dimeric myosin V; the motor domain is shown in red, light chains associated with the neck region are shown in magenta. (d) Schematic overview of the Kinesin-1 dimer; the motor domain is shown in blue; flexible regions in the coiled-coil stalk are shown as bulges.

The structures of kinesin and myosin uncovered an unexpected relationship between the two. The central portion of the myosin head, which harbors the nucleotide binding pocket, is virtually identical in structure to the core of the kinesin motor domain, despite a lack of significant sequence homology [9]. In addition, both structures show similarities to the G-proteins in the region surrounding the nucleotide, the so-called switch I and switch II motifs. In all three proteins the switch regions shift position in response to nucleotide hydrolysis, and thus are instrumental in the catalytic process. These findings are consistent with a common evolutionary origin of myosin, kinesin and G-proteins [11].

Dynein bears no structural relationship whatsoever to kinesin and myosin (Figure 20.3). The dynein heavy chain is a member of the AAA$^+$ ATPases, a highly diverse superfamily of proteins involved in a bewildering spectrum of cellular activities [12]. In its C-terminal portion it contains six AAA$^+$ modules (see Figure 20.1), apparently arranged in a ring [13]. The first four modules exhibit an intact P-loop motif (signifying a nucleotide binding site), while in the other two the P-loops are highly degenerate [7]. Only the first highly conserved AAA domain is believed to hydrolyze ATP, and thus represents the catalytic site. Curiously, the microtubule binding site is located at the end of a unique ~10 nm stalk [14] that extends from the ring at the side opposite of the first AAA unit (Figure 20.3). Thus, the microtubule motors kinesin and dynein are of different evolutionary origin and operate by different mechanisms.

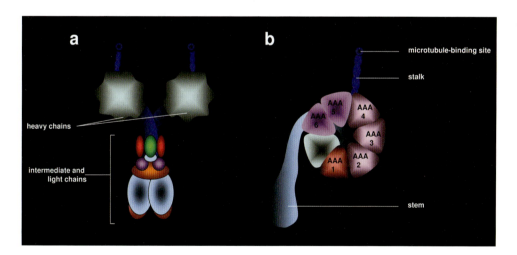

Fig. 20.3 Schematic overview of the dynein motor. (a) Organization of the cytoplasmic dynein dimer. (b) Domain organization of a single heavy chain; the six AAA-domains are arranged in a ring supplemented by a seventh domain of unknown function. A coiled-coil stalk with the microtubule binding site at its tip is extending between the AAA units 4 and 5.

20.3
Initial Events in Force Generation

Conformational changes within the motor domain are the key to an understanding of molecular motors. A consensus is beginning to emerge that force generation and movement begin with small, structural changes in the active site, triggered by ATP hydrolysis, that are translated into a larger conformational change of a mechanical amplifier. This process requires intramolecular coordination of the ATP and filament binding sites. Interruption of this coordination by, for example, mutations of amino acids that flank the nucleotide binding pocket uncouples ATP hydrolysis from track binding [15, 16].

In both kinesins and myosins (Figure 20.4), ATP binding and hydrolysis generate a shift in the position in two structural elements, termed switch I and switch

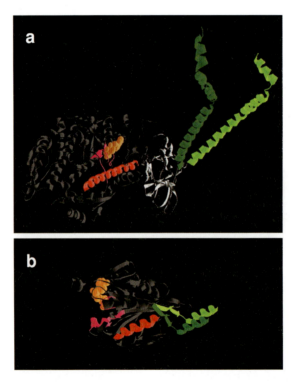

Fig. 20.4 Intramolecular communication in the motor domains of myosin (a) and kinesin (b). ATP is shown as a space-filling model in orange. ATP hydrolysis causes a shift in the position of the switch regions (magenta), which is communicated by the relay helix (red) to the neck region of myosin or the neck-linker of kinesin, respectively, causing a shift in the position of these structural elements (light and dark green).

II (based on the G-protein nomenclature), flanking the nucleotide binding pocket [17, 18]. In myosins, these conformational changes are communicated to the C-terminal region of the head via the so-called relay helix that passes by the actin binding face and ends in a moveable element, the converter domain [19]. Further C-terminal, a mechanical amplifier consisting of an α-helix stabilized by the myosin light chains (see Figure 20.2) originates in the converter domain (Figure 20.4a). According to the crystal structures obtained in different nucleotide states, the amplifier acts as a lever arm that appears to undergo a large conformational change, a rotation through ~70° [20]. This swing of the lever arm is believed to be the ultimate cause of a working stroke. In the myosin II motor, with its two light chains bound to the so-called IQ motifs of the myosin neck region, the size of this working stroke is ~5 nm [21]. If the working stroke of myosins is indeed determined by the length of the neck region, then motors with shorter levers should be slower while those with longer necks should move faster. In agreement with this model, modification of the neck length (fewer or more light chain-binding IQ motifs) either slows down or accelerates the motor both *in vitro* [22, 23] and *in vivo* [24]. Although the different conformational states of the lever correlate nicely with different nucleotide states and thus allow to model the power stroke, all the crystal structures are of motor domains in solution. The "true" power stroke takes place in association with actin, however, so that caution must be exercised in the interpretation of the structures seen.

Kinesins do not possess a rigid lever arm, but they have an analogous element – the neck linker – which consists of approximately 10 amino acids at the C-terminus of the motor domain. The neck linker has been proposed to shift its position in response to nucleotide hydrolysis, being docked along the motor domain in the ATP state or mobile in the ADP state [25]. The information on the state of the active site is relayed via the switch II region to the so-called switch II helix, which is equivalent to the relay helix of myosin (Figure 20.4b). In kinesin, this helix may undergo a length change and a small but significant rotation, which is suggested to tighten the motor domain on the microtubule surface [26] and therefore may be an integral part of the relay mechanism. Interestingly, a twist of the relay helix relative to the rest of the motor domain may also occur in myosins [27]. Thus, the structural elements involved in sensing and transmitting hydrolysis-dependent changes are similar in kinesin and myosin, but the last step of translation into a large-scale conformational change apparently involves the swing of a rigid body (the lever arm) in myosin and the repositioning of a flexible element (the neck linker) in kinesin [28].

How dyneins transmit a conformational change that occurs in the first AAA unit to a microtubule binding site located 20 nm away is presently unclear. As stated above, this is due to the lack of an atomic resolution structure of the motor domain. However, painstaking analysis of dynein molecules by electron microscopy suggests a rolling rotation of the ring/stalk unit relative to the rest of the molecule (Figure 20.5) [29]. This rotation is associated with ATP hydrolysis and release of the products, and shifts the position of the microtubule-binding site at the end of the stalk by about 10 nm. Although a shift in the position of the stalk

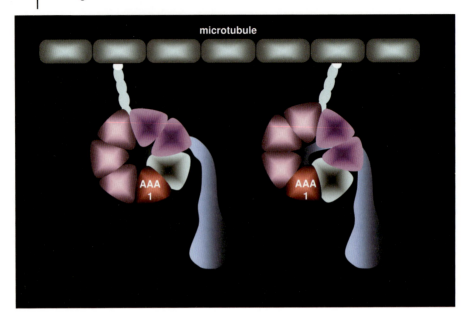

Fig. 20.5 Pre- and post-power stroke conformation of the dynein heavy chain. Rolling rotation of the ring structure causes a shift in the position of the stalk that binds to the microtubule. (Schematic diagram based on the studies of Burgess et al. [29].)

superficially resembles a swinging lever, its cause and transmission mechanism bear no resemblance to that of myosin motors.

20.4
Stepping, Hopping, and Slithering

The question remains as to how these structural changes are translated into movement along the track. There is no generic answer, as a spectrum of behaviors is observed. Several motors, including kinesin-1 or mammalian myosin V, can move along their respective tracks for long distances without dissociating, taking hundreds of steps in the process. This stepping mode, which is referred to as "processive" movement, most likely evolved as a result of a physiological requirement for long-range movement driven by only one or a few motor molecules. Processive motors tend to be dimeric. However, dimerization is neither necessary nor sufficient for continuous movement. Some monomeric motors (e.g., class 3 kinesins, myosin IXb) can move for long distances without dissociating, while certain dimeric motors are non-processive (e.g., myosin II, members of the kinesin-14 family that move in the direction opposite to all other kinesins).

The "duty ratio" is a useful concept by which to explain the difference between processive and non-processive motors. This ratio describes the fraction of the time that a motor domain spends attached to its track. A single step can be envisioned to consist of a working stroke – when the head is bound to the track – and a recovery stroke during the detached phase. For example, a duty ratio of 0.1 indicates that the motor domain spends 10% of its ATPase cycle attached to the track. Processive motors must have a large duty ratio, otherwise the chance to diffuse away from the track during the detached phase would be too high. Better yet is the use of two heads that alternate in their binding to the track so that one head is bound at all times.

A single molecule of conventional kinesin can take over 100 steps of 8 nm, which corresponds to the spacing of tubulin dimers, without dissociating from the microtubule [30], and one molecule of ATP is hydrolyzed per step [31]. These basic features are maintained even in a fast kinesin, the motor domain of which can hydrolyze approximately 260 ATPs per second [32]. Studies with single-headed motors clearly established a requirement for two heads in processive behavior [33]. The hydrolysis cycles of the two heads must be coordinated in such a way that one of the heads is bound to the microtubule while the other is free to move, and the two heads must be kept out of phase [34, 35]. Current models envision a "hand-over-hand" cycle, where the free head moves past the bound head to find a new binding site. So-called "inchworm models", where kinesin is limping along the track with one head always in the lead and the other always trailing [36], apparently can be ruled out [37]. The duty ratio must be at least 0.5, and a phase must exist in which both heads are bound to the microtubule. Since in the crystal structure of dimeric kinesin the two heads are separated by no more than ~5 nm [38], binding of both heads to two tubulin subunits spaced 8 nm apart would require considerable rearrangement of the domains adjacent to the head (i.e., neck linker and/or neck). Indeed, the mobility of the neck-linker in monomeric constructs attached to microtubules is consistent with the existence of a "two heads bound" state [25]. Unraveling of the neck coiled-coil would be another possibility to accommodate such a state, but experimental modifications of the neck only result in minor changes in the run length and apparently do not affect head-head coordination [39].

The paradigm for a processive myosin – mammalian myosin V – is unique in that the neck has six IQ motifs, signifying six light chain-binding sites (see Figure 20.1). Thus, in theory myosin V could take large steps if the longer neck does indeed function as a lever. A number of elegant studies are in agreement with this view. The step size appears to be ~36 nm, which corresponds to the pitch of the actin helix [40]. A rate of movement of ~0.5 μm s^{-1} and an ATPase activity of 13 per second are consistent with the hydrolysis of one molecule of ATP per 36-nm step [41].

The next question is whether this large step is taken in one giant sweep, or are there substeps? This question seems justified since, for example, the kinesin step [42] or the working stroke of myosin I [43] may be composed of two substeps. The working stroke appears to be ~25 nm, and is itself composed of two phases of 20 nm and 5 nm. The missing 11 nm to complete the 36-nm stride is proposed to be contributed by thermally driven diffusion [44]. Thus, myosin V may have a significant Brownian ratchet component in its movement. In other motors, a Brownian ratchet type of movement may prevail, though this is controversial. Nonetheless, it may explain the behavior of a chimera of a short-neck myosin V with only one IQ motif and a smooth muscle myosin tail. This chimera can take relatively large steps, but has a broad distribution of step sizes and exhibits many backward steps [45].

One other intriguing motor is myosin VI, which moves towards the minus end of actin filaments due to the action of a unique insertion near the motor domain [46]. It can work either as a non-processive monomer or a processive dimer, depending on the task in the cell [47]. Although it only has one IQ motif – and therefore a short neck, or lever arm – it can take 36-nm steps when dimeric [48]. Here, the short power stroke component appears to serve primarily as a bias for directionality, while the Brownian movement component contributes a larger fraction of the step. The movement is hand-over-hand, however, implying the presence of a flexible element next to the lever to accommodate such a large step size [49].

For many years, it was believed that the long-range movement of single motors requires dimers, the heads of which are tightly coordinated. It therefore came as a surprise when a monomeric kinesin-3 (KIF1A) [50], a monomeric myosin (class IXb) [51] and a monomeric dynein (inner arm dynein c) [52] were suggested to move along their tracks as monomers for long distances. However, the characteristics of these monomer movements differ from the typical processive movements of dimeric motors. The best-studied of these motors, KIF1A, can stay bound to microtubules for several seconds and diffuses back and forth, with a net movement towards the microtubule plus end. The key to this behavior seems to be the presence of a second microtubule binding site that contains several positively charged residues ("K-loop") which can interact with the negatively charged C-terminus of tubulin ("E-hook"). Thus, the K-loop acts as a tether, while the power stroke of KIF1A provides the push that biases diffusion towards the microtubule plus end [50]. However, other members of the KIF1 family that also possess the ominous K-loop fail to show processive motility [53], leaving some doubts as to the universality of this otherwise plausible mechanism.

Motors that have a low duty ratio and lack a tether to the track are non-processive. The two best-studied non-processive motors are myosin II and the kinesin-14 motor ncd. Both are dimeric, but the two heads do not cooperate. In muscle, processive movement is unnecessary because the myosin molecules operate independently and are part of a large paracrystalline arrangement, the sarcomere. While myosin II is non-processive because the two heads go their separate ways, the reason for the lack of processivity of ncd may be just the opposite: they are too tightly associated with the neck coiled-coil [54]. Ncd lacks a flexible neck linker, and so far no evidence for uncoiling of the neck that would accommodate a two-heads-bound state has been obtained. Interestingly, however, the neck of ncd undergoes a lever arm-like rotation of the neck to produce its one-step movement [55].

A general conclusion emerging from studies on processive motors is that moving along the track may entail both a mechanical component and a diffusive component, with different motors using different proportions of each. Most dimeric, processive motors rely on conformational changes and tight coupling, with a smaller contribution from diffusional searching, although in myosin VI the opposite seems true. Monomeric processive motors appear to have a large contribution from diffusion, which makes their manner and form of processive behavior markedly different from that of dimeric motors. In both, the diffusional component seems to be supported by secondary "tethering" sites that enhance motor performance.

20.5
Directionality

Most cell biologists would have been rather comfortable with the idea that a given superfamily of motors moves in one direction only. However, this comforting thought was shattered when a kinesin-like protein, *Drosophila* ncd [56], and a member of the myosin VI family [57], were found to move towards the minus ends of microtubules and actin filaments, respectively. So dynein is the last hope for a unidirectional motor superfamily.

Analysis of the minus-end kinesins and myosins has offered important insights not only in the basis of reversed directionality, but also into motor mechanochemistry in general. In kinesins, the chief reason for reversed directionality is the placement of the motor domain. While the atomic structures of plus or minus kinesin motor domains are similar, minus-end motors have the motor domain at the C-terminus following the neck region, while in plus-end motors it is at the N-terminus. This placement, which so far is characteristic of all minus-end kinesins, alters the head–neck interaction which, in turn, is the determining factor for directionality. When motor domains are swapped, the resulting chimeras adopt the direction of movement specified by the neck (e.g., [58]). The movement of these chimeras is very slow, and this points to an intrinsic plus-end bias, even in a minus-end motor domain [59]. Convincing evidence for the overriding importance of the neck region in directional determination came from the analysis of a point mutant in the ncd neck that completely lacks directionality, switching stochastically between plus-end and minus-end movement [60].

The reversed polarity of the minus-end-directed myosin VI motor is attributed to a unique 53 amino-acid insertion in the myosin VI converter domain. Surprisingly, atomic structures of the myosin head reveal this insertion to be a modified light chain-binding domain (IQ-motif) [46] that is proposed to reverse the direction of the lever arm swing. Artificial levers accomplish the same trick, demonstrating that the same conformational change in the motor domain can be redirected with appropriate head–neck linkages [61].

20.6
Forces

Molecular motors have evolved to generate forces that can be used to the cell's advantage. In order to measure the forces that single motor molecules can generate, two variants of a laser trap set-up have been widely used. In one system, a microbead carrying one motor is held in a laser trap while the motor moves on its cytoskeletal track attached to a coverslip (Figure 20.6a). As the motor with its bead attempts to escape from the laser trap, it works against an increasing load until it stalls. The stall force is the maximum force that the motor can develop. This set-up can be used with processive motors. In order to measure the forces of non-processive motors such as myosin II, an actin filament suspended between two beads in a double laser trap is lowered onto a myosin molecule attached to a third bead fixed on the surface of a microscopic slide (Figure 20.6b). As the motor interacts with the actin filament, the pull exerted on one of the beads that hold the actin filament is determined. It transpires that the forces of all motors measured so far lie in the range of 1 to 10 pN [62–64], and are sufficient to move even large objects such as vesicles or organelles through viscous cytoplasm.

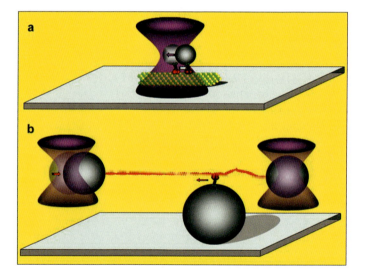

Fig. 20.6 Laser trap set-ups to determine motor activity. The waist of the focused laser beam is shown in magenta. (a) A processive kinesin motor attached to a bead is moving on a microtubule fixed to a coverslip. As the bead is being moved out of the center of the laser trap, retaining optical forces increase until the force exerted by the motor equals the retaining force and the motor stalls in its movement. (b) Set-up to measure the force of a non-processive myosin. An actin filament (red) stretched between two beads suspended by two laser traps is lowered onto a third, fixed bead carrying one motor on its surface. The shift in the position of the left bead caused by the action of the motor is determined.

Laser traps with feedback control that apply a constant external load on the motor allow the determination of load-dependent transitions in the stepping cycle (e.g., [65, 66]).

20.7
Motor Interactions

Molecular motors are involved in a wide variety of cellular activities, including the transport of cellular constituents (e.g., organelles, vesicles, molecular complexes), cell division, translocation, sensory transduction, and development. These activities require specific interaction of the motor(s) with their target sites mediated by regions outside the core motor domain and involving accessory molecules or interacting complexes. The questions of how motors link up to their cargoes, and how transport is regulated, are perhaps the least well understood aspects of motor physiology, because every transport process has its own specific type of motor–cargo interaction, and few generalizations are possible. However, a number of principles have emerged. Some motors, such as certain members of the monomeric myosin I or kinesin-3 families, may bind directly to the phospholipid bilayer. In most cases, however, motors interact with integral membrane proteins either directly or via a (sometimes rather complex) machinery of linker proteins. In the kinesin-1 family, for example, a particularly prominent role is played by a pair of accessory proteins termed kinesin light chains which, in animals at least, figure prominently as linkers to membrane-associated proteins or protein complexes [67]. Cytoplasmic dynein, an organelle transporter towards the microtubule minus end, interacts via one of its associated polypeptides with an activator complex, dynactin, which itself is composed of more than a dozen subunits. An emerging principle for both the recruitment to cargo and regulation of motors is an association with small G-proteins of the Rab family, used by a subset of members of all three motor superfamilies [68]. A prime example is found in melanophores, where pigment granules called "melanosomes" are transported by all three classes of motors (cytoplasmic dynein, kinesin-2, myosin V). Here, kinesin-2 and myosin V cooperate in centrifugal transport, during which myosin V is recruited to melanosomes by Rab27a via the Rab-binding protein, melanophilin, which then binds myosin V [69]. The factors that control the activity of Rab27a may therefore control the recruitment, and thus the activity, of this melanosome motor. During centripetal transport the number of *active* dyneins – but not the total number of dyneins – increases [70]. The mechanisms that govern this switch in dynein activity are poorly understood. In general, the factors that determine the recruitment, activation, inactivation, and unloading of a motor during transport are a fertile ground for further research. Binding to non-motor domains, protein factors, and post-translational modifications all appear to play a role (for examples, see Ref. [71]), but the complete cycle from cargo binding to release is not yet fully understood for any of the transport processes studied to date.

20.8
Outlook

Some 15 years ago, the general belief (or at least hope) was that myosin is for contraction and movement, dynein is for cilia and flagella, and kinesin is for organelle transport. How things have changed! To date, molecular motors are implicated in a bewildering spectrum of cellular activities. In addition to the three "classical" roles, we now count among their tasks unexpected functions such as signaling, RNA localization, and sensory transduction, and we are beginning to appreciate their implications in cellular homeostasis, basic developmental processes, and a growing number of diseases. Because several motors – and even entire classes or families – have not yet been characterized, the full spectrum of cellular roles has yet to be appreciated. Whereas the general principles of motor chemomechanics have been elucidated, large gaps still exist in our knowledge of the biochemical and biophysical details. Recent progress in the design of nanomechanical devices to measure atomic-scale movements and forces has indeed been amazing, and at present there is no end in sight. Spatiotemporal resolution and sensitivity are still improving and, in combination with the rational design of motor mutants, this will allow us to determine the details of motor physiology. Moreover, as yet only a subset of motors has been characterized functionally, and the functions suggested for some of these need still to be confirmed. Many motors – and particularly those in plants – are known only by sequence, and hence this is a fertile playground for the "cell biological hunter–gatherer". Finally, the question of whether cytoplasmic motors will find applications in nanobiotechnological devices remains a challenge for the foreseeable future.

Acknowledgments

The author thanks Günther Woehlke for preparing Figure 20.4. These studies have been supported by the Deutsche Forschungsgemeinschaft and the Fonds der Chemischen Industrie.

References

1 Kühne, W. (1864) *Untersuchungen über das Protoplasma und die Contractilität*. Engelmann Verlag, Leipzig.

2 Gibbons, I. R., Rowe, A. J. (1965) Dynein: a protein with adenosine triphosphatase activity from cilia. *Science* **149**, 424–426.

3 Brady, S. T. (1985) A novel brain ATPase with properties expected for the fast axonal transport motor. *Nature* **317**, 73–75.

4 Vale, R. D., Reese, T. S., Sheetz, M. P. (1985) Identification of a novel force-generating protein, kinesin, involved in microtubule-based motility. *Cell* **42**, 39–50.

5 Foth, B. J., Goedecke, M. C., Soldati, D. (2006) New insights into myosin evolution and classification. *Proc. Natl. Acad. Sci. USA* **103**, 3681–3686.

6 Lawrence, C. J., Dawe, R. K., Christie, K. R., Cleveland, D. W., Dawson, S. C.,

Endow, S. A., Goldstein, L. S., Goodson, H. V., Hirokawa, N., Howard, J. (2004) A standardized kinesin nomenclature. *J. Cell Biol.* **167**, 19–22.

7 Asai, D. J., Wilkes, D. E. (2004) The dynein heavy chain family. *J. Eukaryot. Microbiol.* **51**, 23–29.

8 Rayment, J., Rypniewski, W. R., Schmidt-Base, K., Smith, R., Tomchick, D. R., Benning, M. M., Winkelmann, D. A., Wesenberg, G., Holden, H. M. (1993) Three-dimensional structure of myosin subfragment-1: a molecular motor. *Science* **261**, 50–58.

9 Kull, F. J., Sablin, E. P., Lau, R., Fletterick, R. J., Vale, R. D. (1996) Crystal structure of the kinesin motor domain reveals a structural similarity to myosin. *Nature* **380**, 550–555.

10 Sablin, E. P., Kull, F. J., Cooke, R., Vale, R. D., Fletterick, R. J. (1996) Crystal structure of the motor domain of the kinesin-related motor ncd. *Nature* **380**, 555–559.

11 Kull, F. J., Vale, R. D., Fletterick, R. J. (1998) The case for a common ancestor: kinesin and myosin motor proteins and G proteins. *J. Muscle Res. Cell Motil.* **19**, 877–886.

12 Ogura, T., Wilkinson, A. J. (2001) AAA+ superfamily ATPases: common structure – diverse function. *Genes Cells* **6**, 575–597.

13 Samso, M., Koonce, M. P. (2004) 25 Angstrom resolution structure of a cytoplasmic dynein motor reveals a seven-member planar ring. *J. Mol. Biol.* **340**, 1059–1072.

14 Koonce, M. P., Tikhonenko, I. (2000) Functional elements within the dynein microtubule-binding domain. *Mol. Biol. Cell* **11**, 523–529.

15 Yun, M., Zhang, X., Park, C. G., Park, H. W., Endow, S. A. (2001) A structural pathway for activation of the kinesin motor ATPase. *EMBO J.* **20**, 2611–2618.

16 Murphy, C. T., Rock, R. S., Spudich, J. A. (2001) A myosin II mutation uncouples ATPase activity from motility and shortens step size. *Nat. Cell Biol.* **3**, 311–315.

17 Takagi, Y., Shuman, H., Goldman, Y. E. (2004) Coupling between phosphate release and force generation in muscle actomyosin. *Philos. Trans. R. Soc. Lond. B. Biol. Sci.* **359**, 1913–1920.

18 Marx, A., Muller, J., Mandelkow, E. (2005) The structure of microtubule motor proteins. *Adv. Protein Chem.* **71**, 299–344.

19 Geeves, M. A., Fedorov, R., Manstein, D. J. (2005) Molecular mechanism of actomyosin-based motility. *Cell Mol. Life Sci.* **62**, 1462–1477.

20 Houdusse, A., Szent-Gyorgyi, A. G., Cohen, C. (2000) Three conformational states of scallop myosin S1. *Proc. Natl. Acad. Sci. USA* **97**, 11238–11243.

21 Spudich, J. A. (2001) The myosin swinging cross-bridge model. *Nat. Rev. Mol. Cell Biol.* **2**, 387–392.

22 Uyeda, T. Q., Abramson, P. D., Spudich, J. A. (1996) The neck region of the myosin motor domain acts as a lever arm to generate movement. *Proc. Natl. Acad. Sci. USA* **93**, 4459–4464.

23 Ruff, C., Furch, M., Brenner, B., Manstein, D. J., Meyhofer, E. (2001) Single-molecule tracking of myosins with genetically engineered amplifier domains. *Nat. Struct. Biol.* **8**, 226–229.

24 Schott, D. H., Collins, R. N., Bretscher, A. (2002) Secretory vesicle transport velocity in living cells depends on the myosin-V lever arm length. *J. Cell Biol.* **156**, 35–39.

25 Rice, S., Lin, A. W., Safer, D., Hart, C. L., Naber, N., Carragher, B. O., Cain, S. M., Pechatnikova, E., Wilson-Kubalek, E. M., Whittaker, M. (1999) A structural change in the kinesin motor protein that drives motility. *Nature* **402**, 778–784.

26 Kikkawa, M., Sablin, E. P., Okada, Y., Yajima, H., Fletterick, R. J., Hirokawa, N. (2001) Switch-based mechanism of kinesin motors. *Nature* **411**, 439–445.

27 Schliwa, M., Woehlke, G. (2001) Molecular motors. Switching on kinesin. *Nature* **411**, 424–425.

28 Vale, R. D., Milligan, R. A. (2000) The way things move: looking under the

hood of molecular motor proteins. *Science* **288**, 88–95.

29 Burgess, S. A., Walker, M. L., Sakakibara, H., Knight, P. J., Oiwa, K. (2003) Dynein structure and power stroke. *Nature* **421**, 715–718.

30 Svoboda, K., Schmidt, C. F., Schnapp, B. J., Block, S. M. (1993) Direct observation of kinesin stepping by optical trapping interferometry. *Nature* **365**, 721–727.

31 Hua, W., Young, E. C., Fleming, M. L., Gelles, J. (1997) Coupling of kinesin steps to ATP hydrolysis. *Nature* **388**, 390–393.

32 Kallipolitou, A., Deluca, D., Majdic, U., Lakämper, S., Cross, R., Meyhofer, E., Moroder, L., Schliwa, M., Woehlke, G. (2001) Unusual properties of the fungal conventional kinesin neck domain from *Neurospora crassa*. *EMBO J.* **20**, 6226–6235.

33 Hancock, W. O., Howard, J. (1999) Kinesin's processivity results from mechanical and chemical coordination between the ATP hydrolysis cycles of the two motor domains. *Proc. Natl. Acad. Sci. USA* **96**, 13147–13152.

34 Hackney, D. D. (1994) Evidence for alternating head catalysis by kinesin during microtubule-stimulated ATP hydrolysis. *Proc. Natl. Acad. Sci. USA* **91**, 6865–6869.

35 Hackney, D. D. (1995) Highly processive microtubule-stimulated ATP hydrolysis by dimeric kinesin head domains. *Nature* **377**, 448–450.

36 Hua, W., Chung, J., Gelles, J. (2002) Distinguishing inchworm and hand-over-hand processive kinesin movement by neck rotation measurements. *Science* **295**, 844–848.

37 Kaseda, K., Higuchi, H., Hirose, K. (2003) Alternate fast and slow stepping of a heterodimeric kinesin molecule. *Nat. Cell Biol.* **5**, 1079–1082.

38 Kozielski, F., Sack, S., Marx, A., Thormahlen, M., Schonbrunn, E., Biou, V., Thompson, A., Mandelkow, E. M., Mandelkow, E. (1997) The crystal structure of dimeric kinesin and implications for microtubule-dependent motility. *Cell* **91**, 985–994.

39 Romberg, L., Pierce, D. W., Vale, R. D. (1998) Role of the kinesin neck region in processive microtubule-based motility. *J. Cell Biol.* **140**, 1407–1416.

40 Rief, M., Rock, R. S., Mehta, A. D., Mooseker, M. S., Cheney, R. E., Spudich, J. A. (2000) Myosin-V stepping kinetics: a molecular model for processivity. *Proc. Natl. Acad. Sci. USA* **97**, 9482–9486.

41 De La Cruz, E. M., Wells, A. L., Rosenfeld, S. S., Ostap, E. M., Sweeney, H. L. (1999) The kinetic mechanism of myosin V. *Proc. Natl. Acad. Sci. USA* **96**, 13726–13731.

42 Nishiyama, M., Muto, E., Inoue, Y., Yanagida, T., Higuchi, H. (2001) Substeps within the 8-nm step of the ATPase cycle of single kinesin molecules. *Nat. Cell Biol.* **3**, 425–428.

43 Veigel, C., Coluccio, L. M., Jontes, J. D., Sparrow, J. C., Milligan, R. A., Molloy, J. E. (1999) The motor protein myosin-I produces its working stroke in two steps. *Nature* **398**, 530–533.

44 Veigel, C., Wang, F., Bartoo, M. L., Sellers, J. R., Molloy, J. E. (2002) The gated gait of the processive molecular motor, myosin V. *Nat. Cell Biol.* **4**, 59–65.

45 Tanaka, H., Homma, K., Iwane, A. H., Katayama, E., Ikebe, R., Saito, J., Yanagida, T., Ikebe, M. (2002) The motor domain determines the large step of myosin-V. *Nature* **415**, 192–195.

46 Menetrey, J., Bahloul, A., Wells, A. L., Yengo, C. M., Morris, C. A., Sweeney, H. L., Houdusse, A. (2005) The structure of the myosin VI motor reveals the mechanism of directionality reversal. *Nature* **435**, 779–785.

47 Buss, F., Spudich, G., Kendrick-Jones, J. (2004) Myosin VI: cellular functions and motor properties. *Annu. Rev. Cell Dev. Biol.* **20**, 649–676.

48 Rock, R. S., Rice, S. E., Wells, A. L., Purcell, T. J., Spudich, J. A., Sweeney, H. L. (2001) Myosin VI is a processive motor with a large step size. *Proc. Natl. Acad. Sci. USA* **98**, 13655–13659.

49 Ökten, Z., Churchman, L. S., Rock, R. S., Spudich, J. A. (2004) Myosin VI walks hand-over-hand along actin. *Nat. Struct. Mol. Biol.* **11**, 884–887.

50 Okada, Y., Hirokawa, N. (2000) Mechanism of the single-headed processivity: diffusional anchoring between the K-loop of kinesin and the C terminus of tubulin. *Proc. Natl. Acad. Sci. USA* **97**, 640–645.

51 Inoue, A., Saito, J., Ikebe, R., Ikebe, M. (2002) Myosin IXb is a single-headed minus-end-directed processive motor. *Nat. Cell Biol.* **4**, 302–306.

52 Sakakibara, H., Kojima, H., Sakai, Y., Katayama, E., Oiwa, K. (1999) Inner-arm dynein c of *Chlamydomonas* flagella is a single-headed processive motor. *Nature* **400**, 586–590.

53 Rogers, K. R., Weiss, S., Crevel, I., Brophy, P. J., Geeves, M., Cross, R. (2001) KIF1D is a fast non-processive kinesin that demonstrates novel K-loop-dependent mechanochemistry *EMBO J.* **20**, 5101–5113.

54 Sablin, E. P., Case, R. B., Dai, S. C., Hart, C. L., Ruby, A., Vale, R. D., Fletterick, R. J. (1998) Direction determination in the minus-end-directed kinesin motor ncd. *Nature* **395**, 813–816.

55 Endres, N. F., Yoshioka, C., Milligan, R. A., Vale, R. D. (2006) A lever-arm rotation drives motility of the minus-end-directed kinesin Ncd. *Nature* **439**, 875–878.

56 Walker, R. A., Salmon, E. D., Endow, S. A. (1990) The *Drosophila* claret segregation protein is a minus-end directed motor molecule. *Nature* **347**, 780–782.

57 Wells, A. L., Lin, A. W., Chen, L. Q., Safer, D., Cain, S. M., Hasson, T., Carragher, B. O., Milligan R. A., Sweeney, H. L. (1999) Myosin VI is an actin-based motor that moves backwards. *Nature* **401**, 505–508.

58 Henningsen, U., Schliwa, M. (1997) Reversal in the direction of movement of a molecular motor. *Nature* **389**, 93–96.

59 Endow, S. A., Waligora, K. W. (1998) Determinants of kinesin motor polarity. *Science* **281**, 1200–1202.

60 Endow, S. A., Higuchi, H. (2000) A mutant of the motor protein kinesin that moves in both directions on microtubules. *Nature* **406**, 913–916.

61 Tsiavaliaris, G., Fujita-Becker, S., Manstein, D. J. (2004) Molecular engineering of a backwards-moving myosin motor. *Nature* **427**, 558–561.

62 Finer, J. T., Simmons, R. M., Spudich, J. A. (1994) Single myosin molecule mechanics: piconewton forces and nanometre steps. *Nature* **368**, 113–119.

63 Molloy, J. E., Burns, J. E., Kendrick-Jones, J., Tregear, R. T., White, D. C. (1995) Movement and force produced by a single myosin head. *Nature* **378**, 209–212.

64 Hirakawa, E., Higuchi, H., Toyoshima, Y. Y. (2000) Processive movement of single 22S dynein molecules occurs only at low ATP concentrations. *Proc. Natl. Acad. Sci. USA* **97**, 2533–2537.

65 Schnitzer, M. J., Visscher, K., Block, S. M. (2000) Force production by single kinesin motors. *Nat. Cell Biol.* **2**, 718–723.

66 Purcell, T. J., Sweeney, H. L., Spudich, J. A. (2005) A force-dependent state controls the coordination of processive myosin V. *Proc. Natl. Acad. Sci. USA* **102**, 13873–13878.

67 Gyoeva, F. K., Sarkisov, D. V., Khodiakov, A. L., Minin, A. A. (2004) The tetrameric molecule of conventional kinesin contains identical light chains. *Biochemistry* **43**, 13525–13531.

68 Jordens, I., Marsman, M., Kuijl, C., Neefjes, J. (2005) Rab proteins, connecting transport and vesicle fusion. *Traffic* **6**, 1070–1077.

69 Wu, X. S., Rao, K., Zhang, H., Wang, F., Sellers, J. R., Matesic, L. E., Copeland, N. G., Jenkins, N. A., Hammer, J. A., III. (2002) Identification of an organelle receptor for myosin-Va. *Nature Cell Biol.* **4**, 271–278.

70 Levi, V., Serpinskaya, A. S., Gratton, E., Gelfand, V. I. (2006) Organelle transport along microtubules in *Xenopus* melanophores: evidence for cooperation between multiple motors. *Biophys. J.* **90**, 318–327.

71 Schliwa, M. (2003) *Molecular Motors*. Wiley-VCH, Germany.

21
Biologically Inspired Hybrid Nanodevices

David Wendell, Eric Dy, Jordan Patti, and Carlo D. Montemagno

21.1
Introduction

The exploration and exploitation of biologic modes of design and self-assembly are now studies of both greater urgency and ease, as the tools for manipulation and visualization of nanoscale materials become increasingly available. Engineering biologically inspired nanoscale devices encompasses a wide variety of research, from current nanomaterials such as gecko tape, self-cleaning glass, and artificial shark skin [1], to the mechanics of how biological molecules such as proteins, enzymes, DNA and RNA can function as analogous man-made structures [1–50]. In this chapter, we will briefly examine a sample of these technologies and follow up with current related research. We will discuss what can be gleaned from these emerging technologies, focusing primarily on biomimetic protein-based devices. Finally, we will present our research efforts in the area of biocomputation and extend this discussion to the prospects of future applications.

Engineering hybrid nanoscale devices requires the concurrent application of technology from a variety of fields. Incorporating varying levels of organization is central to creating functional biomimetic materials. The common thread among biological nano-hybrid devices is the need to exploit natural self-assembly schemes that have evolved over the millennia to build complex structures. Often, utilizing such a self-assembly scheme is sufficient, but optimum form may not always follow natural function, and producing devices which convert chemical energy to electricity, or light into chemical energy such as ATP can improve on Nature's design through engineering. Here, we show two bodies of work, including protein-based devices and cellular power generation, the common theme being a fusion of biologic molecules with synthetic structures to produce nanoscale hybrid devices.

Nanobiotechnology II. Edited by Chad A. Mirkin and Christof M. Niemeyer
Copyright © 2007 WILEY-VCH Verlag GmbH & Co. KGaA, Weinheim
ISBN: 978-3-527-31673-1

21.2
An Overview

21.2.1
A Look in the Literature

Biologically inspired synthetic architecture and devices that can both sense and respond on the nanoscale can be found in a number of areas, including optics, microfluidics, and device computation. For example, the waxy nanotextured islands of the lotus leaf have inspired a range of materials from low-drag fluidic channels to self-cleaning glass [2]. Similarly, the microtextured riblets of shark skin has led to low-drag surfaces for turbulent fluid flow situations [3]. In optics, a new hybrid apposition/superposition lens system borrows much of its design inspiration from the compound lens structure found in insects such as the dragonfly [4]. DNA nanotechnology has exploited the Watson–Crick pairing and self-assembly of DNA and RNA in a variety of settings [5–8]. Indeed, it has been used to create small structures in two and three dimensions, including Sierpinski triangles, nanotubes, cubes, octahedrons, nanowires, and many others. Futhermore, RNA has been exploited not only for its self-assembly, but also for its ribozyme and gene-splicing abilities [9]. Other biological materials chosen for nanoscale imitation include bone, shell, and sponge spicules, all selected for their strength and toughness [10]. The durability of many of these rigid materials is derived from sacrificial bonds and "hidden length" within their supporting and connective materials [11]. One biologically inspired commercial design has led to the production of a new type of adhesive, commonly referred to as "gecko tape".

The nanoscale fibrillar structures on the feet of insects and geckos are the principal components which endow them with such extraordinary adhesive wall-walking properties. The fibrillar structures take on a hairy appearance that varies in size and surface density between species. The size – and thus the body weight – of the animal has dictated the evolutionary path of adhesive design, yielding smaller insects with large (several microns) adhesive structures, and geckos with fine nanometer-scale hairs (Figure 21.1) [12]. Concurrently, decreased fibrillar size has coincided with an increase in hair density, with geckos having the highest density of any species studied to date [1, 12]. Recent reports have shown that the origin of the hairy structures stickiness lies in the molecular adhesion produced by van der Waals' interactions of the hairs with the surface [13]. While van der Waals' forces are relatively weak molecular interactions, it is the extremely high fibrillar density that provides the strong cumulative adhesive force capable of supporting a gecko's body weight.

Although the above-described materials are remarkably different, they do share a common fundamental inspiration and similar, design scale. While it is not always necessary to borrow design from Nature, in some instances it can prove useful to do so, especially on the nanoscale, as Nature has already evolved so many efficient designs. Considering the cell and cell membrane as examples, we seek to harness the protein tools within them to make hybrid biomimetic nanoscale

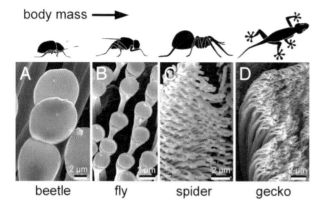

body mass

A B C D

beetle   fly   spider   gecko

Fig. 21.1 Fibrillar attachment: size distribution (from Ref. [12].)

devices. We have begun to do this by coupling light-driven proteins with enzymatic ones, ion channels and connexin; and autocontractile cells with piezoelectric materials for implantable power scavenging.

21.2.2
Membrane Proteins and their Native Condition

The possession of lipid membranes is a defining characteristic of living organisms, as these provide compartmentalized functionality. Lipid bilayers provide a semipermeable barrier, allowing the organism to interact with its environment and, at the same time, to regulate the passage of ions, glucose, DNA and other essential solutes in and out of the cell. The ever-increasing body of structural information on membrane proteins is providing insight into cell membrane architecture. The classic Singer–Nicholson fluid mosaic model of evenly distributed membrane proteins has been updated to include segregated regions of protein with variable membrane thickness [14]. The new picture still maintains membrane fluidity, but paints a patchwork of areas with specific protein function, with other areas devoid of protein. Furthermore, each island of proteins will take on a different appearance, as the shape of embedded membrane protein can vary dramatically [14].

In the past, lipid membranes have served as useful platforms for *in-vitro* protein reconstitution in basic structural and mechanistic studies of membrane proteins [15–17]. However, given that certain parameters such as durability and chain length cannot easily be controlled in lipid systems, polymer membranes offer a compelling approach towards fabricating customizable, rugged devices based on inherent protein functionality. They are versatile in that the block compositions can be easily tailored, depending on the dimensions required to support protein integration.

In an attempt to increase vesicle stability and lifetime, the lipid membranes can be replaced with polymer. In fact, a triblock copolymer, with hydrophilic ends and a hydrophobic core, can provide an environment very similar to lipid. Using a polymer originally slated for contact lenses, Meier and colleagues began using an ABA triblock composed of polymethyloxazoline-polydimethylsiloxane-polymethyl-oxazoline (PMOXA-PDMS-PMOXA) that contains properties similar to that found in lipid bilayers [17], and have shown functional membrane protein incor-poration [18]. In addition to ABA, an ABC copolymer composed of polyethylene oxide-polydimethylsiloxane-polymethyloxazoline (PEO-PDMS-PMOXA) has been recently introduced which reportedly allows oriented protein incorporation [19]. In our laboratory, using poly-2-ethyl-2-oxazoline (PEtOz) and polydimethylsilo-xane (PDMS), we have synthesized a PEtOz-PDMS-PEtOz block copolymer with similar thickness/permeability to a lipid bilayer [20]; these are both important considerations for protein function.

Several recent scientific reviews have focused on amphiphilic block copolymers, and their use as nanocontainers, nanoshells, and scaffolding for artificial mem-branes [21–25]. Unlike polymer, lipid membranes are optimized for fluidity rather than stability, and must rely on the ever-present cellular machinery to maintain dynamic assembly, turnover, and reorganization. Consequently, amphi-philic block copolymer membranes can provide a robust and durable synthetic housing for imbedded membrane proteins that does not require constant upkeep. Moreover, it is proteins such as F_0F_1 ATPase that are particularly useful as ATP is essential for many biological processes. By engineering systems with membrane proteins such as F_0F_1 ATPase, bacteriorhodopsin, ion channels and other criti-cally functional membrane proteins, it is possible to build devices on the nano-scale with hybrid, novel functions. Certainly such devices are not limited to membrane-bound proteins. Much more structural information is available for cy-toplasmic proteins, making them a more practical resource for engineering life processes into nanoscale devices.

21.3
The Protein Toolbox

21.3.1
F_0F_1-ATPase and Bacteriorhodopsin

Nanoscale hybrid systems attempt to integrate biological and synthetic compo-nents. For our purposes, this usually involves packaging proteins in an *in-vitro* environment where they can continue to function. One model system with which our group has conducted extensive studies is that of coupled bacteriorhodopsin (BR) and F_1F_0-ATP synthase (ATP synthase) in an artificial membrane [26]. BR and ATP synthase are both proteins which integrate into the cell membrane, spanning across it.

BR, a membrane-bound protein originally isolated from the purple membranes of *Halobacterium halobium* [27], utilizes solar radiation to translocate protons across the membrane, forming a proton gradient. Subsequently, the F_0F_1-ATPase

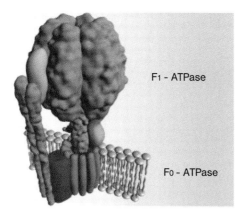

F1 - ATPase

F0 - ATPase

Fig. 21.2 The protein structure of F_0F_1-ATPase, showing protein orientation in the lipid membrane.

complex utilizes the gradient to move the protons into the cell, and synthesizes ATP [27, 28].

ATP synthase, a ubiquitous enzyme which drives the energy transport chain, consists of two separate portions: (i) F_0, the hydrophobic, membrane-bound portion that is responsible for proton translocation; and (ii) F_1, the hydrophilic portion that is responsible for ATP synthesis and hydrolysis. As protons flow through the F_0, the γ subunit rotates clockwise and ATP is synthesized. The a, b, and c subunits of the F_0 portion form the channel which allows protons to flow through the membrane (Figure 21.2). The nucleotide binding and catalytic sites are located on the three α and three β subunits of the F_1-ATPase, respectively [29]. The γ subunit is centrally located within the $\alpha_3\beta_3$ hexamer, and rotates as a function of ATP synthesis/hydrolysis.

Artificial systems consisting of BR and F_0F_1-ATPase in liposomes have been used to demonstrate the light-driven production of ATP [30, 31]. Inversely, hydrolysis of ATP is the process that we use to couple mechanical energy to nanofabricated devices, as demonstrated previously [32] (Figure 21.3). When F_1 is free in solution, the hydrolysis of ATP results in counterclockwise rotation of the γ subunit.

Aside from the fact that the F_1-ATPase motor protein simply looks like a motor, a rotor surrounded by a stator, it is actually capable of producing an astounding amount of torque for its size. The diameter of the rotor is approximately 1 nm, while the diameter of the entire motor is only about 10 nm.

By incorporating both F_0F_1-ATPase and BR into an artificial system, such as a liposome, it is possible to reproduce this process *in vitro*. Liposomes with diameters on the scale of 100 nm are easily formed in bulk from purified lipids by selective solvent removal, using rotary evaporation. Their structure resembles a small cell, with the interior partitioned from the bulk by a lipid bilayer. BR and ATP synthase can be incorporated by partially solubilizing the liposomes with de-

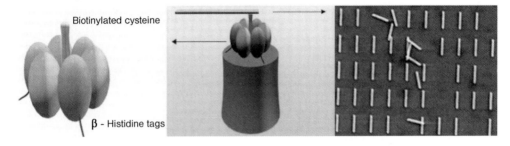

Biotinylated cysteine

β - Histidine tags

Fig. 21.3 F_0F_1-ATPase-based nanopropeller (from Ref. [32]). F_0F_1-ATPase was engineered with histidine tags for placement on nickel posts, approximately 80 nm in diameter. The nanopropeller (length 750–1400 nm, diameter 150 nm) is shown in the illustration to the right.

tergent, adding protein to the mixture, and then dialyzing the solution. As the detergent diffuses away, the proteins left behind assume the lowest energy position, which results in membrane integration. Changes in pH in the interior of the vesicles and production of ATP can be monitored by fluorometric and luminescence techniques. Extending this concept one step further, we have demonstrated such protein insertion and ATP production in ABA polymersomes [42].

21.3.2
Ion Channels and Connexin

The chemical energy produced by BR and F_0F_1-ATPase can be used to power other enzymatic reactions such as ion pumps and cytoskeleton rearrangement. Reconstituting proteins in polymer membrane allows compartmentalization of these reactions and the potential to establish stable ion gradients.

Congruently, we have begun to apply the artificial membrane system to voltage-gated ion channels and connexin in the pursuit of something termed an "excitable vesicle" (EV). In principle, EVs are nanoscale depolarizing units that mimic neurons in their ability to produce and transmit an action potential. By incorporating ion channels and connexin proteins into vesicles, we are attempting to distill the essential elements of excitable cells and to create an artificial neural network capable of creating, distributing and utilizing action potentials.

Neurons can be described as electrochemical units which receive, process, and transmit information in the form of electrical and chemical signals. The simplest networks of neurons are found in invertebrates; for example, with no central brain, flatworms have a collection of reflex neurons distributed throughout their body, that forms a neural net. In its most basic form, the EV is similar to a simplified, miniature neuron – a vesicle that contains voltage-gated sodium (Na^+) and potassium (K^+) ion-channels, and connexin incorporated into the membrane. Connexin is a member of a large family of homologous membrane proteins which form the channels for electrical and metabolic communication between

Excitable Vesicles

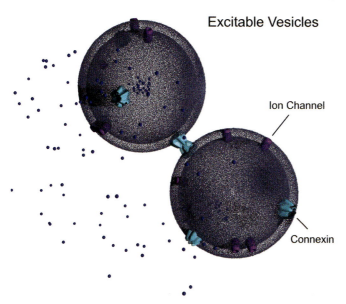

Ion Channel

Connexin

Fig. 21.4 Schematic diagram of an excitable vesicle (EV), showing docked connexin proteins (green), ions (blue) and ion channels (purple). The drawing is not to scale.

cells known as "connexons" or "gap junctions". Self-assembly and unique gating properties allow the protein to be used in a modular fashion to selectively control and tailor the electrical signal propagation between cells. Currently, we are reconstituting ion channels and connexin into polymer vesicles, and tests of protein function are ongoing. In the future, Na^+/K^+ pumps will be required to re-establish the resting membrane potential and allow multiple depolarizations. A schematic diagram of an EV is shown in Figure 21.4.

Computer simulations investigating polymersome volume and ion concentrations have been critical in design parameters such as vesicle size and protein concentration. Simulations show that the ion channels can cause the EV to depolarize briefly, generating an electrical pulse roughly equivalent to that of a biological neuron (Figure 21.5) [33]. Sodium channels provide the outflow of charged ions, which produces a positive potential across the membrane, while potassium channels restore the negative resting potential. The Na^+/K^+ pump ATPase in these simulations [34, 35] re-establishes the resting ion concentrations across the vesicle membrane and enables the EV to be formed without a resting potential and to depolarize multiple times. We have verified ion channel and connexin function via planar black lipid membrane experiments, and are currently integrating these proteins into polymersomes. EV construction is ongoing, and while ion channels are still performing their physiological roles they will be integrated into a hybrid nanoscale system [33].

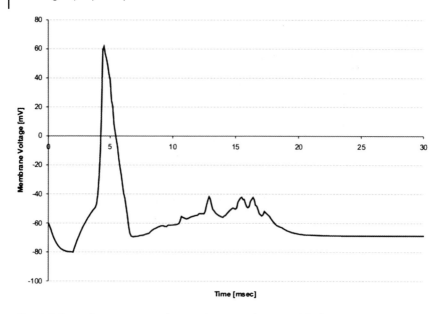

Fig. 21.5 A sample action potential output from a stochastic model of a 600 nm-diameter lipid vesicle. A small stimulus current (0.3 pA) was applied at 2 ms. There was no Na/K pump activity. "Leaks" were allowed only from the gating action of ion channels. The system was very sensitive to initial conditions and channel conductance values. (Illustration courtesy of J. Isobe.)

Thus far, our protein polymer/lipid vesicles represent only a model system, and it is important to recognize the myriad processes that occur at the membrane interface. Receptor binding, signaling, and even cell motility all rely on interactions with membrane bound proteins. Many of these operations can potentially be reproduced in this artificial setting, making it a platform capable of harnessing the work of functional membrane proteins as they are discovered. However, in order to create useful devices, the preserving protein function and vesicle integrity must be made more robust, and our group – along with other laboratories – has begun to address this issue [36]. Nevertheless, the polymer vesicle system remains a platform which is capable of harnessing the work of functional membrane proteins as they are discovered.

21.4
Harvesting Energy

On a modestly larger scale, we have also investigated the benefits of fusing the myosin-powered actin contractile machinery of cardiomyoctyes with a piezoelectric transducer. The aim of this device is to have the cells convert chemical energy into electrical power via charge separation within the piezoelectric material.

The proliferation of increasingly smaller mechanical devices has challenged engineers to develop power sources the weight and volume of which are not disproportionately larger than the systems they power. Traditional power supplies such as batteries have limited lifetimes, and may not be a viable choice if replacement or recharging of the battery is necessary, as in implantable devices [37]. The search for "self charging" and integrated power supplies will be especially pressing in the near future as more implantable micro-electro-mechanical systems (MEMS) devices are produced.

The development of efficient microscale transducers and generators has fostered development in the area of energy harvesting from ambient sources such as vibrations, thermal gradients, and light, for the creation of electricity. Identically, we are developing a power supply that is capable of converting chemical energy into electrical energy through the use of living tissue and piezoelectricity. Our muscle-powered piezoelectric microgenerator integrates neonatal cardiac myocytes, with a PVDF-TrFE-coated cantilever bridge, with piezoelectricity properties capable of converting force transduction into electrical energy. This type of generator is particular suited for implantable devices as it can use the body's own chemical energy to power the cellular machinery. Applicable devices include pacemakers, deep brain-stimulating electrodes, and wireless sensors, to name a few.

21.5
Methods

So far, we have discussed a wide variety of hybrid devices, but we will now present a summary of the methodology actually used to create our biomimetic and cellular-MEMS constructs.

21.5.1
Muscle Power

Previously, we have demonstrated the fusion of biotic cells with abiotic MEMS devices by culturing cardiac myocytes onto silicon (Si) cantilevers (Figure 21.6) [38]. The key to this innovative approach is to separate the MEMS fabrication from the cell culture work and to interface them together using a thermoresponsive polymer. Before conducting any cell culturing, the MEMS devices are fabricated according to traditional surface micromachining techniques. When the devices have been released and characterized, a thermoresponsive polymer, poly-N-isopropylacrylimide (PNIPAAm), is deposited on the MEMS structure; this provides a solid sacrificial substrate for cellular adhesion that is biocompatible.

PNIPAAm has been previously considered as an intelligent substrate for harvesting intact cells [39]. PNIPAAm is a solid at temperatures greater than 32 °C, but undergoes a solid–liquid phase transition as it is cooled; moreover, it can also dissolve in the surrounding medium, without having any harmful effects on biological components. We have applied this technology to the development of a

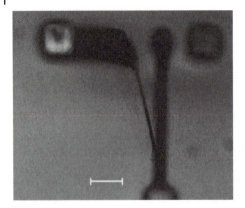

Fig. 21.6 A silicon cantilever is deflected laterally due to the force of contraction, 48.6 μN, of the attached self-assembled cardiomyocyte bundle. Scale bar = 100 μm.

muscle-powered piezoelectric microgenerator (Figure 21.7). Piezoelectric materials have been increasingly used for power generation as energy harvesting has become more prevalent. Polyvinylidene fluoride (PVDF) and its copolymer PVDF (polyvinylidene fluoride)-TRFE (trifluoroethylene) are of great interest because of their strong piezoelectricity and elasticity. PVDF is a semicrystalline homopolymer, which was shown to have a strong piezoelectric coefficient [40]. Its copolymer PVDF-TrFE crystallizes to a much greater extent than PVDF, yielding a higher remnant polarization, lower coercive field, and much sharper hysteresis. After applying poling with a high electric field, the coulombic interactions between injected and trapped charges and oriented dipoles in crystals make the

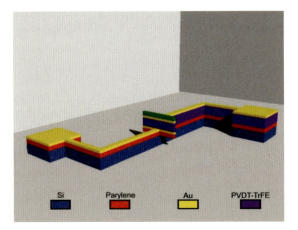

Fig. 21.7 Schematic of the cardiomyoctye piezoelectric generator. The green portion indicates the area of cardiac myocyte assembly.

polarization stable [41]. When a PVDF-TrFE film is stretched in the transverse direction, the dipole crystalline structures realign due to the applied stress and an internal field is created, causing charges to build up on the electrodes, producing the voltage potential.

The external stress to bend the PVDF-TrFE film is supplied by ventricular myocytes isolated from neonatal rats. Unlike mature muscle tissue, which must be dissected and incorporated onto a MEMS device by hand using crude interfaces, the neonatal ventricular myocytes selectively grow onto the device and differentiate into tissue bundles *in situ*. Additionally, ventricular myocytes are autorhythmic and therefore do not require any external stimulus to initiate muscle contraction. Once differentiated and networked into a tissue bundle, the heart muscle begins to contract spontaneously. At this point, the culture medium temperature is lowered below 32 °C (the PNIPAAm transition temperature), which allows the polymer to dissolve into the medium, thus releasing the structure. The structure is now free to bend according to the bending moment applied by the attached muscle bundle, thus generating electricity.

21.5.2
ATPase and BR Devices

Coupled ATPase-BR vesicles require membrane formation and proper protein integration and orientation. By first dissolving the polymer in ethanol and then adding it drop-wise to water, vesicles on the order of 100 nm are formed spontaneously [42]. Varying the concentration of these three components produces a wide range of polymer morphologies. An extensive study using transmission electron microscopy (TEM) has determined which combinations yield micelles, vesicles, or crystallized structures. Furthermore, vesicle size distribution can be somewhat controlled by manipulating these parameters, resulting in vesicles as large as 1 μm (Figure 21.8). Various applications require different vesicle sizes, often de-

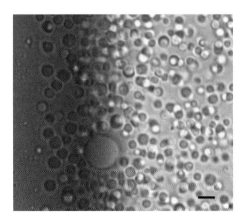

Fig. 21.8 ABA triblock polymer vesicles in water. Scale bar = 5 μm.

pending on whether it is necessary for individual vesicles to be visible with light microscopy. Larger vesicles are usually desirable for light microscopy, since object resolution is ultimately limited by diffraction, leaving the sub-wavelength-sized specimen obscured.

During polymer vesicle formation, purple membrane (the cell membrane from *Halobacterium salinarium*, with an extremely high BR concentration) was incorporated and protein function tested. Protein can be added while the polymer is dissolved in ethanol, and will spontaneously enter the membrane when the mixture is added to water or aqueous buffer. Changes in pH were assayed using the fluorescence shift of the dye pyranine (8-hydroxy-1,3,6-pyrenetrisulfonate), which indicated the successful light-driven pumping of protons. These studies indicate that the polymer can fulfil the role of the cell membrane by isolating the compartment from the outside world, limiting ion transport, and stabilizing membrane proteins.

Coupling this proton gradient to ATP synthase reproduced the ATP-generating behavior of the liposomes (Figure 21.9). Combining these proteins in a stable polymer system provides a solar-powered biological energy source, capable of driving additional biological reactions. ATP synthase, in a purified form, was also incorporated in the same manner as BR, prior to vesicle formation. In this case, the majority of the BR was oriented so that acidification of the liposome interiors occurred. The formation of a proton gradient facilitated the function of ATP synthase.

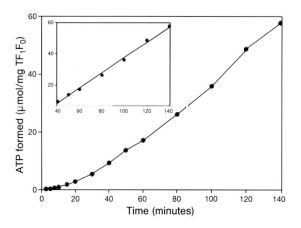

Fig. 21.9 Coupled ATP production from F_0F_1-ATPase and bacteriorhodopsin (BR) in lipid vesicles (from Ref. [26]). Samples contained 56 nM (30 µg mL^{-1}) F_1F_0-ATPase, 5.3 µM (137 µg mL^{-1}) BR, and 3.3 mM liposome. Experiments were performed at 40 °C in dialysis buffer (20 mM Mops, pH 7.3, 50 mM Na$_2$SO$_4$, 50 mM K$_2$SO$_4$, 2.5 mM MgSO$_4$, 0.25 mM DTT, 0.2 mM EDTA). The inset shows the linear regression analysis of steady-state ATP synthesis from 40 to 140 min. The activity calculated from the slope was 490 nmol min^{-1} mg^{-1} F_0F_1-ATPase.

As the protons exited the vesicles via ATP synthase, ATP was synthesized in the bulk solution. A standard luciferase assay was used where the enzyme luciferase catalyzes the consumption of ATP by luciferin, which in turn generates light. After 1 h, ATP production of approximately 1 μmol mg^{-1} ATP synthase was recorded [43]. The fact that ATP synthase can be incorporated to function from purified subunits suggests that many other proteins would be compatible with this system. Consequently, we are currently seeking to utilize this ATP production system and to integrate other membrane proteins with block copolymer membranes.

The initial investigations to address the stability issue, by reconstructing the ATP synthesis process in the laboratory, was described in Section 21.2.2. However, a further challenge is to package the liposomes/polymer vesicles in such a way that they can create useful devices, since desiccation and protein stability become problematic over time.

One approach to this problem (again using the BR-ATP synthase system) is to incorporate vesicles into a sol-gel matrix [44]. Whilst in previous studies the vesicles or simple proteins had been incorporated into a sol-gel, this was the first time that a complex synthesis system had been encapsulated within the polymer matrix. Briefly, a sol-gel is a colloidal suspension of silica particles which has been gelled, forming a solid but porous structure. The gel structure prevents desiccation and also provides a more rigid stabilized system when compared to a purely aqueous environment. The structure of the sol-gel also appears to prevent protease access and keeps vesicles trapped within the matrix. The process of trapping enzymatic reactions within a sol-gel requires certain steps which should be applicable to other, similar systems. For this studies proteoliposomes were used, as lipid systems have been more thoroughly explored than their polymer vesicle descendents.

First, proteoliposomes are formed with incorporated membrane proteins. As in other experiments, no special steps are needed to orient the proteins (they tend preferentially to insert with a particular orientation, most likely due to size/ structural differences between the two ends of the protein). The sol-gel is prepared using a low-alcohol solution route. Tetramethylorthosilicate (TMOS) is added to water and hydrolyzed under acidic conditions, providing the sol. Polyethyleneglycol (PEG) is then added to the sol, preventing premature gelation and providing a protecting environment for the liposomes. At this point, the methanol generated in the hydrolysis reaction is evaporated and buffer is added to the proteoliposomes. The sol-gel proteoliposome mixture (proteogel) is finally cast, being used to form either monoliths or films.

The initial results have indicated that vesicles in the sol-gel remained active for weeks, demonstrating behavior similar to that observed with vesicles in solution. A pyranine-based assay recorded a pH change with light exposure, indicating light-driven BR pumping of protons. Additionally, with BR-ATP synthase vesicles, ATP production was observed over time. When incubated with a protease, the proteogel retained protein function, whereas an equivalent liquid solution showed a marked decrease.

21.5.3
Excitable Vesicles

Extensive investigations have been conducted patterning neurons on microelectrodes. Indeed, the growing of neurons on electrodes in defined geometries has applications in neurolelectric circuits, drug testing, and biosensors [45]. The primary disadvantage of using fully differentiated central nervous system neurons is that almost none divides, thus limiting device integration capabilities. Two other disadvantages include harvesting difficulties and the mixture of cell types, making neuronal-based electrical recordings somewhat dubious as the glial cells continue to grow and can modify synaptic transmission between cells. Investigations into integrating neurons with silicon devices has largely been dedicated to microelectrode readings in, or simulations of, excitable tissue [46–49]. *In-vitro* analysis has included electrical readings from both cultured neurons and whole-brain slices, for example of guinea pig and ferret thalamocortical neurons [50, 51]. In both cases, cultured cells must be harvested from the animal specimen, sometimes creating a mixture of both glial and neuronal cell types, but ultimately creating cultures with a limited life span.

The use of EVs can eliminate these difficulties, as they can be tailored to specific dimensions with desired protein concentrations. Using biomimetic polymer (as mentioned above) also makes their construction more robust then the lipids of the cell membrane. Preliminary studies involved modeling the excitable behavior of EVs using prokaryotic channel proteins, and to date the first ion channels purified have been the sodium channel NaChBac and potassium channel KvAP [52, 53], both from *Escherichia coli*. Connexin 43 (Cx43) has also been purified from insect cells using a baculorvirus expression system [54]. As with F_1F_0-ATPase and BR, the proteins are presently being reconstituted into lipid and polymer vesicles for testing.

21.6
Outlook

In this chapter we have discussed several different nanoscale hybrid devices and their associated construction methodology. While assembly is ongoing, the potential for protein and cellular devices focused on energy conversion is excellent. As we continue to discover novel ways of fusing various synthetic and natural components, more functionally varied devices will follow. Clearly, the variety in these studies bespeaks the possible diversity in this area.

The proteogel system shows potential for prolonging life and preserving the function of nanoscale protein-vesicle systems. Indeed, it may enable these systems to be used in environments which are normally hostile to hybrid biosystems. The demonstration of ATP production is a good model system as it is an essential component for almost all biological reactions. This packaging system would likely be applicable to many other biological nanosystems, where the components must be hydrated and protected. If there is a drawback, it is that the ATP produced is distributed throughout the sol-gel matrix.

Similarly to the proteopolymersomes in adaptability, the hybrid muscle generator system is capable of fusing cardiac myocytes with any number of MEMS structures. The beauty of the myocyte generator is that the cardiac myocytes, if derived from stem cells, would be biocompatible with the host. In addition, as long as the myocytes remain viable there would be no need to replace or recharge the generator's contents, as is the case with current power supplies for implantable devices. At this stage of development, the energy generated is still too low to be useful for macroscale applications; however, as device size scales down so do power requirements, and in such scenarios a fusion of power generation to the device itself can provide an elegant solution to a complex problem.

A network of EVs would offer a means of biological computation with extensive utility in many areas, and offer tremendous research opportunities in areas such as neuroscience, mathematics, biophysics and computer processor design. EV networks may be used to examine complex emergent system fundamentals such as network dynamics and system evolution; they may also provide insight into Cristof Koch's "Neuronal Correlates of Consciousness" [55], a currently unmapped series of neural events linked to consciousness. One complication for studying these correlates, specifically, when examining the individual cells responsible for signal generation, is the low resistivity of neuronal tissue to current flow and the shielding effects of the skull [55]. Investigations involving individual neurons and their neighbors also lack the totality of the extremely interconnected system network, and the distribution of one such series of events.

A recently proposed method for recording an ensemble of neuronal firing events without the need for invasive microelectrodes involves fluorescently responsive ion channels [56]. Indeed, studies conducted by Ataka and Pieribone [57] have brought this idea closer to reality by producing a modified form of the green fluorescent protein within a voltage-gated sodium channel. The protein construct was capable of reporting depolarizations via a change in fluorescence on a 2-ms time scale [57]. Furthermore, photosensitive ion channels have been constructed which provide the possibility of an optically controlled gating mechanism [58]. Together, it is conceivable that protein constructs such as these could provide an optical method both to direct input and to gather output from an EV network or biological neural net in an extremely efficient manner, without the need for hard-wired electrodes.

These previously described cellular and protein-based devices share both the promise for novel functionality and the need for integration sensitive to biological molecules. While our current studies have several aims, we believe that they mirror the prodigious mixture of research involved in engineering biologic devices at the nanoscale.

Acknowledgments

The authors thank Evan Brooks for his critical reading of the manuscript. The National Science Foundation and NASA-CMISE are gratefully acknowledged for financially supporting these studies.

References

1 Forbes, P. (2006) *The Gecko's Foot: Bio-inspiration: Engineering New Materials from Nature.* W. W. Norton, New York.

2 Cho, S. K., Moon, H., Kim, C.-J. (2003) Creating, transporting, cutting, and merging liquid droplets by electrowetting-based actuation for digital microfluidic circuits. *Journal of Microelectrochemical Systems* **12**, 70–80.

3 Ball, P. (1999) Engineering shark skin and other solutions. *Nature* **400**, 507–509.

4 Lee, L., Szema, R. (2005) Inspirations from biolofical optics for advanced photonic systems. *Science* **310**, 1148–1150.

5 Chworos, A., Severcan, I., Koyfman, A. Y., Weinkam, P., Oroudjev, E., Hansma, H. G., Jaeger, L. (2004) Building programmable jigsaw puzzles with RNA. *Science* **306**, 2068–2072.

6 Rothemund, P. W. (2006) Folding DNA to create nanoscale shapes and patterns. *Nature* **440**, 297–302.

7 Winfree, E., Liu, F., Wenzler, L. A., Seeman, N. C. (1998) Design and self-assembly of two-dimensional DNA crystals. *Nature* **394**, 539–544.

8 Seeman, N. C., Lukeman, P. S. (2005) Nucleic acid nanostructures: bottom-up control of geometry on the nano-scale. *Rep. Prog. Phys.* **68**, 237–270.

9 Guo, S., Tschammer, N., Mohammed, S., Guo, P. (2005) Specific delivery of therapeutic RNAs to cancer cells via the dimerization mechanism of phi29 motor pRNA. *Hum Gene Ther.* **16**, 1097–1109.

10 Mayer, G. (2005) Rigid biological systems as models for synthetic composites. *Science* **310**, 1144–1147.

11 Fantner, G. E., Oroudjev, E., Schitter, G., Golde, L. S., Thurner, P., Finch, M. M., Turner, P., Gutsmann, T., Morse, D. E., Hansma, H., Hansma, P. K. (2006) Sacrificial bonds and hidden length: unraveling molecular mesostructures in tough materials. *Biophys. J.* **90**, 1411–1418.

12 Gao, H., Yao, H. (2004) Shape insensitive optimal adhesion of nanoscale fibrillar structures. *Proc. Natl. Acad. Sci. USA* **101**, 7851–7856.

13 Autumn, K., Liang, Y. A., Hsieh, S. T., Zesch, W., Chan, W. P., Kenny, T. W., Fearing, R., Full, R. J. (2000) Adhesive force of a single gecko foot-hair. *Nature* **405**, 681–685.

14 Engelman, D. M. (2005) Membranes are more mosaic than fluid. *Nature* **438**, 578–580.

15 Sauer, M., Haefele, T., Graff, A., Nardin, C., Meier, W. (2001) Ion-carrier controlled precipitation of calcium phosphate in giant ABA triblock copolymer vesicles. *Chem. Commun. (Camb)* 2452–2453.

16 Meier, W., Nardin, C., Winterhalter, M. (2000) Reconstitution of channel proteins in (polymerized) ABA triblock copolymer membranes. *Angew. Chem. Int. Ed. Engl.* **39**, 4599–4602.

17 Nardin, C., Meier, W. (2002) Hybrid materials from amphiphilic block copolymers and membrane proteins. *J. Biotechnol.* **90**, 17–26.

18 Nardin, C., Widmer, J., Winterhalter, M., Meier, W. (2001) Amphiphilic block copolymer nanocontainers as bioreactors. *Soft Matter* **4**, 403–410.

19 Stoenescu, R., Graff, A., Meier, W. (2004) Asymmetric ABC-triblock copolymer membranes induce a

directed insertion of membrane proteins. *Macromol. Biosci.* **4**, 930–935.

20 Choi, H.-J., Brooks, E., Montemagno, C. D. (2005) Synthesis and characterization of nanoscale biomimetic polymer vesicles and polymer membranes for bioelectronic applications. *Nanotechnology* **16**, S143–S149.

21 Antonietti, M., Forster, S. (2003) Vesicles and liposomes: a self-assembly principle beyond lipids. *Adv. Mater.* **15**, 1323–1333.

22 Soo, P. L., Eisenberg, A. (2004) Preparation of block copolymer vesicles in solution. *J. Polymer Sci. B: Polymer Physics* **42**, 923–938.

23 Kita-Tokarczyk, K. Grumelard, J., Haefele, T., Meier, W. (2005) Block copolymer vesicles, using concepts from polymer chemistry to mimic biomembranes. *Polymer* **46**, 3540–3563.

24 Discher, D. E., Eisenberg, A. (2002) Polymer vesicles. *Science* **297**, 967–973.

25 Hamely, I. W. (2005) Nanoshells and nanotubes from block copolymers. *Soft Matter* **1**, 36–43.

26 Hazard, A., Montemagno, C. (2002) Improved purification for thermophilic F1F0 ATP synthase using n-dodecyl beta-D-maltoside. *Arch. Biochem. Biophys.* **407**, 117–124.

27 Vsevolodov, N. (1998) *Biomolecular Electronics: An Introduction via Photosensitive Proteins.* Birkäuser, Boston.

28 Nicholls, D. G., Ferguson, S. J. (1992) *Bioenergetics 2.* Academic Press, New York.

29 Kinosita, K., Jr., Yasuda, R., Noji, H., Ishiwata, S., Yoshida, M. (1998) F1-ATPase: a rotary motor made of a single molecule. *Cell* **93**, 21–24.

30 Richard, P., Pitard, B., Rigaud, J. L. (1995) ATP synthesis by the F0F1-ATPase from the thermophilic Bacillus PS3 co-reconstituted with bacteriorhodopsin into liposomes. Evidence for stimulation of ATP synthesis by ATP bound to a noncatalytic binding site. *J. Biol. Chem.* **270**, 21571–21578.

31 Pitard, B., Richard, P., Duñarch, M., Girault, G., Riguard, J. (1996) ATP synthesis by the F0F1 ATP synthase from thermophilic Bacillus PS3 reconstituted into liposomes with bacteriorhodopsin. *Eur. J. Biochem.* **235**, 769–778.

32 Soong, R. K., Bachand, G. D., Neves, H. P., Olkhovets, A. G., Craighead, H. G., Montemagno, C. D. (2000) Powering an inorganic nanodevice with a biomolecular motor. *Science* **290**, 1555–1558.

33 Isobe, J., Qu, Z., Patti, J., Wendell, D., Choi, H., Montemagno, C. (2005) Preliminary studies on the effect of size on the action potential of an excitable vesicle. *WSEAS Trans. Systems* **5**, 347–352.

34 Santos, H. L., Lamas, R. P., Ciancaglini, P. (2002) Solubilization of Na,K-ATPase from rabbit kidney outer medulla using only C12E8. *Braz. J. Med. Biol. Res.* **35**, 277–288.

35 Cohen, E., Goldshleger, R., Shainskaya, A., Tal, D. M., Ebel, C., le Maire, M., Karlish, S. J. (2005) Purification of Na+,K+-ATPase expressed in *Pichia pastoris* reveals an essential role of phospholipid-protein interactions. *J. Biol. Chem.* **280**, 16610–16618.

36 Graff, A., Sauer, M., Van Gelder, P., Meier, W. (2002) Virus-assisted loading of polymer nanocontainer. *Proc. Natl. Acad. Sci. USA* **99**, 5064–5068.

37 Glynne-Jones, P., Beeby, S. P., White, N. M. (2001) Towards a piezoelectric vibration-powered microgenerator. *IEE Proc. Science Measurement Technol.* **148**, 68–72.

38 Xi, J. Z., Schmidt, J. J., Montemagno, C. D. (2005) Self-assembled microdevices driven by muscle. *Nat. Mater.* **2**, 180–184.

39 Yamada, N., Okano, T., Sakai, H., Karikusa, F., Sawasaki, Y., Sakurai, Y. (1990) Thermoresponsive polymeric surfaces – control of attachment and detachment of cultured cells. *Makromol. Chemie-Rapid Commun.* **11**, 571–576.

40 Kawai, H. (1969) Piezoelectricity of poly (vinylidene fluoride). *Jpn. J. Appl. Physics* **8**, 975–976.

41 Harrison, J. S. (2001) *Piezoelectric Polymers.* ICASE Report. NASA, Vol. No. 2001, pp. 1–26.

42 Choi, H.-J., Lee, H., Montemagno, C. D. (2005) Toward hybrid proteo-polymeric vesicles generating a photoinduced proton gradient for biofuel cells. *Nanotechnology* **16**, 1589–1597.

43 Choi, H.-J., Montemagno, C. D. (2005) Artificial organelle: ATP synthesis from cellular mimetic polymersomes. *Nano Lett.* **5**, 2538–2542.

44 Luo, T.-J. M., Soong, R., Ester, L., Dunn, B., Montemagno, C. (2005) Photo-induced proton gradients and ATP biosynthesis produced by vesicles encapsulated in a silica matrix. *Nat. Mater.* **4**, 220–224.

45 Offenhausser, A., Vogt, A. K. (2004) Defined Networks of Neuronal Cells in Vitro. In: Niemeyer, C., Mirkin, C. (Eds.), *Nanobiotechnology.* Wiley-VCH Weinheim, Germany, pp. 66–76.

46 Muthuswamy, J., Okandan, M., Jackson, N. (2005) Single neuronal recordings using surface micro-machined polysilicon microelectrodes. *J. Neurosci. Methods* **142**, 45–54.

47 Biran, R., Martin, D. C., Tresco, P. A. (2005) Neuronal cell loss accompanies the brain tissue response to chronically implanted silicon microelectrode arrays. *Exp. Neurol.* **195**, 115–126.

48 Liu, S. C., Douglas, R. (2004) Temporal coding in a silicon network of integrate-and-fire neurons. *IEEE Trans. Neural Netw.* **15**, 1305–1314.

49 Buzsaki, G. (2004) Large-scale recording of neuronal ensembles. *Nat. Neurosci.* **7**, 446–451.

50 Debay, D., Wolfart, J., Le Franc, Y., Le Masson, G., Bal, T. (2004) Exploring spike transfer through the thalamus using hybrid artificial-biological neuronal networks. *J. Physiol. (Paris)* **98**, 540–558.

51 Yu, Z., Xiang, G., Pan, L., Huang, L., Xing, W., Cheng, J. (2004) Negative dielectrophoretic force assisted construction of ordered neuronal networks on cell positioning bioelectronic chips. *Biomed. Microdevices* **6**, 311–324.

52 Santacruz-Toloza, L., Perozo, E., Papazian, D. M. (1994) Purification and reconstitution of functional Shaker K+ channels assayed with a light-driven voltage-control system. *Biochemistry* **33**, 1295–1299.

53 Correa, A. M., Bezanilla, F., Agnew, W. S. (1990) Voltage activation of purified eel sodium channels reconstituted into artificial liposomes. *Biochemistry* **29**, 6230–6240.

54 Bao, X., Reuss, L., Altenberg, G. A. (2004) Regulation of purified and reconstituted connexin 43 hemichannels by protein kinase C-mediated phosphorylation of Serine 368. *J. Biol. Chem.* **279**, 20058–20066.

55 Koch, C. (2004) *The Quest for Consciousness: A Neurobiological Approach.* Roberts and Company Publishers, Englewood Co, USA.

56 Pieribone, V., Gruber, D. F. (2005) *A Glow in the Dark: The Revolutionary Science of Biofluorescence.* The Belknap Press, Cambridge, MA.

57 Ataka, K., Pieribone, V. A. (2002) A genetically targetable fluorescent probe of channel gating with rapid kinetics. *Biophys. J.* **82**, 509–516.

58 Horn, R., Ding, S., Gruber, H. J. (2000) Immobilizing the moving parts of voltage-gated ion channels. *J. Gen. Physiol.* **116**, 461–476.

Index

Nanobiotechnology II. Edited by Chad A. Mirkin and Christof M. Niemeyer
Copyright © 2007 WILEY-VCH Verlag GmbH & Co. KGaA, Weinheim
ISBN: 978-3-527-31673-1

Related Titles

Niemeyer, C. M., Mirkin, C.A.

Nanobiotechnology

Concepts, Applications and Perspectives

2004
ISBN-13: 978-3-527-30658-9
ISBN-10: 3-527-30658-7

Willner, I., Katz, E.

Bionanomaterials

Synthesis and Applications for Sensors,
Electronics and Medicine

2007
ISBN-13: 978-3-527-31454-6
ISBN-10: 3-527-31454-7

Kumar, C. S. S. R. (ed.)

Nanomaterials for Biosensors

2007
ISBN-13: 978-3-527-31388-4
ISBN-10: 3-527-31388-5

Kumar, C. S. S. R. (ed.)

Nanodevices for the Life Sciences

2006
ISBN-13: 978-3-527-31384-6
ISBN-10: 3-527-31384-2

Kumar, C. S. S. R. (ed.)

Biofunctionalization of Nanomaterials

2005
ISBN-13: 978-3-527-31381-5
ISBN-10: 3-527-31381-8

Prasad, P. N.

Nanophotonics

2004
ISBN-13: 978-0-471-64988-5
ISBN-10: 0-471-64988-0

Goodsell, D. S.

Bionanotechnology

Lessons from Nature

2004
ISBN-13: 978-0-471-41719-4
ISBN-10: 0-471-41719-X

UNIVERSITY OF STRATHCLYDE

1 4 JUN 2007

UNIVERSITY LIBRARY